Fractography

of

Glasses and Ceramics III

Related titles published by the American Ceramic Society:

Experimental Techniques of Glass Science
Edited by Catherine J. Simmons and Osama El-Bayoumi
© 1993, ISBN 0-944904-58-0

For information on ordering titles published by the American Ceramic Society, or to request a publications catalog, please call 614-890-4700, or write to Customer Service Department, 735 Ceramic Place, Westerville, OH 43081.

Fractography of Glasses and Ceramics III

▼▲▼

Edited by

J.R. Varner
New York State College of Ceramics
at Alfred University

V.D. Fréchette
New York State College of Ceramics
at Alfred University

G.D. Quinn
National Institute of Standards and Technology

▲▼▲

Ceramic Transactions
Volume 64

Published by
The American Ceramic Society
735 Ceramic Place
Westerville, Ohio 43081

ce on the *Fractography of Glasses and Ceramics, held at*
,..y 9–12, 1995.

For information on ordering titles published by the American Ceramic Society, or to request a publications catalog, please call 614-890-4700.

Printed in the United States of America.

1 2 3 4–00 99 98 97 96

ISSN 1042-1122
ISBN 1-57498-007-6

Contents

Preface

These are the proceedings of the Third Alfred Conference on the Fractography of Glasses and Ceramics, held at Alfred University July 9–12, 1995. In the five years since the second conference, many things have happened in the field of fractography of glasses and ceramics, and more people are using fractography in their work. Quite simply, it was time to meet again; past participants were asking us to hold another conference. The objectives of this and the previous two conferences (1986 and 1990) were to review the progress in understanding fundamental principles of fractography of brittle materials; to highlight applications of fractography in research, development, production, testing, and product failure analysis; and to promote discussion among diverse users of fractography of glasses and ceramics.

We believe these objectives were met. The nearly 80 participants heard 27 presentations, nearly all of which are collected in these proceedings. Discussion was lively and in-depth. There was a blend of seasoned fractographers and newcomers, people from academia/research facilities and from industry, explorers of fundamental phenomena and practitioners using fractography as a tool. The papers collected in these proceedings demonstrate progress in this field. For example, it is clear that quantitative fractography in its various forms is continuing to increase in use, importance, and sophistication. New techniques of characterization of fracture processes and surfaces are being tried. More people are using fractography as a tool, particularly in strength and fracture mechanics testing of polycrystalline ceramics.

One measure of the success of the conference is the large number of participants who, when taking their leave at the end of the meeting, asked if we planned to do this again. We are encouraged by this response, and we do indeed plan to have the Fourth Alfred Fractography Conference in about four years.

The papers are arranged in the order in which they were presented. When the program was put together, there was an attempt to group papers covering similar topics together, but there was considerable overlap, making it difficult to have sharp distinctions.

We gratefully acknowledge the financial support provided by the U.S. Army Research Office and by the U.S. Department of Energy.

Special thanks go to the Alfred people who worked so hard to make the participants feel comfortable during their stay here. We take pleasure in singling out Mrs. Marlene Wightman for our expression of gratitude. Marlene made all of the local arrangements, managed registration, and kept an eye on the budget. Throughout all of this, she maintained her marvelous sense of humor, making our task so much more enjoyable.

J.R. Varner
V.D. Fréchette
G.D. Quinn

CERAMIC FRACTURE MODE-INTERGRANULAR VS TRANSGRANULAR FRACTURE

Roy W. Rice, Consultant
5411 Hopark Dr.
Alexandria, Va. 22310

Abstract

Available data on intergranular fracture (IGF) vs transgranular fracture (TGF) of ceramics is summarized and significantly extended. At 22 C, where there is most data, TGF is normally dominant. IGF generally increases with decreasing grain size (G, mainly at G≤ 1-10 μm), increasing grain boundary phase content and the occurrence of: 1) slow crack growth, 2) mist, hackle, and crack branching, and 3) (mainly finer, substantial, grain boundary) porosity, and possibly with increasing elastic anisotropy. Possible effects of grain orientation, stress rate and character, as well as microstructural stresses from thermal expansion anisotropy (TEA) are discussed. At higher temperatures, there is a general shift to more IGF, especially with more grain boundary impurities, finer G, and probably higher elastic anisotropy. This shift often starts with IGF only at the fracture origin, and may not commence until temperatures of the order of 1500 C or more in some materials. While IGF is often attributed to weaker grain boundaries (implying lower strengths), it is also often associated with fine grain size, and thus the highest strengths at lower temperatures. IGF vs TGF reflects not just grain boundary strength (as often emphasized), but a balance of this versus the fracture toughness for grain fracture (usually via cleavage, which may also entail the multiplicity of cleavage planes). Several factors may interact to shift differing balances in different materials, e.g. some increase in IGF at larger G in TiB_2 with high TEA, but more IGF increase in Al_2O_3 at larger G with less TEA, and no IGF increase in BeO and all TGF in B_4C having similar TEA to Al_2O_3.

INTRODUCTION

The fracture mode of polycrystalline ceramics, i.e. the degree of trans-vs intergranular fracture is an important, but often complex, aspect of their mechanical behavior. From a pragmatic standpoint, a higher degree of transgranular fracture (TGF) generally facilitates fractography, especially identification of fracture origins. More fundamentally, fracture mode, or changes in it, can be important clues to the mechanism(s) of failure. Thus toughening mechanisms such as crack branching, deflection, or bridging may reflect some fracture mode changes, e.g. the latter often preferentially occurs with intergranular fracture (IGF). Similarly, the occurrence and extent of slow crack growth often occurs preferentially by IGF. Fracture mode may also be effected by the macroscopic state of fracture stress as well as local microstructural stresses from thermal expansion anisotropy (TEA), elastic anisotropy (EA).

Despite the importance of fracture mode, it has received only limited attention, often not being even mentioned in many papers dealing with fracture. However, there is some, usually limited, data available which collectively is helpful. This paper summarizes literature data and significantly extends it via presentation of the present author's previously published[1-6] and unpublished fracture mode data. The focus is on fracture under typical uniaxial tensile stress, which is the bulk of the available data. It is first shown that the data supports the generally accepted trend for the overall fracture mode to have substantial TGF at 22 C, with IGF progressively increasing at sufficiently high temperatures. However, microstructure, composition, and location on the fracture surface, as well as fracture environment are all shown to be factors that can shift the fracture mode on a local scale or over the entire fracture. Then, the much more limited data for other macroscopic stress states and on possible effects of microstructural stresses due to TEA and EA are considered.

DATA PRESENTATION AND DISCUSSION

Uniaxial Tensile Fracture Mode vs Grain Size and Temperature

Al_2O_3: Consider first data for Al_2O_3, which is most extensive, but often variable. Rice's[2-6] data for the room temperature fracture mode from flexure testing of dense Al_2O_3 (hot pressed without additives), while showing considerable variation, clearly shows substantial, often predominant, TGF from average grain sizes of ~1 μm to ~5-10 μm, then decreasing as grain size (G) further increases (Fig.1). This trend is generally consistent with his observations on a variety of commercial and other experimental aluminas, though they generally are nearer, or above, the higher TGF level shown in Fig.1. Except for Lucalox, which showed

predominantly IGF, i.e. near the lower bounds of the range of Fig.1 (at G ~30 μm), at least part of the lower TGF portion of the trend is more commonly associated with porosity, especially finer intergranular porosity. On the other hand, fusion-cast (larger grain, e.g. 50-200 μm, including ~99% purity) Al_2O_3 refractories had substantial, often predominant, TGF (Fig.2), well above the upper bound shown in Fig.1. Note that while such fused bodies have considerable porosity, it is generally heterogenous distributed and much of it consists of pores whose size is similar to, or much greater than, G.

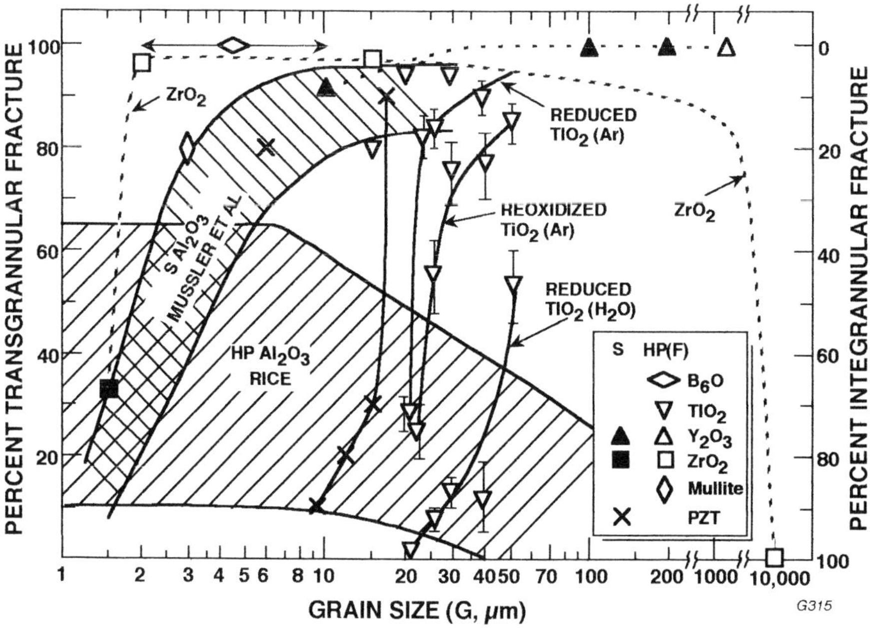

Figure 1) Trans- and inter- granular fracture of oxides as a function of grain size (G) at 22 C. Cross hatched regions show the range found in flexure testing of this author's hot pressed Al_2O_3[3,4] and in K_{IC} (notch beam, NB) testing of dense sintered alumina[19]. Data of other investigators shown as discussed in the text (some estimated from their micrographs), HP=hot pressed, S=sintered.

Subsequent observations of this author[6] on the same hot pressed alumina at finer G (with increasing porosity, P) shows 50% or more TGF on the general fracture area (but not at the fracture origins, as discussed later) at G~1 μm (P~2%, primarily intergranular). IGF increased as G decreased, i.e. to nearly 100% IGF by G ~ 0.3 μm (~16%, fine, intergranular P). In the latter case, larger grains (e.g. ~ 0.5 μm) were generally fractured transgranularly. These results strongly indicate a transition to predominantly TGF at G > ~ 1 μm. This is generally consistent with observations on another hot pressed alumina (from AVCO, with MgO additions), except more IGF at < 5 μm[7]. This alumina also had isolated larger grains (e.g. ~ 100 to over 500 μm) which showed mixed TGF/IGF, but with more TGF at the lower end of this range and more IGF at the higher end.

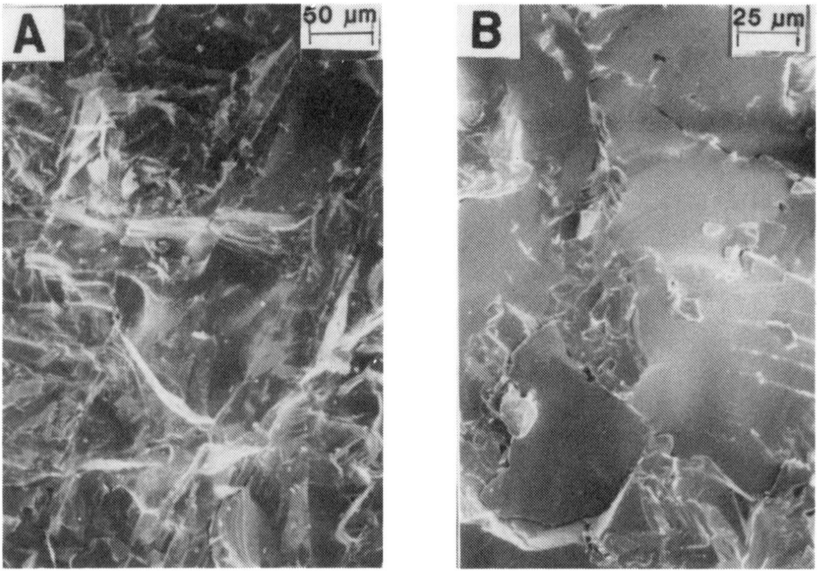

Figure 2) Sample fractographs of 99% pure, fusion cast Al_2O_3 refractory, which is also representative of other less pure, fusion cast Al_2O_3 refractories.

The overall trend, as well as the variations, of Fig.1 are also consistent with the literature data of others for dense sintered or hot pressed Al_2O_3[8-17]. Thus, Spriggs et al[8] noted predominant TGF at G=1-2 μm, but generally predominant IGF by G= 150-250 μm in flexure tests of similar hot pressed Al_2O_3. Sedlacek and Jorgensen[9] found ~ 0 % TGF (i.e. ~ 100 % IGF) in their hoop tension testing of Lucalox rings having G ~ 10, 20, 30, 40, and 50 μm, i.e. in good agreement with this author's Lucalox data noted above. Similarly, Graham and Alers[10] found ~0% TGF on Lucalox fractures (G~20 μm) at 22 C. However, these Lucalox results are in some contrast to some other Lucalox results. Johari and Parikh's[11] Lucalox data (at 22, 400, 700 , and 1000 C), while not quantitative, showed mixed TGF and IGF, but more IGF, especially as G increased over the 5-30 μm range , i.e. consistent with the trend for TGF to decrease with increasing G at larger G, as shown in Fig.1. This is reinforced by their observation of more TGF of the finer than the larger grains on a given fracture. Samuels et al's[12] more limited, qualitative Lucalox results showed similar trends: <50% TGF at -196 C for G~20 μm. Substantial, often predominate TGF is shown for other hot pressed and sintered aluminas (G commonly 5-10 μm) by other investigators[13-17]. TGF of large grains in Al_2O_3 refractories, similar to Fig.2 has also been reported[18].

The above results from flexural or tensile testing at 22 C are also generally consistent with fracture mode data from fracture toughness testing at 22 C. Most extensive are those of Mussler et al[19] on pure alumina sintered without or with additives (MgO or Y_2O_3) for notched beam (NB) tests at 22 C. These showed <20-30% TGF at G < 1.5 μm, but rising rapidly to 80-90 % TGF as G increased to 5-30 μm (Fig 1). Their data thus shows a shift to larger G for the IGF-TGF transition and TGF extending to higher levels that this author's hot pressed alumina data. However, their higher TGF level is reasonably consistent with this authors data on a number of other aluminas, and other data variability, e. g. for Lucalox, as noted above. With regard to the G for the IGF-TGF transition, Mussler et al used a measurement of G that more heavily weighted larger grains (based on their areas) than most measurements, e.g. by a factor of 2-3. Thus their data on the TGF rise as G increases at finer G is consistent with that of this author when differences in G measurement are taken into consideration. (This important issue of G measurement will be discussed further later.)

Other crack propagation data at 22 C support the trends of Fig.1 and results presented above. Thus Cappola and Bradt[20], using the NB test reported ~20% TGF of hot pressed Al_2O_3 (G~1 μm, with 0.5-10% intergranular porosity, P, which they reported had no effect on fracture energy). Hayashi et al[21] reported primarily IGF in their NB test of sintered Al_2O_3 (G~2-70 μm, with P~ 4 & 5% respectively). Heuer[22] noted a high percentage of TGF (often > 50%) in Al_2O_3 hot pressed to < 1% P with G ~ 1 μm whether tested with or without a notch. Similarly Gutshall and Gross[23] reported mixed TGF-IGF of aluminas (including Lucalox) tested via the double cantilever beam (DCB) method at 25 C, with TGF being dominant in larger grain material (30 and 45 μm). Congleton et al[24] report ~ 45% TGF in crack propagation studies of Lucalox at -196 C, i.e. much higher than the author's above results, again showing variation in TGF-IGF levels. Veldkamp and Hattu[25] showed ~ 70% TGF in NB tests of a pure, sintered Al_2O_3 (P~ 1 %, G~ 6 μm), again with TGF predominant on the larger grains and IGF predomi-nant on the finer grains on of a given fracture. Wieninger et al[26] noted that fractures of their pre-cracked beams of a finer G (~3 μm, P~1%) Al_2O_3 for their R-curve studies at 22 C showed mainly IGF in the lower crack velocity (precrack) region transitioning to more TGF at higher crack velocity. In contrast to this, they noted that a similar high purity alumina with larger G (~20-40 μm, P~3 %) with much more R-curve effect showed " a crushed and split appearance" (implying more TGF) on both slow and fast fracture regions. Salem et al[27] also showed substantial, often dominant, TGF in 96% Al_2O_3 bodies of both random and preferred orientation. While over all fracture differences from preferred orientation are discussed, variations in the substantial TGF of their nominally 10-20 μm bodies are not clear.

All Al_2O_3 investigations showed IGF predominant at sufficiently high test temperatures. Thus, Spriggs et al[(8)] reported little change in the dominance of TGF from flexure tests of 1-2 µm Al_2O_3 at 1000 C, but IGF becoming extensive and total at respectively 1200 and 1350 C. The relative decrease in TGF generally correlated with strength reductions with increased test temperature. On the other hand their large G (150-250 µm) Al_2O_3 had modest levels of TGF at all test temperatures. Johari and Parikh[(11)] similarly showed the considerable TGF of their finer G Lucalox decreased at higher temperatures (especially by and above 1000 C). Gruver et al[(13)] showed possible modest decreases in the substantial TGF of their 96% (commercial sintered, G~7 µm, P~5%) Al_2O_3, as well as in their hot pressed Al_2O_3 (P ~0%, G ~3 μm) in flexure testing as low as 800 C. DeWith[(17)] showed TGF decreasing in two dense sintered aluminas (G ~30 µm) substantially by 1100 C, and more by 1500 C. However, larger (eg. 100-200 µm) grains in bodies with bimodal G continued to show substantial TGF at such test temperatures. Hayashi et al[(21)] reported no significant changes in the NB tests of their Al_2O_3 at 1000 or 1500 C, i.e. the fracture mode remained mainly IGF. However, Congleton et al[(24)] noted IGF increasing to 70% at 600 C in their crack propagation studies of Lucalox(G~15 μm), from ~ 45% at -196 C, but with the TGF dropping back to 60% where cracks branched. Gutshall and Gross[(23)] reported 100% IGF in their DCB tests of 3 Lucaloxs (G ~20, 30, and 45 µm) at 1500 C, but some TGF in other commercial aluminas. Samuels et al[(12)] showed ~100 % IGF on Lucalox fractures at 1600 C. Recently White and Hay[(28)] reported very modest increases in IGF from R-curve tests at 22 to 800 C in a 99.7% pure commercial alumina (G~18 μm). They noted that about half of the larger grains of ~30 μm, and all of the grains >65 μm failed transgranularly at 22 C, with the change at 800 C being only for G>65 μm, where the portion of TGF dropped by ~1/2.

Other Single Oxides: Bentle and Kniefel[(29)] reported mainly IGF of dense hot pressed BeO at 25 C and higher temperatures. However, their fractographs at 25 C (G~18 μm) and 1500 C (G~ 38 µm) show respectively substantial (eg. > 50%) and limited, or no, TGF. Gutshall and Gross[(30)] reported mainly IGF in DCB tests of a commercial (>99% pure, G~2-8 µm, P~ 2%) at 22 C. Lynch and Bradt[(31)] showed nearly 100% TGF in a dense (P ~1-3%, G~13-28 μm) commercial BeO in work of fracture tests at 22 C. Thus the limited BeO data is consistent with a transition from mainly IGF at G, ~5 µm, to mainly TGF at larger G .

Rice[(32)] has shown mainly (eg.> 75%) TGF in -196 and 22 C flexure testing of CaO hot pressed without or with (typically 2%) LiF addition (commonly mostly removed by subsequent annealing) for G 20-200 μm. Recrystallization of CaO crystals by hot extrusion[33] or press forging[34] (giving G respectively 40-300 and 20-120 μm) both gave 100% TGF in 22 C flexure tests. The significant <100> preferred orientation parallel to the tensile stress in the hot extruded CaO gave much flatter fractures. Tests at 1315 C(2400 F) gave increased IGF, to 100% in bodies hot pressed with LiF, 20-90% in bodies hot pressed without LiF and to at most a few percent in press forged crystals, where IGF occurred almost exclusively at the 1-2 grain boundaries where failure initiated[4,32]. The recrystallized crystal results for CaO are very similar to those for such MgO (discussed below), but probably reflect even less propensity for localized IGF initiation.

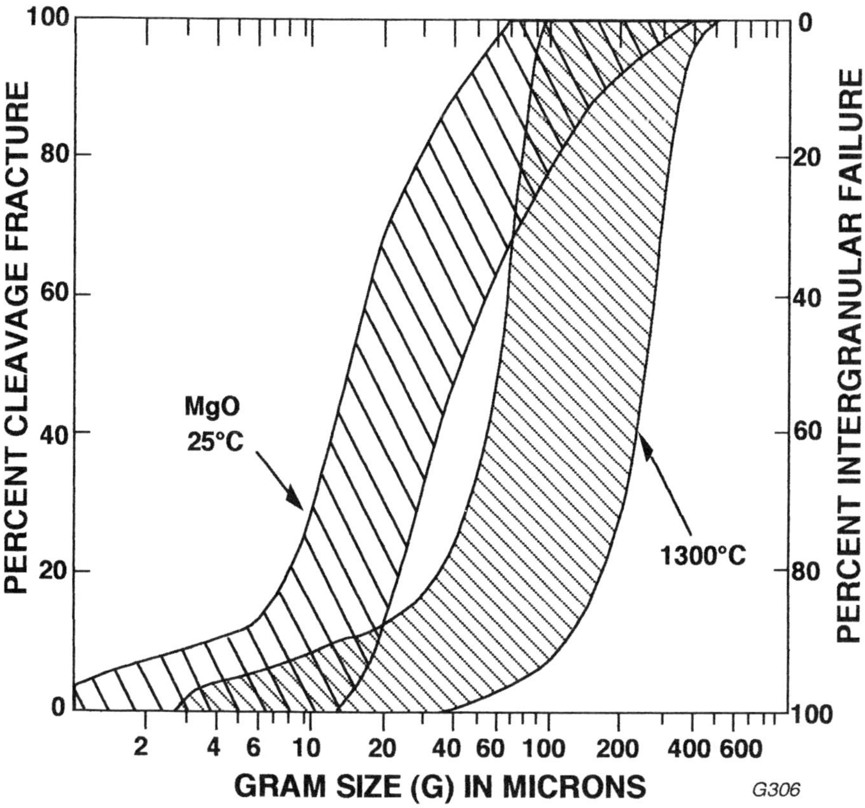

Figure 3)Trans- and inter-granular fracture of MgO versus grain size (G) from flexure tests at 22 and 1315 (2400 F) C[1,3,4]. Note the shift of the transition to all TGF to larger grain size at 1315 vs 22 C (see also Fig.4).

This author's MgO made by hot pressing without or with LiF[1](much, or all of this, commonly being removed by annealing for all but the finest G bodies), showed TGF dominated fracture, except at the finer G at 22 C[1] (Figs.3&4). The

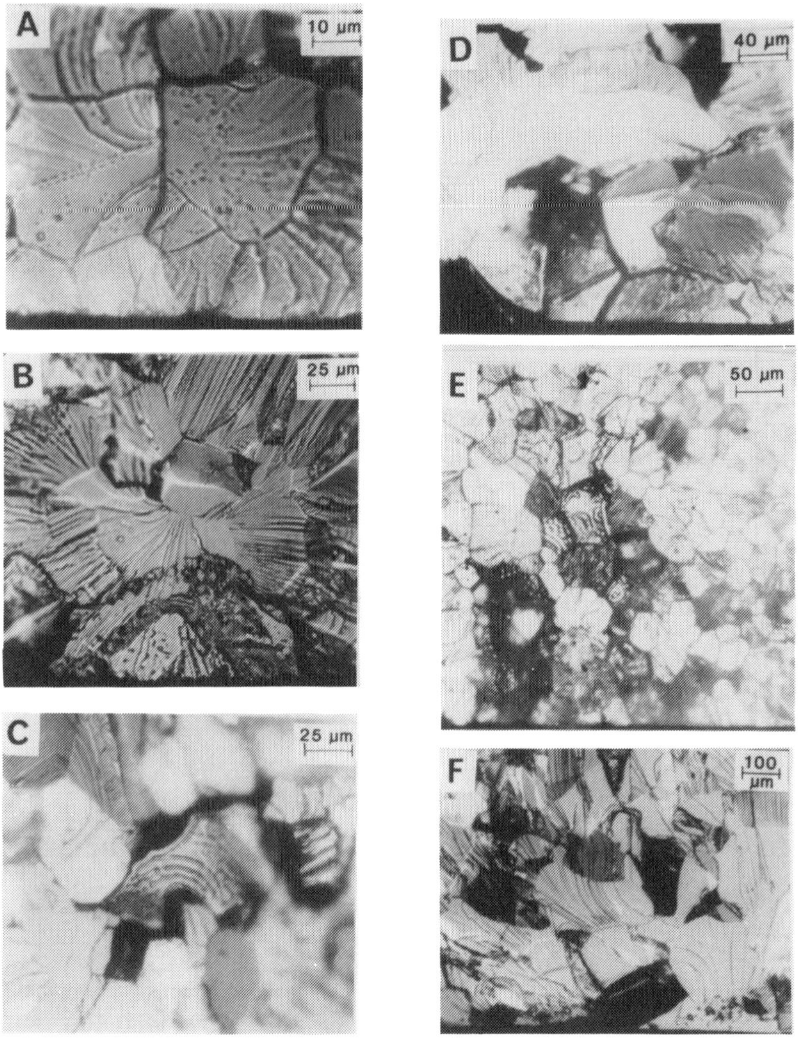

Figure 4) Fracture mode at, and around, fracture origins, of bend bars from different polycrystalline hot extrused MgO billets. Optical fractographs (with resultant <100> axial extrusion texture normal to the photos) from hot pressed billets tested at 22 C: A), B), and C), and 1315 C: D) and E), and from MgO crystal billets tested at 1315 C: F). Origins (usually inside of the tensile surface due to surface work hardening)[3,33,35] were from grain boundaries, whose exposure on the fracture (often darker areas) increased with test temperature.

presence of residual (mostly intergranular) porosity and impurities or additives shifted the IGF-TGF transition to somewhat larger G. The IGF-TGF transition was also shifted to higher G at higher test temperatures. As with CaO, MgO recrystallized from single crystals by hot extrusion[33,35] or press forging[34] (giving G 15-70 μm) showed higher TGF (and greater retention of it at higher temperatures than for hot pressed bodies, Fig.4). Like CaO, hot extruded MgO had much flatter fractures (reflecting the significant <100> axial texture parallel to the tensile axis in flexure testing). The flatter TGF allowed specific fracture origins to be determined (as with such CaO). Thus, while all fractures initiated from grain boundaries (often from slip bands ending on the boundary, eg. Fig.4A), there were distinct fracture differences depending on the type of extrusion billet material. While some specimens from hot pressed billets failed from boundaries at a high angle to the fracture surface so that both grains abutting the boundary fractured transgranularly (Fig.4A), most failures originated from boundaries exposed on the fracture surface. Some of these exposed boundaries appeared relatively clean (Fig.4B), but many had substantial impurities, residual porosity, or both[1](eg. Fig.4C). Specimens from extruded (hence recrystallized) crystals almost universally had fractures like that of Fig.4A.

At 1315 C fracture origins of hot extruded MgO specimens from hot pressed billets were sometimes from 1 or 2 boundaries exposed on the fracture surface, often with a few other such exposed boundaries nearby (Fig.4D), but more commonly from several such exposed boundaries (Fig.4E), or from so many of such boundaries that the origin could not be determined. In contrast to this, recrystallized crystal specimens sometimes had origins from similarly exposed, but usually single, isolated, generally nearly featureless boundaries at 1315 C (2400 F), and universally at 1540 C(2800 F). Tests of the same hot pressed MgO that had not been hot extruded showed more IGF (and rougher fracture) than their hot extruded counter parts. Increased TGF may reflect preferred orientation effects, cleaner grain boundaries (from recrystallization), or both.

More limited MgO literature data sopports Fig.3. Graham and Alers[10] reported 100 % IGF in dense (AVCO) hot pressed MgO (G ~23μm). Vasilos et al's[36] more qualitative flexure results of hot pressed MgO showed mainly IGF at 22 C, increased at 700 C, and became 100% at 1200 C for G=1 or 3 μm, but some TGF in larger G (to 200 μm) at all test temperatures. Similarly Yasuda et al[37] showed mixed TGF-IGF in flexure tests of their hot pressed MgO (G~10-200 μm) at 22 C (with TGF possibly decreasing some at larger G), and mostly IGF at 1000 C.

This author has observed essentially 100% TGF in 22 C flexure test of hot pressed B_6O (G~2-10 μm, P~0 to >10%) and of sintered CeO_2 (G~30 μm, P~7% ,mainly large, heterogeneously distributed, intergranular pores). Ownby and Jungquist[38] showed considerable (often >50%) TGF in sintered Cr_2O_3(mainly dense bodies with a few percent P and G>10-20 μm). Spriggs et al's[39] micrographs showed ~0 and 20- 40% TGF at 22 C on hot pressed NiO fractures with G ~0.5 and ~2 μm respectively. (Note, their finer G material was weaker contrary to the almost universal trend for strength to increase with decreasing G. The lower strength is attributed to the use of much higher pressures, and lower temperatures, in hot pressing the finer G bodies and the probable greater entrapment of anion impurities at grain boundaries.)

Room temperature data of this author for other single oxides is shown in Fig.1, along with data of others, which by themselves, or combined, support the broader G trend. Most extensive are data of Minnear and Bradt(40) from K_{IC} and slow crack growth (double torsion) tests of hot pressed TiO_2 (with or without various annealing, P ~2-10%). These show the common TGF increase with increasing G in both K_{IC} and slow crack growth tests, but somewhat less TGF for reoxidized TiO_2, and much less for slow crack growth. This author's flexure results for TiO_2 hot pressed with additions of ~1/4% of 1 of 4 oxides (giving P ~10%, G~20 μm), and of sintered TiO_2(41) (G~30 μm) are reasonably consistent with the above K_{IC} results. The amount, as well as the size, of pores are probable factors in the variations within and between various tests. The occurrence of substantial TGF in TiO_2 is also supported by Kirchner and Gruver's(42) observation that microcracks in their hot pressed TiO_2 (G~30-140 μm) were commonly transgranular.

Earlier(43,44) and more recent flexure tests of Y_2O_3 (sintered to transparency with or without additions of ThO_2 or La_2O_3) showed ~100% TGF across the range of G available (10-200 μm, Fig.1), and are consistent with other recent tests(45). DCB crack propagation studies of some of the same Y_2O_3 bodies showed 100% TGF, and an apparent tendency for fracture to avoid grain boundaries(46). Also, sizeable (e.g.~1 cm fragments from skull melts of Y_2O_3, with or without La_2O_3 additions, (giving G at least a few hundred μm) showed mainly, if not exclusively, TGF. Data of Desmaison-Brut et al(47) for vacuum sintered and HIPed Y_2O_3 is basically consistent with, and extends the above data, except for their as-sintered bodies having essentially all IGF. However, air annealed of their sintered material (P ~ 1.2%, G ~ 15 μm) gave essentially 100% TGF at 22 C. Similarly, their as- HIPed materials (G~ 0.6 μm) had substantial, but < 100%, TGF, and air annealing of this also increased the TGF. Their sintered material fractured at 1400 C also had ~100 % TGF, while their HIPed material showed some reduction in TGF at 1000 C and mostly IGF at 1400 C.

Earlier(2,33,48) and more recent data of this author and colleagues on various dense, sintered, hot pressed, or hot extruded (not textured) ZrO_2, fully or partially stabilized with Y_2O_3, or partially stabilized with CaO or MgO, all showed high TGF (G ~2-60 μm, Fig.1). Flexural failure originating from grain boundary surfaces approximately in the plane of fracture (often with some associated, mostly intergranular, porosity) have been observed in such larger G partially(4,48)

Figure 5) Flexure fracture of commercial (Diamonite Products) TZP at 22 C; ~ 1/3 TGF, indicating a transition to IGF at finer G (µm) in ZrO_2(see also Figs.1,6).

and fully stabilized ZrO_2 bodies (often with the rest of the fracture being mostly, or all, TGF. On the other hand directional solidification of skull melts of a variety of partially and fully stabilized ZrO_2 compositions yielding large columnar grains (~½ to many cm in diameter, with substantial preferred orientation) have nearly complete, or total intergranular separation, i.e. 100% IGF[(49)]. The extent and ease of such separation, which tends to increase with the size, especially diameter, of the columnar grains (crystals) clearly shows a transition to IGF at quite large G (e.g. mm to cm or more in scale. The degree of boundary separation in such fused ZrO_2 is clearly more than in similar columnar, fused MgO, which in turn may have more boundary separation than similar fused CaO. At the other end of the G range, this author's study of TZP materials (G~ 1-2 µm, strength >700 MPa) showing only of the order of 1/3 TGF indicates a transition to IGF at fine G (Fig.5), as for other materials (Fig.1). This transition is also indicated by all IGF in the finer, but not the coarser, grains of another TZP with bimodal G (Fig.6). ZrO_2 data of Adams et al[(50)] for bodies made via the same processing as the larger G ZrO_2 of the present author also showed mainly TGF in flexure tests at 22 C. Again residual porosity (mainly intergranular) appears to be a major factor limiting TGF. The predominance of TGF decreased with increasing test temperature, e.g. being mixed TGF-IGF at 1000 C, and ~0 % TGF at 1500 C.

Similarly, Rice et al's flexure testing of ZrO_2 partially stabilized with CaO (2.9 %, G ~20 μm), showed ~100 % TGF at 22 C, and 100 % IGF at 1300 C[33].

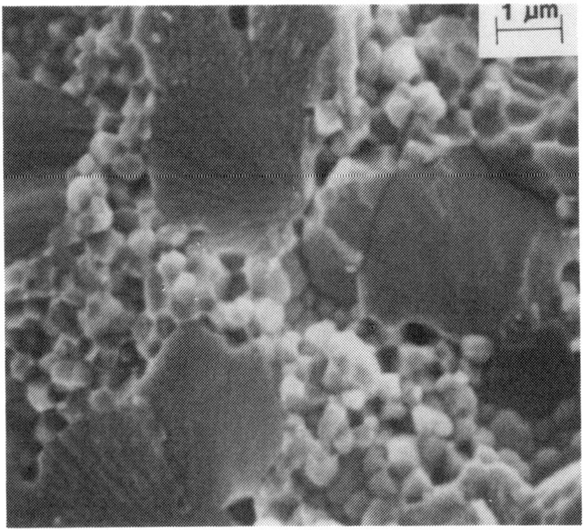

Figure 6) SEM fractograph of ZrO_2-8 wt.% Y_2O_3 with bimodal grain structure showing 100 %TGF of the larger grains and 100% IGF of the smaller grains at 22 C. See also Figs. 9, 12, and 14 for similar trends with much less extreme grain size differences.

Predominant TGF has been observed in the immediate vicinity of fracture origins, e.g. in the fracture mirror area, but ~ 100% IGF further out on the fracture surface, i.e. in the mist-hackle area and beyond in hot pressed ThO_2 (+3% Y_2O_3, G ~2 μm) flexure tested at 22 C[51]. Again the limited IGF in the area of TGF was mainly for grains < 2 μm, e.g. 0.5-1 μm, with TGF in this area being essentially universal for larger grains (G ~2 μm or more). The TGF was presumably mainly {111}cleavage, since this has been reported to be the dominant fracture of ThO_2 crystals[51,52], as is commonly the case for other fluorite structure crystals, including UO_2[53]. Daniel and Tagahasi[53] showed substantial TGF at 22 C in their sintered UO_2 (G >20 μm). Kennedy and Bandyopadhyay[54] showed TGF decreasing from ~100 to ~20 % with increasing severity of thermal shock in sintered UO_2 (P ~5% , G~5 μm). Roberts and colleagues[55,56] reported predominantly TGF in flexure tests of their sintered UO_2 (P ~6-16%, G ~10-20 μm) at 500 C, but TGF at 1250 C decreased from~90% to ~10% as P increased from 6 to 16 %. Evans and Davidge[57] reported TGF decreased from ~30 % in flexure tests of their G ~8 μm at 22 C to <5% at 1100 C, while TGF was >60% at all test temperatures to1600 C in their G=25 μm sintered UO_2 (both bodies had ~3%P).

Schulson et al's[58] observed essentially 100 % TGF of ice at -25 to -50 C (i.e. from > 80 to > 90 % of it's absolute melting temperature !) for G ~ 1 to 10 mm (i.e. ≥1000 μm) despite a few percent of mainly finer, grain boundary porosity.

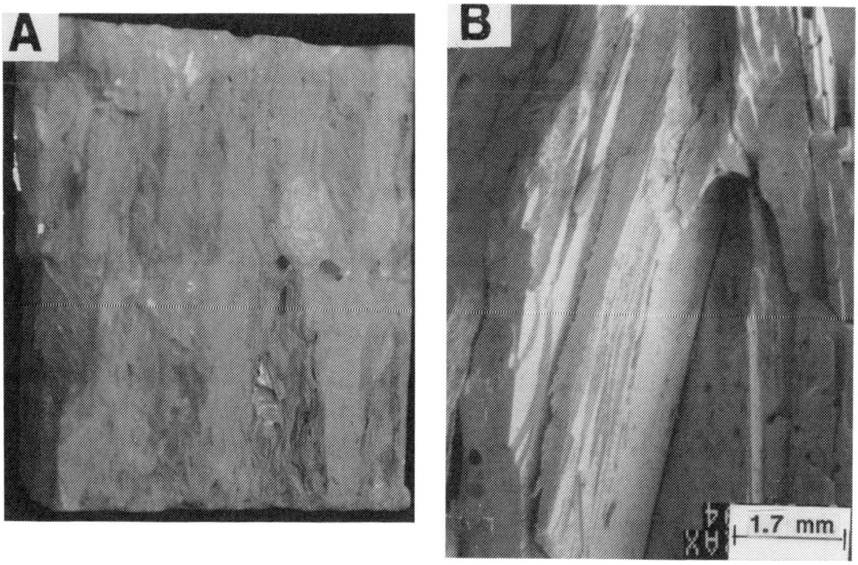

Figure 7) Fracture of fusion cast beta-alumina refractory. A) lower magnification SEM of complete crossection of the 5 cm (2") thick brick showing essentially 100 % TGF, mainly by cleavage of the large, columnar grains whose growth was normal to the brick (and hence also two photo) sides, extending ~ ½ way through the brick and being ~ 0.6-1.2 cm wide. B) higher magnification SEM, mostly of one grain, showing predominant cleavage, which occured despite several percent porosity (mostly intragranular and not visible in this photo).

Mixed Oxides: TGF predominates at 22 C over a broad range of beta alumina compositions, processing and grain sizes[59-62], including fused cast β-Al_2O_3 refractories consisting of long platey grains (Fig.7). Much more limited data shows mainly TGF at 22 C in mullite (G~3 μm, Fig.1)[63]. Substantial, possibly exclusive TGF is reported in crystallized 3BaO $5SiO_2$ glass in room temperature fracture[64]. In earlier crystallization stages the bodies consist of spherulitic grain colonies . The ~5 μm, or less, diameter spherulites all fractured transversely (and it is implied that the much smaller grains making up the spherulites also fractured transgranularly). Later stages of crystallization involved the development of lath-shaped grains (<5 μm in size), which clearly had TGF (and increased flexural strengths from 93 to 160 MPa and K_{IC} from ~0.2 to 0.5 $MPam^{1/2}$).

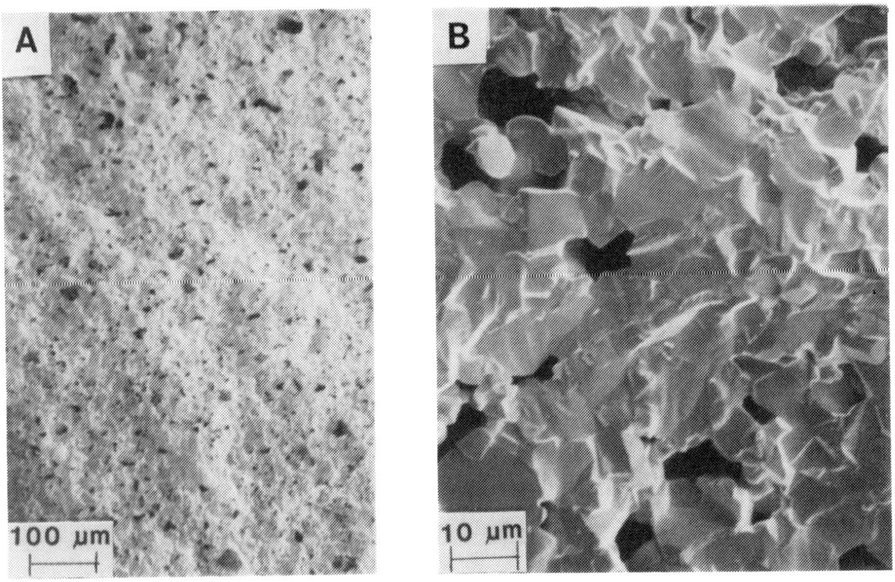

Figure 8) Fracture of porous $MgAl_2O_4$ at 22 C. A) lower magnification SEM showing larger pores, which are >G. B) higher magnification SEM of the sample showing finer pores (< or ~ G), and nearly 100%TGF despite the presence of both pore populations. Note limited areas of IGFof smaller grains.

TGF dominates fracture of $MgAl_2O_4$ at 22 C over a range of stoichiometries and G, and with (e.g. Fig.8) or without porosity or additives, (Fig. 9A). There is a possible trend for some increase in IGF at fracture origins at large G (eg.>50 μm)[44]. A pronounced deviation is greatly increased, typically dominant IGF in dense (transparent) $MgAl_2O_4$ hot pressed with LiF additions (Fig.9B). Rhodes et al[65] had reported similar effects in their dense $MgAl_2O_4$ hot pressed (at 1200 or 1300 C for 20 min.) with LiF additions had ~100% IGF in the fine grain (G ~0.2-0.3 μm) matrix and mixed IGF-TGF in the larger (G ~3-5 μm) grains of the bimodal grain size. Conversely, their material hot pressed to transparency without LiF (at 1400 C for 90 min.) exhibited nearly 100% TGF (and the highest bend strength) at 22 C. Graham and Alers[10] also showed 100 % TGF in dense, transparent $MgAl_2O_4$ (made without LiF). White and colleagues[66-68] reported predominantly IGF in DCB tests of large G(~50-100 μm) $MgAl_2O_4$ hot pressed to transparency with LiF. (This and their observation of TGF occurring mainly in isolated larger grains, e.g. >100 μm, is consistent with this author's observations on the same material, Fig. 9B). They further noted that the IGF increased overall above 800 C, and that the threshold G for TGF to occur increased to >150 μm.

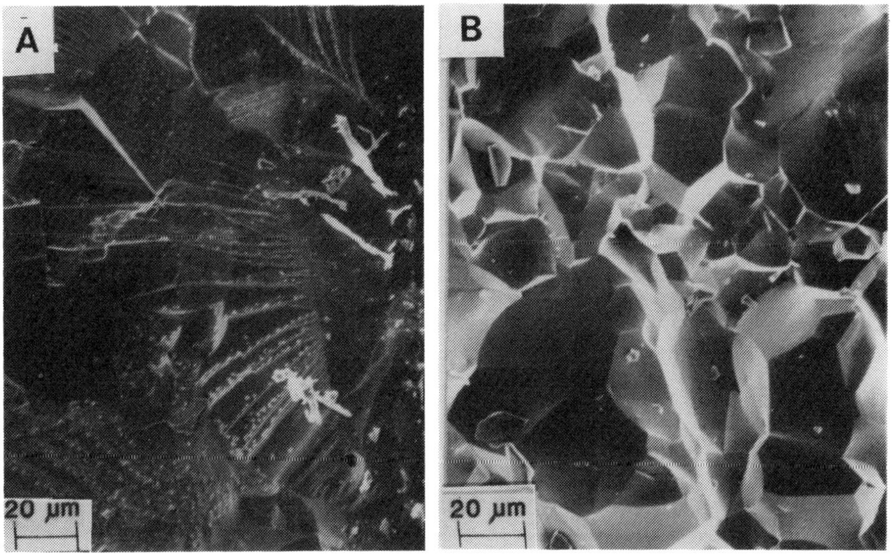

Figure 9) Comparison of dense (transparent) $MgAl_2O_4$ made without and with additives and fractured at 22 C. A) SEM of material sintered with out additives showing 100% TGF, including the fracture origin (center, right. The white spots and longer, irregular objects are artifacts). B) SEM of sample hot pressed with LiF addition, with almost 100% IGF, and only TGF of larger grains.

Veldkamp and Hattu[25] reported ~50% TGF in a dense Ni-Zn ferrite (G ~5-10 μm), and mainly TGF in a similarly dense strontium hexaferrite (fractured in various directions relative to it's preferred orientation) in K_{IC} tests at 22 C. (They also showed mainly TGF in a dense Al_2O_3, G ~5 μm). In contrast to the above ferrite results, Beauchamp and Monroe[69] reported fast fracture in Mn-Zn ferrites was predominantly IGF, while SCG was mostly TGF, especially with G ~45 μm, and somewhat less in finer G (35 μm) with more grain boundary (e.g. Ca) phase.

Pohanka and colleagues[2,70-72] have shown predominantly TGF in a variety of commercial and experimental $BaTiO_3$ bodies (e.g. for sonar applications) with G ~5-200 μm. Similarly, commercial PZT's (G~5 μm) showed mainly TGF[7,73]. More recently Kim and Kim[74] have shown a smooth, but substantial, increase in TGF, from ~ 10% at G=9.5 μm to ~20% at G ~12 μm, ~30% at G=15 μm, and to ~ 90% at G=17 μm in PZT (Fig.1) sintered to ~3% P (mostly at grain boundaries, but with pore size increasing with sintering time and temperature, and hence also with G). Meghirhi[75] showed sintered $MgNb_2O_4$+0.1 $PbTiO_3$(P ~3-5%, G ~2-15 μm) had mainly IGF at G=5 μm, and mainly TGF at G=15 μm (addition of a few percent $Pb_5Ge_3O_{11}$ mainly reduced G and often increased P some).

Kuszyk and Bradt[76] showed some probable TGF in fine (~5 μm) G $MgTi_2O_5$, but predominantly IGF at larger G, consistent with spontaneous microcracking commencing at G~3 μm (since such cracking often involves substantial IGF). This author has also observed mainly IGF in Al_2TiO_5-based bodies (with several % P) above their critical G for microcracking.

Kim and Kim[77] report significant changes in fracture mode of hot pressed PLZT with annealing. The as-hot pressed fracture mode was ~ 100 % IGF, while that of this material annealed (in air or in a PbO atmosphere at 900 C) was ~ 100 % TGF. The annealing had no effect on G (~4 μm), but changed the grain boundaries from relatively flat as- hot pressed to wavy or undulating after annealing. These changes correlated with the K_{IC} increasing from 0.85 ± 0.2 to 1.3 ± 0.3 $MPam^{1/2}$. Similar effects of annealing on the grain boundary character and resultant fracture mode changes were shown by Chang et al[78] in strontium calcium ferrite.

Nonoxides: Less refractory, dense transparent materials, developed for IR windows, provide useful fracture data, at 22 C and larger G. Thus, this author and colleagues' observations on transparent, CVD ZnSe (G ~ 50-100 μm)[3,79] and ZnS (G ~ 5-30 μm) showed almost exclusive TGF in flexure and other fracture. Similarly large G (~100-200 μm) transparent CaF_2 showed 100% TGF.

This author's extensive observations on B_4C processed from a variety of powders, densified by various organizations (mostly by hot pressing) without or with subsequent heat treatment, all gave essentially exclusive TGF in flexure at 22 C across the G range studied (~2 -100 μm, e.g. see ref. 44). Literature data[80-83] is consistent with this, showing predominantly TGF at least down to G~2 μm, even with ~ 15% P(mostly intergranular from sintering)[82]. Finer G and higher P gives mostly TGF, e.g. for G~1-2 μm. Effects of P on IGF appear to be modest at 22 C, again indicating a TGF-IGF transition at G~ 1 μm. Literature data also shows continued TGF for typical G as test temperatures increased to 1200 C[80].

This authors observations on flexural fracture of several CVD SiC bodies at 22 C showed essentially exclusive TGF over a range of grain diameters (1-20 μm) and shapes (including long tabular grains). This is consistent with Gulden's [84] observations of essentially exclusive TGF of CVD β-SiC (G ~1-15 μm, $P < 0.1\%$) in flexure at 22 C. This is also consistent with mainly TGF in dense SiC bodies sintered or hot pressed with (nonoxide) additives tested in flexure at 22 C (G ~ 3 μm, with tabular grains to ~ 500 μm long)[85], and similar tensile frac-ture of sintered α-SiC fibers[86]. Kim and Lee[87] showed that their SiC made by hot pressing mixtures of alpha SiC powder, 20 % polycarbo-silane, and 0.5-2 % of B, C Al, or AlN (singlely or in combinations of 2) giving G~1-2 μm,had 22 C flexural fractures of predominantly TGF when B was present (i.e. with 1% B, 0.5 % B + 1% AlN, or 1% B + 2 % C), but mainly IGF when B was not (i.e. with

1.3 % Al, 2 % AlN, or 1.3 % Al + 2 % C). The dominance of TGF in the absence of Al-based additives (that commonly yield Al_2O_3) is also consistent with mainly TGF in 22 C flexure testing of reaction processed SiC, until a transition to mainly IGF occurs as the actual bond area between SiC grains decreases independent of grain or pore sizes (commonly nearly the same size as G)[86].

Fracture toughness testing of reaction processed SiC also showed mainly TGF[88], but mainly IGF of SiC hot pressed with Al_2O_3 additions[89]. The indicated fracture mode differences, especially with nonoxide vs oxide, additives are supported by Takada and Nakamura's[90] observations of ~100% TGF in SiC sintered with 1% B (G ~10-20 μm) vs mixed TGF and IGF for sintering with 1% Al (part, or all, of this ends up as Al_2O_3). This is also shown by Lee and Kim's[91] observation of mixed TGF/IGF in K_{IC} tests at 22 C of SiC sintered (G ~1-10 μm) with $Al_2O_3 + Y_2O_3$ additions, but more IGF than with nonoxide additions. Their micrographs indicate more IGF as G decreases, consistent with Orange et al's[92] predominantly TGF in dense sintered alpha or beta SiC fractured in either NB or WOF tests at 22 C. The occurrence of mainly IGF with oxide grain boundary phases and mainly TGF with no, or nonoxide, grain boundary phases in SiC is corroborated by results of Lee et al[93] for their SiC bodies fabricated with either Al_2O_3+ Y_2O_3 or with B+C. They also observed substantial R-curve effects in the oxide processed bodies with mainly IGF, and no significant R-curve effect with the nonoxide processed bodies with mainly TGF. Some different results are also indicated in Kodama and Miyoshi's[94] fracture toughness tests of SiC hot pressed or HIPed from pyrolyzed polycarbosilane derived powders. They showed their finest (~ 0.2 μm) and largest (~ 1.4 μm) G bodies with ~50% TGF had lower K_{IC} (~2.5 and 3.5 $MPam^{1/2}$ respectively) than their ~ 0.7 μm body with the maximum K_{IC} (~5 $MPam^{1/2}$) with mainly IGF.

Results of SiC studies at elevated temperature also showed variable results. Gulden's[84] dense CVD β-SiC retained essentially exclusive TGF to the maximum test temperatures (1200-1400 C, with flexure strength increasing above 900 C attributed to the onset of some plasticity). However, material with ~ 1 % P went from predominantly TGF at 22 C to predominantly IGF at high temperatures. Prochaska and Charles[85] reported substantial TGF still occurring on fractures of their SiC at 1600 C. On the other hand Bradt et al[95] and Henshall et al[96] reported fracture toughness/energy tests showing transitions from mainly TGF by respectively 1200 and <1000 C. Again, however, the difference appears to be due to use of Al-based (probably Al_2O_3) additions in the latter case.

Other tests support and extend the above results. Faber and Evans[97] report TGF in SiC sintered with B, (G ~5 μm), and IGF in SiC hot pressed with Al_2O_3 or ZrO_2 (G ~ 5 and ~2 μm respectively) at 22 C. Takeda et al[98] report mainly TGF in SiC hot pressed with Be (G ~ 10 μm), ~100% TGF with B (G~10 μm), and ~ 100% IGF with Al (G ~2 μm). These trends are supported by the survey of

Niihara and Hirai[99] of SiC, showing mainly TGF at 22 C in CVD bodies and those with out oxide phases, and mainly IGF in bodies with oxide boundary phases. They also reported that IGF generally increased with test temperature, but that CVD bodies or those made with nonoxide additions, frequently retained substantial TGF to at least 1400-1600 C.

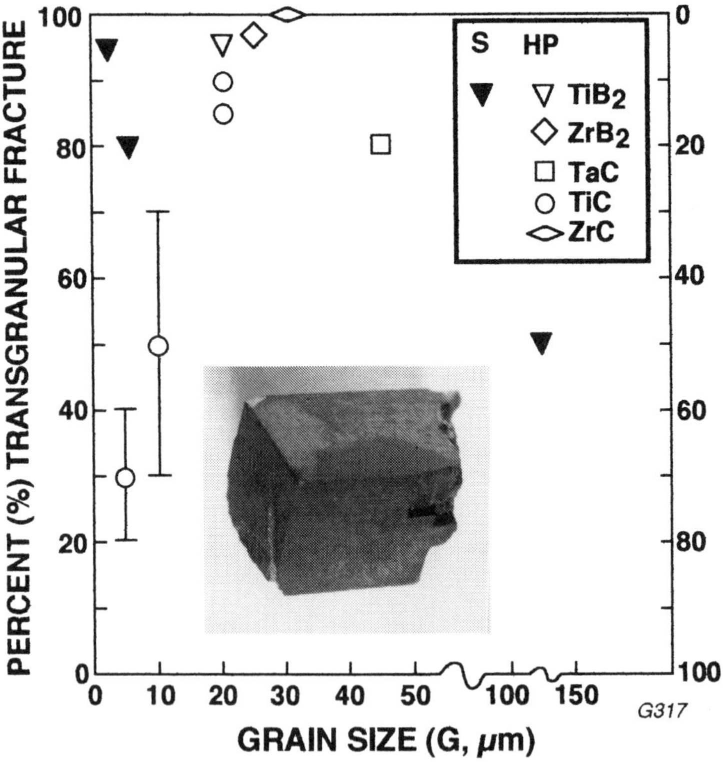

Figure 10) Plot of fracture mode versus grain size (G) for nonoxide ceramics flexure tested at 22 C. S=sintered and HP=hot pressed. Compare with Fig.1, especially the counter trend of Al_2O_3 and TiB_2 for more IGF at larger G. Insert-cleaved TiC crystal (~½ cm on a side) showing cubic, {100}, cleavage.

More limited data of this author for hot pressed TaC, TiC, and ZrC (G~20-45 μm) and of Leipold and Nielsen[100] for TaC (all cubic, NaCl structure materials with known or expected {100} cubic cleavage, e.g. Fig.10 insert) showed very predominant TGF (Figs.10,11). Harrison's[101] 22 C flexure data for hot pressed TiC (P~ 0.7-2.2%, G ~ 4-12 μm), while uncertain in specific IGF/TGF values showed a definite IGF to TGF transition with increasing G, which is consistent with the currant author's data (Fig.10), as well as that of Chermant et al[102] (G ~

6-20 µm, who also observed less TGF in compressive fracture at 22 C). Similarly Miracle and Lipsitt[103] showed mainly TGF in 22 C flexure testing of TiC hot pressed to C/Ti ratios of 0.93 to 0.66(G ~14 to 22 µm respectively). Kecskes et al[104] reported ~100% TGF for TiC made by hot pressing (G ~ 25 µm) and by dynamic compaction of hot bodies from self propagating synthesis (SPS) with a C/Ti ratio=1. Both increasing impurities and decreasing the C/Ti ratio (which was accompanied by increasing P) introduced some IGF.

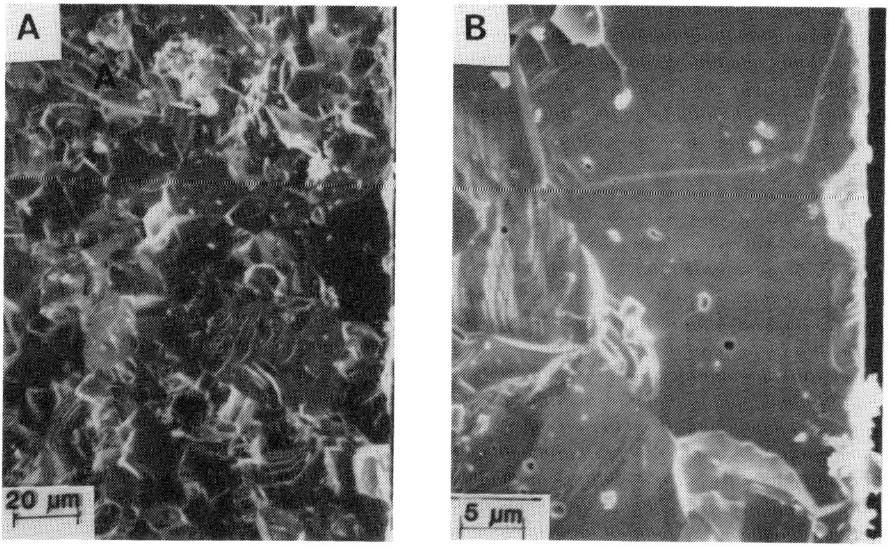

Figure 11) Fracture of hot pressed TiC at 22 C. A)lower magnification SEM showing essentially 100 % TGF. B) higher magnification SEM of the origin, showing all TGF there also. Note fracture origin at right center of photos.

LaSalvia et al [105]reported similar and different fracture mode behavoir in TiC made by reaction synthesis of Ti+C during high rate forging (yielding a few to several percent of mixed inter- and intra- granular porosity and G ~ 15 to 50 µ). Compressive fracture over strain rates of 10^{-4} to 10^{-2} sec.$^{-1}$ resulted in essentially 100 % TGF, but thermal stress fracture during rapid cooling was ~ 100 % IGF.

Miracle and Lipsitt[103] noted that all 4 of their TiC stoichiometries showed strength decreases associated with transitions to mainly IGF at 1100-1200 C (but at lower temperature for C/Ti =0.66). (Several pores, a fraction of the boundary facet size, were revealed by IGFon most grain boundaries. TGF at lower temperatures also showed some intragranular pores). Chermant et al[102] reported mainly IGF in pure WC (G ~6 µm) in both flexure and compression at 22 C.

Flexure tests at 22 C by this author of sintered and hot pressed TiB_2(P ~ 0.5-1.5 %) showed TGF quite dominant at finer G, but clearly mixed TGF/IGF at larger G (Fig.10). Thus TiB_2 showed a counter trend to that of other nonoxides at moderate and larger G similar to that of Al_2O_3 (Fig.1), but shifted to higher TGF. The higher degree of TGF in TiB_2 is corroborated by Baumgartner and colleague[106-108] showing nearly 100% TGF in both strength and K_{IC} tests of purer material (G ~1.5 -10 μm, the source of much of the TiB_2 tested by this author). They noted that the small amount of IGF observed was for the finer grains on the fractures, implying a transition to IGF at finer G(~1 μm)[106]. They did not directly report the fracture mode of their larger G (weaker) TiB_2, but noted it was microcracked [106], which would support greater IGF at larger G. A lower purity material (from carbothermic oxide reduction, apparently leaving some oxide boundary impurities) showed distinctly less TGF. Itoh et al[109] corroborate nearly 100% TGF in hot pressed TiB_2, but with this decreasing at finer G (~1-2 μm) on a given fracture and in bodies with such average G. Kecskes et al's[104] results for hot pressed TiB_2 and dynamically compacted -SPS TiB_2 showed mainly TGF in dense, purer TiB_2 (G ~25 μm), but increasing IGF with the presence of modest amounts of grain boundary Fe impurity or porosity (some of which also reduced G to ~10-15 μm).

Crack growth tests of some of the above TiB_2 materials in molten Al[107,108,110] still showed 100% TGF in the purer material, but considerable SCG in the less pure material. Tests of the purer, finer G, material at higher temperature (in an inert atmosphere) showed very little reduction in TGF until >1000 C. Flexure tests by this author of a hot pressed ZrB_2 at 22 C showed ~100% TGF, i.e. at least as much, and probably higher than for TiB_2 of comparable G (Fig.10).

Studies by O'Day and Letherman[110] and by this author and colleagues[112] of dense AlN sintered or hot pressed with CaO or Y_2O_3 additions (G<~5-15 μm) typically showed mainly IGF, with some tendency for an increase in the small amount of TGF at the larger G values. Alexandre et al[113] showed mainly TGF in ZrN (G~10 μm) HIPed from various powders, then fractured at 22 C.

Mc Kinney et al's substantial Si_3N_4 data[114] closely parallels that of SiC. Thus CVD Si_3N_4 from at least 3 different sources, having G ~20-200 μm, showed essentially exclusively TGF in various tests at 22 C, as is also typically the case for RSSN (G <1 μm). On the other hand Si_3N_4 made with oxide additions showed substantial, often dominant IGF, but with probable increasing TGF with G (Fig.12), which is also indicated by other literature data, e.g. of Himsolt et al[115]. Greskovitch and colleagues[116,117] showed mixed TGF-IGF in fracture of Si_3N_4 hot pressed with mainly $SiBeN_2$ plus some SiO_2, G~1-2 μm (with TGF of the larger, and IGF of the smaller, grains) at 22 C. Their sintered- HIPed material (G~3-5 μm) had mainly TGF at 22 C, but had slow crack growth via IGF at the fracture origins at 1400 and 1500 C, surrounded by TGF fast fracture[116]. Electron microscopy studies of this material shows very little grain boundary phase consistent

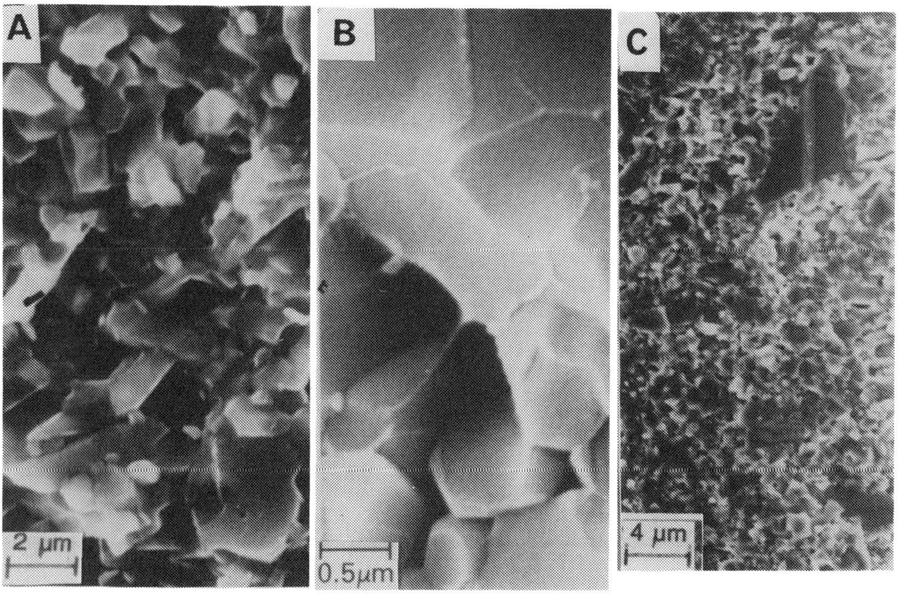

Figure 12) Comparison of hot pressed Si_3N_4 fractured at 22 C A) SEM showing typical mixed mode fracture B) higher magnification SEM showing IGF with higher grain boundary oxide content, and C) SEM of earlier hot pressed (pre- HS 110) showing TGF of an isolated larger grain..

with the much greater TGF. These results are supported and extended by Niihara and Hirais'[98] CVD Si_3N_4 showing ~100% TGF in larger grain, crystalline bodies, but transitioning to mainly IGF with diameters (of columnar grains) ~1 µm at 22 C. Their amorphous material apparently had conchoidal fracture (e.g. along growth cone boundaries). Grain boundary porosity or amorphous phases in their crystalline α-Si_3N_4 contributed to IGF. Their survey showed mainly IGF in Si_3N_4 hot pressed with MgO at 22 C and higher, but apparently mainly TGF at 77 K.

Fracture Mode Effects of Location on the Fracture, Test, and Material Variables

Location on the Fracture: The focus of the preceding section was on the overall fracture mode. However, while not studied extensively, data clearly shows that the fracture mode is not necessarily uniform over the fracture surface. Thus, for

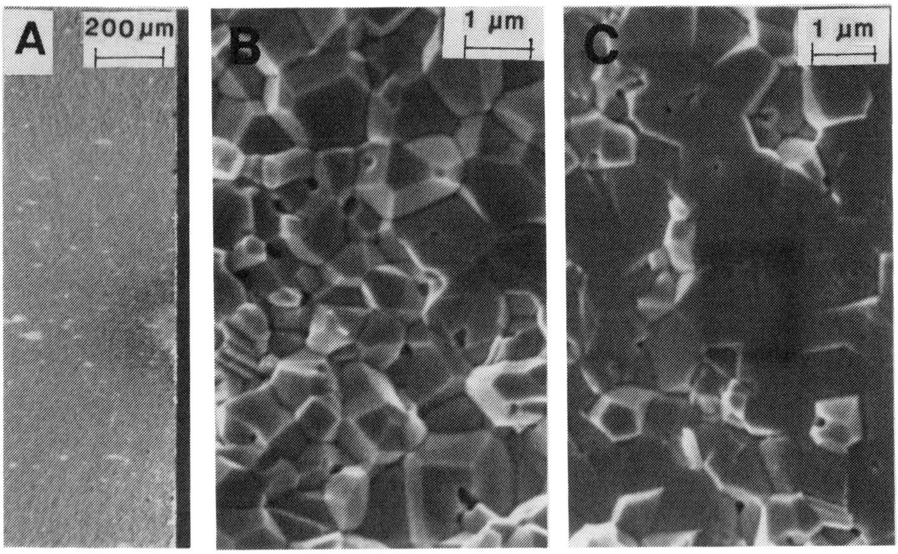

Figure 13) Preferential intergranular slow crack growth in hot pressed Al_2O_3 at 22 C. A) lower magnification SEM showing the complete flexure fracture and the fracture origin (right center of photo-the white, ~ semicircular region). B) higher magnification SEM of the origin (white) area; note all IGF. C) higher magnification SEM just outside the failure causing flaw, i.e. in the mirror region-darker region around the origin flaw in A). Note substantial TGF.

example failure initiation from 1 or a few grain boundary surfaces at 22 C, or as the initiation of the TGF-IGF transition at higher temperatures was indicated in the previous section for CaO and MgO (Fig.4). While some observations of machining flaw fracture origins showed no obvious change in fracture mode between the flaw and the surrounding fracture(44) this has not been examined in detail, except for slow crack growth (SCG) studies. Thus Mecholsky(111) showed that SCG (due to H_2O) in dense, fine grain MgF_2 was entirely IGF, with subsequent predominantly transgranular catastrophic fracture. Similarly, McKinney et al(129) showed SCG in (large G, CVD) ZnSe occurred by IGF in contrast to subsequent catastrophic TGF. Mecholsky(120) also reported more IGF in SCG vs fast fracture of Al_2O_3 at 22 C. Rice(6) showed that fine G (~0.7-4 μm) hot pressed Al_2O_3 (P ~ 0.5-16.5 %) failing from machining flaws, again showed SCG via IGF at 22 C. Subsequent fast fracture had some definitive TGF, except at the finest G (P~16.5 %), which all fracture was exclusively intrergranular (Fig.13). Rhodes et al(121) reported SCG in dense MgO (G~25-45 μm) being all IGF, with the extent of SCG apparently increasing with decreasing purity. However again note the

dominant at >1000 C, i.e. apparently more TGF at lower temperatures, but a similar TGF-IGF shift with increasing temperature as with flexure testing. Nash[131] reported that a higher purity (99.7%) Al_2O_3 gave mainly IGF, while a 97.5% Al_2O_3 (both with average G a few µm, with isolated grains to ~25 µm) gave mainly TGF in compressive test in air or physiological media at 22 C. Both showed considerable twinning at >50% of failure stress, especially from pores and in large grains, and with greater frequency in the lower purity material. Keski[132] reported mainly IGF in compressive testing of Bi_2O_3 (P ~ 4%, G ~5-10 µm) at 22 C. Lankford[133] showed mainly TGF to at least 500 C in compressive failure of commercial sintered α-SiC(G ~10 µm) at various strain rates.

Langdon and Pask[134] reported that dense MgO hot pressed with or without LiF additions(G~10-80 µm) had respectively mainly TGF and IGF at<1000 C (at 1200 C specimens were plastic). Patterson and Weaver[135] reported MgO hot pressed with 3% LiF (then annealed at 1300 C, giving dense bodies with G=10-15 µm) failed in compression by extensive IGF at 22 C. However, MgO sintered without any additive at 1700 C (P~3%, G ~30 µm) had mainly TGF (via {100} and {110} cleavage). Thus, while there may be some (possibly important) differences in fracture mode in compressive vs tensile failure, overall they show quite similar behavior, i.e. with G, impurities, and temperature. More limited studies of B_4C[136], TiB_2 and SiC[137], and AlN[138] failed in ballistic testing also indicate overall similarity of fracture mode to that found in flexure, as well as compressive[137], failure of the corresponding material at 22 C. However, some fracture changes were noted as the actual projectile point of impact were approached, e.g. within a few cm, in both B_4C and AlN. The changes were not clearly specified in the B_4C, but were reported as predominantly TGF near the impact site in AlN, then transitioning to the normal IGF with increasing distance from the impact site.

There are indications of other loading-temperature effects. Thus, Kirchner et al's[14] studies of fine grain, hot pressed Al_2O_3, showed quenching (e.g. from up to 1700 C) substantially reduced the amount of TGF in subsequent flexure testing at 22 C of the resultant strengthened specimens. Similarly, Kennedy and Bandyopadhyay[53] showed quenching of UO_2 from substantially lower temperatures (e.g. 700 C for thermal shock testing) also reduced the amount of TGF. More extreme was the essentially 100 % IGF of TiC in thermal shock vs 100 % TGF compressive in fracture[105].

Stucke and Wronski[139] reported on 3 MgO and 2 Al_2O_3 bodies as well as a UO_2 body fractured in flexure or diametral compression while under varying levels of hydrostatic pressure at 22 C.Two porous MgOs (P ~ 10 and 12 %,with G~20-30 µm) showed substantial IGF with and without superimposed hydrostatic pressure. The dense MgO (hot pressed with LiF, G~3-4 µm) showed complete IGF. The UO_2 (P~2%, G~10 µm) showed essentially 100% TGF with or without the pressure. Similarly, the Lucalox (G~ 90 μm) showed no effect of pressure on the 10-30% TGF, while the other alumina (P~ 2%, G ~10 μm) showed some possible effect of pressure to increase the substantial TGF (>50%) seen without pressure. Thus, while the superimposed hydrostatic pressure had definitive effects on the failure stress in some cases, it had, at best, an uncertain effect on fracture mode. Quinn et al[140] report predominant TGF for indent flaws in TZP (G~ 0.5 μm) versus mainly IGF for surrounding failure (which is consistant with limited observations of this author on other materials, e.g. earlier Si_3N_4).

Material Variables: Key material variables are composition (stoichiometry and the phase(s) present), microstructure (i.e. the grains, pores and second phases), and the crystal structure. Porosity can play a role, mainly in enhancing IGF. This appears particularly pronounced at elevated temperatures, where grain boundary pores enhance IGF in both tensile and compressive fracture, as noted earlier. There is less effect of porosity on fracture mode at lower temperatures. However, the uncertain and variable effects of porosity on fracture mode, in at least some of these cases, is probably due in part to the number and size of pores relative to G and to pore location. Thus, fewer, but larger, intergranular pores probably have less effect in enhancing IGF (e.g. the $MgAl_2O_4$ of Fig.8) than the same amount of much finer intergranular porosity. Pore location must play an important role, with intragranular pores contributing little if any to IGF, but may aid TGF (this may be a factor in some UO_2 TGF[54]). Clearly both pore location in the grains and the size and concentration of pores in the grains are likely to be factors.

The main effects of grain size have been presented earlier, showing it to have one of the most significant effects on fracture mode (e.g. Figs. 1,3). Thus, broad support was given for increasing IGF at finer G, typically at G < 1-5 μm, though with considerable variation in the IGF increase with decreasing G. Much variation is related to grain size issues discussed earlier, and below. Some increased IGF at fine G is probably extrinsic due to residual boundary porosity, impurities, or both, commonly resulting from the lower temperatures and finer powders used to get fine G. Anion impurities, which can be particularly prevalent in hot pressed oxide bodies[141] may often be an important factor in IGF in them. However there may also be an intrinsic effect based on TEA, and especially EA, discussed below. At and above G~1-5 μm, TGF is, or becomes, dominant as G increases. In a few cases TGF may decrease with increasing G at larger G (e.g. Al_2O_3, Fig.1, and TiB_2, Fig.10), or at very large G (e.g. ZrO_2, Fig.1), which may also be true of other materials since so little data exists at such large G values. Some of the decrease in TGF at large G in materials such as Al_2O_3 and TiB_2 may also be extrinsic due to increasing grain boundary impurities at larger G. However, there are arguments for intrinsic IGF at larger G due to TEA and EA, as discussed below.

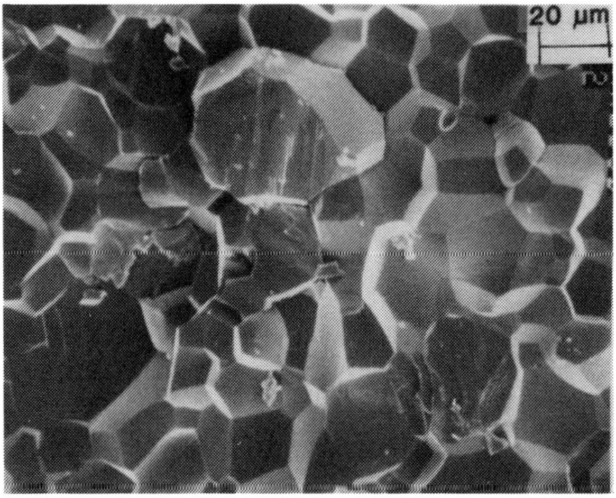

Figure 14) Predominantly IGF in hot pressed AlN at 22 C. SEM shows mainly IGF, but some isolated TGF, mainly of grains larger than average.

One factor related to G, is the occurrence of slip and twins. These, especially twins, can influence fracture, e.g. as shown in Lankford's[129,130] and Nash's[125] compressive and Becher's flexure[142] studies of Al_2O_3. Whether the extensive twinning in B_4C is a factor in it's extreme TGF (e.g. relative to Al_2O_3, with a very similar Young's modules and TEA as Al_2O_3) is unknown, in part because cleavage information on B_4C (and BeO, which also has similar TEA) is lacking. While slip and twinning probably play a role in specific cases, general G trends are due to broader effects discussed below. Clearly closely related factors are grain : 1)size distribution, 2) orientation, and 3) shape, especially shape, for which information is limited.

The effects of grain size distribution appear to universally be consistent with the effects of average G. Thus, in bodies showing mixed fracture mode with decreasing TGF with decreasing average G, finer grains on a given fracture will show more IGF than the larger grains, which will have (often much) more TGF than the finer G matrix.(e.g. Figs. 6, 9). Thus, the preferential TGF of larger grains in bodies in which an increase in TGF with increasing G is not known (due to the lack of larger grain bodies for test, e.g. AlN, Fig.14) implies that TGF will increase at larger G in that material. On the other hand, in materials such as Al_2O_3, where both TGF and IGF occur, but with the latter often increasing with increasing average G at larger G, IGF is also observed more often in larger isolated grains or clusters of them. This appears to be especially common for isolated or clustered larger grains in Al_2O_3 that are platey in nature (Fig.15).

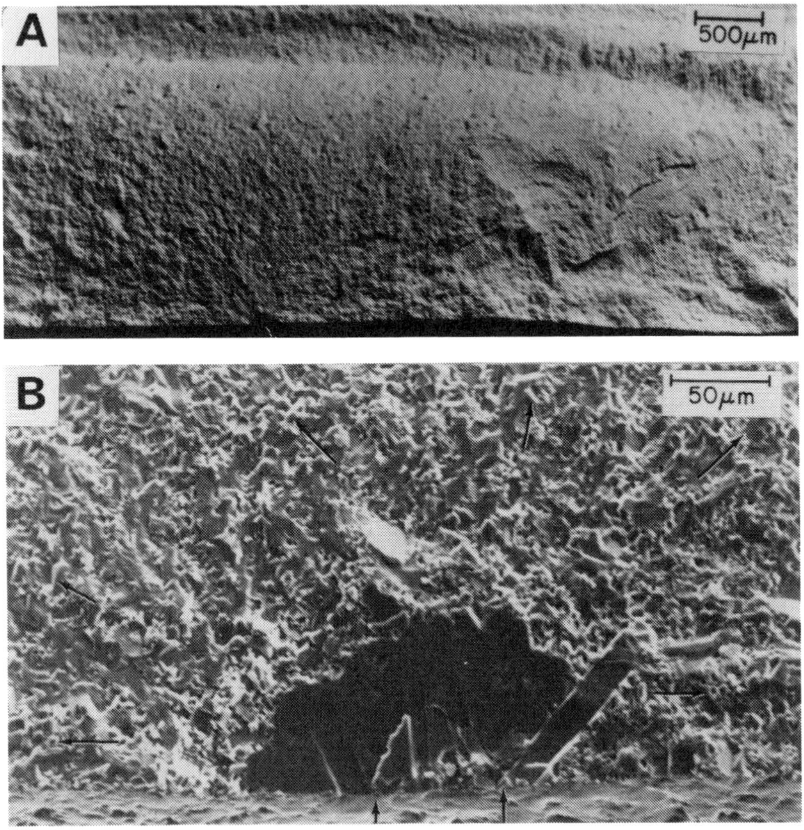

Figure 15) Intergranular fracture (IGF) of larger, especially platey, grains at fracture origins in hot pressed Al_2O_3. Note also transgranular fracture of a smaller large grain, and of many of the grains of the main population.

Preferred grain orientation in bodies can have measurable effects on mechanical properties, including those involving fracture, e.g. K_{IC} and strength, and hence possible effects on fracture mode. Thus, hot extruded CaO and MgO (Fig.4) with resultant <100> axial textures had more TGF than hot pressed bodies of comperable density and grain size. However, how much of this greater TGF is due to the orientation and how much to cleaner grain boundaries (formed by recrystallization) is not known. Limited studies of preferred orientation , e.g. of Salem et al[27] are insufficient to discern possible fracture mode differences. Similarly, while Iwasa et al[144] have shown marked effect of preferred grain orientation on K_{IC} at 22 C in commercial Ba hexaferrite (P ~ 5%, G< 1 μm), and shown definite effects on the overall fracture character, the effects on actual fracture mode (which appears to be mixed IGF/TGF) are not clear. Veldkamp and Hattu[25], while also showing substantial K_{IC} effects and some effect on fracture topography due to preferred grain orientation in a sintered Sr hexaferrite (P ~5%, G~5 μm) fracture both perpendicular and parallel to the C-axis texture, showed essentially 100 % TGF for both orientations. Limited data for jades of varying composition, grain size, and preferred orientation showed a transition from mainly TGF at larger G to mainly IGF with finer G (~1 μm dia.), very elongated, and highly oriented grains (which also had higher strengths and toughness)[143].

As noted earlier, measurement of G is an important issue, especially when there is a change in fracture mode as G changes. It is less criticalt when there is no significant change in fracture mode with G. While many studies do not even address G, despite it's frequent importance, many that do address G give little or no information on it's measurement (or variation). Thus many use the linear intercept method, but do not state whether a factor was used to convert the linear intercept to a "true" grain size. This commonly makes 50% or more difference in values, and may make larger, e.g. 100 % or more differences[145]. Others may measure grain diameter on a fracture or polished surface, which results in similar, or larger differences. As recommended elsewhere[19,145], measurement of average grain diameters is best to better reflect the effects of grain size distribution and shape, in particular to account for the greater fracture area occupied by larger grains since fracture is an area generating process. The average grain diameters can readily be weighted to reflect the grain areas on the fracture. Measurement should be on fracture, rather than polished, surfaces since the former will reflect any possible bias in the fracture process with regard to grain size or shape.

While improved measurement of G is very important, improved measurement of the amount of transgranular and intergranular fracture is also important, particularly when there is a significant mix of fracture modes. Many of the measurements are not particularly precise, often being only visual estimates that are variable by 10-20% or more. Truly quantitative results are fairly laborious to obtain, and these are still only qualitative or at best semiquantatative with regard to the impact of grain size distribution on IGF vs TGF on a given fracture. An important step would be computer automation of fracture analysis, preferably in combination with stereological analysis. The work of Yang et al[146] in using software to interpret SEM line scan profiles to give TGF/IGF data is promising. They show good correlation of their computer results with visual analysis for both an alumina and an alumina-SiC whisker composite. They incidently also showed more TGF in the regions of fast, versusstable fracture, in both materials, consistent with results presented earlier.

A major extrinsic factor is grain boundary phases, such as oxide grain boundary phases in nonoxides as noted earlier. The role of oxide grain boundary phases in oxides, e.g. Al_2O_3, is more complex. Some of these phases, e.g. CaO, MgO, and SiO_2, may enhance IGF and SCG (via IGF). However, inter-actions between impurities, additives, and the body material can be complex, e.g. de With[17] showed reduced strength and toughness in Al_2O_3 with MgO+CaO additions vs with MgO alone due in large part to grain size and distribution (e.g. bimodal) changes, but similar SCG for the bodies with MgO+CaO and with CaO only. Al_2O_3 with only MgO additions (e.g. Lucalox) appears to often have more IGF than other aluminas, as shown earlier, and indicated by results of Taylor et al[147] who showed some Ca boundary segregation with MgO additions. Thus, the variability of IGF in Lucalox may reflect variability in residual boundary phases, especially of MgO and CaO. However, some, e.g. SiO_2-based phases prevalent in many commercial aluminas may possibly enhance TGF (e.g., as noted earlier in Nash's compressive tests[131]), for example by relaxing microstructural stresses (discussed below). Additives or impurities may also impact fracture mode by effecting G. Thus, Koyoma et al[148] showed additions of CaO+SiO_2 to Al_2O_3 increasing G from ~2 to ~20 μm with fracture mode going from 100% IGF to mainly TGF.

It is well established that grain boundary phases that exhibit high temperature flow greatly enhance IGF at elevated temperatures. However, the effect of these phases also depends on the extent of boundary coverage, which in turn depends on the amount of the phase, it's degree of boundary wetting, as well as on grain size, shape and probably orientation. The better high temperature strength of Lanxide Al_2O_3 vs commercial 99% Al_2O_3 is attributed to the presence of molten Al in the processing of the former preventing the formation of SiO_2-based phases[149]. Similarly Crandall et al's[3,150] data showing much less, or no, decrease in high temperature strengths of Al_2O_3 hot pressed with a few percent SiO_2 additions vs no additions may reflect effects of the reducing processing environment on SiO_2 stability (hence presence) and wetting (hence distribution). The well known strain rate sensitivity of strength of such bodies with softening grain boundary phases implies a possible strain rate sensitivity of the resultant fracture mode that has apparently not been examined.

Another composition issue is stoichiometry. As noted earlier stoichiometry changes due to processing in reducing environments (e.g. hot pressing) and re-oxidizing by annealing can have substantial effects. These include: 1) moderate shifts to more TGF at finer G for hot pressed TiO_2[40] (Fig.1), and 2) more extreme, opposite changes (i.e. from ~ 100 % IGF in partially reduced, to ~ 100 % TGF on air annealing) of Y_2O_3[47], PLZT[77], and a strontium calcium ferrite[78].

The issue of second phases distributed in other fashions than as grain boundary phases gets into the area of particulate and other composites. This is potentially a large subject in itself, but with limited data available, a brief outline of which follows. This author has observed considerable TGF in many particulate and whisker composites fractured in flexure at 22 C. Much of the TGF was in the matrix (commonly Al_2O_3, G~1-10 μm), but frequent TGF of the dispersed phase (eg, TiC) is seen. Thus, laboratory and commercial Al_2O_3-TiC (G ~2-5 μm, strength >700 MPa) showed >50% TGF at 22 C, while Al_2O_3-ZrO_2 bodies (G ~2-5 μm, strength ~700 MPa) commonly showed 10-30% or more TGF, i.e. similar to, possibly more than in, bodies of either phase alone with the same G. Again, in both cases the TGF preferentially occurred in the larger grains on the fracture surface. These results are similar to those for ThO_2, again hot pressed with (3 wt % Y_2O_3), but with addition of 25 m/o ZrO_2[(51)] (with some reduction in grain size). Krell and Blank[(151)] also reported considerable TGF in their Al_2O_3 - TiC composites fractured at 22 C. The ready cleavage of some of these dispersed phases (e.g. TiC) is probably a factor in the TGF of the Al_2O_3 matrix in these composites. However, there clearly also is substantial TGF in the Al_2O_3 matrix with SiC whiskers, which frequently fracture both transversely and along the

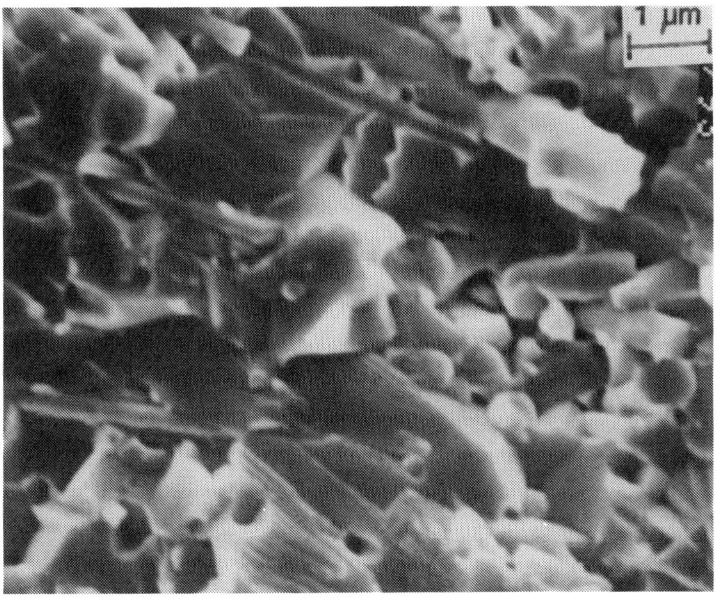

Figure 16) Flexure fracture of commercial (Cercom) Al_2O_3-SiC whisker composite at 22 C. Note substantial TGF in the Al_2O_3 and considerable transverse fracture of the whiskers, as well as considerable interfacial fracture (i.e. IGF) around part of many SiC whiskers.

Al_2O_3-SiC interface (Fig.16). The degree of TGF in the matrix of these composites is typically higher than in the matrix alone (of the same G). This is clearly shown by work of Baek and Kim[152] where the matrix alone (G~6 μm) had predominantly IGF, but with 20% SiC whiskers the matrix (with ~ the same G) had mainly TGF. Similarly, Dauskardt et al[153] reported essentially 100% TGF in the Al_2O_3 matrix (G ~2 µm) of Al_2O_3-SiC_W composites fractured in fatigue tests. Thus TGF is typically enhanced by the presence of the disbursed phase.

WC-Co composites, which have been extensively studied, corroborate both the variability and basic trends for other ceramic materials. Thus, earlier studies of flexure fracture at 22 C by Gurland and Bardzil[154] showed 50-75% TGF of WC grains for G ~1-4 μm, while Murray[155] showed, 10% for similar G material in K_{IC} tests. However, Hara et al[156] reported a transition from IGF to TGF as G increased. Pickens [157], Pickens and Gurland[158], and Roebuck and Almond[159] conducted their own studies and surveyed the WC-Co literature. The latter reported that TGF increased with G and the amount of Co. Pickens agreed, but was more quantitative, indicating more IGF for G <2 μm and more TGF for G=2 μm and above. He also notes more TGF of larger grains in a given body (which thus can vary TGF with G distribution), and cited possible effects of crack velocity and of G measurements. Comparison with the one data point for pure WC[102] suggests that WC-Co has more TGF at comparable G.

Effects of Material Properties: The three material properties expected to be important factors in fracture mode are fracture toughness, elasticity, and thermal expansion, with anisotropies of these being particularly important. Thus the absence or occurrence and K_{IC} level of preferred cleavage planes, the occurrence and level of TEA, and the level of EA are prime parameters. A summary of these properties and the typical fracture mode at 22 C for materials where most, or all, of the properties are known (Table 1) shows there is no single factor comprehensively explaining fracture mode behavior. However, substantial insight in understanding this behavior can be gained from considering the above three factors and their interactions.

The presence or absence of cleavage planes is clearly a major factor in the overall fracture mode, as well as in complicating correlations with other factors (discussed below) in view of uncertainties in them. Thus while many associate IGF with weak grain boundaries, which can clearly be important, IGF may also reflect the absence of easy cleavage. Thus, for example, more TGF in MgO than in Al_2O_3 probably reflects the easy {100}, and probably some {110}, cleavage in MgO and the absence of similarly easy cleavage in Al_2O_3, Fig.17. The easy cleavage on both {100} and {110} in $MgAl_2O_4$ is probably a major factor in it's extensive TGF (in the absence of residual boundary phases), despite it's fairly high EA (e.g. > 6%,Table 1). Similarly the extensive TGF in materials such as CaO, TiC

Table I. Comparison of Fracture Modes and Material Characteristics[a]

Mat.	Cleavage	Av. $\Delta\alpha$[b]	Max $\Delta\alpha$[c]	EA (%)[d]	Fracture Mode[e]
Al_2O_3	none pronounced, if any	0.3	1.1	1.7	Mixed
BeO	uncertain, possibly basal	0.7	1.1	0.4	Mixed
CaO	{100}primary,{110}secondary	0	0	0	TGF
MgO	{100}primary,{110}secondary	0	0	2.3	TGF
TiO_2	{110}primary,{100}secondary	1.3	1.9	11	TGF
ZrO_2 (cubic)	Probably mainly {111}	0	0	6-10[f]	TGF
$MgAl_2O_4$	{100}primary, {110} also primary or close secondary	0	0	6.9	TGF
β-Al_2O_3	basal	0.2-0.7[f]	0.3-1.1[f]	--	TGF
TiB_2	unknown	2	3	~6.3	Mixed
ZrB_2	unknown	--	--	--	Mixed
B_4C	unknown	0.6	0.9	--	TGF
TiC	{100}primary	0	0	0.1-0.9[f]	TGF
ZrC	{100}primary	0	0	0.3-1[f]	TGF

a) Data from refs.160-163; -- shows lack of data.
b) Difference between extreme crystallographic and polycrystalline thermal expansions and c) the two extreme crystallographic expansions (typically to 1000 C). Note, cubic materials intrinsically have 0 values.
d) EA = elastic anisotropy; after Chung and Bussem[160]; 1% EA means the maximum/minimum Young's modules ratio in the crystal is ~1.2, ~ 1.4 at 2% EA , and ~2 at 7% EA.
e) At moderate to large G (eg.>10 μm).
f) Depends on composition.

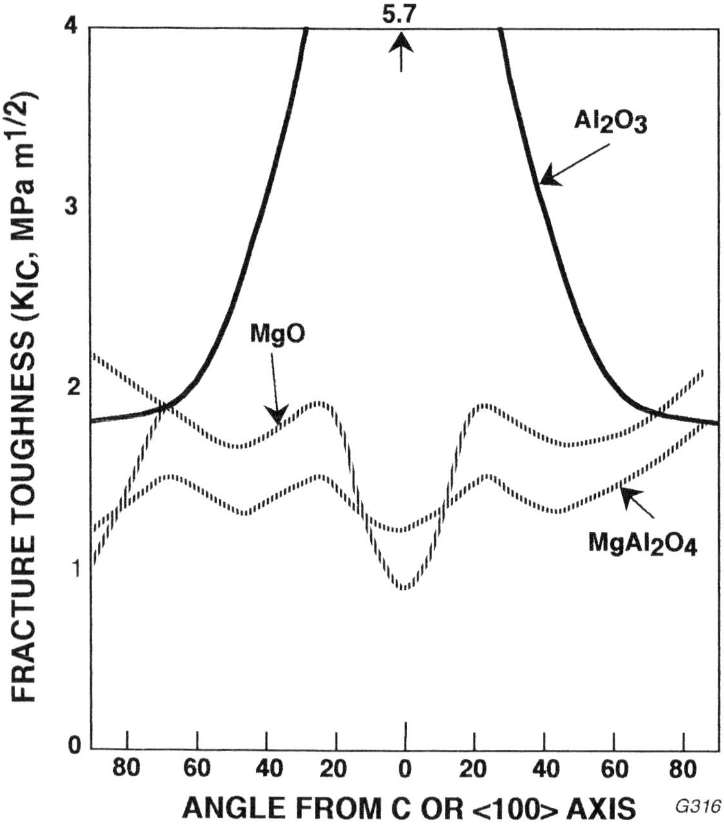

Figure 17) Semiqunatitative plot of the fracture toughness of Al_2O_3, MgO, and $MgAl_2O_4$ crystals as a function of crystal orientation, i.e. for fracture on a plane normal to a line having the angle plotted relative to the C or <100> axes. This is semiquantatative first, since for MgO only the {100} K_{IC} is known ({110} fracture is known to be much less frequent and extensive, but it's K_{IC} isn't known, other than it must be a fair amount higher than that for {100}). Second, while the K_{IC} for both plane sets is known for $MgAl_2O_4$, the transition, as a function of orientation, between the two planes is not known, nor is it for MgO.

and ZrC is probably much more due to their pronounced (i.e. low K_{IC}){100} cleavage (Fig. 10) than to stronger grain boundaries. Similar considerations apply to the relative trends amongst noncubic materials. Thus, TiB_2 and TiO_2 have more TEA and EA than Al_2O_3 and BeO, but Al_2O_3 has the most IGF. This probably reflects an increasing ease of cleavage in the order: Al_2O_3, BeO, TiB_2, and TiO_2 to balance probable effects of their anisotropies. Similarly, the dominance of TGF in B_4C at lower temperatures and moderate to large G must reflect easy cleavage or other easy fracture in it, not just its lower TEA. In materials that twin, e.g. Al_2O_3 and B_4C, preferred fracture paths along twin-matrix interfaces, reported in Al_2O_3[123,135] may be an added factor and complication (e.g. introducing the dependance of twining on G), but is uncertain in its scope, which depends on the extent of twining. The level of the K_{IC} of preferred cleavage (or other fracture) planes is probably most important, but the number of such planes must also play some role. Thus, while beta-Al_2O_3 has TGF, presumably due mainly to cleavage of it's single basal plane, the multiple cleavage of $MgAl_2O_4$, i.e. {100} and {110} giving respectively 3 and 4 independent cleavage planes, is probably an important factor in TGF occurring so extensively (in the presence of boundary porosity, but absence of boundary phases) in spite of it's high EA. Similar cleavage of TiO_2 is probably a factor in it's high TGF despite high EA and TEA. While cleavage (or other easy fracture) is clearly very important, it does not explain all behavior. Other key needs are understanding the: 1) decreasing TGF at respectively larger and very large G of Al_2O_3 and B_4C, and of ZrO_2, 2) frequent decrease of TGF at fine G at 22 C in most, possibly all, materials, and 3) increasing IGF, especially at fracture origins (eg. of CaO and MgO) at higher temperatures, despite surrounding TGF.

The presence and ease of cleavage also plays a role in local fracture mode variations, but with an even more critical tradeoff with grain boundary character in the case of SCG. Thus, for bodies having substantial TGF overall, SCG via TGF appears to reflect substantial SCG for single crystal cleavage or fracture surfaces, and SCG via IGF reflecting less, or no, single crystal SCG. Hence, SCG by 100% IGF in MgO reflects, at least in part, the absence of SCG on {100} (and presumably {110}) crystal cleavage planes[4,164]. On the other hand similar levels of substantial SCG on both {100} and {110} crystal cleavage planes of $MgAl_2O_4$ is consistent with SCG via TGF in polycrystalline $MgAl_2O_4$[164,165]. Similar SCG via TGF in CeO_2 and TiO_2[166] implies similar single crystal SCG. SCG via IGF in MgF_2 and Al_2O_3 (but more TGF in fast fracture), implies more limited single crystal SCG, though there clearly is some in sapphire.

Thermal expansion anisotropy (TEA), while occurring only in noncubic materials, is potentially important in the IGF-TGF balance in them because of its generation of added stresses and their concentration near, or at, grain boundaries. Spontaneous cracking from TEA, which begins at a minimum G value and is a function of the Young's modules, the TEA, and grain boundary fracture energy, increases in frequency and extent with increases of G beyond the minimum G[161]. Much, but not all, of the resultant microcracking is along grain boundaries. Such cracking probably also occurs in stressing a body to failure as a result of the superposition of applied and TEA stresses, which is likely to be a factor in the IGF of noncubic materials. Thus, the increasing IGF at larger G noncubic Al_2O_3 and TiB_2 is probably due, at least in part, to TEA stresses. The ~0% IGF in B_4C, TiO_2 and β-Al_2O_3, which have similar, or greater, TEA probably reflects the availability of easy cleavage (and possible effects of twins in B_4C). TEA stresses in materials susceptible to SCG , e.g. Al_2O_3, may be a factor in SCG via TGF, and in more IGF in mist, hackle, and crack branching. TEA stresses inherently decrease with increasing temperature (disappearing at the stress relief or processing temperatures, but then increasing above the processing temperature). Thus TEA is not likely to be a major factor in IGF at high temperatures, but is probably an increasing factor as temperatures decrease below stress relief temperatures (eg. below 1000 C). Where TEA is of possible significance, it's effects are probably complex in view of it's dependance on processing temperatures, with the opportunity for intrinsic or extrinsic (due to boundary phase) stress relaxation at high temperatures complicating it's processing relation. Local, as well as global, variations in preferred grain orientation can be another serious complication, as is interaction with EA. Some or all of these may be factors in the (anomalously) high TGF of large grains in Al_2O_3 refractories.

Elastic anisotropy (EA), though generally neglected, is probably an important factor in fracture mode. It enhances or reduces applied stress at grain boundaries, similar to the occurrance of TEA stresses there. However, EA, which occurs in both cubic and noncubic material, while low to moderate in many material, is moderate to substantial in a number of noncubic materials and is often higher in many cubic materials[(163)] (Table I). Complications of defining the role of EA in effecting fracture mode arise from the lack of data and it's complex nature and interaction with other factors, e.g. TEA (which it can reinforce, counteract, or both). However, it seems probable that EA aids IGF locally and globally (similar to TEA). This may be favored at larger G, e.g. as indicated for ZrO_2, since, as for TEA, the volume of strain energy driving fracture of each grain should scale with the volume of the grain involved, but the fracture energy requirements scale only with the grain fracture area. However, EA may also aid IGF at finer G not only because of the greater number of grains in the high stress area near the crack tip, but also because more choices of grains should give more opportunity for fracture along a path of net boundary tensile stresses from EA near the crack tip. Thus, Rice[(4,5)] suggested that EA can contribute to IGF at finer G by enhancing such fracture shortly ahead of the crack tip (Fig. 18). Besides interaction with TEA, another complication is local or global preferred grain orientation in bodies. EA has been suggested as a factor in the generally modest fracture energy and toughness maxima as a function of G that often occur in cubic ceramics[(167,168)], and has been correlated with increasing grain boundary cracking and spalling around hardness indents, (along with boundary phases)[(169)].

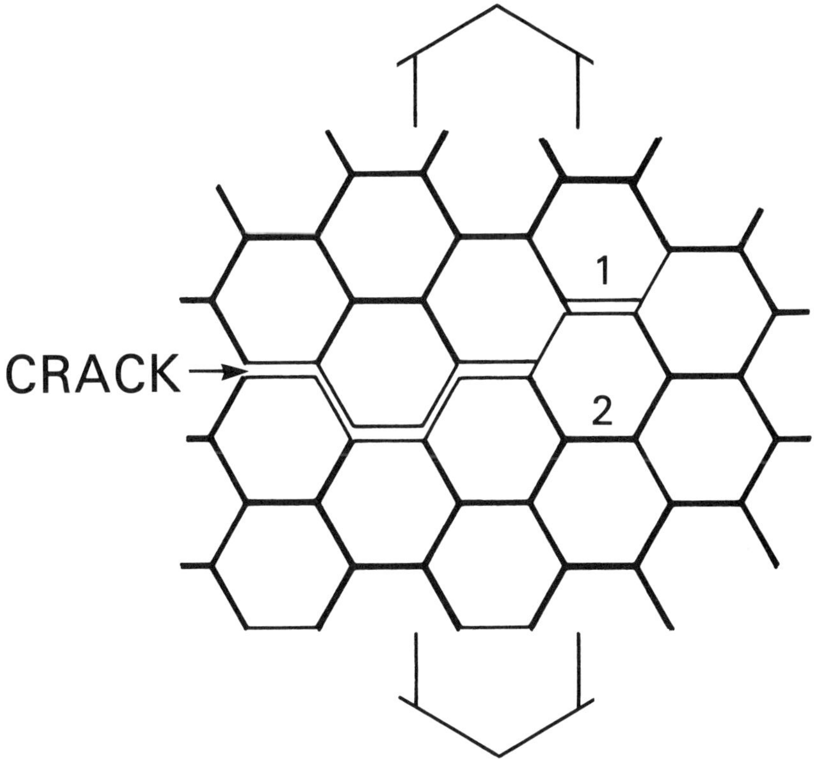

Figure 18) Schematic of how grain boundary stresses from EA and TEA may aid IGF. A crack, as shown (from the left) generates stresses on grain boundaries ahead of it due to EA (or experiences TEA stresses in noncubic materials). While varying levels of tensile or compressiven boundary stresses occur, boundaries with higher levels of tensile stress (eg. 1) will fracture, even somewhat beyond the immediate crack tip. Many boundaries with compressive stresses (eg.2) are readily by-passed. However there are two counter trends: 1) the occurrence of cracking from EA stresses probably increases with increasing G in a similar fashion to cracking from TEA stresses, but 2) the level of enhanced tensile (or compressive) boundary stresses from EA rapidly decrease ahead of the crack, giving higher average EA stress levels on smaller versus larger grains which each have their boundaries the same distance from the crack. (TEA stresses do not depend on distance from the crack tip and hence on G, so they can extend to large G.) Also, smaller grains give more opportunities for cracks to avoid EA boundary compressive stresses (as sketched above), and thus presumably giving more opportunity for EA effects to manifest themselves.

An important factor in sorting out effects of EA on fracture mode is it's effects on the temperature dependance of fracture mode, which is of importance itself. EA changes with increasing temperature can be to increase or decrease substantially, or very little, depending on the particular material[163]. However, several important ceramics such as CaO, MgO, and ZrO_2 show substantial increases in EA with temperature, which correlates with the IGF increase of these materials with increased temperature after effects of grain boundary impurities are minimized (eg. in recrystallized crystals). Similarly, the degree of intergranular separation of large columnar grains of CaO, MgO, and ZrO_2 on cooling from skull melting appears to correlate with their levels of EA at higher temperatures. While the data base is very limited, both basic principles and some data show that effects of EA must be considered. Thus, EA is probably a factor that aids IGF where it is high, with the effect and level required probably being interactive with other factors such as grain boundary impurity type, level, and distribution, as well as competition with cleavage.

SUMMARY AND CONCLUSIONS

The overall as well as the local degrees of TGF or IGF are important indicators of the fracture conditions and hence an important factor in fractography and understanding mechanical behavior. Despite this importance, attention to fracture mode has been limited, similar to the neglect of fractography itself. However, sufficient data exists to show overall trends, e.g. mainly overall TGF at lower temperatures, as well as probable or possible effects of other variables. Variations in results reflect the impact of other factors, as well as limited characterization and the need for more understanding. Trends and research needs are outlined in TableII for specific parameters, followed by a general summary of major trends and issues.

Table II. Summary of Effects of Parameters on TGF/IGF and Their and Status/Study Needs

A)Microstructure:

1) Grain Size(G): Considerable data shows increasing G generally increases TGF, especially at lower temperature, but exceptions (e.g.Al_2O_3 and TiB_2) occur./More and better data needed, especially on coupling with other parameters.

2) Grain Shape: Possible effects probably heavily impacted by orientation of elongated grains and cleavage planes relative to stress./Some study needed as part of more comprehensive effort.

3) Grain Orientation: Limited data shows some uncertain effects, reflecting

impact of it's interaction with other parameters./Study needed, as above.

4) Pores: Intergranular pores can enhance IGF, mainly when pore size <<G, but must depend on pore concentration and the balance with cleavage, and probably with intragranular pores. Open porosity may enhance SCG. Intragranular pores probably enhance TGF if there are sufficient pores and cleavage (or other easy fracture) is feasible. Pore size relative to G and location in grains probably factors besides amount of porosity./Better and more comprehensive data needed along with modeling.

5) Second Phases: Grain boundary phases can greatly enhance IGF, especially at high temperature, but effects can be quite variable, including possible increased or decreased SCG, or increased TGF, e.g. by limiting TEA. Dispersed particles commonly extend matrix TGF to substantially finer G (and frequently exhibit substantial TGF themselves)./Much study needed, especially of grain-boundary composition, wetting and boundary coverage, as well as prarticle size, shape, orientation, and location relative to the grains.

B) Test Parameters:

1) Temperature: Clearly increasing it increases IGF, but rates and extents can depend greatly on other factors, especially boundary phases, and probably grain orientation and EA. Effects of temperature <22 C, e.g. via changes in TEA and EA are not known (other than reducing SCG and thus IGF)/Need substantially more detailed studies.

2) Environment: Main effect is via SCG, which often occurs by IGF (but not universally, a case of SCG via TGF with fast fracture by IGF is reported), and may be aided by open intergranular pores./Much more study needed.

3) Compressive, or Superimposed Hydrostatic, Stress: Some possible effects of such stresses (e.g. in flexure) reducing some the extent of TGF, but no large effects know./Some study useful.

4) Multiaxial Stress: Limited data indicates limited, or no, effect./Some study of possible coupling with other parameters (e.g. TEA and EA) useful.

5) Strain Rate: Ditto above.

C) Material Parameters:

1) Composition: Can be important in determining properties such as cleavage, TEA, and EA./Study recommended in this context,

2) Crystal Structure: Ditto above.

3) Cleavage: Important in TGF vs IGF, especially at lower temperature./Much more crystal cleavage and K_{IC} data needed along with broader experimental and theoretical study.

4) EA: Some materials have high EA at lower temperature, and EA can also increase substantially with temperature. Appears to be a possibly important factor in higher IGF, especially at very large G of ZrO_2 and at higher temperature of some cubic materials. /More data and study (experimental and theoretical)

needed.

5) TEA: Probably a factor in IGF of some noncubic materials (eg, Al_2O_3 and TiB_2) at lower temperatures. Effects should disappear on reaching the processing or stress relief temperatures (but then increase again with increases above the prossing temperature)./Some further experimental and theoretical study useful.

D) Location on Fracture: Is a factor, e.g, some reduction of TGF at mist, hackle, and branching, and hence of possible broader significance to fractography, both in practice and in basic understanding./Further study recommended.

This survey corroberates two generally accepted observations about fracture mode: 1) at low to moderate temperatures the overall fracture mode is mainly TGF, except at fine grain sizes, and possibly(typically very) large grain sizes, and 2) at sufficiently high temperatures the predominant mode becomes IGF. IGF is seen not simply as a result of grain boundary weakness as proposed by some, but as a balance between grain boundary fracture and transgranular fracture (mainly the ease of grain cleavage). Factors shifting this balance have greatest effect where there is a relatively close balance between these two factors. An important, but unexpected, factor found in this survey is that of stoichiometry. Though seldom examined, the four cases where it has been noted, show measur-able, often extreme effects. Since the effects are not always in the same direction, this may indicate their effects are via changing other properties such as TEA or EA, since these could have differing changes with changing stoichiometry. However, changes in grain boundary character with at least some of these stoich-iometry changes is clearly a complicating factor.

Much of the common increase of IGF at finer grain size at lower temperatures is probably intrinsic due to TEA and EA stresses at grain boundaries. Also, at very fine grain size the crack deflection caused by IGF may become negligible, which combined with the larger volume fraction of the more disordered grain boundary structure may favor IGF. (The latter may also slow, or even reverse, the trend for strength to increase with decreasing grain size.) However, some IGF at finer grain sizes may be extrinsic due to lower processing temperatures and finer powders used to obtain fine grain sizes leaving residual boundary, especially anion, impurities. These intrinsic and extrinsic factors thus both lead to the general trend for TGF to increase with increasing grain size. This occurs both as the grain size increases from body to body, as well as with the larger grains in a given body (where TGF increases with the average grain size), i.e. fracture of larger and smaller grains in a given body follows the trend for the effect of average grain size on fracture mode. As grain size increases at very large grain sizes, e.g. as found in skull melts, some materials such as ZrO_2 which have substantial EA may again exhibit considerable, to complete, transition back to IGF.

There are also a few cases, e.g. Al_2O_3 and TiB_2, where IGF increases at intermediate to larger grain sizes within the normal range of grain size (e.g.. at $G \geq 10\mu m$) at low and moderate temperatures. Again some of this may be extrinsic due to greater concentration of impurities at grain boundaries. However, TEA and possibly EA are likely to be intrinsic factors that also enhance extrinsic effects of deleterious impurities. Intergranular porosity aids IGF some at lower temperatures (and substantially more at higher temperatures). It appears to have most effect at finer pore and grain sizes, i.e. mainly accelerate the onset of IGF, and is most effective when cleavage is more difficult. Conversely, intragranular porosity may aid TGF, depending on the amount and size and location of such pores relative to the grain size, and the ease of cleavage. Results show other parameters such as the stress state (e.g. tension vs compression), stress rate (e.g. ballistic impact), thermal shock, etc., and other microstructural characteristics such as grain shape and orientation can be (usually modest) factors in fracture mode. However, they need more study to make them more useful tools of fractography. Again boundary phases can have complex effects. Thus, for example, LiF addition to $MgAl_2O_4$ and oxide additions to Si_3N_4 and SiC enhance IGF and crack bridging, but may decrease strength and toughness ($MgAl_2O_4$), or introduce slow crack growth. Also, in ceramic composites, especially particulate composites, the second phase often enhances TGF of the matrix, and possibly the dispersed phase. Whether this is an intrinsic effect, or is extrinsic, e.g. reflecting higher processing temperatures, but with finer grains (due to the presence of the dispersed phase) and hence less anion impurities, is not known. However, this raises questions of the role of bridging commonly attributed to such composites since bridging is typically most effective with IGF.

The above are trends for the overall fracture, which may often have a relatively uniform fracture mode. However, there can be, sometimes substantial, local variations of fracture mode, again especially near grain sizes, or other parameters near a transition in fracture mode . Thus, slow crack growth due to environmental effects often occurs partly or totally by IGF, especially at moderate to finer grain sizes, and thus is an important factor in IGF. However, there is at least one case where slow crack growth occurs by TGF, sometimes exclusively so. This reinforces the basic role of the trade off between TGF and IGF. Though observations are again limited in scope, it is also clear that some bodies having considerable TGF will have some increase in IGF at and beyond the onset of mist, hackle, and crack branching. If stress character varies over a fracture, this may change fracture mode over the fracture. Thus, some increase in TGF on the compressive side of flexure fractures has been observed. Also, at least some indent flaws have much more TGF than the surrounding fracture. Fracture following thermal shock has increased IGF, in at least some materials.

While extensive consistency exists between different data sets, especially for over all trends, considerable variations can occur, e.g. as noted for some Al_2O_3. Key sources of variations are both the measurements and the material. Most TGF/IGF levels are estimated giving considerable, e.g. 10-20%, variation in the estimated mode where one mode is not very dominant. Even where actual measurements are made, their accuracy and specific basis (e.g. whether the value reflects the percent of grains or of the fracture area having the mode reported) are often uncertain. Another source of serious measurement differences is of G, which again is of significance primarily when the mode is changing with G. Material variations that can lead to fracture mode variations, e.g. due to variations in the amount, type, and location of other phases or impurities, with some grain boundary phases being particularly important. Other microstructural variations such as the amount, and size and location (especially relative to G) of porosity as well as variations of G can also vary fracture mode, again especially where the mode is in, or near, a state of transition.

The transition to mainly IGF at high temperatures is well known, but poorly understood from a fundamental stand point. Clearly grain boundary phases play an important, often dominant, but variable, role. Thus, while some materials show extreme IGF due to some boundary phases (e.g. from LiF additions to $MgAl_2O_4$), some show less effect e.g. of LiF additions to MgO). Also, some impurities can limit some, especially boundary, stresses, and their effects also depend on their amount and processing temperature and boundary wetting. EA, which increases with increasing temperature for many important ceramics may often be a factor in such increased IGF at higher temperatures, along with frequent contributions of grain boundary phases to such IGF. Intergranular porosity, especially that (often much) smaller than the G has greater effect on IGF at higher temperatures. However, some dense materials, apparently mainly with clean grain boundaries (and probably low EA), can have mainly TGF to 60 to > 80 % of their absolute melting temperatures. Materials with considerable TGF at moderate to higher temperatures commonly begin the transition to IGF at the fracture origin, with surrounding fracture remainding TGF. Further increase in test temperature (or presumably a decrease in strain rate) will expand the extent of IGF at, and around, the origin, and introduce new, often nearby, areas of IGF that also grow with increased test temperature. There may also be some changes at < 22 C, in addition to effects of suppressing SCG, e.g. due to changes in TEA and EA.

REFERENCES

1) R. W. Rice,"Strength and Fracture of MgO", *Proc. Brit. Cer. Soc.* No.20, pp 329-363 (June 1972).

2) R. W. Rice,"Fracture Topography of Ceramics" Surfaces and Interfaces of Glass and Ceramics, *Materials Science Research*, V. 7 (V.D.Frechette, W.C.

LaCourse, and V.L.Burdick eds.) pp 439-472, Plenum Press, New York (1974).
3) R. W. Rice,"Microstructural Dependance of Mechanical Behavior of Ceramics", *Treatise on Materials Science and Technology*, V.11, Properties and Microstructure (R.K.Mac Crone, ed.) pp 191-381, Academic Press, New York (1977).
4) R. W. Rice,"Ceramic Fracture Features, Observations, Mechanisms, and Uses" *Fractography of Ceramic and Metal Failures ASTM STP 827* (J. J. Mecholsky Jr. and S. R. Powell Jr. eds.) pp 5-103, ASTM (1984).
5) R. W. Rice,"Perspective on Fractography", *Advances in Ceramics*, V. 22, Fractography of Glasses and Ceramics' (J. R. Varner and V. D. Frechette eds.) pp 3-56, Am. Cer. Soc., Westerville, Oh. (1988).
6) R. W. Rice,"Porosity Effects on Machining Direction-Strength Anisotropy and Failure Mechanisms", *J. Am .Cer Soc.*, 77(8), pp 2232-36 (1994).
7) R. W. Rice,"Specimen Size-Tensile Strength Relations for a Hot Pressed Alumina and Lead Zirconate/Titanate", *Am. Cer. Soc. Bull.*, 66(5), pp 794-98 (1987).
8) R. M. Spriggs, J. B. Mitchell, and T. Vasilos,"Mechanical Properties of Pure, Dense Aluminum Oxide as A Function of Temperature and Grain Size", *J. Am. Soc.*, 47(7), pp 323-27 (1964).
9) R. Sedlacek and P. J. Jorgensen,"Processing of Ceramics-Surface Finishing Studies", *SRI Final Tech. Rept.* for U.S. Navy Cont. N00019-70-C-0179, Sept. (1971).
10) L. J. Graham and G. A. Alers ,"Microstructural Aspects of Acoustic Emission Generation in Ceramics", *Fracture Mechanics of Ceramics*, V.1 (R. C. Bradt, D. P. H. Hasselman, and F. F. Lange, eds.) pp 175-88, Plenum Press, New York (1974).
11) O. Johari and N. M. Parikh"Scanning Electron Fractographic Study of Aluminum Oxide as a Function of Testing Temperature and Grain Size", ibid, pp 399-420.
12) J. M. Samuels, M. R. Hoover Jr., L. Tarhay, G. G. Johnson Jr., H. A. McKinstry, and E. B. White, "Quantitative SEM and Raster Profilimeter Analysis of Fracture Surfaces", ibid, pp 421-429.
13) R. M. Gruver, W. A. Sotter, and H. P. Kirchner,"Fractography of Ceramics", Cer. Finishing Co. Rept. for ONR Contract No. N00019-73-C-1356,11/22/1974.
14) H. P. Kirchner, R. M. Gruver, and R. E. Walker,"Strengthening Hot-Pressed Al_2O_3 by Quenching", *J. Am. Cer. Soc.* 56(1), pp 17-21 (1973).
15) H. P. Kirchner, R. M. Gruver, and D. M. Richard,"Fractographic Investigation of Subcritical Crack Growth at Inherent Flaws in Polycrystalline Ceramics", Cer. Finishing Co. Rept. for ONR Contract No. N00019-77-C-0328 (Nov.1978).
16) G. J. Bansal, W. H. Duckworth, and D. E. Niecz,"Strength-Size Relations in Ceramic Materials: Investigation of an Alumina Ceramic", *J. Am. Cer. Soc.*,

59(11-12), pp 472-478 (1976).
17) G. DeWith,"Fracture of Translucent Alumina: Temperature Dependance and Influence of CaO Dope", *J. Mat. Sci.*, V. 19, pp 2195-2202 (1984).
18) C.-C. Chou, I-L. Cheng, K.-J. Chen, and Y.-C. Ko,"Fracture Energy and Thermal Stress Resistance Parameter of High Alumina Brick", *Am. Cer. Soc. Bull.*, 65(7), pp 1042-46 (1986).
19) R. Mussler, M. V. Swain, and N. Claussen,"Dependance of Fracture Toughness on Alumina on Grain Size and Test Technique", *J. Am. Cer. Soc.*, 65(11), pp 566-72 (1982).
20) J. A. Cappola and R. C. Bradt,"Effects of Porosity on Fracture of Al_2O_3", *J. Am. Cer. Soc.* 56(7), pp 392-3 (1973).
21) K. Hayashi, K. Gotoh, and T. Nishikawa,"Grain Size Dependance of Fracture Toughness in Polycrystalline Alumina", *J. Cer. Soc. Japan, Intl .Ed.*, V.99, pp 604-8 (1991).
22) A. H. Heuer,"Trans granular and Intergranular Fracture in Polycrystalline Alumina", *J. Am. Cer. Soc.*, 52(9), pp 510-11 (1969).
23) P. L. Gutshall and G. E. Gross,"Observations and Mechanisms of Fracture in Polycrystalline Alumina", *Eng. Fract. Mech.*, V.1, pp 463-71 (1969).
24) J. Congleton, N. J. Petch, and S .A. Shiels,"The Brittle Fracture of Alumina Below 1000 C", *Phil. Mag.*, V.19(106), pp 795-807 (1969).
25) J. D. B. Veldkamp and N .Hattu,"On the Fracture Toughness off Brittle Materials", *Phillips J. Res.*, 34, pp 1-25 (1979).
26) H. Wieninger, K Kromp, and R. F. Pabst,"Crack Resistance Curves of Alumina and Zirconia at Room Temperature", *J. Mat. Sci.*, V. 21, pp 411-18 (1986).
27) J. A. Salem, J. L. Shannon Jr., and R. C. Bradt,"Crack Growth Resistance in Textured Alumina", *J. Am. Cer. Soc.*, 72(1), pp 20-27 (1989).
28) K. W. White and J. C. Hay,"Effect of Thermoelastic Anisotropy on the R-Curve Behavior of Monolithic Alumina", *J. Am. Cer. Soc.*,77(9), pp 2283-88 (1994).
29) G. G. Bentle and R. M. Kniefel,"Brittle and Plastic Behavior of Hot Pressed BeO", *J. Am. Cer. Soc.*, 48(11), pp 570-77 (1965).
30) P. L. Gutshall and G. E. Gross,"Fracture Energy of Polycrystalline Beryllium Oxide", *J. Am. Cer. Soc.*, 51(10), p 602 (1968).
31) J. F. Lynch and R. C. Bradt,"Work of Fracture of BeO", *J. Am. Cer. Soc.*, 56(4), pp 228-29 (1973).
32) R. W. Rice,"CaO: II Properties", *J. Am. Cer., Soc.*, 52(8), pp 428-36 (1969).
33) R. W. Rice, J. G. Hunt, G. I. Freidman, and J. L. Sliney,"Identifying Optimum Parameters of Hot Extrusion", Final Boeing Co. Report for NASA Contract NAS-7-276 (8/1958).
34) R. W. Rice,"Effect of Hot Extrusion, Other Constituents, and Temperature on

the Strength and Fracture of Polycrystalline MgO", *J. Am. Cer. Soc.*, 76(12), pp 3009-18 (1993).
35) R. W. Rice,"Deformation, Recrystallization, Strength, and Fracture of Press Forged Ceramic Crystals", *J. Am. Cer. Soc.*, 55(2), pp 90-97 (1972).
36) T. Vasilos, J. B .Mitchell, and R. M. Spriggs,"Mechanical Properties of Pure, Dense Magnesium Oxide as a Function of Temperature and Grain Size", *J. Am. Cer. Soc.* 47(12), pp 606-10 (1964).
37) K. Yasuda, S.-D. Kim, Y. Kanemichi, Y. Matsuo, and S .Kimura,"Influence of Grain Size on Fracture Toughness of MgO Sintered Bodies", *J. Cer. Soc. Japan, Int. Ed.*, V.98, pp 110-15 (1991).
38) P. D. Ownby and G. E. Jungquist,"Final Sintering of Cr_2O_3", *J. Am. Cer. Soc*, 55(9), pp 433-36,1972.
39) R. M. Spriggs, L. A. Brissette, and T. Vasilos,"Pressure Sintered Nickel Oxide", *Am .Cer. Soc. Bull.*, 43(8), pp 572 77 (1964).
40) W. P. Minnear and R. C. Bradt,"Stoichiometry Effects on the Fracture of TiO_2", *J. Am. Cer. Soc.*, 63(9-10), pp 485-90 (1980).
41) S. M. Neirman,"The Formation, Sintering, and Electrical properties of Monodispersed TiO_2 particles Doped with Ta and Coated with $SrCO_3$", MIT Cer. Proc. Lab. Report No.57 (5/1986).
42) H. P. Kirchner and R. M. Gruver," Strength-Anisotropy-Grain Size Relations in Ceramic Oxides", *J. Am. Cer. Soc.*, 53(5), pp 232-36 (1970).
43) R. W .Rice, S. W. Freiman, and J. J. Mecholsky Jr.,"The Dependance of Strength-Controlling Fracture Energy on the Flaw-Size to Grain-Size Ratio", *J. Am. Cer. Soc.*, 63(3-4), pp 129 (1980).
44) R. W. Rice, "Machining Flaws and the Strength Grain Size Behavior of Ceramics", *Science of Ceramic Machining and Surface Finishing II* (B. J. Hockey and R. W. Rice eds.), pp 429-452, Nat. Bureau of Stand. Spec. Pub. 562, Wash., D.C. (1979).
45) D. C. Harris, G. A. Hayes, N. A. Yeager, L. D. Sawyer, R. C.Scheri, M. E. Hills, K. R. Hayes, S. E. Homer, Y. L. Tsai, and J. J. Mecholsky Jr.,"Mechanical Strength of Hemispheric Domes of Yttria and Lanthana-Doped Yttria", *J. Am. Cer. Soc.*,75(5), pp 1247-53 (1992).
46) C. Cm. Wu., R. W. Rice, and P. F.Becher,"The Character of Cracks in Fracture Toughness Measurements of Ceramics", *Fracture Mechanics Methods for Ceramics, Rocks, and Concrete* (S. W. Freiman and E. J. Fuller Jr. eds), ASTM STP 745, pp 137-40, ASTM (1982).
47) M. Desmaison - Brut, J. Montintin, F.Valin, and M. Boncoeur, "Influence of Processing Conditions on the Microstructure and Mechanical Properties of Sintered Yttrium Oxide", *J. Am. Cer. Soc.*, 78 (3), pp 716-22 (1995).
48) R. W. Rice, K. R. McKinney, and R. P. .Ingel,"Grain Boundaries, Fracture, and Heat Treatment of Commercial Partially Stabilized Zirconia", *J. Am. Cer.*

Soc., 64(12), pp C175-77 (1981).
49) R. P. Ingel, R. W. Rice, and D. Lewis,"Room-Temperature Strength and Fracture of ZrO_2-Y_2O_3 Single Crystals", *J. Am. Cer. Soc.*, 65(7), pp C109-9 (1982).
50) J. W. Adams, D. C. Larson, R. Ruh, and K. S. Masdiyasni, "Thermomechanical Properties of Zyttrite Fully Stabilized Y-ZrO_2 Ceramics", sub. to *J. Am. Cer. Soc.*
51) R. Brezny and R. W. Rice, unpublished data.
52) A. Gilbert, "Deformation and Fracture of Thoria", *Phil. Mag.* A, 12 (115), pp 139-44 (7/1965).
53) J. L. Daniel and S. Takahashi,"Fracture of Uranium Dioxide", *Proc.1st Mat. Conf.on Fracture*, V.3, pp 1987-97, Jap. Soc. for Str. of Mat. (1966).
54) C. R. Kennedy and G. Bandyopadhyay,"Thermal-Stress Fracture of UO_2", *J. Am. Cer. Soc.*, 59(3-4), pp 176-77 (1976).
55) R. F. Canon, J. T. A. Roberts, and R. J. Beals,"Deformation of UO_2 at High Temperature", *J. Am. Cer. Soc.*, 54(2), pp 105-112 (1971).
56) J. T. A. Roberts and Y. Ueda,"Influence of Porosity on Deformation and Fracture of UO_2", *J. Am. Cer. Soc.*, 55(3), pp 117-24 (1972).
57) A. G. Evans and R. W. Davidge,"The Strength and Fracture of Stoichiometric Polycrystalline UO_2", *J. Nuc. Mat.*, V.33, pp 249-60 (1969).
58) E. M. Schulson, I. Baker, C.D. Robertson, R. B. Bolon, and R. J. Harnimon," Fractography of Ice", *J. Mat. Sci. Let.* V 8, pp 1193-94 (1989).
59) A. V. Virkar and R. S. Gordon,"Fracture Properties of Polycrystalline Lithia-Stabilized β-Al_2O_3", J. *Am .Cer. Soc.*, 60(1-2), pp 58-61 (1977).
60) W. J. McDonough, D. R. Flinn, K. H. Stern, and R. W. Rice,"Hot Pressing and Physical Properties of Na Beta Alumina", *J. Mat. Sci.*, V.13, pp 2403-12 (1978).
61) R. Stevens"Strength and Fracture Mechanisms in Beta-Alumina", *J. Mat. Sci.*, V.9, pp 934-40 (1974).
62) D .C. Hitchcock and L. C. DeJonghe,"Fracture Toughness Anisotropy of Sodium β-Alumina", *J. Am. Cer. Soc.*, 66(12), pp C204-205 (1983).
63) R. W. Rice and J. J. Mecholsky Jr.,"The Nature of Strength Controlling Machining Flaws in Ceramics", *Science of Ceramic Machining and Surface Finishing II* (B. J. Hockey and R. W. Rice eds.), pp 351-378, Nat. Bureau of Stand. Spec. Pub. 562 , Wash.,D.C. (1979).
64) S. W. Freiman, G. Y. Onada, and A. G. Pincus,"Mechanical Properties of 3BaO $5SiO_2$ Glass-Ceramic", *J. Am. Cer. Soc.*, 57(1), pp 8-12 (1974).
65) W. H. Rhodes, P.L.Berneburg, and J.E.Niesse,"Development of Transparent Spinel($MgAl_2O_4$)",AVCO Corp. Final Rept. DAAG-46-69-C-0113 for US Army Contract AMMRC-CR-70-19 (10/1970).
66) K. W. White and G. P. Kelkar,"Evaluation of the Crack Bridging Mechanism

in a $MgAl_2O_4$ Spinel", *J. Am. Cer. Soc.*, 74(70), pp 1732-34 (1991).
67) A. Ghosh, K. W. White, M. G. Jenkins, A. S. Kobayashi, and R. C. Bradt,"Fracture Resistance of a Transparent Magnesium Aluminate Spinel", *J. Am. Cer. Soc.*, 74(7), pp 1624-30 (1991).
68) K. W. White and G. P. Kelkar,"Fracture Mechanisms of a Coarse-Grained, Transparent $MgAl_2O_4$ at Elevates Temperatures", *J. Am. Cer. Soc.*, 75(12), pp 3440-40 (1992).
69) E. K. Beauchamp and S. L. Monroe,"Effect of Crack-Interface Bridging on Subcritical Crack Growth in Ferrites", *J. Am. Cer. Soc.*, 72(7), pp 1179-84 (1989).
70) R. C. Pohanka, S. W. Frieman, and B. A. Bender,"Effect of the Phase Transformation on the Fracture Behavior of $BaTiO_3$", *J. Am. Cer. Soc.*, 61(2),pp 72-75 (1978).
71) B. E. Walker Jr., R. W. Rice, R. C. Pohanka, and J. R. Spann,"Densification and Strength of BaTiO3 with LiF and MgO Additives", *Am. Cer. Soc. Bull.*, 55(3), pp 274-278 (1976).
72) R. C. Pohanka, R. W. Rice, and B. E. Walker Jr.,"Effect of Internal Stress on the Strength of $BaTiO_3$", *J. Am. Cer. Soc.*, 59(1-2), pp 71-4 (1976).
73) B. K. Molnar and R. W. Rice,"Strength Anisotropy in Lead Zirconate Titanate Transducer Rings", *Am. Cer. Soc. Bull.*, 52(6), pp 505-9 (1973).
74) S.-B. Kim, D.-Y. Kim, J.-J. Kim, and S.-H. Cho,"Effect of Grain Size and Poling on The Fracture Mode of Lead Zirconate Titanate Ceramics", *J. Am. Cer. Soc.*, 73(1), pp 561-63 (1990).
75) M. H. Megherhi,"The Effect of Additives, Microstructure , and Macrostructure, on the Fracture Behavior of Lead Magnesium Niobate-Based Ceramic Actuators", M. S. Thesis in Ceramic Science, Mat. Res .Lab., Pen. State U. (8/1988).
76) J. A. Kuszyk and R. C. Bradt,"Influence of Grain Size on Effects of Thermal Expansion Anisotropy in $MgTi_2O_5$", *J. Am. Cer. Soc.*, 56(8), pp 420-23 (1973).
77) J.-J. Kim and D.-Y. Kim,"Change in the Fracture Mode of PLZT Ceramics by Chemically Induced Grain-Boundary Migration", *J. Am. Cer. Soc.*, 71 (5(, C228-29 (1988).
78)E. K.Chang, Z. Y. Wang, D. M. Smyth, and M. P. Harmer,"Wavy Boundaries and Fracture Modes in Ceramics", 88th Annual Meeting Ceramographic Exhibit, *J. Am. Cer. Soc.*, 69 (10), back cover (1986).
79) S. .W. Freiman, J. J. Mecholsky Jr., R. W. Rice, and J. C. Wurst,"Fracture of ZnSe and As_2S_3 Laser Window Materials", *Proc. 4th Conf. on Infrared Laser Window Materials* (C. R. Andrews and C. L. Strecker eds.), pp 698-715, US Air Force (1/1975).
80) G. DeWith,"High Temperature Fracture of Boron Carbide: Experiments and Simple Theoretical Models ", *J. Mat. Sci.* , V.19, pp 457-66 (1984).
81) B. Champagne and R. Angers,"Mechanical Properties of Hot-Pressed B-B_4C

Materials", *J. Am. Cer. Soc.*, 62(3-4), pp 159-53 (1979).
82) K. Niihara, A. Nakahira, and T. Hirai,"The Effect of Stoichiometry on Mechanical Properties of Boron Carbide", *J. Am. Cer. Soc.*, 67(1), pp C13-14 (1984).
83) K. A. Schwetz and W. Grellner,"The Influence of Carbon on the Microstructure and Mechanical Properties of Sintered Boron Carbide", *J. Less - Com. Metals*, V. 82, pp 37-47 (1981).
84) T. D. Gulden,"Mechanical Properties of Polycrystalline β-SiC", *J. Am. Cer. Soc.* 52(11), pp 585-90 (1969).
85) S. Prochaska and R. J. Charles,"Strength of Boron-Doped, Hot Pressed Silicon Carbide", *Am. Cer. Soc. Bull.*, 2(12), pp 885-91 (1973).
86) G. V. Srinivasan and V. Venkateswaran,"Tensile Strength Evaluation of Polycrystalline SiC Fiber", *Cer. Eng. & Sci. Proc.*14(7-8), pp 563-72 (1993).
87) Y.-W. Kim and J.-G. Lee,"Effect of Polycarbosilane Additions on Mechanical Properties of Hot Pressed Silicon Carbide", *J. Mat. Sci.*,V. 27, pp 4746-50 (1992).
88) C.-B.Lim and T.Iseki,"Strength-Porosity-Microstructure Relations in Porous Reaction-Sintered SiC with Free Silicon Removed", *J. Mat. Sci.*, V.25, pp 4203-10 (1990).
89) J. A. Cappola and R. C. Bradt,"Measurement of Fracture Surface Energy of SiC", *J. Am. Cer. Soc.*, 55(9), pp 455-60 (1972).
90) Y. Takeda and K. Nakamura,"Effects of Additives on Microstructure and Strength of Dense Silicon Carbide", *Proc.23rd Jap. Cong. on Mat. Res., Jap. Soc. Mat. Sci.*, Japan, Koyto , pp 215-19 (1980).
91) S. K. Lee and C. H. Kim,"The Influence of Polytypes of Starting Powders on the Microstructure and Fracture Toughness of Sintered SiC with Al_2O_3-Y_2O_3 Additions', *J. Am. Cer. Soc.*, 77(6), pp 1655-58 (1994).
92) G. Orange, H. Tanaka, and G. Fantozzi,"Fracture Toughness of Pressureless Sintered Silicon Carbide: A Comparison of K_{IC} Measurement Methods", *Cer. Intl.*,V. 13, pp 159-65 (1987).
93) S. K. Lee, D. K. Kim, C. H. Kim, “Flaw-Tolerance and R-Curve Behavoir of Liquid-Phase-Sintered Silicon Carbides with Different Microstructures”, *J. Am. Cer. Soc.*, 78 (1), pp 65-70 (1995).
94) H. Kodama and T. Miyoshi,"Study of Fracture Behavior of Very Fine-Grained Silicon Carbide Ceramics", *J. Am. Cer. Soc.*, 73(10), pp 3081-86 (1990).
95) R. C. Bradt, S. D. Hartline, J. A. Cappola, G. Q. Weaver, and R. A. Alliegro,"Fracture of Commercial Silicon Carbide", *SiC-1973* (R. C. Marshall, J. W. Faust Jr., and G. E. Ryan eds.), pp 420-27, U. South Cariliona Press, Columbia, S.C. (1964).
96) J. L. Henshall, D. J. Rowcliffe, and J. W. Edgington,"K_{IC} and Delayed Fracture Measurements on Hot-Pressed SiC', *J. Am. Cer. Soc.*, 62(1-2), pp 36-41 (1979).

97) K. T. Faber and A. G. Evans,"Intergranular Crack-Deflection Toughening in Silicon Carbide", *J. Am. Cer. Soc.*, 66(6), pp C94-6 (1983).
98) Y. Takeda, K. Nakumura, K. Maeda, and M. Ura,"Effects of Elemental Additives on Densification, Microstructure, Strength, and Thermal Conductivity of Silicon Carbide Ceramics", *Adv. Cer. Mat.*,1(2), pp 162-5 (1986).
99) K. Niihara and T. Hirai,"Fractography of Chemical Vapor-Deposited Si_3N_4", *J. Mat. Sci.*, V 13, pp 2385-93 (1978).
100) M. H. Leipold and T. H. Nielson,"The Mechanical Behavior of Tantalum Carbide and Magnesium Oxide", JPL-NASA Tech. Rept. 32-1201 (12/1967).
101) W. B. Harrison,"Mechanical and Electrical Properties of TiC", 2nd Interim Tech. Rept. for US Army Contract DA-11-022-ORD-3441 (2/1965).
102) J. L. Chermant, A. Deschanvres, and F. Osterstock,"Toughness and Fractography of TiC and WC", *Fracture Mechanics of Ceramics*, V.4,(R . C. Bradt, D. P.H.Hasselman, and F. F. Lange eds.) pp 891-901, Plenum Press, New York (1978).
103) D. B. Miracle and H. A. Lipsitt,"Mechanical Properties of Fine-Grained Substochiometric Titanium Carbide", *J. Am. Cer. Soc.*, 66(8), pp 592-7 (1983).
104) L. J. Kecskes, T. Kottke, and A. Niiler,"Microstructural Properties of Combustion-Synthesized and Dynamically Consolidated Titanium Boride and Titanium Carbide", *J. Am. Cer. Soc.*, 73(5), pp 1274-82 (1990).
105) J. C. LaSalvia, L. M. Lothar, and M. A. Meyers,"Densification of Reaction-Synthesized Titanium Carbide by High-Velocity Forging", *J. Am. Cer. Soc.*,75 (3). pp 592-602 (1992).
106) H. R. Baumgartner and R. A. Steiger,"Sintering and Properties of Titanium Diboride Made from Powder Synthesized in a Plasma-Arc Heater", *J. Am. Cer. Soc.* , 67(3), pp 207-212 (1984).
107) H. R. Baumgartner,"Mechanical Properties of Densely Sintered High-Purity Titanium Diborides in Molten Aluminum Environments", *J. Am. Cer. Soc.*, 67(7),pp 490-7 (1984).
108) H. R. Baumgartner,"Subcritical Crack Velocities in Titanium Diboride under Simulated Hall-Heroult Cell Conditions", *Am. Cer. Soc. Bull.*,63(9), pp 1172-5 (1984).
109) H. Itoh, S. Naka, T. Matsudaira, and H. Hamamoto,"Preparation of TiB_2 Sintered Compacts by Hot Pressing", *J. Mat. Sci.*, V 25, pp 533-36 (1990).
110) R .C. Dorward,"Aluminum Penetration and Fracture of Titanium Diboride", *J. Am. Cer. Soc.*, 65(1), p C6 (1982).
111) M. E. O'Day and G. L. Leatherman,"Static Fatigue of Aluminum Nitride Packaging Materials", *Intl. J. Microcircuts and Elect. Pack.*,16(1), pp 41-7 (1993).
112) R. W. Rice, K. R. McKinney, C. Cm. Wu, S. W. Freiman, G. S. White, J. H. Enloe, and L. E. Dolhert,"Slow Crack Growth in AlN",to be published.

113) N. Alexandre, M. Desmaison-Brut, F. Valin, and M. Boncoeur,"Mechanical Properties of Hot Pressed Zirconium Nitride", *J. Mat. Sci.*, V.28, pp 2385-2390 (1993).
114) K. R. McKinney, B. A. Bender, R. W. Rice, and C. Cm. Wu,"Slow Crack Growth in Si_3N_4 at Room Temperature", *J. Mat. Sci.*, V.26, pp 6467-72 (1991).
115) G. Himsolt, H. Knoch, H. Huebner, and F. W. Kleinlein," Mechanical Properties of Hot-Pressed Silicon Nitride with Different Grain Structures", *J. Am. Cer. Soc.*, 62 (1-2), pp 29-32 (1979).
116) J. A. Palm and C. D. Greskovitch,"Thermomechanical Properties of Hot-Pressed $Si_{2.9}Be_{0.1}N_{3.8}O_{0.2}$ Ceramic", *Am. Cer. Soc. Bull.*, 59(4), pp 447-52 (1980).
117) C. D. Greskovitch, W. D. Pasco, and G. D. Quinn,"Thermomechanical Properties of a New Composition of Sintered Si_3N_4", *Am. Cer. Soc. Bull.*, 63(9), pp 1165-70 (1984).
118) J. J. Mecholsky Jr.,"Intergranular Slow Crack Growth in MgF_2", *J. Am. Cer. Soc.* 64(9), pp 563-66 (1981).
119)K. R. McKinney, J. J. Mecholsky Jr.,and S. W. Freiman,"Delayed Failure in Chemically Vapor Deposited ZnSe", *J. Am. Cer. Soc.*, 62(7), pp 336-40 (1970).
120) J. J. Mecholsky Jr.,Private communication 1993.
121) W. H. Rhodes, R. M. Cannon Jr., and T. Vasilos,"Stress-Corrosion Cracking in Polycrystalline MgO", *Fracture Mechanics of Ceramics*, V.2 (R. C. Bradt, D. P. H. Hasselman, and F. F. Lange eds.) pp 708-733, Plenum Press, New York (1974).
122) D. M. Kotchick and R. E. Tressler,"Surface Damage and Environmental Effects on the Strain-Rate Sensitivity of the Strength of Sapphire and Silicon Carbide Filaments", *J. Mat. Sci.*, V.10, pp 608-10 (1975).
123) A. G. Evans and F. F. Lange,"Crack Propagation and Fracture in Silicon Carbide", *J. Mat. Sci.*, V.10, pp 1659-64 (1975).
124) K. D. McHenry, T. Yonushonis, and R. E. Tressler,"Low-Temperature Subcritical Crack Growth in SiC and Si_3N_4", *J. Am. Cer. Soc.*, 59(5-6), pp 262-64 (1976).
125) R. W. Rice,"The Difference of Mirror to Flaw Size Ratios Between Dense Glass and Polycrystals", *J. Am. Cer. Soc.*, 62 (9-10), pp 533-35 (1979).
126) C. Cm. Wu, K. R. McKinney, and R. W. Rice,"Zig-Zag Crack Propagation in $MgAl_2O_4$ Crystals", *J. Mat. Sci.Let.*, 14, pp 474-77 (1995).
127) R. W. Rice and C. Cm. Wu,"Slow Crack Growth in $MgAl_2O_4$ Single- and Poly- Crystals", *J. Mat. Sci.*, 14, pp 723-27 (1995)..
128) P. Chantikul, S. J. Bennison, and B. R. Lawn,"Role of Grain Size in the Strength and R-Curve Properties of Alumina", *J. Am. Cer. Soc.* 73 (8), pp 2419-27 (1990).
129) J. Lankford,"Compressive Strength and Microplasticity in Polycrystalline Alumina", *J. Mat. Sci.*,V.12, pp 791-96 (1977).

130) J. Lankford,"Compressive Strength and Damage Mechanisms in Ceramic Materials", So.W. Res. Inst. Interim. Rept. for ONR Contract N00014-75-C-0668 (9/10/1980).
131) A. Nash,"Compressive Failure Modes of Alumina in Air and Pathological Media", *J. Mat. Sci.*, V.18, pp 3571-77 (1983).
132) J. R. Keski,"Bismuth Oxide Ceramics", *Am. Cer. Soc. Bull.*, 51(6), pp 527-31 (1972).
133) J. Lankford,"Uniaxial Compressive Damage in -SiC at Low Homologous Temperatures", *J. Am. Cer. Soc.*, 62 (5-6), pp 310-12 (1979).
134) T. G. Langdon and J. E. Pask,"Effect of Microstructure on Deformation of Polycrystalline MgO", *J. Am. Cer. Soc.*, 54(5), pp 240-46 (1971).
135) M. S. Paterson and C. W. Weaver,"Deformation of Polycrystalline MgO Under Pressure", *J. Am. Cer. Soc.*, 53(8), pp 465-71 (1970).
136) R. N. Katz and W. A. Brantly,"Fractography of High Boron Ceramics Subjected to Ballistic Loading", *Ceramics in Severe Environments* (W .W. Kriegel and H. Palmour III eds.), pp 271-82, Plenum Press, New York (1971).
137) C. Tracy, M. Slavin, and D. Vichnicki,"Ceramic Fracture During Ballistic Impact", *Advances in Ceramics, V.22, Fractography of Glasses and Ceramics* (V. D. Frechette and J. R. Varner eds.), pp 195-306, Am. Cer. Soc., Westerville, Oh. (1988).
138) W. B. Brubaker, and R. Hoffman,"Evaluation of Low-Cost Aluminum Nitride Armor Tiles", Dow Chemical Co.(Midland, Mich.) Report cera 1992.
139) M. S. Stucke and A. S. Wronski,"The Dependance of the Strength of Several Oxide Ceramics on Hydrostatic Pressure", *Proc. Brit. Cer. Soc.*, 25, Mech. Prop. Cer.(2), pp 139-49 (5/1975).
140) G. D. Quinn, J. J. Kubler, and R. J. Gettings;"Fracture Toughness of Advanced Ceramics by Surface Cracks in Flexure (SCF) Method : A VMAS Round Robin", VMAS Rept. # 17, NIST (1994).
141) R. W. Rice,"The Effect of Gaseous Impurities on the Hot Pressing and Behavior of MgO, CaO, and Al_2O_3", *Proc. Brit. Cer. Soc.*, pp 99-123 (3/1969).
142) P. F. Becher,"Fracture-Strength Anisotropy Of Sapphire", *J. Am. Cer. Soc.*, 59(1-2), pp 59-61 (1976).
143) C. Cm. Wu, K. R. McKinney, and R. W. Rice,"Strength and Toughness of Jade and Related Natural Fiberous Materials", *J. Mat. Soc.* ,V 25, pp 2170 - 74 (1990).
144) M. Iwasa, E. C. Liang, and R. C. Bradt,"Fracture of Isotropic and Textured Ba Hexaferrite", *J. Am. Cer. Soc.*, 64 (7), pp 390-3 (1981).
145) R. W. Rice,"Ceramic Tensile Strength-Grain Size Relations: Grain Sizes, Slopes, and Branch Intersections", to be published.
146) W.-J. Yang, C. T. Yu, and A. S. Kobayashi,"SEM Quatification of Transgranular vs Intergranular Fracture", *J. Am. Cer. Soc.*,74 (2), pp 290-95

(1991).
147) R. I. Taylor, J. P. Coad, and A. E. Hughes,"Grain-Boundary Segregation in MgO-Doped Al_2O_3", *J. Am. Cer. Soc.* 59(7-8), pp 374-76 (1976).
148) T. Koyama, A. Nishiyama, and K. Niihara,"Effect of Grain Morphology and Grain Size on the Mechanical Properties of Al_2O_3 Ceramics", *J. Mat. Sci.*,V 29, pp 3949-54 (1994).
149) M. K. Aghajanian, N. H. Macmillan, C. R. Kennedy, S. J. Luszcz, and R. Roy,"Al_2O_3-Al Ceramic Composite Materials", *J. Mat. Sci.*, V 24, pp 658-70 (1989).
150) W. B. Crandall, D. H. Chung, and T. J. Gray,"The Mechanical Properties of Ultra-Fine Grain Hot-Pressed Alumina", *Mechanical Properties of Engineering Ceramics* (W.W.Kriegel and H.Palmour III eds.), pp 349-76, Wiley(Interscience), New York (1961).
151) A. Krell and P. Blank,"Inherent Reinforcement of Ceramic Microstructures by Grain Boundary Engineering", *J. Eur. Cer. Soc.*, V.9, pp 309-322 (1992).
152) Y. K. Baek and C. H. Kim,"The Effect of Whisker Length on the Mechanical Properties of Alumina-SiC Whisker Composites", *J. Mat. Sci.*, V 24, pp 1589-93 (1989).
153) R. H. Dauskardt, B. J. Dalgliesh, D. Yao, R. O. Ritchie, and P. F. Becher,"Cyclic Fatigue-Crack Propagation in a Silicon Carbide Whisker-Reinforced Alumina Composite: Role of Load Ratio", *J. Mat. Sci.*,V 28, pp 3258-66 (1993).
154) J. Gurland and P. Bardzil,"Relation of Strength, Composition, and Grain Size of Sintered WC-Co Alloys", *J. Metals*, pp 311-15 (2/1955).
155) M. J. Murray,"Fracture of WC-Co Alloys: an Example of Spatially Constrained Crack Tip Opening Displacements", *Proc. Roy. Soc. Lond.A*, 356, pp 483-508 (1977).
156) A. Hara, T. Nishikawa, and S. Yazu,"The Observation of the Fracture Path in WC-Co Cemented Carbides Using a Newly Developed Replica Method", *Planseeberichte fur Pullver Metallurgie*, V.18, pp 28-43 (1970).
157) J. R. Pickens,"The Fracture Toughness of Tungsten Carbide-Cobalt Alloys as a Function of Microstructural Parameters", Brown U. Report PhD Thesis and Technical Report for NSF Grant GH-42182 (4/1977).
158) J. R. Pickens and J. Gurland,"The Fracture Toughness of WC-Co Alloys Measured on Single-Edged Notched Beam Specimens Precrracked by Electron Discharge Machining", *Mat. Sci. & Eng.*,V.33, pp 135-42 (1978).
159) B. Roebuck and E. A. Almond,"Deformation and Fracture Processes and the Physical Metallurgy of WC-Co Hardmetals", *Intl. Mat. Revs.*, 33(2), pp 90-110 (1988).
160) D. H. Chung and W. R. Bussem,"The Elastic Anisotropy of Crystals", *Anisotropy in Single-Crystal Refractory Compounds*, V.2, (F. W. Vahldiek and S.

A. Mersol), pp 217-45, Plenum Press, New York (1968).
161) R. W. Rice and R. C. Pohanka,"Grain-Size Dependance of Spontaneous Cracking in Ceramics", *J. Am. Cer. Soc.*, 62(11-12), pp 559-63 (1979).
162) R. W. Rice,"Effects of Thermal Expansion Mismatch Stresses on the Room-Temperature Fracture of Boron Carbide", *J. Am. Cer. Soc.* 73(10), pp 3116-18 (1990).
163) R. W. Rice,"Possible Effects of Elastic Anisotropy on Mechanical Properties of Ceramics", *J. Mat. Sci. Let.*, V.13, pp 1261-66 (1994).
164) R. W. Rice, C. Cm. Wu, and K.R. McKinney,"Fracture and Fracture Toughness of Stoichiometric $MgAl_2O_4$ Crystals at Room Temperature", submitted for publication.
165) R. W. Rice and C. Cm. Wu,"Slow Crack Growth in $MgAl_2O_4$ Single- and Poly-Crystals", *J. Mat. Sci.Let.*,V.14, pp 723-27 (1995).
166) C. Cm. Wu and R. W. Rice,"Slow Crack Growth in Oxide and Nonoxide Ceramics", to be published.
167) R. W. Rice,"Elastic Anisotropy and the Grain Size Dependance of Ceramic Fracture Energies", *J. Mat. Sci.*, V. 19, pp 1267-71 (1984).
168) R. W. Rice,"Grain Size and Porosity Dependance of Fracture Energy and Toughness of Ceramics at 22 C", to be published.
169) R. W. Rice, C. Cm. Wu, and F. Borchelt,"Hardness-Grain Size Relations in Ceramics", *J. Am. Cer. Soc.*, 77(10), pp 2539-53 (1994).

THE VAMAS FRACTOGRAPHY ROUND ROBIN: A PIECE OF THE FRACTOGRAPHY PUZZLE

Jeffrey J. Swab
U.S. Army Research Laboratory-Materials Directorate
Watertown, MA 02172

George D. Quinn
National Institute of Standards and Technology
Gaithersburg, MD 20899

ABSTRACT

Seventeen agencies, representing eight nations, participated in a VAMAS (Versailles Project on Advanced Materials and Standards) sponsored round robin exercise on fractography of advanced ceramics. Our objectives in conducting this exercise were to identify and rectify any erroneous, ambiguous or weak areas in Military Handbook 790 "Fractography and Characterization of Fracture Origins in Advanced Structural Ceramics", (MIL HDBK-790), to solicit constructive criticism and new ideas, and to support the development of an ASTM (American Society for Testing and Materials) standard practice. The exercise was divided into three topics. Topic #1 addressed the detection and interpretation of machining damage on photographs of the fracture surfaces of ceramic specimens. Topic #2 dealt with the fractographic analysis of ceramic specimens. Topic #3 was optional, and asked the participants to perform fractography on a ceramic material of their choice.

The results from Topic #1 showed that there are problems in detecting and interpreting machining damage in advanced ceramics. These problems stemmed from legitimate differences in interpreting the various manifestations of machining damage, or alternatively, from the participants' unfamiliarity with the nature of subsurface machining damage. Topic #2 indicated that the guidelines and the characterization scheme outlined in MIL HDBK-790 are adequate to completely characterize fracture origins in advanced ceramics but some refinements are necessary. There was a good-to-excellent consensus in origin characterization in many cases. The instances where concurrence was not forthcoming indicated where improvements to the handbook should be made and highlighted the key steps that are integral to a proper fractographic evaluation.

INTRODUCTION

The fundamentals of fractography of ceramics and glasses are well documented (1-7) but analysis can be interpretative and dependent on the fractographic experience level of the analyst. MIL HDBK-790 (and its subsequent ASTM standard practice) attempt to rationalize and guide fractographic analysis of ceramics which will lead to more consistency and clarity in the interpretation and characterization of fracture origins.

The fractographic analysis of ceramic specimens or components is akin to assembling a jigsaw puzzle or being a detective attempting to solve a crime. In the case of a puzzle: if a few key pieces are missing, then the picture may not be understood, as illustrated in Figure 1. As a detective the fractographer must look beyond the fracture surface and investigate all clues and properly combine them in order to appropriately characterize the fracture origin. In both instances the fractographer must be cognizant of the various pieces of fractography-related information and how they fit together in order to reach the appropriate conclusion.

This paper will provide a brief synopsis of the round robin exercise and use examples from the exercise to discuss the various pieces of the fractography puzzle. Examples will also be used to highlight "clues" which can assist in the proper characterization fracture origins in advanced ceramics materials.

SYNOPSIS OF THE ROUND ROBIN EXERCISE

A comprehensive discussion and complete compilation of the results from the VAMAS round robin exercise can be found in Reference 8. This report identifies areas of concern which require further study, provides amendments to MIL HDBK-790 and evaluates each of the round robin participants fractographic analysis.

Topic #1: Characterization of Machining Damage

Objective: The objective of this topic was to characterize machining damage in a variety of advanced structural ceramics. The participants were asked to use the SEM photographs provided to characterize the origin's identity, location, and size, and to mark and measure the size of the associated fracture mirror. A short write-up was furnished for each material. This included information such as: material composition; processing, machining, and exposure history; microstructural data; a fracture toughness value; and the room temperature flexure strength of the particular specimen. Each participant received three sets of six photographs of the fracture surfaces of a flexure bar from three different ceramic materials: Set #1 - Zirconia/Alumina; Set #2 - Silicon Nitride; Set #3 - Alumina.

Identity: The consensus of the participants was that machining damage was the fracture origin in all three materials. There were however some participants who questioned the organizers machining damage identification on some of the photograph sets. These disagreements may be due to a failure to understand how machining damage can appear in ceramics, a lack of experience observing this type of flaw, or an error on the part of the organizers. In general the organizers had the benefit of examining many specimens of each material. This more comprehensive examination enabled the organizers to recognize key features in machining damage that differentiate it from alternate types of fracture origins.

Location: Most participants reported the location of the origin in all three sets as "surface" in agreement with the organizers and the guidelines of MIL HDBK-790. Several reported that the origins were "near surface", however. These respondents may have felt that the subsurface crack(s) were better described by such a characterization, but this may be merely a difference of semantics. The organizers feel that machining damage is inherently a surface- or edge-distributed fracture origin.

Flaw Size: The methods of reporting flaw size widely varied between participants. Some reported a diameter or radius. Still others reported a flaw depth or a 2-dimensional value (depth x width). Even with these differences in how the size was reported, there was some participant-to-participant consistency for two of the three photograph sets.

Mirror Size: Many participants were not familiar with fracture mirror analysis, including the use of mirror constants. Most had difficulty measuring the mirror size because the boundaries were difficult to delineate and the feature was not distinct. The general outline of the mirror could be seen at lower magnifications, however. As with the flaw size, the participants reported mirror size in many different ways.

Topic #2: Characterization of Fracture Origins in Specimens

Objective: The objective of this topic was to locate and characterize fracture origins in fractured flexure specimens and determine the effectiveness of the characterization scheme in MIL HDBK-790. Each participant received both mating halves of the primary fracture surface of six specimens. A short description of the material and the test conditions which led to fracture (including the fracture stress and a fracture toughness value) were furnished for each specimen. Individual agencies received different sets of six specimens, but in each category all the specimens were from identical batches and had the same origin type. This was confirmed by the organizers who examined and characterized all fracture origins before sending them to the participants. Together the participating agencies evaluated a total of 102 fracture origins. The six ceramics and their origin types are listed in Table 1.

Table 1
CERAMIC MATERIALS ANALYZED IN TOPIC #2

Specimen #	Material	Origin Type*
1	Hot-pressed Al_2O_3 w/SiC whiskers	Large Grain (LG^V)
2	Sintered (99.9%) Al_2O_3	Handling Damage (HD^S)
3	Sintered ZrO_2	Pore (P^V)
4	Reaction Bonded SiC	Pit (PT^S)
5	Hot Isostatically Pressed Si_3N_4	Machining Damage (MD^S)
6	Sintered TiB_2	Porous Seam or Porous Region (PS^V or PR^V)

* In the code scheme of MIL HDBK-790, each origin type is represented by an acronym. The superscript S or V attached to each acronym indicates that the origin type is inherently surface- or volume-distributed.

Summary: The results from this topic showed that the characterization scheme in the handbook contains the necessary attributes for the characterization of fracture origins in advanced structural ceramics. The general attribute scheme of identity, location, and size for an origin worked well, but some clarification and refinements are needed regarding the particulars of each attribute. The exercise also revealed that the guidelines provided can enable an inexperienced fractographer to identify and properly characterize fracture origins present in this class of ceramics.

Table 2 lists the ratios of concurrence in interpretation of the participants with the organizers for the six specimens of Topic #2. It is clear that there was mixed success; in some instances there was a good consensus and in other specimens, there was not. There was a good consensus on the location of the origin in all six specimens and the overall characterization of specimen 3 (Pore in Zirconia). In the latter instance this may be a result of pores being a common anomaly in many ceramics and it may be safe to assume that most of the participants had seen pores before. One reason for the lack of a consensus for some of the other specimens appears to be that the participants had only one specimen of each material to analyze while the organizers had seventeen or more specimens available for analysis. Nonetheless, some participants ignored some of the available information at their fingertips, i.e., the mating half of the primary fracture surface, the external surfaces of the specimen, or the specimen's past history.

Table 2
CONCURRENCE WITH ORGANIZERS' EVALUATION

Specimen #	Identity	Location	Size
1: LG^V in Al_2O_3/SiC	6/17 = 35%	12/17 = 71%	7/17 = 41%
2: HD^S in Al_2O_3	5/17 = 29%	14/17 = 82%	-----
3: P^V in ZrO_2	14/17 = 82%	14/17 = 82%	10/17 = 59%
4: PT^S in SiC	5/17 = 29%	15/17 = 88%	4/17 = 24%
5: MD^S in Si_3N_4	11/17 = 65%	10/17 = 59%	4/17 = 24%
6: PS^V or PR^V in TiB_2	8/17 = 47%	14/17 = 82%	8/17 = 47%

---- Size values not determined by organizers.

Round Robin Conclusions

1) The guidelines and characterization scheme in MIL HDBK-790 are adequate for complete characterization of fracture origins in advanced ceramics, but some refinements are needed.
2) Fracture mechanics was not used enough by the participants in this exercise to assist in the characterization of fracture origins.
3) In many instances, fractographers failed to use all available information about the material and its history during the characterization.
4) Characterization of the fracture origin size was difficult in many instances.
5) Characterization of origins from photographs or a single specimen can be misleading.
6) Fractographers must examine the mating halves of the primary fracture surface.
7) Fractographers must examine the external surfaces or the specimen or component, especially if the origin is surface- or edge-located.
8) Few participants were familiar with fracture mirror analysis. There is a need for new mirror constants for today's advanced ceramic materials.

THE FRACTOGRAPHY PUZZLE

Background Information

A material or specimens' background information (processing history, exposure or testing conditions, etc.) provides vital clues to the types of fracture origins which may limit the strength of the material. This is one piece of information that was overlooked or ignored in many instances in the round robin exercise.

An example is specimen 4 from Topic #2. The specimens provided to the participants were reaction-bonded polycrystalline silicon carbide (SiC). Prior to room-temperature strength testing, the specimens were thermally shocked between 1,000 and 10,000 times from ≈ 1350°C to room temperature. Quenching was done during each cycle with a jet of compressed air and took ≈ 15 seconds. All fracture origins were clearly connected to the specimen surface.

Only five participants agreed with the organizers that the identity of this origin type was a pit (PT^s). Eight of the twelve participants who labeled the origin as something else did not take the thermal history of the specimen into account. Two of these participants provided a label of surface void (SV^s) which may indicate that they did not understand the difference between pits and surface voids. Pits appear similar to the examples of surface void given in MIL HDBK-790, see Figure 2. Pits are cavities created at the surface (exterior) of a specimen or component during the reaction or interaction between the material and the environment, e.g., corrosion or oxidation. Surface voids are cavities created at the surface as a consequence of the reaction or interaction between the material and the processing environment, e.g., a surface reaction layer. The bars used for specimen 4 were all machined from a large billet, therefore surface voids can not be the origin type. The appropriate identity may have been reported in some cases had one or both of these been taken in account.

Fracture Surfaces

The interaction between the fracture origin and its associated fracture surface makes it imperative that the fractographer analyze both of the mating halves of the primary fracture surface. Although the features on these mating fracture surfaces are typically mirror images of each other, there are times when slight differences may result. Thus, restricting the fractographic analysis to only one half of this surface may lead to an erroneous characterization of the fracture origin. Several times in Topic #2 it appeared that an examination of both the mating halves of the primary fracture surface may have resulted in the participant providing a different identity or location for a particular origin.

An agglomerate is an example of an origin where failure to examine both mating halves of the primary fracture surface can lead to an errant origin identity. Many times when agglomerates are the strength-limiting flaw, one half of the fracture surface will have a depression which is matched on the mating half of the fracture surface by a protrusion of the same size. Examination of only one of the mating halves of the primary fracture surface could lead to an origin identity of a porous region or pore. In both specimens 5 and 6 of Topic #2 there were instances where the participants' identity did not agree with that of the organizers due to a limited analysis. In these instances the identity of the fracture origin was much clearer on the mating half of the fracture surface that the participant did not examine. Figure 3 is an example of such a case. The participant provided an agglomerate label which appeared to be based on only one half of the primary fracture surface, Figure 3A. Examination of both halves, Figures 3B and 3C, show that the origin is a porous region.

Figure 4 shows how the origin location can differ when the analysis is restricted to one half of the primary fracture surface. In this case the participant based the location solely on Figure 4A and reported a volume-location. In Figure 4B the origin is in contact with the tensile surface, thus a surface-location would appear to be more appropriate.

Other Specimens

There is no substitute for the information that can be gathered from actually viewing the fracture surfaces from a *set* of specimens of the same material. The round robin participants were at a distinct disadvantage in this case since they only received photographs of the fracture surfaces (Topic #1) or one specimen of each material (Topic #2).

Characterization of a fracture origin from photographs can be compromised by the fractographic experience level of the photographer, the restricted viewpoint, or the quality of the photographs. Some of the misinterpretations of machining damage in Topic #1 can be directly attributed to the fact that *only* SEM photographs were used in the analysis. A couple of participants marked striations on the tensile surface as the fracture origin rather than the subsurface cracks, Figure 5. The SEM photographs do show some striations which were deeper than those around them. Had the participants had the chance to analyze all the machined surfaces of the specimen or a group of identically machined specimens, as the organizers had, it would have become clear that these few "deep" striations were not unusual. A fracture mechanics calculation showed that the striations are too small to be the genuine fracture origins for this specimen.

Fractography of a "single" specimen of a material can also be misleading. The origin in one specimen may not clearly represent the dominant fracture origin type in the material or it may not be possible to completely characterize the origin in that particular specimen. In many instances the dominant origin type may not be become clear until many specimens of the same material with the same origin type have been examined. In general the organizers had the benefit of examining many specimens of the materials used in the exercise. This more comprehensive examination enabled the organizers to recognize key features common to the strength-limiting flaw in a given material.

Other Specimen Surfaces

Specimen surfaces other than the fracture surfaces can also contain value fractographic information. The specimens selected for specimen 2 in Topic #2 were known to have failed due to handling damage (HD^s) because the organizers intentionally scratched one of the 4mm x 50mm faces. The scratch was imparted on to the surface with a diamond indenter using only finger pressure, to simulate gross damage due to misuse of the specimen, Figure 6. These faces were subsequently placed in tension and all specimens fractured from surface or subsurface cracks associated with the scratch.

Only five participants agreed with the organizers that handling damage was the strength-limiting flaw, but several reported identities such as crack or machining damage. A closer examination of the tensile surface would have revealed that the damage was really a gouge and not a crack, and that the severity and wavy nature of the scratch along the tensile surface was more indicative of handling damage. The remaining participants incorrectly labeled the origin as something else probably because they did not examine the tensile surface.

An examination of the external surfaces of the SiC specimens (specimen 4), discussed above, would have revealed the excessive pitting that had occurred due to the thermal cycling. Even if the analyst did not have the information about the past history of the specimen it would have been apparent that pits had been created due to some reaction/interaction with an environment.

Fracture Mechanics

The organizers sense that a number of participants were not familiar with fracture mechanics. Only five participants reported* that they used some form of fracture mechanics to assist them in the size characterization of the flaws and mirrors examined, despite the fact that the organizers (by inclusion of the fracture stress and a fracture toughness value for each specimen in both topics) hinted that a size estimate should have been calculated by the participants. The exercise showed that the determination of the size of a fracture origin or fracture mirror is complicated for a variety of reasons. Even so, the use of fracture mechanics can provide valuable information about these sizes. A detailed discussion of how to use fracture mechanics as an aid to fracture analysis is in reference 8. MIL HDBK-790 contains a brief explanation, but evidently it was not adequate.

Flaw Size: Fracture mechanics can be used to estimate the size of the machining-related fracture origins seen in Topic #1, since these origins tend to be 2-dimensional and assume a semicircular or semielliptical shape (9, 10). The best correlation between the size predicted by fracture mechanics and that determined from fractographic analysis was in Set #2 (Silicon Nitride) and Set #3 (Alumina). The lack of correlation in Set #1 (Zirconia/Alumina) may be due to: 1) R-curve behavior; 2) The heat treatment the specimen received prior to strength testing (This may have reduced or eliminated any residual stresses and/or a changed acuity of the crack tip).; or 3) Crack nesting - Multiple cracks of similar size and shape can shield each other from experiencing the full stress intensity that is assumed for a single, stand alone crack.

The origins characterized in Topic #2 typically had complex geometries, making it difficult to use a single value to represent the size. Even with this complication, fracture mechanics can assist in the proper size characterization. (It is not unreasonable to approximate the origin size with circular or elliptical penny-shaped cracks.) If the origin is located near the surface, and the measured size is smaller than that estimated by fracture mechanics, it may indicate that the ligament of material between the surface and the origin should be considered part of the origin size. If the size and shape of the origin can be measured with confidence, then fracture mechanics can be used to estimate the toughness of the material. This calculated toughness value can then be compared to independently measured toughness values and possibly corroborate the identity of the origin. If there is a large discrepancy, then the fractographer should reassess their characterization of the fracture origin. A detailed discussion in how to use rudimentary fracture mechanics concepts to estimate or corroborate flaw sizes has now been added to the draft ASTM standard practice. Factors which can complicate the comparison of the fracture mechanics origin size estimate and the measured origin size are discussed in Appendix 2 of reference 8.

Mirror Size: Mirror constants and the mirror size-to-origin size ratio (11) are valuable aids in the fractographic interpretation of either mirror or origin sizes. Conceptually, it would appear to be straightforward to measure fracture mirrors since they are essentially circular or semicircular, and are centered on all (or part of) the fracture origin in uniformly stressed specimens or components, but in practice there are a number of complications. Unlike in glasses, these fracture features may not be easy to see or measure in polycrystalline ceramics. Non-uniform stress states and free surface effects can cause deviations from the assumed circular or semicircular shape.

The results from Topic #1 showed that few participants were familiar with fracture mirror analysis. The organizers determined two mirror size ranges, one for each technique, for

* Several participants subsequently stated that they had used fracture mechanics but had not stated such in their reports.

the three specimens in Topic #1 and compared the participants values to these ranges. In Set #1 - Zirconia/Alumina, there was a very good agreement as only two participants' values were outside either range. The agreement was not as good in Set #2 - Silicon Nitride, as the values from five participants were outside the mirror constant range and six were outside the range estimated by the ratio method.

Suprisingly the easiest fracture "mirror" to measure (Set #3 - Alumina) was the one that gave the participants the most problems. Many participants were fooled by the appearance of the fracture features and labeled the "smooth" region of transgranular fracture as the mirror, see Figure 7. Fracture mechanics was used to estimate the flaw and mirror size in this low strength specimen (σ = 228 MPa). These calculations indicate that the flaw and mirror are quite large. The mirror is in fact larger than the specimen cross-section size. In such instances, a mirror will not form. Only two participants recognized this.

It must be noted that these techniques do have limitations. The mirror constants used to estimate the range for Set #1 and Set #2 may not be appropriate for these materials. The constants used for Set #1 are most likely for a refractory-grade partially-stabilized zirconia. In Set #2 the silicon nitride contains yttria and an elongated microstructure that can be a source of R-curve behavior (12) while the mirror constants in the literature are for a magnesia-doped silicon nitride with negligible R-curve behavior. Additionally, the ratio method relies on accurate measurement of the origin size which is difficult in its own right.

OVERALL SUMMARY

The detective work entailed in fractographic analysis requires experience, patience, and time on the part of the fractographer. Time is an element that is part of every piece of the fractography puzzle. Upon reviewing a draft version of Round Robin report (8) several participants indicated that their analysis may have been better if they had the opportunity to spend more time on the specimens. The organizers concur. We believe that time-constraints caused many of the origin misidentifications seen in the exercise, especially in instances when one of the mating halves of the primary fracture surface or the external surfaces were not examined.

Successful fractography has a number of ingredients: broken pieces, a material conducive to fracture surface interpretation, background information, a skilled fractographer, equipment, adequate time, and a framework or guide for the analysis. This round robin has identified a number of key factors that are essential for the last item. These factors will be incorporated into the developing ASTM standard.

ACKNOWLEDGMENTS

One of the authors (GQ) received partial support for this project from the Ceramic Technology Project, U.S. Department of Energy, Office of Transportation Services, under contract DEAC05-84OR21400 with Martin Marietta Energy Systems, Inc.

REFERENCES

1. Fractography of Glasses and Ceramics, Advances in Ceramics, Vol. 22, eds., J. Varner and V. Frechette, American Ceramic Society, Westerville, OH, (1988).

2. Fractography of Ceramic and Metal Failures, eds., J. Mecholsky, Jr and S. Powell, Jr., ASTM STP 827, Philadelphia, PA, 1984.

3. Concepts, Flaws and Fractography, Fracture Mechanics of Ceramics, Vol 1., eds., R. Bradt, D. Hasselman and F. Lange, Plenum Press, New York, 1974.

4. Failure Analysis of Brittle Materials, Advances in Ceramics, Vol. 28, V. D. Frechette, American Ceramic Society, Westerville, OH, 1990.

5. J.J. Mecholsky, Jr. and S.W. Freiman, "Determination of Fracture Mechanics Parameters Through Fractographic Analysis of Ceramics," Fracture Mechanics of Ceramics Applied to Brittle Materials, ed. S. Freiman, ASTM STP 678, 136-150, (1979).

6. R.W. Rice, "Ceramic Fracture Features, Observations, Mechanisms and Uses," Fractography of Ceramic and Metal Failures, ASTM STP 827, 5-103 (1984).

7. "Failure Analysis" Engineering Materials Handbook, Vol 4, Ceramics and Glasses, 629-673 (1991).

8. J.J. Swab and G.D. Quinn, "Fractography of Advanced Structural Ceramics: Results from the VAMAS Round Robin Exercise," ARL-TR-656, December 1994. (Also published as VAMAS Report #19, NIST, February 1995.)

9. A.G. Evans, M.E. Meyers, K.W. Fertig, B.I. Davis and H.R. Baumgartner, "Probabilistic Models of Defects Initiated Fracture in Ceramics," J. Non-Destruct. Eval., [1] 111-114 (1980).

10. R.W. Rice, "Fractographic Determination of K_{IC} and Effects of Microstructural Stresses on Ceramics," Fractography of Glasses and Ceramics II, Ceramic Transactions, Vol. 17, eds., J. Varner and V. Frechette, Amercian Ceramic Society, Westerville, OH, 509-545 (1988).

11. J.J. Mecholsky, Jr., S.W. Freiman and R.W. Rice, "Fracture Surface Analysis of Ceramics," J. Mat. Sci., 11, 1310-1319 (1976).

12. J.A. Salem and S.R. Choi, "Toughened Ceramics Life Prediction," Department of Energy, Oct.-Nov. Bimontly report, 220-234 (1991).

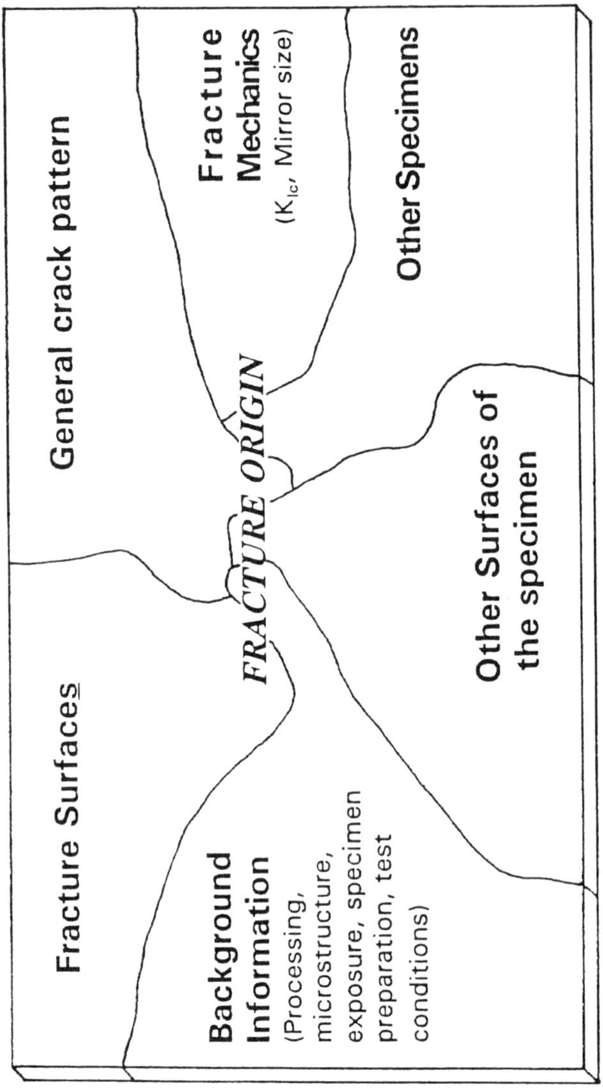

Figure 1. Fractography is like assembling a jigsaw puzzle, if a few key pieces are missing then the picture may not be understood.

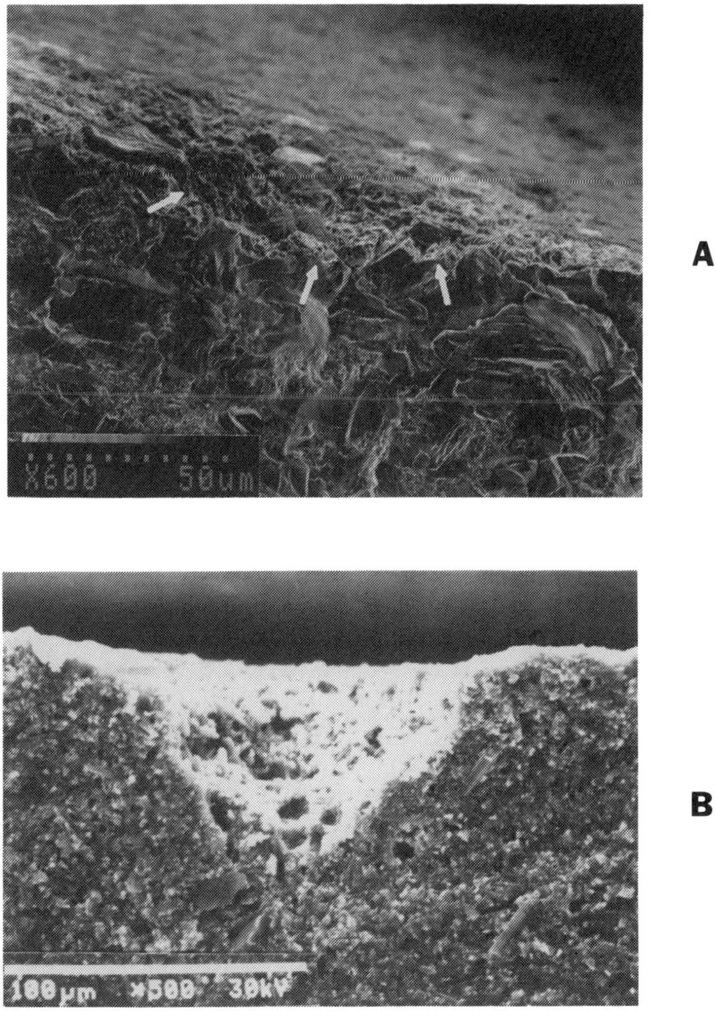

Figure 2. A) Participants' SEM photograph of the pit in specimen 4. Arrows added by the organizers. B) SEM photograph of a surface void from MIL HDBK-790, Figure 33, page 40. Note the similarity.

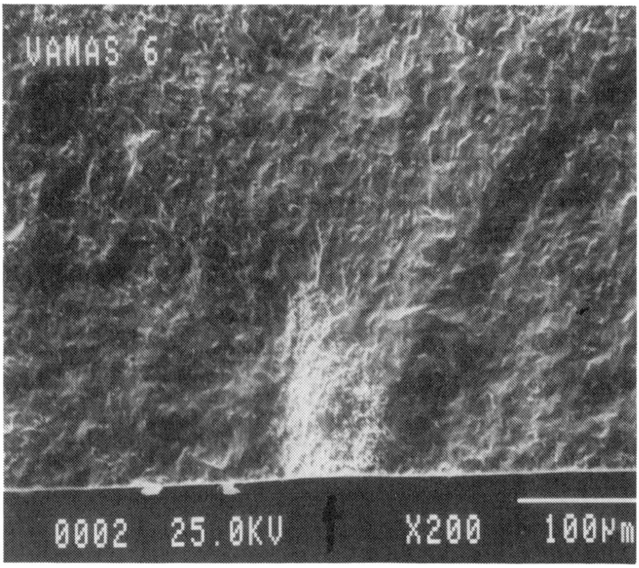

Figure 3A. Participants' SEM photograph of specimen 6. Participant labeled the origin as an agglomerate.

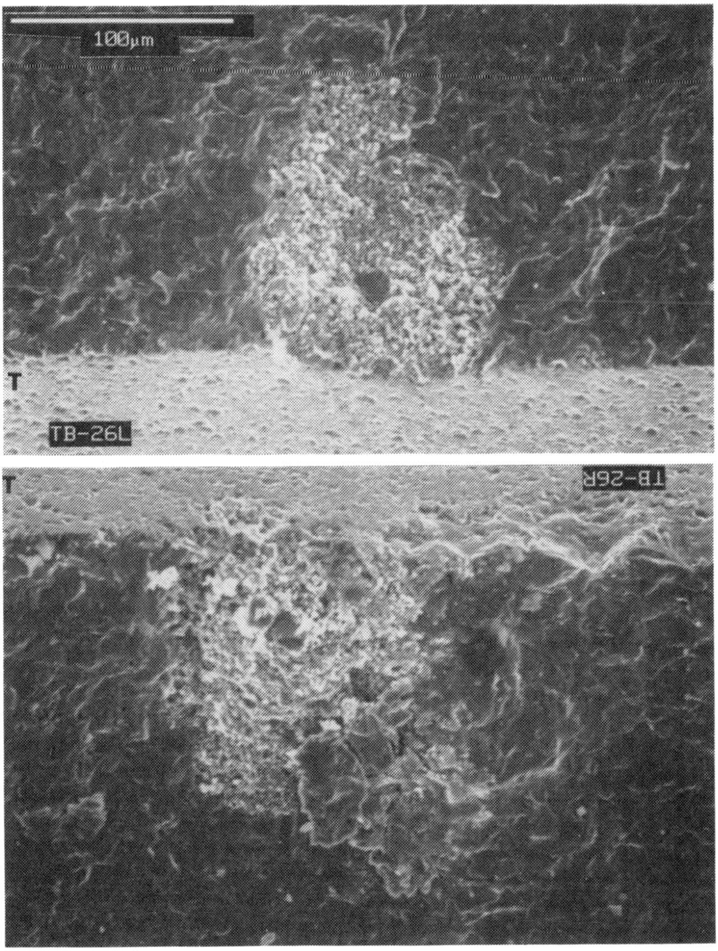

Figure 3B and 3C. Organizers' photographs of the mating halves of the primary fracture surface of specimen 6. The photographs in B) and C) indicate that this origin is a porous region and not an agglomerate as labeled in A). "T" denotes the tensile surface.

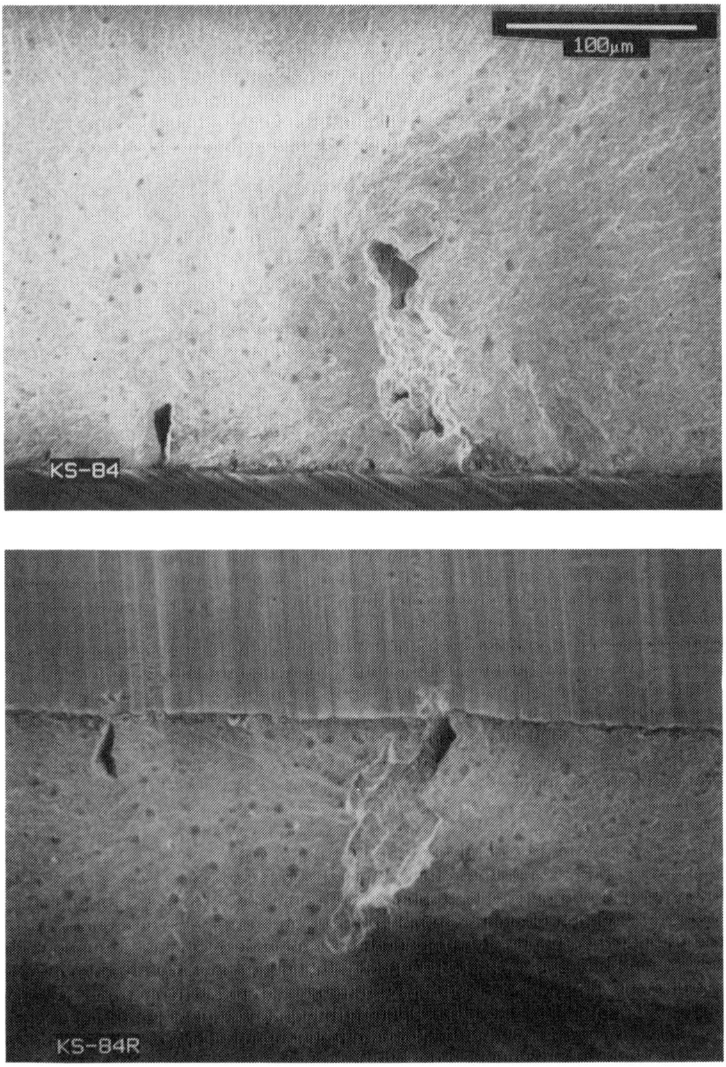

Figure 4. SEM photographs of the mating halves of the primary fracture surface from specimen 3. Examining photograph A) only could lead the fractographer to report the origin as volume-located. B) indicates that the origin is surface-located instead of volume-located. The origin is a volume-distributed pore. Since the pore contacts the machined specimen surface, the origin location is thus characterized as "surface" for this particular specimen.

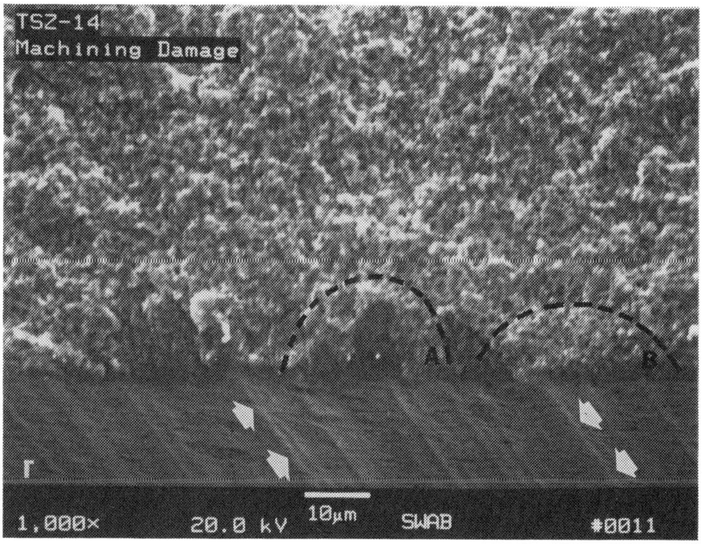

Figure 5. High magnification SEM photograph of one half of the primary fracture surface from Set #1 of Topic #1. Some participants indicated that the striations on the tensile surface (white arrows) may be the machining damage. Organizers outlined two possible semicircular origins (dashed lines.)

Figure 6. Example of Handling Damage in specimen 2: Alumina. "T" denotes the tensile surface.

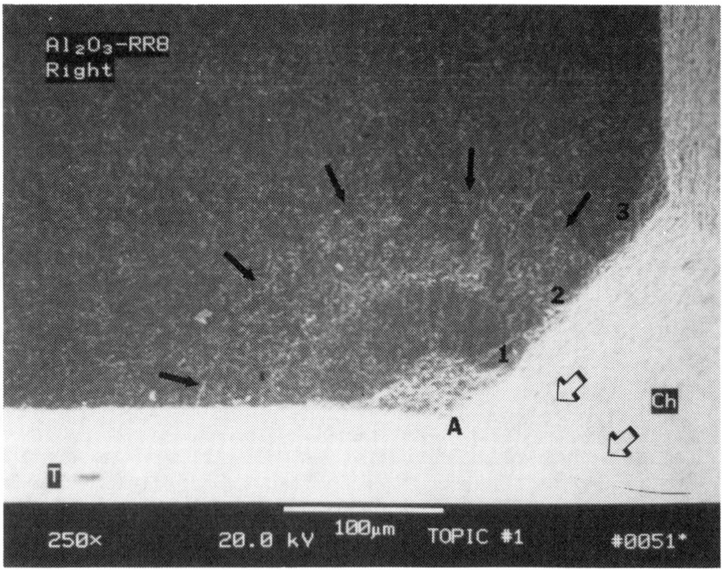

Figure 7. High magnification SEM photograph of one half of the primary fracture surface from Set #3 of Topic #1. Organizers outlined three stages of fracture. Stage 1 is transgranular, Stage 2 is subcritical crack growth and Stage 3 is fast fracture. "T" denotes the tensile surface and "Ch" the chamfer.

STANDARDIZED GUIDELINES FOR FRACTOGRAPHY OF ADVANCED CERAMICS - A VIEW FROM EUROPE

R Morrell
National Physical Laboratory, Teddington, Middlesex, TW11 0LW, UK

ABSTRACT

The development of CEN standards for advanced technical ceramics is briefly reviewed, with an emphasis on fracture tests. The development of a fractography standard is discussed. The subjective nature of the topic poses some problems for appropriate wording. In particular, issues relating to the scope of the standard and the degree of caution that would need to be inserted for the benefit of inexperienced standards users are highlighted, especially areas seen as problems. These include identification of machining damage, identification and use of mirror radius, and multiple fracture origins. Some of these are illustrated with examples of common materials. A flow chart for a standardised procedure is proposed.

INTRODUCTION

The advent of advanced ceramics for wide-ranging applications has led to the need for new industrial standards to be developed which are more relevant to the new engineering uses than those originally designed for some of the more traditional applications, such as whitewares, electrical insulators, etc. In particular, mechanical properties become much more critical, and correspondingly there is a need for a more rigorous approach to mechanical testing and evaluation of materials and products. In Europe (CEN Technical Committee 184), just as in the US (ASTM Committee C28) and Japan, there have been programmes of standardization operating for some years now specifically to address this problem. The relatively independent approaches in the different geographical areas have led to differences, and the issue of harmonisation is now being addressed through the recently formed ISO Technical Committee 206.

The brittle nature of advanced ceramics means that considerable emphasis in standardisation work has been placed in strength, fracture toughness and related areas of testing and analysis. The inclusion of fractography in these programmes is a mark of recognition that as a tool for investigation it has considerable value. This paper reviews the standardisation programme in CEN, with special emphasis on mechanical testing issues, and illustrates the directions in which European thinking on a fractography standard is moving. In particular, the feasible scope of such a standard is discussed, and some of the problem issues are aired, illustrated with examples from the author's own experience.

CEN STANDARDS

The CEN (Comité Européen de Normalisation) standards programme was established in 1989 under a European Commission mandate in order to bring together different contemporary standards activities amongst the member states, in particular the UK, France and Germany, but also encouraging equivalent interests in other EU countries. There are currently nine active countries in a broad programme now into its second mandate period. Under the first mandate 47 standards were produced and agreed by the Technical Committee covering the activities of five Working Groups: Classification and Terminology; Powders; Monolithic Ceramics (includes particulate, whisker and short-fibre composites); Composites (principally long-fibre composites); and Coatings. The full extent of the programme has recently been reviewed [1]. Overall, the emphasis has been on establishing a suite of methods for all aspects required for industrial specification purposes, commencing with basic issues such as density and porosity, through to issues of high-temperature mechanical and thermal testing. By commencing with this broad programme, an even-handed approach has been possible, not biased in any one direction. There is strong industrial interest with support from GROUPISOL, the forum for European manufacturers' federations, and the expert membership of the Working Groups is defined by industrially led national standards committees. As a consequence, a key issue is the industrial usability of the standards being developed. This gives rise to several important factors:

1. Although one target is to improve test practice, the test method must not set too high a target in terms of accuracy of measurement, since this may not be achievable in industrial laboratories.

2. The vast range of materials and products covered by the term advanced technical ceramic causes some uncertainties as to whether the test method really does work. There is a need to pool the experience of the Working

Groups to identify those types of material for which the tests are inappropriate, or need to be modified.

3. Clear step-wise procedures are required in accordance with the CEN and/or ISO formats, with clear definitions as to which aspects are mandatory and which are informative.

4. A clear statement of uncertainties is generally required. These may not be known at the initial drafting stage. Research, especially round-robin activities, are seen to be of paramount importance in providing appropriate evidence.

5. The CEN procedure offers a facility which is different from most other standardization procedures. If a standard is not considered to be definitive to the extent that it can be adopted formally by all member states of CEN, the TC can agree to it being published as a Euronorm Volontaire (ENV), broadly equivalent to an ISO Technical Report. This has a three-year life before decisions have to be made on its formal voting to a full standard, a period during which it can be used experimentally and any issues investigated. This is particularly useful for standards being developed for the first time in a new technical area.

These factors result in a high level of detail being required to account for all recognised issues, but the formalised format assists in providing a clear procedure for industrial laboratories to follow, especially with increasing requirements for recognised quality systems to be employed in industry.

This paper is not the place to review in detail the standards related to mechanical properties. However, some aspects are worth mentioning concerning fracture testing and analysis.

Flexural Strength at Room Temperature

CEN EN 843-1, based on previous DIN and AFNOR standards is broadly equivalent to ASTM C1161 in most respects, although test bar size C (6 x 8 x 80 mm span) is omitted. In preparing this standard, full consideration was given to MIL STD 1942 [2] and the related documentation produced by the US Army [3,4]. Issues related to prescription of machining of test-pieces were fully discussed, and a degree of flexibility over procedures has been allowed to cope with varying circumstances. Fully articulating test jigs are required, unlike the relaxation in

ASTM C1161 for machined test-pieces, because of lack of confidence concerning perfection of test-jig and test-piece manufacture [5], and to allow the testing of as-fired, oxidised or corroded test-pieces. This standard is intended as a baseline method to be referenced in other standards in which geometrical perfection of test-pieces will be absent. A separate standard has been designed for long-fibre composites, ENV 658-3, and also requires articulation. A loading arm to thickness ratio of 10 is specified.

Flexural Strength at Elevated Temperatures

CEN ENV 820-1, again based on a previous DIN standard, extends the room-temperature test to elevated temperatures, subject to pseudo-elastic behaviour in the test-piece. Test jigs may be made from metallic materials provided that the roller hardness remains adequate, e.g. up to typically 400 °C, but recommends high-alumina or silicon carbide rollers for much higher temperatures. Jig articulation is again required. Notes concerning recommended designs and procedures are included to help the inexperienced, as well as warnings concerning interactions between test-pieces and rollers.

Tensile Strength

CEN TC184 has produced a standard for tensile strength of long-fibre composites, which is a simplified version of that adopted by ASTM as C1275. There are plans to produce a method for elevated temperature use. For monolithic ceramics, a decision was made not to develop a tensile test method because of fears that industry would come under increasing unnecessary pressure from customers for its use as a general data collection/QA method, rather than for special purposes. Another view is that it is more cost-effective to evaluate flexural behaviour than tensile behaviour, and in high-strength materials both tests are equivalently surface finish dominated [6].

Statistical Evaluation of Fracture Data

Like ASTM standard C1239, the CEN ENV, which is in its final stages of review, looks at the general situation of a set of equivalently obtained fracture data obtained by any relevant uniaxial tensile stressing method. Apart from using slightly different nomenclature, it is otherwise equivalent to C1239 in approach, analysis method and reporting requirements.

DEVELOPMENT OF A FRACTOGRAPHY STANDARD

Background

As in the US, fractography is seen to be an important tool for analysis of test results and for component failures from the field. In this respect, the analysis of small flexural test-pieces broken under known and well-controlled conditions is rather more straightforward in terms of preparing a procedure than is the evaluation of fragments from other circumstances.

The development and publication of MIL-HDBK-790 [7,8], based on materials development and testing experience, provides a good basis for development of a standard. This document, however, is not in a "standard" format and needs redrafting. In ASTM C28 this has been done by the original authors following a very successful VAMAS international round robin [9] reviewed at this conference [10]. The CEN version will use the same concepts and sources of information, and will take into account most of the practical issues and items of guidance that are essential to good fractographic practice. However, in taking the wide range of factors into consideration, a number of crucial questions need to be satisfactorily answered. The remainder of this paper discusses these issues.

Overall Objectives

Taking a broad view of fractography, it is useful to examine the applications to which it may be put. Table 1 indicates some of the typical questions that might be asked about a fracture, and that should be addressed in a standard.

Table 1. Why do we do fractography?

Question	Application
Where did it break from?	Engineering
Did it break suddenly or crack slowly?	Engineering
Why did it break from here?	QA, process monitoring
What was the nature of the fracture source?	Material development, QA
What was the stress at fracture?	Design
Did environment or fatigue have influence?	Engineering
Did we do a good test?	Material evaluation
Whose fault?	Commercial, legal (care!)

Taking these individually:

Where did it break from? For test-pieces, e.g. bars or discs, there are some typical patterns of cracks that allow speedy identification of the original fracture source. Some simple rules can be used, such as for biaxial testing in which the primary cracks tend to be continuous, but the secondary ones tend to start abruptly from primary ones. Some of these are illustrated in the draft ASTM standard. For other shapes with multiple fractures (e.g. in a tube burst test), it becomes extremely difficult to identify the position of the originating flaw from the fragments. Allowance has to be made for this eventuality in preparing the standard.

Did it break slowly or suddenly? In other words, did the flaw grow slowly, or was there a sudden increase in mechanical load? Subcritical crack growth (SCCG), except under controlled laboratory fracture conditions, is usually very hard to observe. Its detection relies on seeing a significant and identifiable change in fracture morphology. Some participants in the VAMAS high-temperature fracture toughness test round robin organized by the Japan Fine Ceramic Center [11,12] noted SCCG before fast failure in a single-edge precracked beam test, but others did not. Recognition of some differential degrees of oxidation of the crack face may not always be obvious, depending on the thermal history. In many other cases, especially in materials such as alumina, macroscopic crack morphology seldom changes significantly as the crack develops, perhaps with only a change from transgranular to intergranular fracture as sometimes noted in fracture toughness tests. Distinguishing between the normal subcritical crack growth that occurs before fast fracture, and any effects that were genuinely of a longer-term type due to fatigue is difficult.

Why did it break from here? Having followed the fracture lines back towards the source, interpretation of the feature at the source may or may not be obvious, but clearly involves subjectivity. It may appear differently when observed using different means. For example, for white, semitranslucent materials, optical evaluation using specimens without a metallic coating is difficult as a result of light scattering, e.g. Figure 1. Recourse may need to be taken to SEM examination where the depth of field at higher magnifications is a key factor in unambiguously identifying microstructural defects, but this is normally classed as destructive. The use of fracture mechanics to relate the apparent source with actual fracture stresses is also a helpful aid to correlate with the defect size, but this requires prior knowledge of the applied local stress and the material fracture toughness. In the absence of such information, e.g. in the failure of components, greater care needs to be taken.

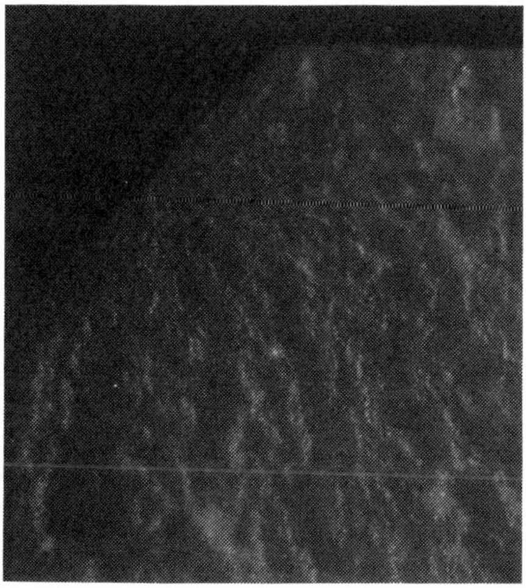

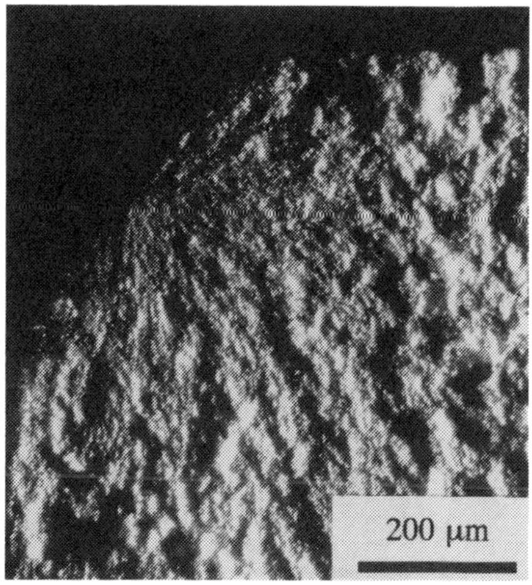

Figure 1. Fractography of a translucent ceramic (3Y-TZP) under grazing incident illumination, with (right) and without (left) coating.

The most difficult fracture origin is machining damage. This produces no obvious heterogeneity in the majority of materials, and in some senses runs the risk of being the default option if nothing else is seen. However, as the VAMAS Fractography Round Robin has shown [9,10], identification of the precise location and size of flaws developed by machining is highly subjective. It is helpful if there is a difference in fracture surface morphology between the machining damage and subsequent macroscopic fracture. A standardised terminology [13,14] will certainly help to provide a unique method of describing the observations made.

What was the nature of the fracture source? Once the source has been positively identified, the nature of any heterogeneity needs to be proved. This generally requires SEM work, especially if the source is small, or needs to be analysed chemically for material development of QA purposes. Cleanliness becomes paramount, but this is usually not possible for fractures in the field.

What was the stress at fracture? The defect cannot always be considered to be a regular shaped Griffith type flaw for direct calculation of strength via a known or assumed toughness. While in some materials this may be usefully done, in those which suffer SCCG before fast fracture, critical flaws are usually much larger than any microstructurally obvious defect. One possibility is to resort to fracture mirror analysis. The mirror size has been shown to be related experimentally to the

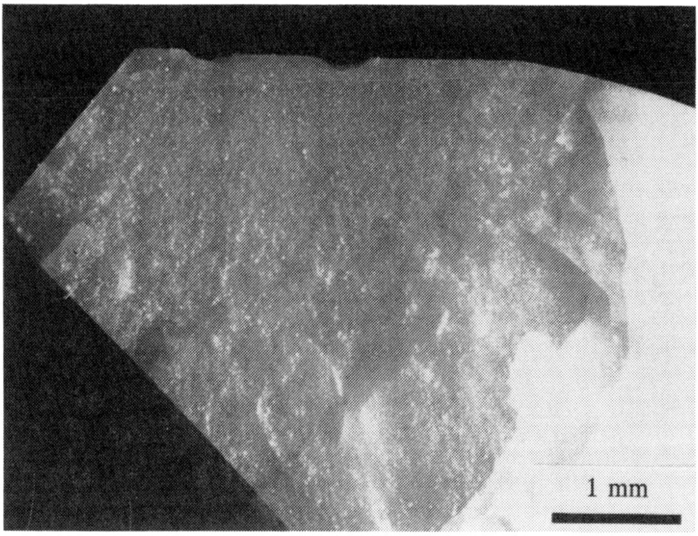

Figure 2. Clear mirror/hackle system in a fine-grained alumina component, optical.

fracture stress, so that if the mirror constant has been determined from test bars, it should be possible to identify the fracture stress in failed components.

An interesting example of this has recently been studied for TZP. The question arising was whether the less favourable strength characteristics of a component of a new design resulted from a ceramic fabrication change, or from the mechanical design change. Using quality-control flexural strength test bars of known fracture stress, fractography was performed to determine the mirror constant. The key problem was positive identification of the mirror size. This particular ceramic material does not seem to give a clearly defined mirror, compared with, for example, a high-alumina ceramic (Figure 2). The roughness of the fracture surface appears to increase progressively with distance from the source. As a consequence, any attempt to measure the mirror radius requires some criterion of roughness to be adopted. If done using light microscopy or SEM this is highly subjective. However, in order to provide the required demonstration it proved possible, using photographs of about x40 magnification, for two people to make independent but reasonably consistent estimates of a radius corresponding with a given agreed level of apparent roughness (that may or may not be equivalent to the mirror). These measurements could be made with a precision of only ± 1 mm on the photograph. The possibility of measuring an apparent mirror size more definitively using laser scanning profilometry is being investigated.

Figure 3 Fracture surface of a TZP test bar showing a lack of a clearly defined mirror (optical, grazing incidence, coated).

The results are plotted in Figure 4 using stress levels corrected for the distance of the fracture origins from the outer tensile surface of the flexural test-piece. The precision limitations resulted in the grouping of the data at each value of radius. From the results, the average mirror constant was deduced as about 10 MPa $m^{1/2}$. The same procedure and the same photographic magnification was used for the components failing at various locations within the unknown stress field. The experimentally measured fracture mirror radii are plotted in Figure 5 on the mean line from Figure 4 in order to estimate the original fracture stress. The results indicate that the component fracture stresses were on average just a little lower than those of the test bars, as would be expected from the volumetric effect. The source of the less favourable component strength is thus the manner in which the components were stressed, not the component fabrication.

This example highlights that for some materials there may be a strong degree of subjectivity in assessing mirror radius. In fact one could ask whether a mirror truly exists, let alone the so-called mist region that is sometimes obvious in high-energy glass fractures. However, the above demonstration shows that some useful observations can be made using self-set criteria, but including such practice as part of the guidance in a standard is problematical.

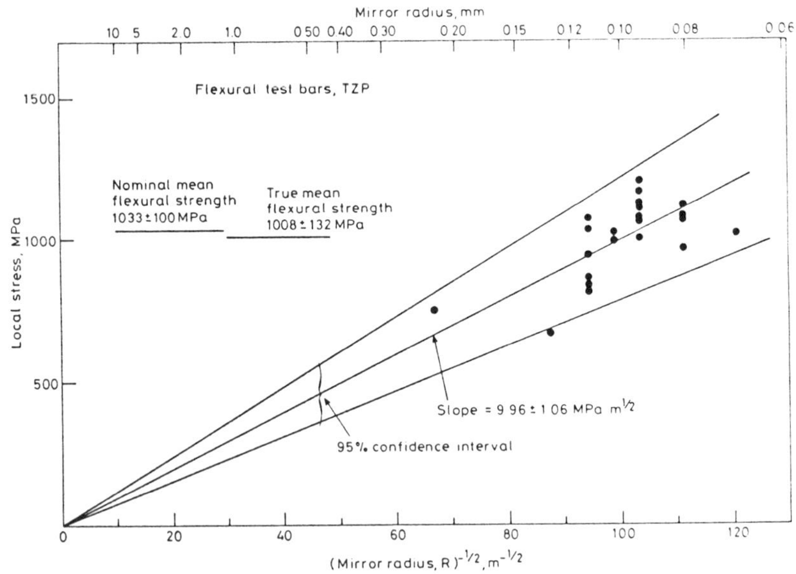

Figure 4. Approximate fracture mirror radius versus fracture stress for 3Y-TZP.

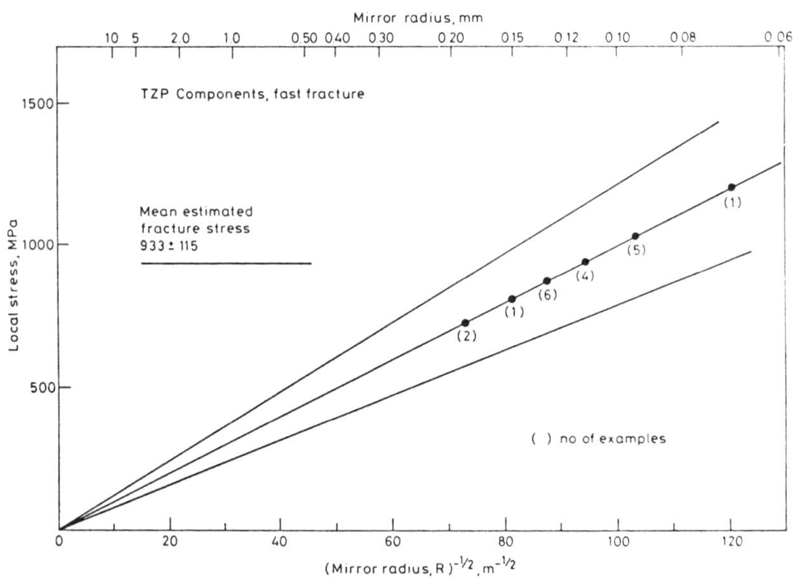

Figure 5. Use of the fitted slope and confidence interval in Figure 4 to define strength of fractured 3Y-TZP components from measured mirror radii.

Did environment or fatigue have influence? Unless the effect of these factors on the fracture surface morphology has previously been investigated and documented, the correct assignment of observed features to these factors is highly subjective. It is seldom the case that fatiguing effects in practice are sufficiently well controlled to have a systematic effect on the fracture surface, e.g. of a series of obvious interruptions in the growth of the flaw resulting from cyclic stress. It is felt that this aspect cannot have specific instruction within a standard.

Did we do a good test? Fractography, even if performed superficially, can clearly assist in deciding whether fracture tests found defects randomly, or whether there was some error in the procedure, either in terms of edge failures resulting from poor machining, or imperfect loading geometry leading to biased failure positions, such as shoulder failures in tensile testing or ring failures in ring-on-ring disc testing.

Whose fault? Using fractography to assist in assigning blame for products considered defective has risks associated, as remarked upon by Frechette [15], and must be considered in developing a standard. It is the present author's opinion that in view of their subjective nature, fractography skills may only be able to draw inferences, so not in all circumstances prove the cause beyond reasonable doubt. An appropriate phrase must be included to ensure that the procedure contained in the standard is merely a guide, for which no legal responsibility can be held.

Flow chart for fractography

A possible approach for a standard is to provide the inexperienced user with a flow chart to follow in a step-wise fashion. It would allow the user to see the sequence of basic steps unhindered by the do's, don'ts and notes of a formally written procedure. This is under discussion in the CEN WG, and is shown in Figure 6. By following the sequence, the user is prompted to avoid missing steps. Each step would be cross-referenced to the relevant paragraph in the standard.

PROBLEM AREAS FOR DEVELOPMENT OF A STANDARD

A standard needs to have a defined scope in terms of the uses to which it is put, and for advanced ceramics this is often uncertain. A standard can fall into disrepute because it is inapplicable to the scope written in it, and thus considerable vigilance and breadth of knowledge must be brought to bear in drafting. Based on personal and collected experience of colleagues, some of the issues are given below as questions.

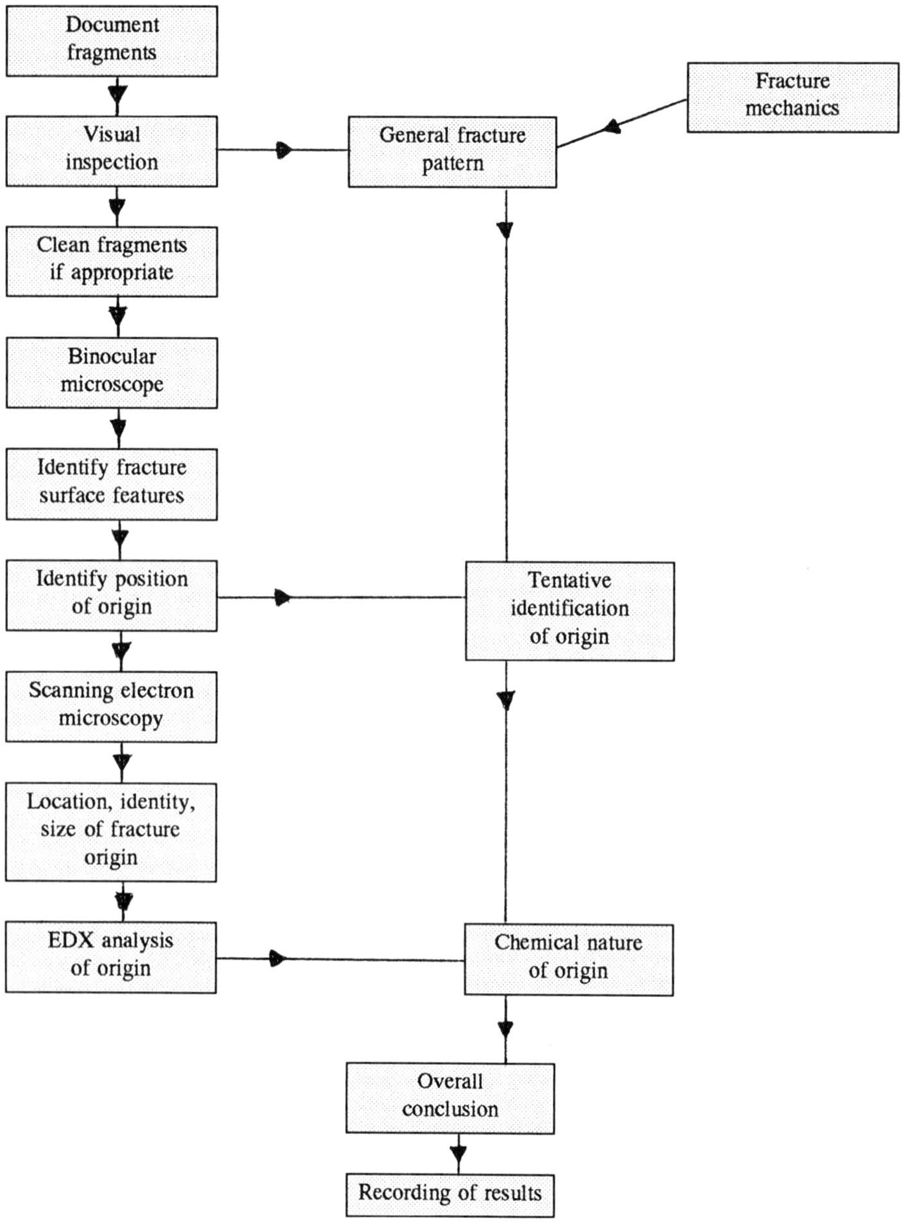

Figure 6. Fractography - a general flow chart for the basic steps

Does standardisation lead to deskilling?

One of the objectives in creating standards is to provide a common starting point and a rationalized reporting procedure for use by any laboratory. However, it can be argued that fractography is a personal skill developed through experience with a variety of subject matter. There is an element of experience-related subjectivity in being able not only to recognise the existence of fracture surface features, but also to make the correct deductions from these observations. As the quality of the available fractographic information declines from the ideal situation, such as a carefully controlled flexural test, this task becomes more difficult, and more reliant on past experience to provide a correct interpretation. Two key questions are:

> Can most cases be reduced to a standardised set of procedural steps?

> How much instructional information should one place in a standard?

If this is done, is there a risk that an inexperienced user will follow the procedure rigidly whereas an expert might adopt a flexible but more appropriate approach?

> Will a standard result in wider-spread usage but poorer quality?

It would certainly be useful to extend a standard's utility beyond test bars but it cannot cover every eventuality, and there is no substitute for experience.

Machining damage

As mentioned above, machining damage seems to be the most difficult area for categorical identification. Often, the damage is a series of partial shallow flaws which may join up on fracture. Interpreting the originating flaw and sizing it seems highly subjective. Even experts can disagree in drawing flaw outlines. As the stress intensity factor calculations for surface flaws demonstrate [16], the shape of the flaw determines whether it is more likely to propagate from its deepest point or from its junction with the external surface, and this in turn will influence the pattern of fracture lines visible in subsequent fractography.

> Can there be a standardised approach to identification and interpretation of machining damage, and to sizing the observed flaws?

SEM contrast issues

SEMs can operate under various of detection modes with different excitation voltages, with different coatings applied [17]. While some conditions are basically suited to topographic observation, identification of flaw sizes can prove more difficult. During the VAMAS Surface Flaw in Flexure Fracture Toughness Round Robin [18,19], initial difficulties observing pre-crack boundaries in TZP were overcome when the observational mode was changed from low voltage without coating to higher voltage with a thin metallic coating, under which conditions clear contrast could be seen between the precrack and surrounding fast fracture, distinguished by a small change in crack face morphology (Figure 7).

> Can we give clear general instructions on SEM observation conditions?

Difficult materials

Some materials give clear fractographic evidence, others do not. Medium to coarse grained alumina is one such material. Optical investigation of fracture origin often reveals no clear pattern of markings. Figure 8 shows an example of a fractured 96% alumina component. The series of holes implies the existence of multiple fracture origins, so identification of the primary origin is difficult. In the SEM, no fracture lines can be seen to guide the observer to the source. However, careful searching revealed a large grain close to one of the suspected initiating points.

In very strong materials, e.g. polished sapphire, the energy dissipated on fracture can be so great that the material tends to shatter as the crack jumps between low energy fracture planes. Fracture origins are invariably lost, so application of fracture mechanics of initiating flaws to explain strength variability is impossible.

> Can we identify the material types for which fractography clearly works, and those for which it is generally difficult to impossible?

Sizing of mirrors

As described above, not all materials give clear mirrors and hackle for fracture mechanical investigation. In some cases this may be because the test-piece or component is not large enough for the crack velocity to reach the hackle condition under the applied stresses. Discussing mirrors and hackle in detail, and providing

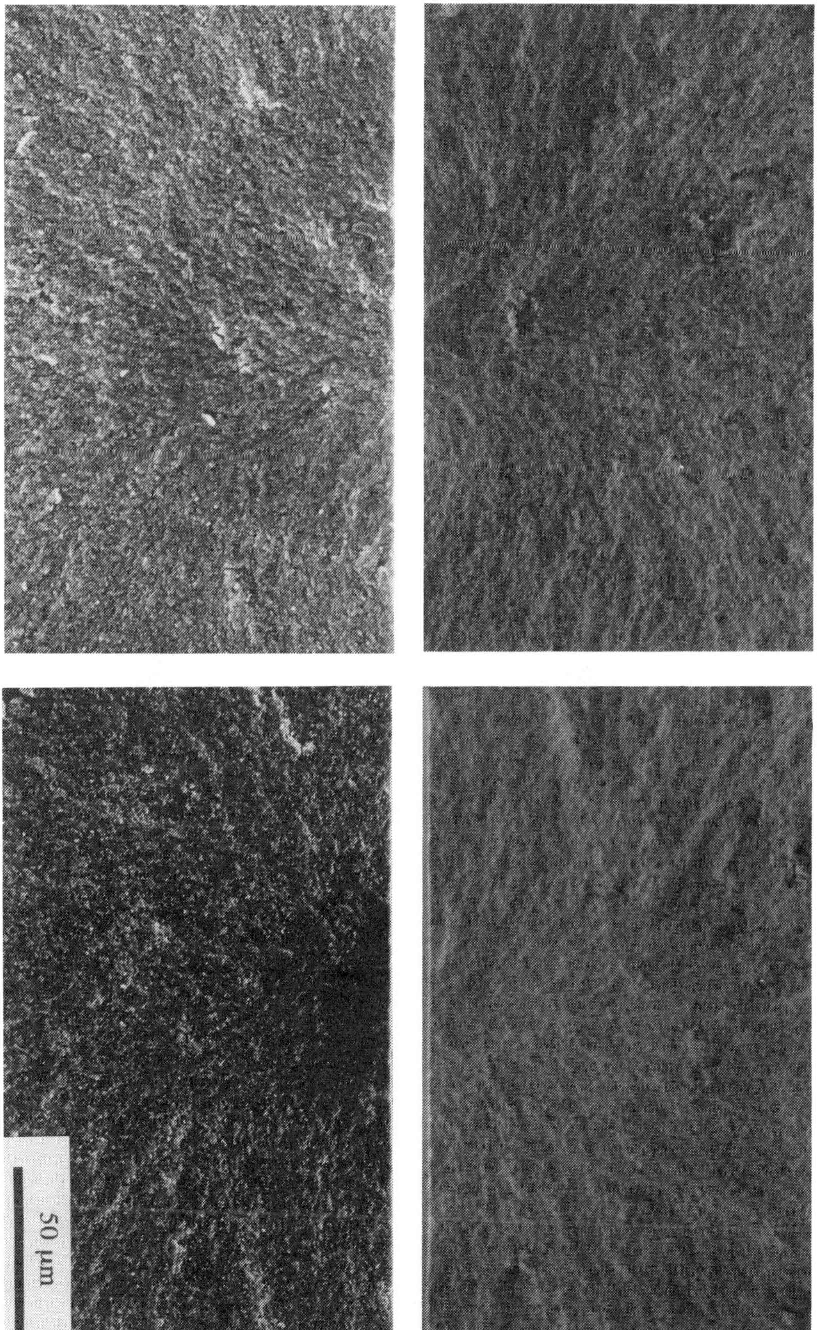

Figure 7. Matching halves of controlled surface flaws in TZP examined under (top) low voltage mode, uncoated, and (bottom) 30 kV with optimised gold/palladium coating thickness to reveal pre-crack region.

(a)

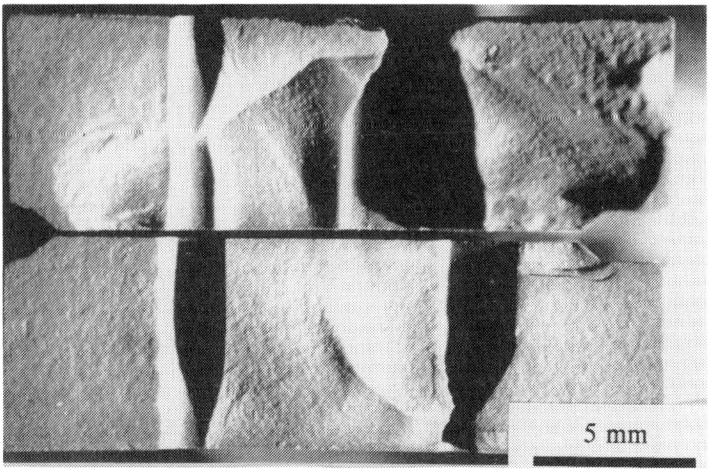

(b)

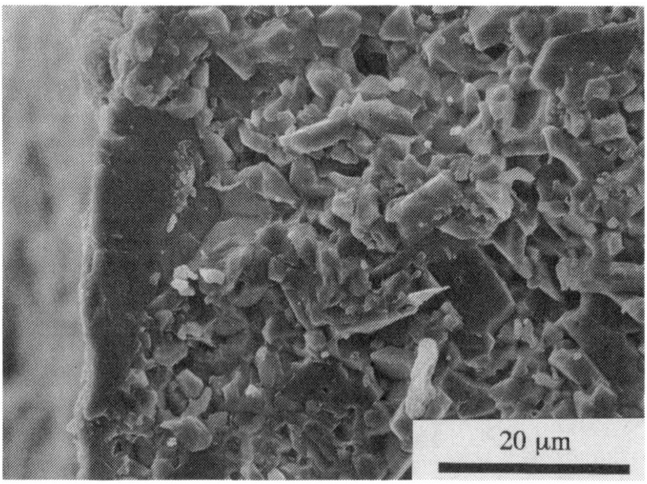

Figure 8. (a) Multiple potential fracture origins in an alumina component with weak fracture markings. (b) Fracture source finally identified as large grain, at left of SEM image, adjacent to the edge of one of the holes.

mirror constants, in a standard may suggest that they should always appear, so it is essential to include some instruction on the expectation of seeing them:

What guidelines should be given concerning the appearance of classical mirrors in polycrystalline materials?

Coping with failures from the field of operation

Incomplete sets of contaminated fragments pose a problem that must also be addressed by a standard. While cleanliness of operation can be emphasized for fractures obtained under laboratory conditions, retrieval from the field usually results in a bag of poorly packed fragments impacting on each other, often covered in dirt and metallic pick-up. A standard may help to educate those wishing to undertake an examination, but the damage is usually done before this point. Some instruction is needed on distinguishing those features which must have occurred subsequent to fracture from those present on the original fracture. Indeed the dirt one might normally wish to remove for clearer observation may itself be the key to the fracture.

How should we instruct on identification of subsequent damage?

CONCLUSION

This paper has reviewed the development of CEN standards with particular reference to those for strength and fracture testing. The extension towards a standard for fractography is now under consideration in Europe, and a number of questions have been raised about the level of instruction that should be included. By focusing on the purposes to which such a standard might be put, a number of questions have been raised concerning observational and interpretational issues which need to be clarified.

A modern standard is frequently used as a documented procedure for quality systems, and thus must have a clear step-wise approach. To assist this a flow chart for the basic steps in fractography is proposed. In addition, there are a number of more difficult areas, such as interpretation of machining damage, non-appearance of fracture mirrors, SEM observation conditions (voltage, imaging mode, coating, take-off angles, etc. to be adapted to each material type), and examination of fractures from field operation. Some instruction needs to be included in a standard for these factors, but the necessary scope and statements remain unclear.

Acknowledgements

The author would like to thank Liam Byrne (NPL) for assistance with SEM studies, and Morgan Matroc Ltd for permission to include some examples of fractographic problems.

References

1. Kübler, J.J., Morrell, R. "European standardisation activities for advanced technical ceramics", ASME Paper 93-GT-158 [1993].
2. MIL-STD 1942 (MR), "Flexural strength of high-performance ceramics at ambient temperature", [1983].
3. Quinn, G.D.; Baratta, F.; Conway, J. "Commentary on US Army Standard test method for flexural strength of high performance ceramics at ambient temperature", US Army Report TR 85-21, US Army Materials Technology Laboratory, Watertown, USA, [1985].
4. Baratta, F., Quinn, G.D., Matthews, W.T. "Errors associated with flexure testing of brittle materials", US Army Report TR 87-35 [1987], US Army Materials Technology Laboratory, Watertown, USA.
5. Quinn, G.D., "Twisting and friction errors in flexure testing", Cer. Eng. Sci. Proc. **13**(7/8) [1992] 319-30.
6. Quinn, G.D., Morrell, R. "Design data for engineering ceramics: a review of the flexure test", J. Amer. Ceram. Soc. **74**(9) [1991] 2037/66.
7. MIL-HBK-790, "Fractography and characterisation of fracture origins in advanced structural ceramics", [1992] 47 pp.
8. Swab, J.J., Slavin, M.J., Quinn, G.D. "Fractography and characterisation of fracture origins in advanced structural ceramics", Cer. Eng. Sci. Proc. **14**(7/8) [1993] 75-83.
9. Swab, J.J., Quinn, G.D., "Fractography of advanced structural ceramics: results from the VAMAS Round Robin exercise", Army Research Laboratory Report ARL-TR-656, 1994, 141 pp.
10. Swab, J.J., "The VAMAS Fractography Round Robin: a piece of the fractography puzzle", this conference.
11. Mizuno, M., Okuda, H., "VAMAS round-robin on fracture toughness of silicon nitride at high temperature", VAMAS Report No. 16, [1993] 64 pp., Japan Fine Ceramics Center, Nagoya, Japan.
12. Mizuno, M., Okuda, H., "VAMAS round-robin on fracture toughness of silicon nitride", J. Amer. Ceram. Soc., in press.
13. Frechette, V.D., "Markings on crack surfaces in brittle materials: a suggested unified nomenclature", in Fractography of Ceramic and Metal Failures, ed. Mecholsky, J.J., Jr, Powell, S.R., Jr, ASTM STP 827, [1984] 104-9.
14. ASTM Standard C 1256-93 "Standard practice for interpreting glass fracture surface features".
15. Frechette, V.D., "Failure analysis of brittle materials", Adv. Ceram. **28** [1990].

16. Newman, J.C., Jr., Raju, I.S., "An empirical stress-intensity factor equation for the surface crack", Eng. Fract. Mech. **15**(1-2) [1981] 185-92.
17. Healey, J.T., Mecholsky, J.J. "Scanning electron microscopy techniques and their application to failure analysis of brittle materials", In ASTM STP 827 (1984), edited by Mecholsky, J.J., Powell, S.R., Jr, pp. 157-81.
18. Quinn, G.D., Kübler, J.J., Gettings, R.J. "Fracture toughness of advanced ceramics by the surface flaw in flexure (SCF) method: a VAMAS round robin", VAMAS Report No. 17, [1994], NIST, Gaithersburg, USA.
19. Quinn, G.D., Gettings, R.J., Kübler, J.J., "Fracture toughness by the surface crack in flexure (SCF) method: results of the VAMAS Round Robin", Cer. Eng. Sci. Proc. **15**(5) [1994] 46-55.

FRACTURE TOUGHNESS ESTIMATION BY FRACTOGRAPHIC ANALYSIS OF TENSILE SPECIMENS

R. Nathan Katz, Mechanical Engineering Department, Worcester Polytechnic Institute, Worcester, MA 01609
Kyu Cho, Army Research Laboratory, Materials Directorate, Aberdeen Proving Ground, MD 21005

ABSTRACT

Empirical criteria are presented for selection of uniaxial tensile failure origins which will yield valid fracture toughness estimations. Using these criteria estimations of fracture toughness based on fractography of an AlN, two SiC's, a AlN-SiC solid solution and a particulate composite, and a TiB_2 were performed. Comparison of the fracture toughness values estimated from fractography with fracture toughness measured by SEPB method, will be shown. Close agreement is observed between estimates made from fractographic analysis and measurements made on specimens with large artificially induced flaws if the fracture toughness is 4 MPa√m, or below. For materials in which the SEPB fracture toughness is greater than about 4.8 MPa√m, estimates by fractographic analysis significantly underestimate the SEPB fracture toughness. In attempting to rationalize this discrepancy the importance of rigorously applying the empirical criteria was re-emphasized.

INTRODUCTION

Fractography is widely utilized to identify failure initiating flaws and flaw populations in advanced ceramics. However, fractography is seldom used to determine the fracture toughness of ceramics. This is surprising since fractography has long been utilized to estimate the fracture energy of brittle materials, glasses in particular. One such study of the use of fractography to calculate fracture surface energy for a wide variety of ceramics and glasses was conducted by Mecholsky et al [1]. They were able to show a good correlation between the size of the outer fracture mirror in bend specimens and K_{IC} as measured in a double cantilever beam (DCB) test. Rice [2] presented a comprehensive paper on the fractographic determination of K_{IC} in ceramics at the Second Conference on the Fractography of Glass and Ceramics in 1990. He showed that for bend bars with large flaws good agreement was obtained between K_{IC} estimated from fractographic analysis and DCB tests. However, as the flaw sizes decreased there was a marked tendency for an increasing disparity between the fracture toughness or fracture energies estimated by fractographic analysis and DCB measurements.

In the past several years two very useful documents [3,4] have been prepared by researchers at the Army Research Laboratory and the National Institute of Science and Technology, which provide handbook type guidance in the use of fractography, and the relationship between fractography and fracture toughness, but they provide no specific data correlating K_{IC} and fracture origins. Reference 4, contains a very useful appendix which discusses, in detail, various circumstances which might cause discrepancies in estimates of fracture origin size predictions from fracture mechanics, and fractographically observed flaw sizes. Virtually all examples of the use of fracture mechanics to estimate critical flaw sizes or fracture energies/toughness discussed in the above references involved bend tests and the fracture origins were often associated with surface machining damage.

With the growing use of tensile testing in the determination of strength statistics for advanced ceramics, a large number of samples of known uniaxial tensile stress and unambiguous failure origins have become available for analysis. Little work has been reported on using fractography of uniaxial tensile test specimens to estimate fracture toughness. One early work that utilized fractography of carefully lapped uniaxial tensile specimens of HS130* Si_3N_4 (an early version of NC132† material) [5] yielded an estimated K_{IC} of 2.8 MPa√m. No correlation with measured values was made. Recent work by Katz and

* Norton Co., Worcester, MA 01606.
† Norton Co., Worcester, MA 01606.

his coworkers [6] compared K_{IC} estimated by fractographic analysis and those measured by SEPB for an AlN, a 50%/50% AlN-SiC solid solution, and a 25% SiC particulate toughened AlN composite. For all three materials the agreement between estimated and measured K_{IC} values were within 10%. In a subsequent study of two SiC materials it was observed that the SiC with a K_{IC} of 4.9 MPa$\sqrt{m}$ showed the largest deviation (-12%) from measured SEPB values of any material that we had tested [7]. A study by Oishi et al [8] conducted on Y-TZP material, also compared measured K_{IC} (by SEPB) with estimates from fractography. Fractographic estimates were as much as 35% less than the SEPB values.

The purpose of the work reported in this paper was to extend the range of polycrystalline ceramics for which estimates of K_{IC} could be compared to measured values. TiB_2 which has a moderately high K_{IC} and a moderate tensile strength was chosen as a material that would likely produce a reasonable percentage of internal, as opposed to machining induced, failure origins. Discrepancies between estimated and measured values for TiB_2 and other "high" K_{IC} ceramics are rationalized in terms of subcritical crack growth and/or R-curve behavior. Consideration of the consequences of subcritical crack growth re-emphasizes the difference between the failure initiating flaw and the critical flaw, even though the later contains the former. Fractographs showing a hierarchy of concentric flaws in a TiB_2 specimen will be presented to emphasize this point.

EXPERIMENTAL TECHNIQUE

The tensile specimens utilized in this work were right circular cylinders, measuring between 100 and 120 mm in length and 8.8 to 9.2 mm in diameter. 40 mm on each end of the test specimen were inserted into a steel piston and adhesively bonded in place with high strength epoxy[‡] [9-11]. The adhesively bonded specimen-piston assembly is inserted into the pressure chamber of an ASCERA hydraulic Tensile Tester[§]. Pressure is applied via a hydraulic fluid and increased at a rate of approximately 3 MPa per second until the specimen is broken

‡ ARALDITE AV 118, Ciba Geigy Corp., Madison Height, MI 48071.

§ ASCERA Hydraulic Tensile Tester, Robertsfors, Sweeden.

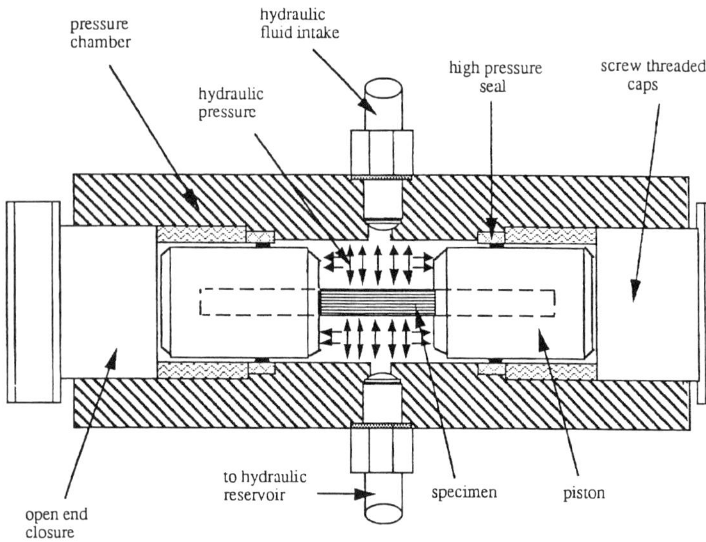

Figure 1. Hydraulic pressure chamber with specimen-piston assembly.

apart by the hydraulic pressure acting against the piston faces. Figure 1 shows a schematic of the pressure chamber with the adhesively bonded specimen - piston assembly in place. The tensile fracture stress is calculated using the pressure at the instant of failure and the geometric parameters of the specimen-piston assembly [9-11]. Although this tensile testing method is inherently self-aligning, a small amount of eccentricity may occur, arising from either the specimen being slightly curved or the epoxy bonding being slightly off-center in the pistons. Procedures to correct for this source of error exist, and have been described by several investigators [9-12]. Where required such corrections have been applied to all tensile data used in estimating fracture toughness reported in this paper. Typically these corrections are in the 3 to 5 % range, and in no case exceeded 10%. The stress is related to the pressure and test geometry by the following equation:

$$\sigma = \sigma_{NOM} + \Delta\sigma = (A - A_S) / A_S \cdot P + \Delta\sigma \qquad (1)$$

where σ_{NOM} is the nominal fracture stress, $\Delta\sigma$ is the correction for eccentricity which is equal to $\sigma_{NOM} \times f$(fracture origin location, specimen eccentricity and diameter), A is the cross sectional area of the piston, A_S is the cross sectional area of the specimen, and P is the pressure at failure, respectively.

Single edge pre-cracked beam (SEPB) tests [13] were performed to measure the fracture toughness of materials from the same lots of materials used for tensile testing. At least six valid [14] SEPB tests and more usually ten to fourteen valid tests were performed for each material utilizing 3x4x25 mm beams. Pre-cracking was accomplished by indenting the 3 mm face of each specimen with a Vickers diamond indentor (with the diagonals parallel to the edges of the beam), and loaded in a double anvil configuration to "pop-in" a straight crack [15]. "Pop-in" and fracture toughness measurements were carried out on Instron model 8502 and 4201 testing machines[**], respectively. Detailed descriptions of the SEPB test procedure are available in references 15 and 16.

Fractographic examination of the fracture surfaces of the tensile bars was carried out in a JEOL 840A SEM[††]. Fracture surfaces were cleaned with acetone and coated with Au-Pd or carbon prior to SEM examination. Fractographic analysis of the tensile failures are conducted prior to carrying out K_{IC} measurements by SEPB. This avoids any subconscious bias in the measurement of the critical flaw

Table I. Materials and Property Data Utilized in This Study.

	Source	T-σ_{CHAR}	m	AGS[a]	Reference
HP TiB_2	Cercom	275	14.9	15μm	This Study
HP AlN	Dow	234	12.7	1.9μm	Ref[6]
HP 50AlN/50SiC	Dow	361	26.4	0.5μm	Ref[6]
HP 75AlN/25SiC	Dow	371	9.9	2[b]& 5[c]μm	Ref[6]
HP SiC Pad B	Cercom	304[d]	9.6[d]	4.0μm	Ref[7]
HP SiC Pad N	Cercom	347[d]	9.6[d]	4.0μm	Ref[7]

HP=Hot Pressed; T-σ_{CHAR}=Tensile Characteristic Strength; m=Weibull Modulus; AGS=Average Grain Size. [a]Based on the company product literature. [b]AGS of AlN. [c]AGS of SiC. [d]Combined test data obtained from specimens machined by two vendors.

[**] Instron Corp., Canton, MA 02021.
[††] JEOL USA, Inc., Peabody, MA 01961.

size for the K_{IC} estimate.

The materials data and their source utilized in this work are listed in Table I. The methodology utilized to estimate fracture toughness from the fractography of failed uniaxial tensile bars is presented in the following section.

CRITERIA FOR ESTIMATION OF K_{IC} BY FRACTOGRAPHIC ANALYSIS

Only a fraction of the fracture surfaces resulting from uniaxial tensile failures are suitable for estimating K_{IC}. Not all samples have an obvious failure origin, this is often the case for samples failing from surface "defects". In a previous paper [6] the authors and their coworkers defined a set of empirical conditions which restrict those samples to be used for K_{IC} estimation. Later in this paper it will be shown that, in fact, these restrictions are necessary for close estimation of K_{IC} (within approximately 10%), and if they are relaxed the difference between the estimated values and the values measured by SEPB tests increase significantly. The criteria for selecting samples for fracture toughness estimates are:

1) The failure must be from an internal flaw.

2) Fracture origins must exhibit fully developed "mirror", mist and hackle regions.

3) The mirror has to be "essentially" circular (samples with ellipsoidal mirrors are occasionally encountered and are not utilized).

4) The mirror, mist, and hackle regions must be close to normal to the specimen axis. (This assures that the onset of unstable fracture occurred under a pure uniaxial stress field.)

5) The largest linear dimension of the failure initiating flaw at the center of the mirror is used as the crack length in the K_{IC} calculation.

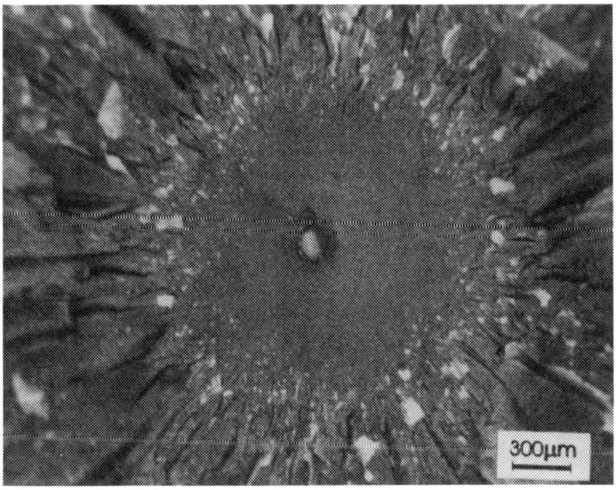

Figure 2. Circular fracture mirror in AlN.

The toughness, K_{IC}, is calculated using the Sneddon closed form solution for a penny shaped crack [17] :

$$K_{IC} = 2\,\sigma_C\,(a_C / \pi)^{1/2} \tag{2}$$

where σ_C is the failure stress and a_C is one half the principal flaw dimension. It is worth noting that criteria 1) and 4) address issues raised by Rice with regard to improving the accuracy of fractographic estimation of K_{IC} [2].

Figure 2, shows a specimen of AlN which exhibits a well defined mirror, mist and hackle, as well as a circular mirror. This typifies an acceptable specimen for K_{IC} estimation. Figure 3, shows a specimen of an AlN-SiC solid solution which has an elliptical mirror, and is therefor unacceptable for use in estimation of K_{IC}. Figure 4, shows a specimen of TiB_2 which has what we refer to as a "pseudo mirror". In this case there is a circular area which suggests a mirror, but upon close examination it can be seen that no mirror, mist, and hackle regions are truly present, only mist and hackle. It will be shown later that such fracture origins are associated with significant underestimation of K_{IC}.

Whether or not a given material has a mirror associated with the

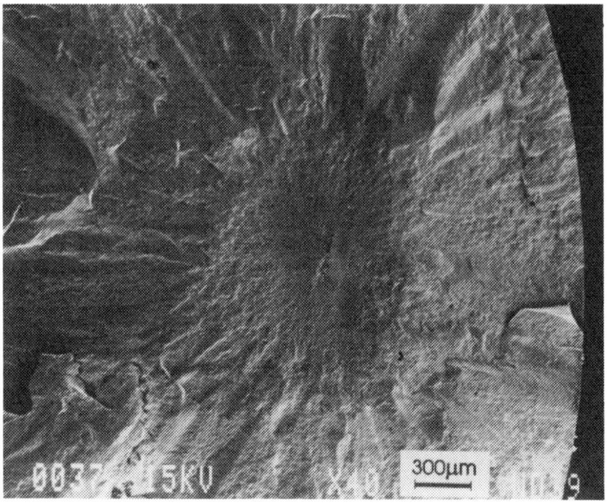

Figure 3. Elliptical fracture mirror in AlN-SiC solid solution unacceptable for analysis

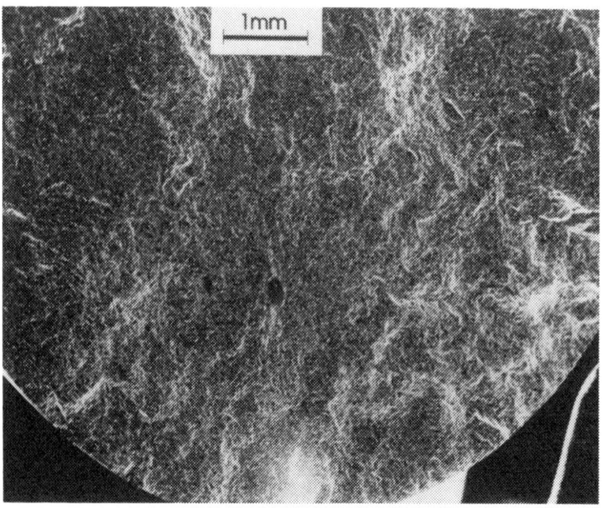

Figure 4. "Pseudo" mirror in TiB_2 shows mist and hackle but no mirror.

fracture origin can be a function of the chemistry, phase content and processing history. For example we have examined the fracture origins

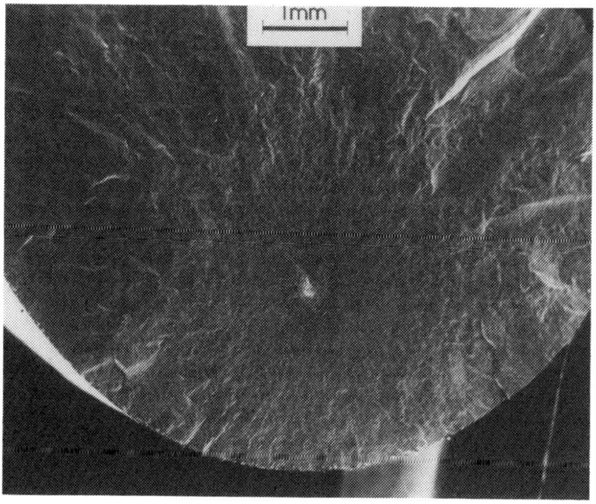

Figure 5. SiC Pad B shows mirror, mist, and hackle.

in three different SiC's. A Dow experimental SiC[‡‡] did not exhibit a mirror or mist region. Cercom Pad B[§§] material did exhibit mirror, mist, and hackle, Figure 5, whereas Cercom Pad N[***] material exhibited no mirror but only mist and hackle, Figure 6.

ESTIMATION OF K_{IC}

Utilizing the above criteria estimates of K_{IC} were made for the six materials listed in Table I. SEPB K_{IC} measurements were also made on these six materials. Figure 7 presents a plot of the fractographic estimations versus the measured SEPB values. Very good agreement between the estimated and measured values is obtained for K_{IC} values determined by SEPB of less than ~ 4.5 MPa$\sqrt{m}$. Around this value the estimates of K_{IC} by fractography begin to substantially underestimate the measured values.

This behavior led us to re-examine the fracture surfaces and calculations that we had used in estimating K_{IC}. Upon re-examination it

‡‡ Dow Chemical Co., Midland, MI 48674.
§§ Cercom, Inc., Vista, CA 92083.
*** Cercom, Inc., Vista, CA 92083.

Figure 6. SiC Pad N shows only mist and hackle.

became apparent that we had "relaxed" criteria 2) and used specimens exhibiting "pseudo mirrors". This clearly emphasized the importance of rigorously applying all of the empirical criteria listed above, in estimating K_{IC} utilizing fractograpy. Rationalizing the observed behavior will be the concern of the balance of the paper.

POTENTIAL SOURCES OF DISCREPANCY BETWEEN ESTIMATED AND MEASURED K_{IC} AT "HIGH" K_{IC}

While the tensile test technique of Oishi et al [8] differed substantially from those used by the present authors and their colleagues, the variation that they found in estimated and measured K_{IC} is similar to that observed by other investigators as one goes from "small" flaws to "large" flaws in Y-TZP. The increase in fracture toughness as the flaw size increases has been related to "micro R-curve behavior" [4]. Rice's data on estimation of K_{IC} as a function of flaw size for TZP and ZTA materials also show evidance of “micro R-curve behavior” [2]. It is reasonable that the difference in fracture toughness encountered between large artificially induced flaws, such as in an SEPB test, and the small naturally occurring flaws used in fractographic estimation,

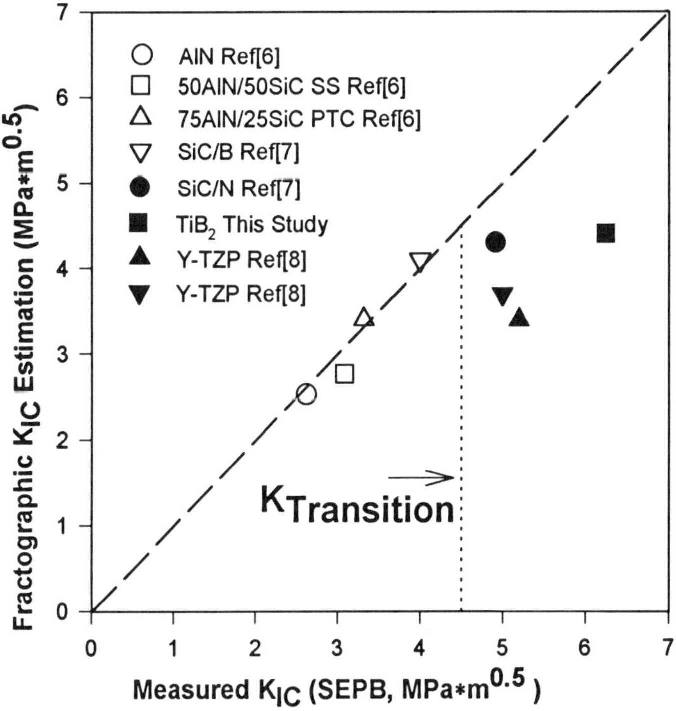

Figure 7. Fracture toughness comparison between fractographic estimates and SEPB measurements.

would be very sensitive to R-curve behavior. Swab and Quinn [4] point out that one cause of observing critical flaws significantly smaller than those predicted from macroscopic fracture toughness tests is the presence of R-curve behavior. As shown in table II, the range of flaws measured on TiB_2 tensile fracture surfaces is significantly smaller than the average flaw anticipated using the average strength and measured fracture toughness. By contrast, in the case of AlN and AlN/SiC solid solution the observed flaw sizes bracket the predicted flaw size, as anticipated. The presence of behavior consistent with a given mechanism, however only means that the mechanism is plausible. Other mechanisms may in fact be responsible for the observed behavior.

Table II. Anticipated Vs Observed Flaw Size.

	AlN	50AlN/50SiC	TiB_2
average σ_C	230 MPa	360 MPa	280 MPa
average K_{IC}	2.62 MPa√m	3.08 MPa√m	6.25 MPa√m
calculated average a_0	~ 100 μm	~ 57 μm	~ 390 μm
observed ranges of a_0	42 - 135 μm	5 - 200 μm	170 - 280 μm

σ_C=Fracture Stress; a_0=flaw size.

An alternative mechanism may be subcritical crack growth without R-curve behavior. Normally one associates subcritical crack growth with slow crack growth at high temperatures or in the presence of water or other reactive environment. Neither situation prevailed in the tensile testing reported in this work. All tests were at room temperature and since the fracture origins were well in the volume, not near the surface, no reaction with the hydraulic pressurization fluid was possible. Therefore, the most plausible explanation for the existence of subcritical crack growth, in TiB_2, is R-curve behavior as described by Bennison and Lawn [18]. To settle the issue an investigation of R-curve behavior in TiB_2 should be undertaken.

THE FRACTURE INITIATING FLAW VERSUS THE CRITICAL FLAW

In ceramic materials that exhibit subcritical crack growth, whether associated with R-curve behavior or some other cause, the fracture initiating flaw will not be the critical flaw for fracture mechanics purposes. An interesting example of this was encountered in one TiB_2 specimen examined in this study. The fracture origin of the specimen shown in Figure 4, is detailed at increasing magnifications. Figure 8a, shows the large grain at the center of the "pseudo-mirror". Figure 8b, shows that this grain itself has a true fracture mirror with a small (~ 20 μm) particle at the center of the mirror. This inclusion showed traces of W, Co, Fe, and Ni (by EDS), which are all consistent with a fragment of WC grinding media. It is probable that this inclusion was the initiating flaw but it is much too small to be the critical flaw.

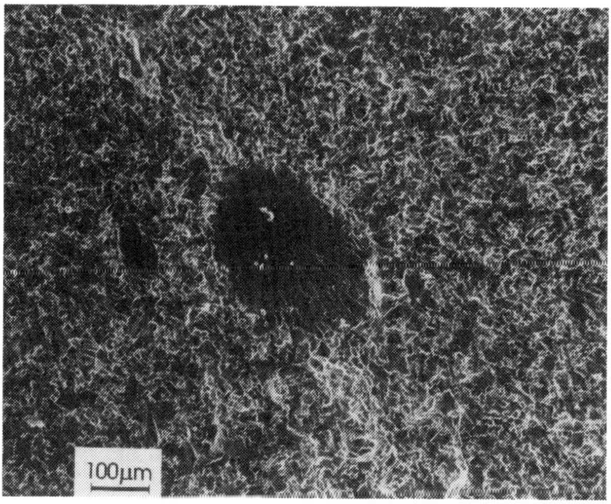

Figure 8a. Large grain at origin of fracture in TiB_2 specimen.

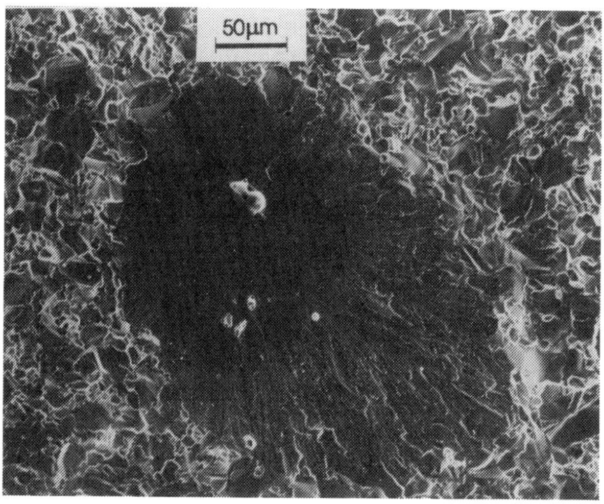

Figure 8b. Particle in large grain at origin of fracture in TiB_2 specimen.

Similarly, the large grain is too small to be the critical flaw predicted by fracture mechanics based on the conventional large flaw toughness measurement. Even the various ellipses that one can imagine seeing, linking large grains surrounding the very large grain in figure

8a are not large enough to be the critical flaw. It is evident that care should be taken not to use the terms "fracture initiating flaw" and "critical flaw" interchangeably. They may, or they may not, be equal.

SUMMARY

We have shown that if the empirical criteria presented in this paper are applied excellent correlation between fractographically estimated and measured K_{IC} can be attained. It is especially critical in applying these criteria to be certain that a true fracture mirror is present. By implication, if the estimated K_{IC} lies on the curve in Figure 7 it can be taken as evidence of a lack of R-curve behavior. Conversely, if the estimated K_{IC} lies substantially below the curve in Figure 7 R-curve behavior may be present. It has also been shown that for "tough" ceramics the terms failure initiating and critical flaw, are not synonymous.

REFERENCES

[1] J.J. Mecholsky Jr., S.W. Freiman, and R.W. Rice, "Fracture Surface Analysis of Ceramics," J. Mat. Sci., **11**, 1310-19 (1990).

[2] R.W. Rice, "Fractographic Determination of K_{IC} and Effects of Microstructural Stresses in Ceramics," in Fractography of Glasses and Ceramics II, Edited by V.D. Frechette and J.R. Varner, American Ceramic Society, pp.509-545, (1991).

[3] "Fractography and Characterization of Fracture Origins in Advanced Structural Ceramics," Military Handbook, MIL-HDBK-790, U.S. Department of Defense, 1 July 1992.

[4] J.J. Swab and G. Quinn, "Fractography of Advanced Structural Ceramics: Results From the VAMAS Round Robin Exercise," ARL-TR-656, Army Research Laboratory, December 1994.

[5] F.I. Baratta, G.W. Driscoll, and R.N. Katz, "The Use of Fracture Mechanics and Fractography to Define Surface Finish Requirements for Si_3N_4," in Ceramics for High Performance Applications, Edited by J.J. Burke, A.E. Gorum, and R. N. Katz, Brook Hill Pub, MA, pp. 445-476, (1974).

[6] R.N. Katz, S. Grendahl, K. Cho, I. Bar-On, and W. Rafaniello, "Fracture Toughness of Ceramics in the AlN-SiC System," Ceram. Eng. Sci. Proc., **15** [5] 877-884 (1994).

[7] K. Cho, R.N. Katz, and I. Bar-On, "Strength and Fracture Toughness of Hot Pressed SiC Materials," Ceram. Eng. Sci. Proc.,**16**, in-press.

[8] M. Oishi, Y. Matsuda, K. Noguchi and T. Masaki, "Evaluation of Tensile Strength and Fracture Toughness of Y-Stabilized Zirconia Polycrystals with Fracture Surface Analysis," J. Am. Ceram. Soc., **78** [5] 1212-16 (1995).

9) J. Nilson and B. Mattsonn, "A New Tensile Test Method for Ceramic Materials," in <u>Proceedings of 2nd International Symposium on Ceramic Materials and Components for Engines</u>, Edited by W. Bunk and H. Hausner, Deutsche Keramische Gesellschaft, Wiesbaden, (1986).

[10] H.P. Lucas, "Direct Tensile Testing of Brittle Materials," Master Thesis, Worcester Polytechnic Institute, Worcester, MA (1991).

[11] H.A. Tautanjii, "The Development of a Cementitious Composites Test Technique and its Application to Carbon Fiber Reinforced Cementitious Composites," Ph.D. Thesis, Worcester Polytechnic Institute, Worcester, MA, (1992).

[12] L. Hermanson, J. Adlerborn, and M. Burstrom, "Tensile Testing of Ceramic Materials - A New Approach," in <u>High Tech Ceramics</u>, Edited by P. Vicenzini, Elsevier Science Pulb., Amsterdam, pp. 1161-68, (1987).

[13] T. Nose and T. Fujii, "Evaluation of Fracture Toughness for Ceramics Materials by a Single-Edge-Precracked-Beam Method," J. Am. Ceram. Soc., **71** [5] 328-333 (1988).

[14] ASTM STD E-399, "Standard Test Method for Plane-Strain Fracture Toughness of Metallic Materials," <u>Annual Book of ASTM Standards</u>, vol. 03.01, pp. 487-511 (1989).

[15] I. Bar-On, J. Beals, G. Leatherman, and C. Murray, "Fracture Toughness of Ceramic Precracked Bend Bars," J. Am. Ceram. Soc., **73** [8] 2519-22 (1990).

[16] G.D. Quinn, J. Salem, I. Bar-On, K. Cho, M. Foley, and H. Fang, "Fracture Toughness of Advanced Ceramics at Room Temperature," J. Res. Natl., Inst. Stand. Technol. **97** [5] 579-607 (1992).

[17] I.N. Sneddon, "The Stress Distribution in the Neighborhood of a Crack in an Elastic Solid," Proc. Royal Soc. Ser. A, **187**, pp. 229-260 (1949).

[18] S.J. Bennison and R.R. Lawn, "Origin Tolerance in Ceramics with Rising Crack Resistance Characteristics," J. Mat. Sci., **24**, pp.3169-75 (1989).

FRACTOGRAPHY AND THE SURFACE CRACK IN FLEXURE (SCF) METHOD FOR EVALUATING FRACTURE TOUGHNESS OF CERAMICS

G. D. Quinn and R. J. Gettings
Ceramics Division
National Institute of Standards and Technology
Gaithersburg, MD 20878

J. J. Kübler
Swiss Federal Laboratories for
Materials Testing and Research
Dubendorf, Switzerland

ABSTRACT

The surface crack in flexure (SCF) method, also known as the controlled surface flaw method, has been used to measure fracture toughness of ceramics and glasses for almost 20 years. New fracture toughness results for a range of ceramics and glasses including alumina, boron carbide, silicon carbide, silicon nitride, titanium diboride, zirconia, glass ceramic, borosilicate crown glass, and a whisker-reinforced alumina are presented in this paper. Some materials are conducive to precrack measurements, while others are not. New techniques for detecting the precracks are presented.

A surprising outcome from a recently concluded Versailles Advanced Materials and Standards (VAMAS) round robin project was that the computed toughness is often *not sensitive* to the exact precrack size measurement. Consistent results were obtained by many laboratories despite different viewing modes and magnifications. The reasons for this consistency and why toughness is insensitive to precrack size is presented.

INTRODUCTION

The surface crack in flexure (SCF) method follows conventional practice for measuring fracture toughness: a specimen is precracked, the specimen is fractured, the precrack is measured, and the toughness is computed form a stress intensity formula for a well-defined crack geometry. Unlike most fracture toughness tests, however, the precrack is very small and must be measured using fractographic methods. A hardness test machine with a Knoop indenter is used to make an indentation in the surface of a common flexure specimen as illustrated in Figure 1. In brittle materials, a semicircular or semielliptical surface crack will form under the indentation. The size of the precrack can be controlled by the indentation load.

The method is also known as the controlled-surface flaw method. Petrovic et al.[1-3] developed this method in the 1970's, primarily using Knoop indentation precracks. Kenny[4] previously had used a row of Knoop indentations as precursors to creating a single-edge cracked flexure specimen. Govila[5,6] and later Kinsman et al.[7] used a single Vickers indentation in flexure specimens. The later work of Petrovic and associates focussed on Knoop precracks since the crack patterns were simpler than those created by Vickers indentation.[1-3] The Knoop indenter is preferred since only one median crack is formed with the Knoop indentation and the precracks are larger than those formed by Vickers indentations at the same load.[8] A critical contribution was Petrovic et al.'s[1-3] recognition that residual stresses from the indentation had to be removed by polishing or annealing in order to obtain a valid fracture toughness result. The SCF method is very similar to a new American Society of Testing and Materials standard for metals: "E-740, Fracture Testing with Surface-Crack Tension (SCT) Specimens[9]" in which surface cracks are introduced by machining followed by fatigue precracking. To maintain consistency with fracture mechanics terminology conventions, we hereafter refer to the method described in this paper for advanced ceramics as the surface crack in flexure (SCF) method.

This paper highlights fractographic techniques used to find and measure the precracks in a variety of ceramics evaluated in a series of related studies[10,11,12] as well as in a Versailles Advanced Materials and Standards (VAMAS) round robin project.[13,14]

PROCEDURE

The materials evaluated and the experimental procedures are described in detail in References 10-14 and only a brief summary of this information is given below.

Figure 1. Schematic of the Surface Crack in Flexure Specimen. A Knoop indenter is used to form a semielliptical precrack in a flexure specimen. The precrack size is exaggerated in the cross section view of the specimen shown on the right. The specimen tilt, also exaggerated, aids the fractographic analysis.

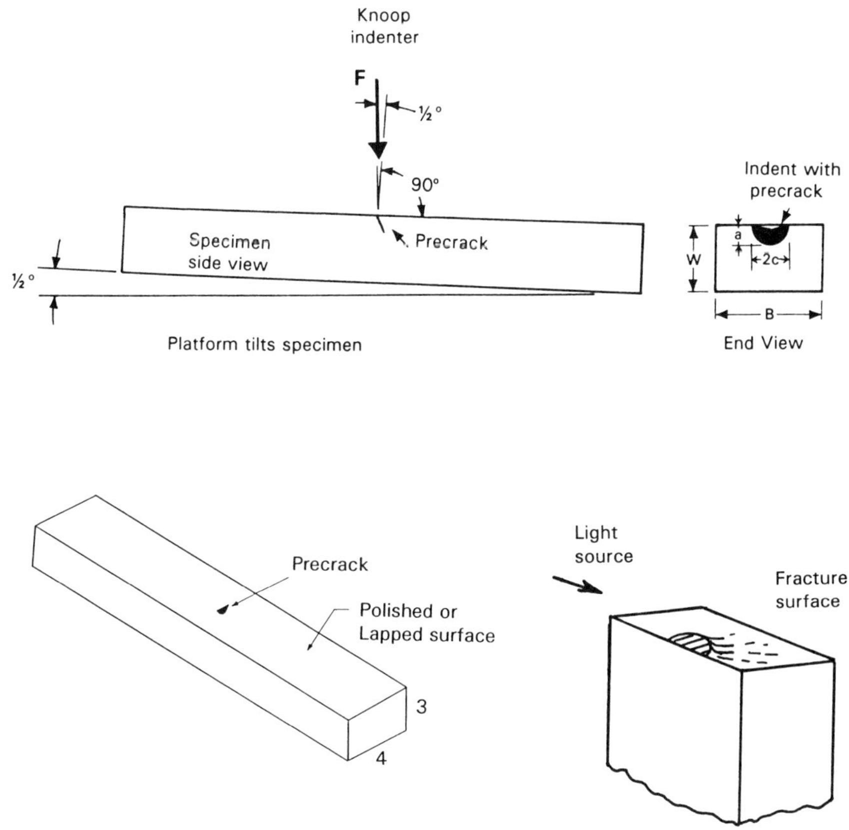

Common flexure specimens were used for all materials evaluated which are listed in Table I. Specimens were indented with a Knoop indenter at different loads depending upon the material. A common indentation test load of 49 N (5 kgf) has been adopted for most of our routine evaluations. Alternate loads are used for some materials if the 49 N load is not satisfactory (eg. the specimens break from natural flaws and a larger precrack is needed) or if fracture toughness versus precrack size data is desired. It is best that the specimen surface that is to be indented is polished and flat so that reproducible precracks can be obtained. A new procedure that we have implemented is to tilt the specimen by 1/2° as shown in Figure 1. The resulting precrack will be at 0 - 3° off normal and therefore is at a slight angle relative to the final fracture surface plane. The precrack is much easier to discern, especially with low-angle incident lighting. Higher specimen tilt angles are not as satisfactory since the precrack may be at steeper angles and may be truncated by the final fracture such that the precrack is not fully exposed. Precracks formed by the 1/2° specimen tilt have a negligible K_{II} loading component and the measured K_{Ic} is not affected by the Mode II loading.[13]

A special precracking procedure using a tilted and canted Vickers indenter was used for the zirconia, since Knoop precracks at loads up to 490 N (50 kgf) were not successful in making an adequately large precrack in this relatively soft but tough material. Figure 2 illustrates the precracking procedure for the zirconia. In this case, the cant caused one of the Palmqvist cracks to be larger than the other. After polishing, a slightly skewed semielliptical precrack remained in the specimen.

After precracking, the residual stresses under the indentation impression must be removed preferably by polishing since annealing, which is an alternative treatment, has risk of precrack healing or crack tip blunting. Dry silicon carbide papers were used in most instances, although diamond wheel grinding was used by some of the round robin participants. Diamond wheel grinding was also used for some of the very hard, tough materials such as titanium diboride and boron carbide. Petrovic et al.[1-3] ascertained that three times the indentation depth (X) was necessary to eliminate the residual stresses under the Knoop indentation, and they recommended that 4X be removed for most materials. In our recent work, we remove 4.3 - 4.5X in order to make the shape of the precrack a shallow ellipse such that the maximum stress intensity factor will be located at the deepest part of the precrack.

Flexure testing to measure flexure strength usually was on standard 20 mm x 40 mm fully-articulating fixtures at a crosshead speed of 0.5 mm/min in accordance with ASTM C 1161.[15] The round robin participants used similar procedures.[14]

TABLE I. Materials tested.[+] Abbreviations are: (ρ) density, (E) elastic modulus, (υ) Poisson's ratio, (HP) hot pressed, (HIP) hot-isostatic pressed, (RB) reaction-bonded, (GPS) gas-pressure sintered. (*) The SNW 1000 specimens had a lower density than that which is normal for this material.

Material	Code and/or Source	Processing	ρ (Mg/m^3)	E (GPa)	υ
Si_3N_4	NC 132, Norton	HP	3.23	320	0.27
Si_3N_4	NCX 34, Norton	HP	3.37	310	0.27
Si_3N_4	147A (MgO), Ceradyne	HP	3.17	291	0.27
Si_3N_4	NBD 200, Cerbec	HIP	3.16	320	0.26
Si_3N_4	NT 154, Norton	HIP	3.23	320	0.27
Si_3N_4	EKasin, ESK	HIP	3.18	315	0.27
Si_3N_4	NCX 5102, Norton	HIP	3.23	--	--
Si_3N_4	EC 141, NTK	GPS	3.22	310	0.26
Si_3N_4	SNW-1000*, Wesgo	sintered	3.06	248	0.27
Si_3N_4	KBI	RB	2.58	200	0.22
Si_3N_4	NC 350, Norton	RB	2.53	180	0.22
Syalon	201, Lucas	sintered	3.24	321	0.29
SiC	NC 203, Norton	HP	3.36	460	0.17
SiC, Si	NC 435, Norton	siliconized	2.99	350	0.18
α-SiC	SA, Carborundum	sintered	3.12	410	0.15
Al_2O_3	Referceram, Asahi	sintered	3.92	--	--
Al_2O_3	AD 94, Coors	sintered	3.51	291	0.23
Al_2O_3	AD 999, Coors	sintered	3.96	386	0.21
Al_2O_3/SiC	ARCO	HP	3.70	401	0.24
Al_2O_3/SiC	ARCO	HP	3.70	401	0.24
B_4C	Ceradyne	HP	2.48	~455	--
TiB_2	Ceradyne	HP	4.51	545	0.11
TiB_2	LANL	sintered	4.55	557	0.13
TiB_2	3120, Osram/Sylvania	GPS	4.64	542	0.12
TiB_2	3122, Osram/sylvania	GPS	4.67	581	0.12
Y-TZP	EMPA	sintered, HIP	6.03	211	0.31
Glass	BK 7, Schott	melt	2.51	82	0.21

[+]Certain commercial materials or equipment are identified in this report to specify adequately the experimental procedure. Such identification does not imply endorsement by the National Institute of Standards and Technology or the Swiss Federal Laboratories for Materials Testing and Research, nor does it imply that these materials or equipment are necessarily the best for the purpose.

Figure 2. A modified indentation procedure can be used for soft materials such as the zirconia used in the round robin project. The specimen was tilted and canted relative to the Vickers indenter.

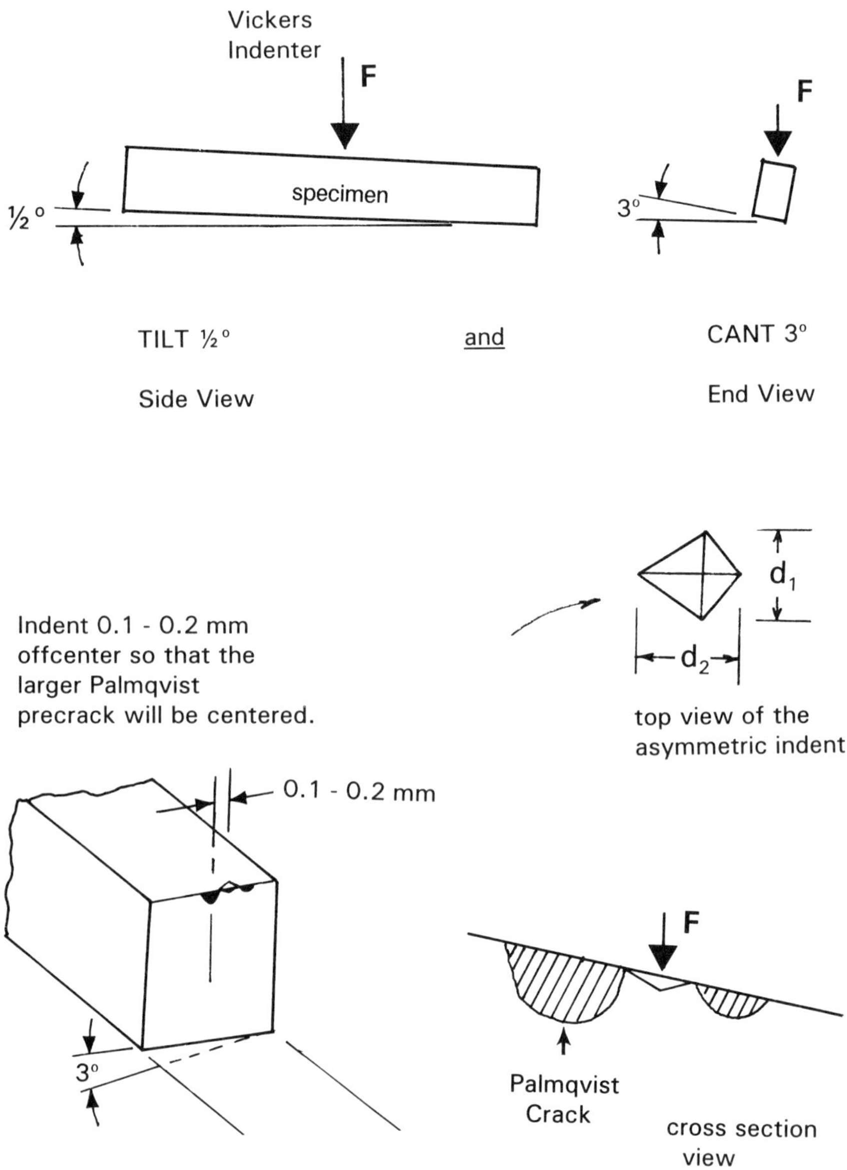

The alumina composite material was tested on a flexure fixture with 10 mm x 20 mm spans.

Precrack sizes and shapes were measured on the fracture surfaces, both with optical microcopy and scanning electron microscopy, usually at magnifications between 100 and 500X. Fractographic inspection techniques and procedures are described below in the Results section. The precracks usually are semicircular or semielliptical in shape after indentation, but the subsequent polishing step to remove the residual stresses alters the shape to a section of a circle or a semiellipse.

Fracture toughness (K_{Ic}) was calculated from the formula for a semicircular or semielliptical surface crack in tension or flexure:

$$K_{Ic} = Y \sigma \sqrt{a} \qquad (1)$$

where:

	Y	is the stress intensity shape factor
	σ	is the flexure strength of the specimen (MPa)
	a	is the crack depth (m)

This simple formula illustrates that only two things need be measured: the stress at fracture and the crack size. Y is dimensionless and is a function of the crack size and shape. The stress intensity shape factors, Y, for semicircular and semielliptical surface cracks in bending are from the empirical equations developed by Newman and Raju[16] which are given in simplified form in Appendix A. Y depends upon the ellipticity ratio (a/c) and the size of the precrack relative to the specimen cross section and takes into account the bending stress fields. The Y factor is not constant and varies along the precrack boundary. *It is important to take this into account and to use the maximum value of Y* which will either be at the deepest point of the precrack, or the point where the precrack intersects the surface. Fracture from the former location is preferred and can be promoted by polishing 4.3 to 4.5X material off the specimen after indentation so as to make the precrack a shallow ellipse. In some materials (that left clear fracture markings) it was possible to confirm that fracture did initiate from the point of the precrack periphery where Y was a maximum. It is also important to calculate the Y maximum for each and every precrack, rather than use a single average value for a set of specimens. The Newman-Raju formulas for Y are widely accepted and it is estimated that they are accurate to within a few percent. ASTM standard, E 740, uses them for the fracture toughness of surface crack in tension specimens.[9]

TABLE II. Results for ceramics with a 49 N (5 kgf) Knoop precrack.

Material	Code or source	Halo	Brightness change	Hackle lines	Fractographic amenity	crack size a, 2c (μm)	$K_{Ic} \pm 1\sigma$ (MPa√m)
Si_3N_4	NCX34	none	slight	slight	difficult	70, 184	6.35 ± .10
Si_3N_4	NBD200	none	slight	clear	moderate	77, 230	5.41 ± .24
Si_3N_4	NT154	none	slight	moderate	moderate	80, 206	5.80 ± .23
Si_3N_4	NCX5102	none	none	moderate	difficult	59, 150	5.36 ± .62
Si_3N_4	EC141	none	slight	slight	difficult	82, 219	5.22 ± .18
Si_3N_4	KBI	none	none	none	not possible	--	--
Si_3N_4	NC350	none	moderate	clear	easy	147, 383	1.65 ± .09
Syalon	201	none	slight	clear	moderate	100,286	4.41 ±..12
SiC	NC203	slight	slight	moderate	moderate	133, 252	4.37 ± .38
SiC, Si	NC435	none	slight	moderate	moderate	140, 340	3.70 ± .27
Al_2O_3	Referceram	none	none	slight	too difficult	--	--
Al_2O_3	AD999	clear	slight	moderate	easy	119, 274	3.39 ± .09
B_4C	Ceradyne	none	clear	clear	easy	163, 300	3.08 ± .07
TiB_2 3120	Osram	none	clear	moderate	easy	121, 312	5.20 ± .41
TiB_2 3122	Osram	none	clear	moderate	easy	120, 304	5.36 ± .50
Glass, dry N_2	BK7	none	clear	none	easy	229, 524	0.96 ± .09
Glass, ambient	BK7	none	clear	none	easy	219, 543	0.88 ± .09

TABLE III Results for materials with various indentation loads were used depending upon the material.[10] (*) denotes that a range of loads were used for the AD999 alumina in order to create different sized precracks. (†) denotes a wide variation of precrack sizes were observed for the grade 147A silicon nitride.

Material	Code or source	Load P (N)	Halo	Brightness change	Hackle lines	Fractographic amenity	crack size a, 2c(μm)	$K_{Ic} \pm 1\sigma$ (MPa√m)
α-SiC, S	Carborundum	108	none	clear	clear	easy	250, 550	3.01 ± .06
SiC, HP	NC203	88	none	slight	moderate	moderate	149, 361	3.85 ± .32
Al_2O_3, S	AD999	29 - 255	clear	slight	moderate	easy	*	3.2 - 4.0
Al_2O_3, S	AD94	108	none	slight	slight	difficult	148, 354	3.78 ± .34
Al_2O_3/SiC	ARCO //	176	--	--	--	difficult	182, 404	6.32 ± .13
Al_2O_3/SiC	ARCO ⊥	176	--	--	--	difficult	206, 382	8.66 ± 1.54
Si_3N_4, HP	147A	108	slight	clear	clear	easy	†	4.70 ± .91
Si_3N_4, S	SNW1000	59	none	slight	moderate	moderate	104, 247	4.66 ± .29
TiB_2 , HP	Ceradyne	230	--	slight	moderate	difficult	288, 689	5.52 ± .61
TiB_2, S	LANL	279	--	--	--	difficult	281, 561	5.14 ± .47

RESULTS

Tables II through IV give the results for the 26 materials evaluated. In each table, the average precrack size is identified as well as the average fracture toughness with an uncertainty of one standard deviation. Three columns in each table categorize certain fractographic features that are described below which were instrumental in detecting and measuring the precracks. Detailed discussions about the fracture toughness results are in References 10-14 and we confine our discussion below to fractographic aspects of the SCF method.

A variety of features, some obvious and some subtle, help delineate the precracks in these materials. From a detailed examination of several hundred precracks in these twenty six materials, it is possible to make some generalizations about what features and fractographic methods enable the precracks to be found and measured with confidence. Certain features make the precrack stand out and appear differently than the surrounding fast fracture zone. Precracks are detectable if:

(a) a ridge or halo outlines them clearly,
(b) the color or contrast of the precrack is different,
(c) the mode of fracture is different (e.g. transgranular versus intergranular),
(d) microhackle lines change direction at the boundary.

The following techniques enhance these features and aid fractographic evaluation:

(a) low angle incident lighting in optical microscopy,
(b) photography of the fracture surface at different magnifications (optical or SEM),
(c) photography with photo pairs taken at the same magnification in the SEM, one with the specimen normal to the detector, the other with the specimen inclined at 15-20°, (either to cause a change in general appearance of the precrack or to create stereo photo pairs),
(d) examination with the SEM in the backscattering mode (optional),
(e) gold coating (prior to SEM examination) with highly-directional (shadow) sputtering

An overall, qualitative assessment of the fractographic amenity (easy, moderately difficult, or not possible) is shown in Tables II through IV. In some materials, the precracks were obvious due to one or more of the above features. In other instances, only careful scrutiny and patient observation enabled the fractographer to detect the precracks. For example, a faint contrast difference between the precrack and surrounding material was sometimes confirmed by slight redirection

TABLE IV. Results from the VAMAS round robin project.[13]

Material	Code or source	Load P (N)	Halo	Bright-ness change	Hackle lines	Fractographic amenity	crack size a, 2c(μm)	$K_{Ic} \pm 1\sigma$ (MPa√m)
Si_3N_4	NC132	24.5	clear	none	clear	easy	53, 150	4.59 ± .37
Si_3N_4	ESK	49	slight	none	moderate	difficult	78, 250	4.95 ± .55
Y-TZP	EMPA	147	none	moderate	moderate	moderate	--	4.36 ± .44

of microhackle lines at the precrack boundary. Either one of these features may not have been sufficiently convincing, but the combination led to a higher level of confidence in the interpretation. The following illustrations demonstrate one or more of the above features that enable precracks to be detected.

Figures 3 and 4 are examples where the precrack has a distinct brightness or contrast difference relative to the fast fracture region. Furthermore, in each case a distinct halo is present. The halo is probably due to the precrack tilt and the halo represents the region where the crack realigned its orientation to be perpendicular to the applied stresses. The hot-pressed silicon nitride specimen is not susceptible to environmentally-assisted slow crack growth (under the lab ambient conditions the specimen was fractured) nor rising R-curve resistance, and all residual stresses had been removed, so the halo cannot be attributed to stable crack extension. The same cannot be said conclusively for the sintered alumina, and the halo could represent one or both of the two former phenomena.

Figure 5 shows a precrack in ESK hot-isostatic pressed silicon nitride photographed by an optical microscope with low-incident angle, highly-directional lighting. The precrack is quite obvious in the lower magnification photo since it reflects the light differently due to the its tilt relative to the fracture surface. The precrack is not as distinct in higher magnification photo, in large part due to the limited depth of field of the microscope at the higher magnification and the uneven plane of the precrack, but a size measurement is nonetheless practical. This material was prone to develop uneven, 3-dimensional precracks as shown in Figure 6. Another example of a brightness change due to the precrack tilt is in Figure 7 which shows a precrack in boron carbide.

An example of a change in fracture mode is shown in Figure 8. The precrack has a transgranular mode of fracture, whereas the fast fracture region is intergranular. The difference manifests itself as a brightness difference in the SEM.

Figure 3. Optical microscope (a) photo of both fracture halves in a NC 132 specimen. A faint halo is evident. The same precrack as viewed by the SEM in shown in (b).

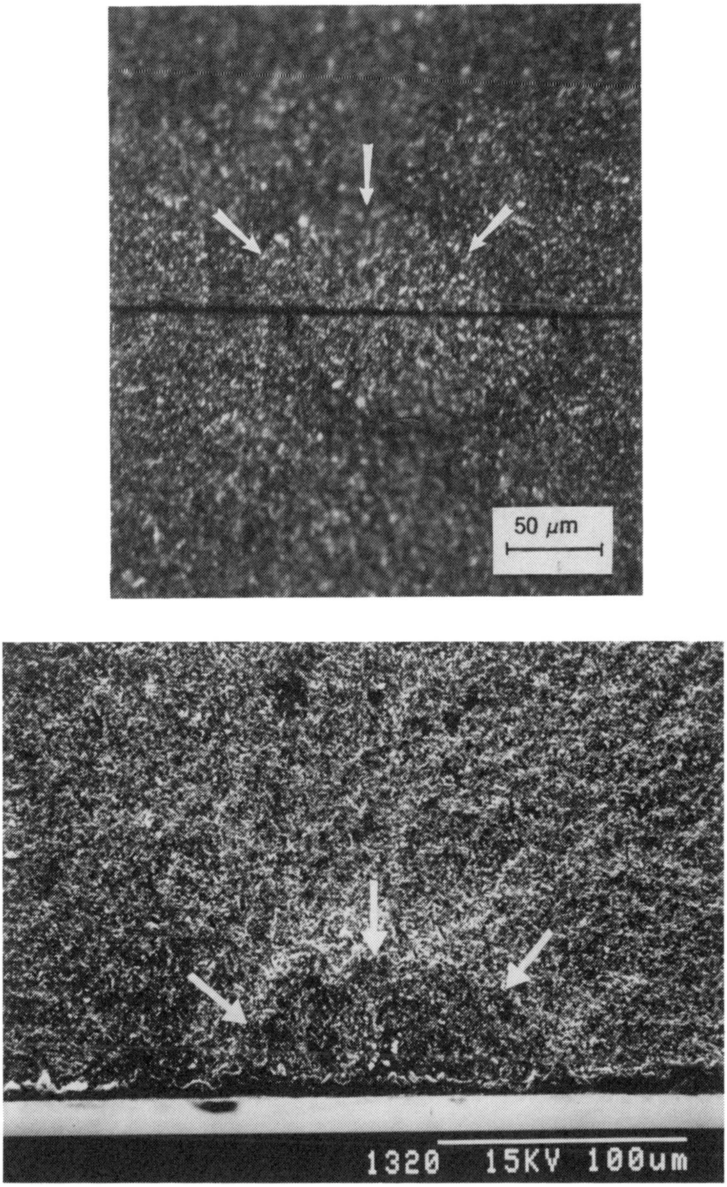

a

b

Figure 4. SEM photos of precracks in: (a), AD 999 sintered alumina, and (b), Pyroceram 9603. The halos help define the precrack. The residual stresses and the Knoop indent have not been removed by polishing in (b) and the damage zone is readily seen.

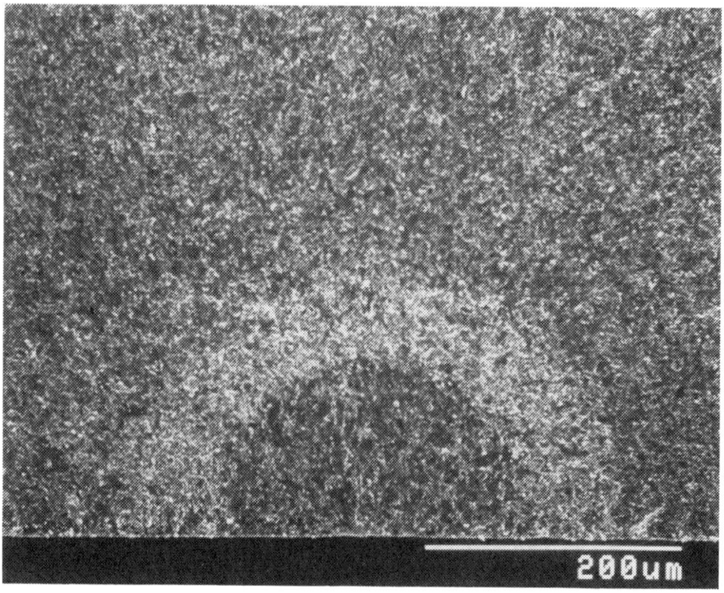

a

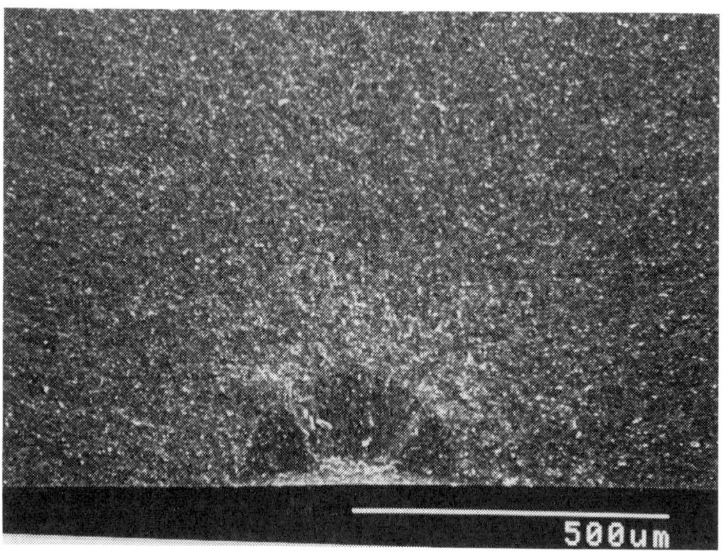

b

Figure 5. Low (a) and high (b) magnification optical microscope photos of a precrack in the ESK HIP silicon nitride.

a

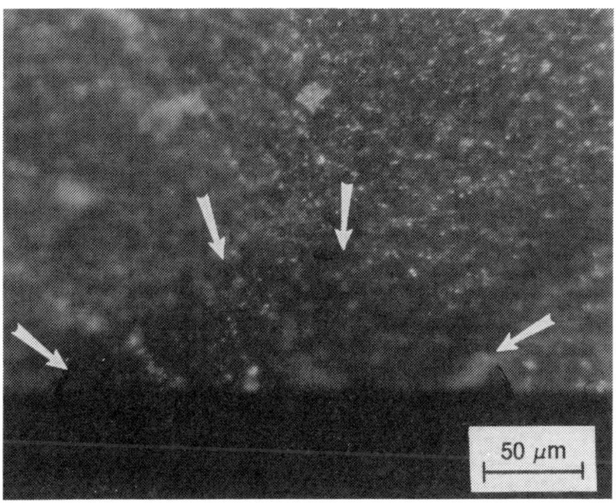

b

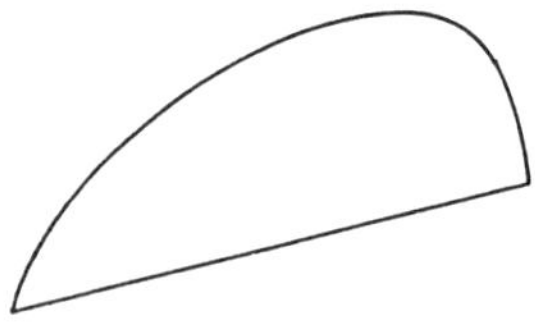

a

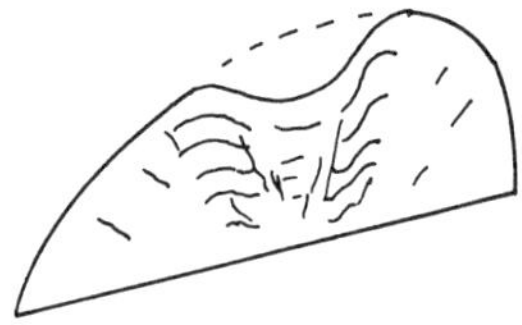

b

Figure 6. Schematic showing that precracks were either flat, (a) as was the case for NC 132 hot-pressed silicon nitride, or uneven and three dimensional, (b) as was the case for the hot-isostatic pressed silicon nitride.

In many materials the combination of low and high magnification photos proved to be helpful. Figure 9 shows two photos of the same precrack in the sintered zirconia. The precrack is fairly obvious by a brightness difference (it is darker) in the low magnification photo, but an accurate size estimate could not be obtained from such a photo. The precrack is less distinct in the higher magnification photo. The precrack is asymmetric was produced by the modified precracking procedure shown in Figure 2. (This precrack should be rejected for analysis since its shape deviates from the semielliptical shape assumed in the fracture toughness equation.) Figure 5 discussed above is another good example of the benefits of using low and high magnification photos together.

On numerous occasions, microhackle lines form inside the precrack such as illustrated schematically in Figure 10. When the specimen is loaded to fracture these microhackle lines may change direction as the crack front moves in a different direction than when the precrack was formed. The net result is that the microhackle lines may bend or be kinked at the precrack boundary. This can be a valuable aid in marking the precrack boundary and can even help determine from which point on the precrack boundary that fracture commenced. In our experience, for materials that show these markings clearly, it was often possible to confirm that fracture commenced from that portion of the precrack boundary where the calculated Y factor was a maximum. Figures 4b, 9b, 11, and 12 show precracks with microhackle redirection.

The backscattering mode of SEM operation is helpful in some instances. Figure 13 shows two photos of the hot-isostatic pressed silicon nitride with different settings. This material had rippled, undulating precracks that were often very difficult to interpret. The backscattering mode tended to make the fracture

Figure 7. Boron carbide precrack which is darker than the surrounding material due to the precrack tilt. (b) is a closeup of the right side of the precrack. The boundary is marked by the arrows.

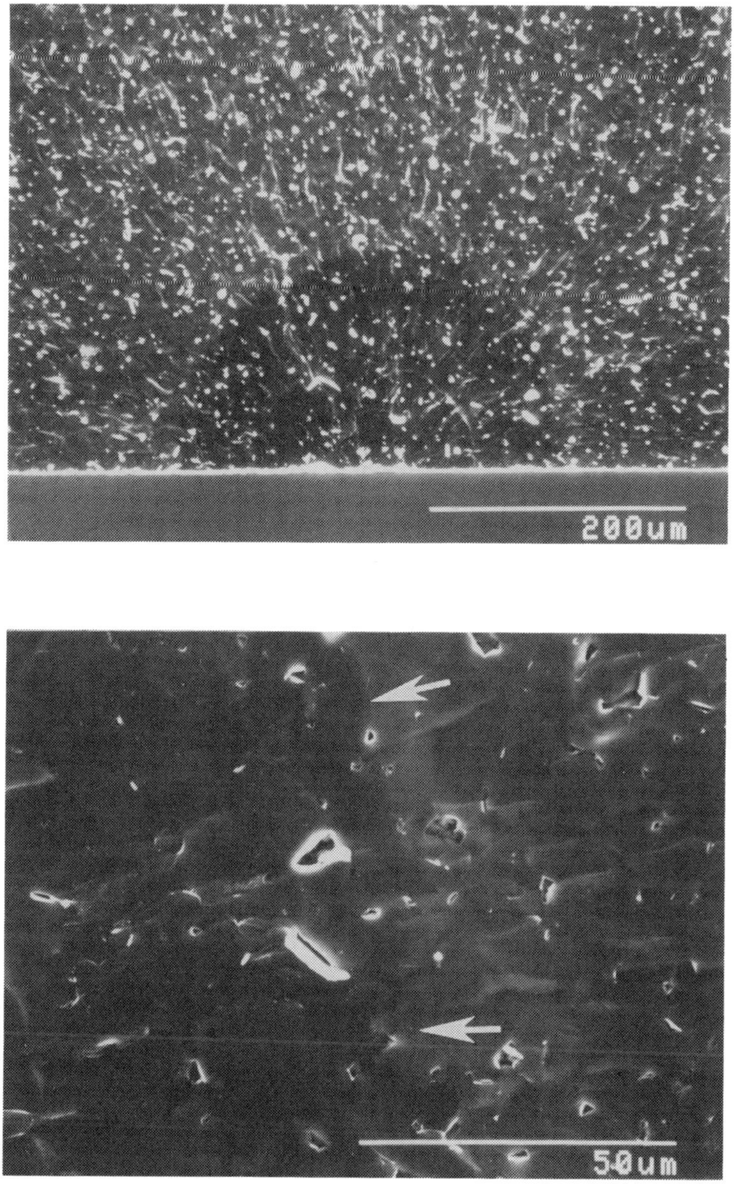

a

b

Figure 8. Precrack in zirconia. (a) shows the entire precrack which is darker than the surrounding material since the precrack fractured transgranularly as shown in the closeup, (b), whereas the fast fracture zone was intergranular, (c).

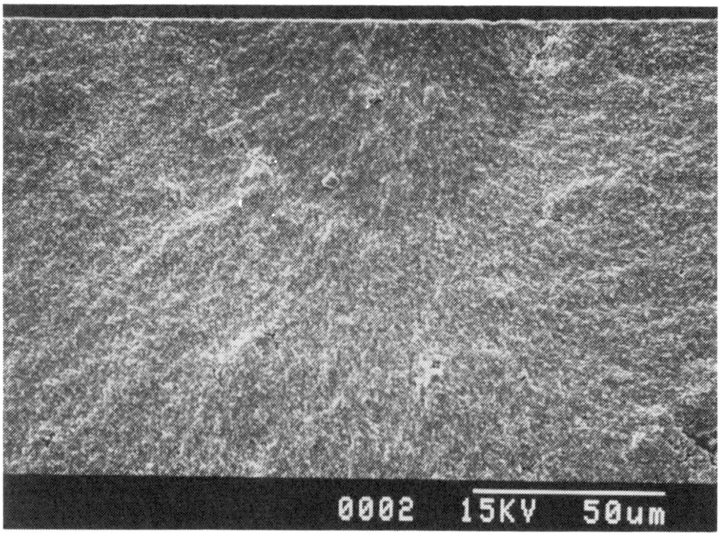

a

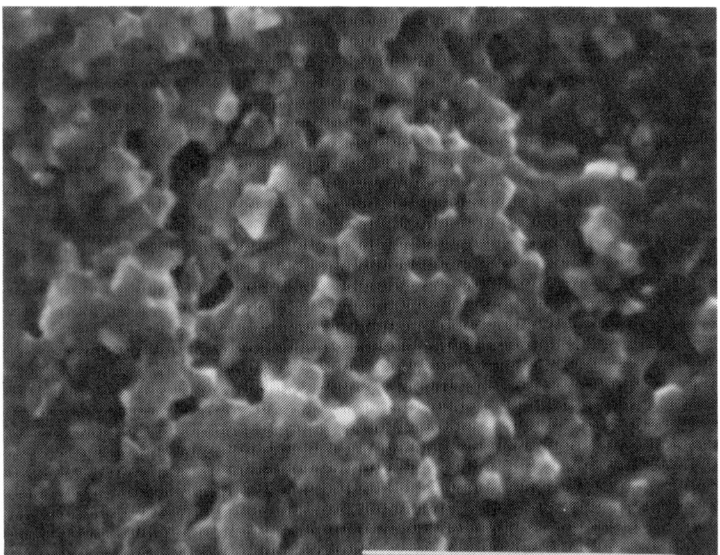

b

8c

Figure 9. Combination low (a) and high (b) magnification photos of a zinconia precrack. The precrack is fairly obvious in the low magnification photo. From laboratory #2 in the VAMAS project.[13]

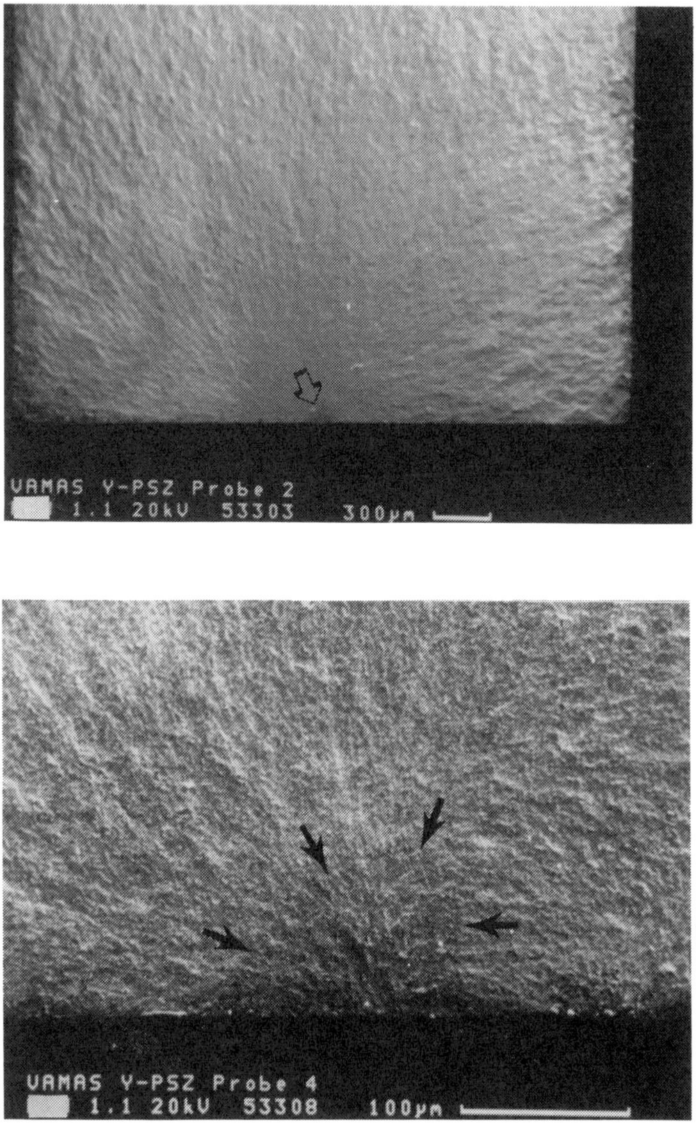

Figure 10. Schematic showing microhackle lines and how they bend or kink at a precrack boundary. Such lines help mark the precrack boundary. They also help interpret the portion of the precrack boundary from which the crack propagated (initiated) during final specimen fracture.

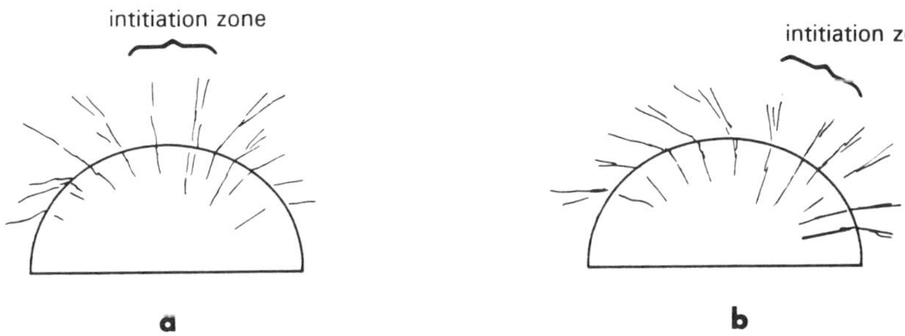

Figure 11. Precrack in NC 132 hot-pressed silicon nitride that illustrates redirection of microhackle lines at the precrack boundary. A highly directional gold/palladium coating was used to prepare the specimen which creates a high-contrast image that accentuates detail. From laboratory #20 in the VAMAS project.[13]

Figure 12. Precracks in NC 350 reaction-bonded silicon nitride that show redirection of microhackle lines at the precrack boundary. (a) shows a specimen that has had the residual stress removed by polishing and the precrack is a shallow ellipse. (b) shows an as-indented specimen which shows the original Knoop impression and the damage zone directly underneath. In (b), the microhackle lines inside the precrack stop at the precrack boundary (arrows) and do not extend into the fast fracture region. Both photos were taken with the top of the specimen tilted back 15° in the SEM.

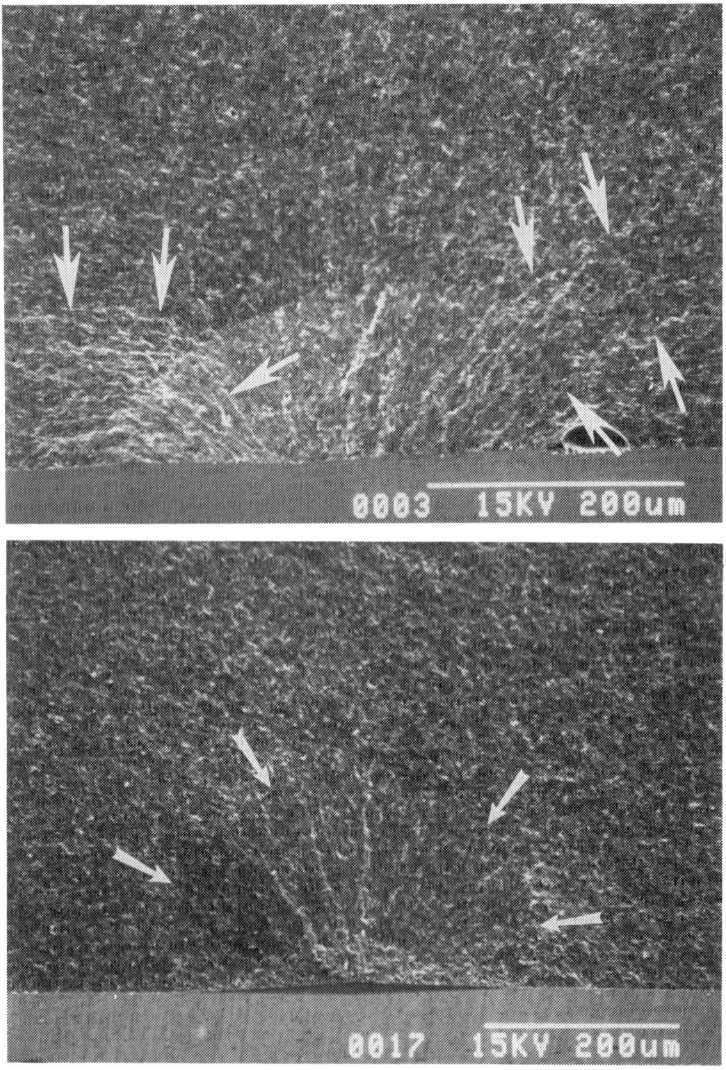

a

b

Figure 13. Backscattered SEM photos of a precrack in the ESK HIP silicon nitride. The backscattered mode tends to flatten the appearance of the precrack, which helps make the boundary a little clearer.

surface appear flatter and the precrack rim stands out somewhat better than on a secondary electron mode photograph.

In general, one can hope that a distinct brightness or contrast difference will highlight the precracks. In instances where this is not the case, and especially in instances where the precracks are uneven, rippled, or undulating, we have discovered that the best fractographic tool that can be employed is stereo SEM photography. Unfortunately, we are unable to illustrate the effect in this paper, but it has proven to be invaluable in our routine work when using the SCF method. A photo is taken of the precrack (or the area we suspect the precrack is in) with the fracture surface normal to the detector. A magnification is used such that the precrack will cover about 40 - 50% of the area of the photograph. This will provide a sufficiently large image of the precrack so that a good size measurement can be made, and some of the hackle outside of the precrack can be viewed. A second photograph is then taken, but with the specimen tilted about so that the deepest part of the precrack is about 15° tilted back relative from the detector. Since the magnifications of these photos are typically rather low for the SEM (200-400X) the tilting and second photo making can often be taken quickly with very little or no refocussing. (The extra work usually takes less than one minute.) After the SEM session is concluded, the photos are viewed in a simple stereo viewer. With a little practice, it is easy to see a three dimensional image. The three dimensional nature of the precrack, the microhackle, and the precrack tilt all reveal themselves very nicely. We strongly recommend this fractographic tool for characterization of precracks on a routine basis.

VAMAS ROUND ROBIN

A twenty laboratory VAMAS round robin investigated whether many different laboratories, with different experience levels with this method, could obtain consistent results.[13,14] Ten specimens each of three materials were delivered to each laboratory. The three materials were progressively more difficult to test by the SCF method. The NC-132 hot-pressed silicon nitride was relatively easy to precrack, polish, test, and fractographically interpret. The hot-isostatic pressed silicon nitride also was easy to precrack, but was more fractographically challenging. The zirconia was optional and precracking and fractographic analysis were challenging for this material. The modified indentation procedure shown in Figure 2 was specified for the zirconia. Only fourteen labs attempted to test the zirconia.

Most of the laboratories had little difficulty in the precracking steps for all three materials, although some labs used universal testing machines instead of conventional hardness machines. The former invariably produced larger and more varied precracks than those made by the latter. Most labs had no problem with the polishing to remove the residual stresses, although a few resorted to diamond grinding. A few laboratories had problems when they removed too much material and obliterated the precracks or reduced their size so that they no longer were the largest flaws in the specimen, in which case, the specimens broke from natural flaws.

The numerical results of this project have been well documented elsewhere,[13,14] and we only present the highlights in this paper. The grand averages and standard deviations for the three materials are listed in Table IV.

The best results were obtained with the NC 132 HPSN as shown in Figure 14. All laboratories had a high success rate (fraction of specimens tested to total numbers tested, and consistency with the overall project results). Thc grand average toughness is shown as a dashed line. The fracture toughness was independent of the specimen orientation used (i.e., whether indents were put into the 3 mm or 4 mm wide faces.) Precracks were 35-55 micrometers deep and 120-160 micrometers wide. The results are in excellent agreement with the preponderance of data in the literature for NC-132 HPSN by a variety of techniques (including Petrovic et al.'s[1-3] results from over 20 years ago) as discussed in detail elsewhere.[12-14]

The hot-isostatic pressed silicon nitride was more challenging fractographically, but consistent results were obtained from sixteen of the eighteen labs that tried it. Precracks were 60-95 micrometers deep, and 210-270 micrometers wide. Every laboratory reported at least some problem in discerning the precracks and success rates were much lower.

The zirconia was rather challenging, but most of the fourteen laboratories that attempted this material obtained acceptable results. Precracks were 20 to 40 micrometers deep and 45 to 80 micrometers wide. Two labs reported no success with the method at all. Four labs sent results where the precracks were apparently marked incorrectly.

All the round robin results were analyzed in accordance with Standard Practice E 691[17] in order to evaluate the precision of the SCF method with the results shown in Table V. The within-lab precision, or repeatability, characterizes the scatter of results obtained with the test method, in the same laboratory, by the same operator, with the same equipment, in the shortest practical period of time, using test specimens taken at random from a single quantity of material that is as

Figure 14. Results from the VAMAS round robin for the NC-132 hot-pressed silicon nitride. Each laboratory's average is shown, as well as ± one standard deviation (error bars), and the number of specimens tested.

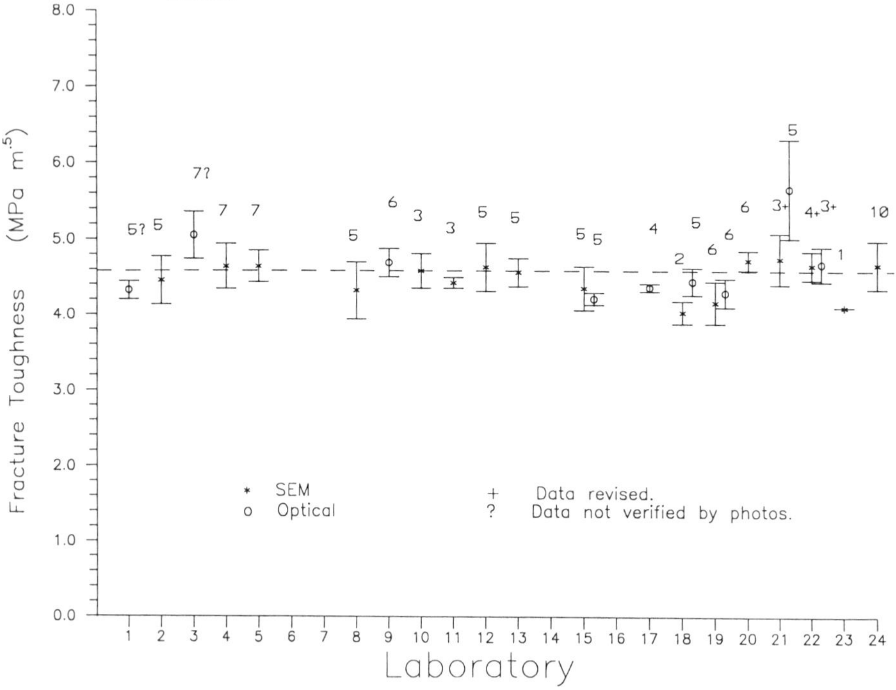

Table V Precision of the SCF method based upon the VAMAS round robin results according to ASTM Standard Practice E691-92.[17]

Material	# Labs †	Total # Spec.	Average ** MPa√m	Std. Dev. ** MPa√m	Repeatability (Within-lab) Std. Dev. MPa√m	Repeatability (Within-lab) COV % *	Reproducibility (Between-labs) Std. Dev. MPa√m	Reproducibility (Between-labs) COV % *
HPSN NC132	19	102	4.56	0.32	0.24	5.4	0.31	6.8
Hipped SN	15	100	5.00	0.48	0.38	7.7	0.45	8.9
Y-PSZ	7	29	4.47	0.31	0.29	6.6	0.29	6.6

* Coefficient of variance.

** Average and standard deviation of all individual test results.

† One outlier data set was removed for each material which accounts for the small difference with the average results listed in Table IV.

nearly homogeneous as possible. The between-lab precision, or reproducibility, characterizes the variability in test results between labs. The standard deviations are all less than 0.5 MPa$\sqrt{m}$ demonstrating an excellent consistency in the results.

The round robin participants concurred that the most difficult part of the SCF method was finding and measuring the precracks. Many of the round robin participants, particularly those who had never tried the SCF method before, had difficulty in locating and measuring the precracks, and felt there was some subjectivity to their fractographic assessments. In most instances, these participants found the fracture mirrors without difficulty, and knew to look in the middle of the mirror to find the precrack. Unfortunately, the precracks either did not stand out well on the photos, or the investigators lacked confidence in marking the subtle features that sometimes are the only markings present. All participants sent representative photos of their marked precracks to the project organizers. Some of the photos were thermal prints which did not have as much clarity as the instant-developing film that most participants used. Other participants reported that they could not see the precrack on their SEM monitors, so they didn't take photos. This was unfortunate since the video monitors often do not show the precracks very clearly. Careful scrutiny of hard-copy photos is essential. Most participants found that patience and multiple views of the precrack area enhanced their success rates.

Many participants reported that they had very good success in finding and characterizing the precracks. They used various techniques to enhance precrack detectability: tilting in the SEM, using the SEM backscattered signal, stereo SEM imaging, shadow gold coating, or using multiple photos with different magnifications. A few labs reported they had no difficulty and produced clear, sharp photos of the precracks. Experience was helpful, but not essential. Laboratory #20, which had complete fractographic success, used a graduate student who had a "golden touch" with the SEM. All photos were sharp and clear, and the precracks were readily apparent. Low-incident angle (shadow) gold coating and multiple photos at different magnifications were the keys to their success. Figure 11 is an example from laboratory 20.

Figure 14 shows that there is no difference in the fracture toughness values obtained whether from optical or scanning electron microscopy was used to measure the precracks. This was a pleasantly surprising outcome and is discussed in more detail the final report of the project.[13] The optical photos taken by the participants were in the same magnification ranges (100-400X) that were used by those who used the SEM, but the former usually were less sharp or clear due to the reduced depth of field. Figure 3 shows SEM and optical microscope photos of the same precrack in the NC 132. The precracks in this material were relatively flat. In contrast, the limited depth of field of optical microscopes created difficulties when photographing the hot-isostatic pressed silicon nitride precracks since they were more rippled or undulated as illustrated in Figure 6. It appeared that the precracks followed microstructural variations in the hot-isostatic pressed silicon nitride.

One surprising conclusion of the round robin was that the computed fracture toughness is *insensitive to the exact precrack boundaries marked*, because the computed fracture toughness is not sensitive to the precrack size measurement! This is partly due to the dependence of the toughness on the square root of the crack size (eq. 1). Thus, a 10% uncertainty (or error) in crack size is diminished to, at the most, a 5% uncertainty in fracture toughness. We also discovered an additional, fortuitous influence of the stress intensity shape factor (Y) on computed toughness. The Y factor often offset errors in the crack size measurement. For example, if the crack depth is overestimated, the corresponding Y is underestimated. The ellipse shape affects the Y factor, which ranges from 1.28 to 1.99 for shallow semicircular and semielliptical surface cracks in bending. It is important that the Y factor be calculated for each and every precrack.

Figure 15 illustrates the influence of an error in the precrack size on the computed fracture toughness. The error in K_{Ic} is shown as a function of three possible errors sources: errors in crack depth, a; errors in crack width, 2c; or errors in both a and 2c. In each case, equation 1 was solved with an assumed error in the crack size and with the Y factor computed for the erroneous crack size estimate. For this numerical simulation, the specimen was assumed to be 4 mm wide and 3 mm high. The initial ("correct") crack depth, a, was set at 50 micrometers and the width was 166 micrometers so that the a/c ratio is 1/1.66. The normalized crack size of 1.0 on the horizontal axis corresponds to a correctly measured precrack. Other values correspond to under- or overestimates of the crack size. If there was no error in K_{Ic}, the normalized K_{Ic} had a value of 1.0. The maximum Y value was used, whether it occurred at the surface or at the deepest point of the precrack. For this particular precrack shape, the maximum stress intensity factor, Y, is always at the deepest point of the precrack.

In every instance, the maximum error is bounded by the error locus for the "both a and 2c" error. This means that the whole crack size is under or over estimated as might happen if the magnification of a photo were wrong. The shape factor Y remains approximately constant and the resulting error in K_{Ic} is half the error in the crack size. Figure 15 shows that errors in a <u>or</u> 2c lead to an even smaller error in K_{Ic}. For example, an underestimate of 20% in the crack depth, a, leads

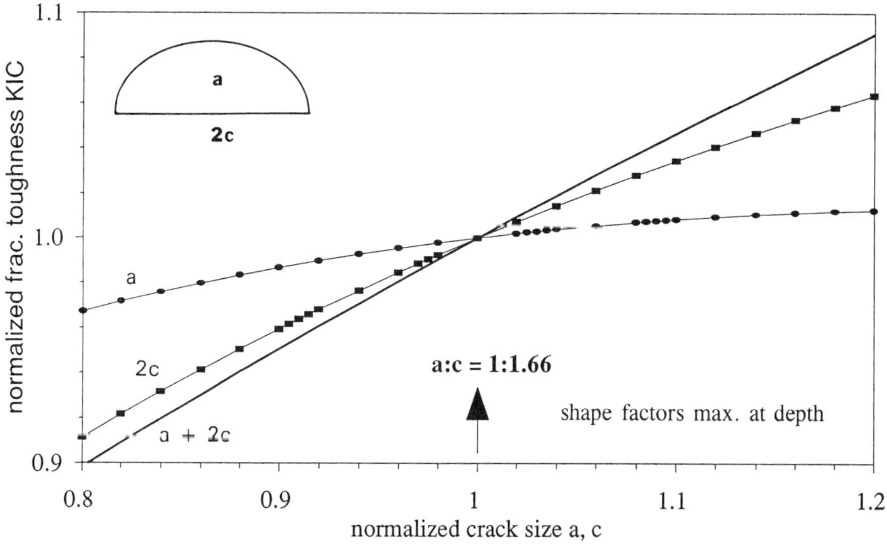

Figure 15. Normalized fracture toughness versus normalized crack dimensions. This figure illustrates the effect of uncertainty in the precrack size measurements upon computed fracture toughness for a 50 μm deep semiellipse with an ellipticity (a/c) of 1/1.66 in a 3 mm thick flexure specimen. Three possible errors in crack size dimension are shown on the horizontal axis: errors in crack depth (a) alone; in crack width (2c) alone; or both depth and width (a+2c).

to an underestimate of K_{Ic} of less than 4%. A detailed discussion of this crack size - stress intensity shape factor interrelationship and graphs for other semielliptical and semicircular precracks are in Reference 13.

The organizers of the VAMAS round robin project were initially surprised that some obviously misinterpreted precrack photos sent by a few participants led to fracture toughness estimates that were not seriously in error. Hackle lines on one side of the precrack (particularly in the zirconia) were erroneously identified as the precrack, or the intersection of the precrack with the specimen surface was misidentified (ie., 2c was overestimated). In general, even if the participants made a completely wrong appraisal, the calculated toughness was usually within a factor of 2 of the correct value. Indeed, the worst reported numbers from some of the inexperienced participants corresponded to their having marked the fracture mirror as the precrack. The computed fracture toughness values were no more than a factor of 2.5 - 3 in error which is a simple consequence of the fact that mirrors are generally 6 - 12 times the size of the fracture origin.

The numerical calculations discussed above lead to the conclusion that fracture toughness uncertainty has only a weak dependence on crack size measurement errors. We believe that this fortuitous insensitivity of fracture toughness on precrack size has gone unrecognized since most users of the SCF method have used an average Y factor for their precracks in a group of specimens, and have not calculated the maximum Y factor for each and every precrack.

The most sensitive input to fracture toughness in the SCF method is the flexure strength measurement, which can be accurately measured to within 1-2%. The stress intensity shape factors for the precracks are expected to be accurate to within a few percent for the instances where failure initiates at the deepest point of the precrack periphery. Precrack sizes can be measured to within 5% with either optical or electron microscopy provided that the material is conducive to fractographic interpretation. Uncertainties in crack size, a and 2c, are alleviated by the offsetting changes of the stress intensity factor coefficient, Y. For a material that fractures from the deepest part of the precrack, and which has a clearly visible precrack, the overall precision of the SCF method for a "fractographically friendly" material is expected to be of the order of $\pm 5\%$ (coefficient of variation). The round robin results for the NC 132 silicon nitride confirm this estimate.

We should note that the round robin did highlight one problem in the fractographic analysis process. The person who took the photographs of the precracks occasionally was not the person who had to interpret them. In a number of instances, the laboratory's chief participant in the round robin gave the fractured specimens to technicians, students, inexperienced or even experienced SEM operators and told the them to "go find the precracks and photograph them." This often led to problems whereby the photograph taker was inexperienced with conducting ceramic fractography and often didn't know what to look for or how to photograph the precracks. The interpreter was left to scrutinize hopelessly inadequate photos. It was no wonder that a few participants complained that the precracks were difficult to find and mark with any degree of confidence. The organizers addressed this problem in the project final report:[13] *"We therefore conclude in general that a necessary prerequisite for the success of the project is that the analyst must have some ceramic fractographic experience. The fractographer must work closely with the SEM or optical microscope operator. It is not enough to ask that the latter person should take a few photos and present them to the analyst later."* On the other hand, there were quite a few participants let inexperienced personnel do the fractography, and nonetheless received excellent photos and obtained completely satisfactory results. The project organizers concluded that if the participating laboratories retried their experiments, with the experience of the round robin behind them and with the benefit of seeing the many precrack photos and interpretations, they probably would have higher confidence and better success.

The consistency of these SCF results over the course of 20 years, the good repeatability that is inherent to the technique, the good agreement with other results from reliable test methods, and the fact that toughness is measured for precracks at the size scale of natural flaws, prompted ASTM Committees C-28, Advanced Ceramics, and E-8, Fracture and Fatigue, to include the SCF method as one of three in a new draft fracture toughness test method standard.

A list of recommended steps to aid in precrack detection is given in Appendix B. In some instances where there is some doubt about whether specimens fracture from precracks, or with a new material for which the fractographer has no experience, it is often helpful to indent and fracture a few specimens *without* removal of the residual stresses. Although the computed toughnesses will be underestimates due to the residual stresses, the fracture surfaces (which will have a distinct Knoop impression as well as the precrack beneath it) will help the fractographer get acquainted with the nuances of the material's fracture surfaces. Some guidelines for interpretation of precracks also are given in Appendix C. Appendix D illustrates why orienting Knoop precracks with a "twist" angle relative to the specimen longitudinal axis usually is unsuccessful in aiding precrack detection.

Future work will focus on making precrack detection and characterization easier, possibly by means of precrack decoration. A few investigators have attempted flaw decoration (e.g. see References 18-22) or thermal tinting[4-6] but there usually is some uncertainty whether the dye penetrant completely reveals the precrack. SCF precracks are rather small and tight closed. There is also a risk that the dye penetrant might blunt or alter the crack tip in an undesirable manner.

We will apply the SCF method to a broader range of materials including those with rising R-curve crack extension resistance. Preliminary findings indicate that the method can generate apparent toughness versus crack size data.[10-13] We also plan to compare techniques by testing some materials by the SCF method as well as single-edged precracked beam and chevron notch test methods.

CONCLUSIONS

The surface crack in flexure method has been refined. Fractographic analysis is an integral step to the test method and techniques to aid the fractographer have been presented. The computed toughness is insensitive to the precrack size measured, a consequence of a beneficial interrelationship between the crack size and shape, and the stress intensity shape factor, Y. It is important to compute the Y for each crack in each specimen (and not to use an average Y factor) in order to benefit from this relationship between Y and the crack size. A round robin project demonstrated that consistent results can be obtained by different laboratories.

ACKNOWLEDGEMENT

Support for this program was furnished by the Ceramic Technology Project, U.S. Department of Energy, Office of Transportation Technologies, under contract DE-AC05-84OR21400 with Martin Marrietta Energy Systems, Inc.

We wish to thank Mr. Lewis Ives of NIST for recommending that we use stereo SEM microscopy. Prof. James Varner of Alfred University, USA and Dr. Chris Debel of RISO Denmark, who participated in the VAMAS round robin, also suggested stereo SEM photography. Dr. David Rowcliffe suggested the gold shadow coating procedure during preparation for the SEM.

REFERENCES

1. J. J. Petrovic, L. A. Jacobson, P. K. Talty, and A. K. Vasudevan, "Controlled Surface Flaws in Hot-pressed Si_3N_4," J. Amer. Ceram. Soc., 58 [3-4] 113-116 (1975).
2. J. J. Petrovic, R. A. Dirks, L. A. Jacobson, and M. G. Mendiratta, "Effects of Residual Stresses on Fracture from Controlled Surface Flaws," J. Amer. Ceram. Soc., 59 [3-4] 177-178 (1976).
3. J. J. Petrovic and M. G. Mendiratta, "Fracture from Controlled Surface Flaws," pp. 83-102 in Fracture Mechanics Applied to Brittle Materials, ASTM STP 678, ed. S. W. Freiman, ASTM, Philadelphia, PA, 1979.
4. P. Kenny, "The Application of Fracture Mechanics to Cemented Tungsten Carbides," Powder Met. 14 [27] 22-38 (1971).
5. R. K. Govila, "Dislocation Etch Pits in Vanadium Carbide Monocrystals," Phil. Mag., 22, 431-436 (1970).

6. R. K. Govila, "Cleavage Fracture of VC Monocrystals," Acta Metal., 20 447-457 (1972).
7. K. R. Kinsman, R. K. Govila, and P. Beardmore, "The Varied Role of Plasticity in the Fracture of Inductile Ceramics," pp. 465-82 in "Deformation of Ceramic Materials," eds. R. C. Bradt and R. E. Tressler, Plenum, NY, 1975.
8. J. J. Petrovic, "Effect of Indenter Geometry on Controlled-Surface-Flaw Fracture Toughness," J. Amer. Ceram. Soc., 66 [4] 277-283 (1983).
9. ASTM E-740-88, "Standard Practice for Fracture Testing with Surface-Crack Tension Specimens," ASTM Annual Book of Standards, Volume 03.01, ASTM, Philadelphia, PA, 1994.
10. C. A. Tracy and G. D. Quinn, "Fracture Toughness by the Surface Crack in Flexure (SCF) Method," Cer. Eng. and Sci. Proc., 15 [5] 837-845 (1994).
11. R. J. Gettings and G. D. Quinn, "Surface Crack in Flexure (SCF) Measurements of the Fracture Toughness of Advanced Ceramics," to be publ. Cer. Eng. and Sci. Proc., 1995.
12. G. D. Quinn, R. J. Gettings, and J. J. Kübler, "Fracture Toughness of Ceramics by the Surface Crack in Flexure (SCF) Method," to be publ. in Fracture Mechanics of Ceramics, ed. D. Munz, Plenum, NY, 1996.
13. G. D. Quinn, J. J. Kübler, and R. J. Gettings, "Fracture Toughness of Advanced Ceramics by the Surface Crack in Flexure (SCF) Method: A VAMAS Round Robin," VAMAS Report #17, National Institute of Standards and Technology, Gaithersburg, MD, June 1994.
14. G. D. Quinn, R. J. Gettings, and J. J. Kübler, "Fracture Toughness of Advanced Ceramics by the Surface Crack in Flexure (SCF) Method: Results of the VAMAS Round Robin," Cer. Eng. and Sci. Proc., 15 [5] 846-855 (1994).
15. ASTM C 1161-90, "Standard Test Method for Flexural Strength of Advanced Ceramics at Ambient Temperature," Annual Book of ASTM Standards, Vol. 15.01, ASTM, Philadelphia, PA, 1994.
16. J. C. Newman, Jr. and I. S. Raju, "An Empirical Stress-intensity Factor Equation for the Surface Crack," Eng. Fract. Mech., 15 [1-2] 185-192 (1981).
17. E 691-92, "Practice for Conducting an Interlaboratory Study to Determine the Precision of a Test Method," Annual Book of ASTM Standards, ASTM, Philadelphia, PA, Vol. 14.02, 1994.
18. R. R. Wills and J. M. Wimmer, "Controlled Surface Flaw-Initiated Fracture in Reaction-Densified Silicon Carbide," J. Amer. Ceram. Soc., 59 [9-10] 437-440 (1976).
19. S. L. Jones, C. J. Norman, and R. Shahani, "Crack-Profile Shapes

Formed Under a Vickers Indent Pyramid" J. Mater. Sci. Let., 6 721-723 (1987).
20. J. J. Hansen, R. A. Cutler, D. K. Shetty, and A. V. Virkar, "Indentation Fracture Response and Damage Resistance of Al_2O_3-ZrO_2 Composites Strengthened by Transformation-Induced Residual Stresses," Comm. Amer. Ceram. Soc., 71 [12] C-501-C505 (1988).
21. S. Mizuno, N. Kobayashi, K. Miyata, "Analysis of Sintered Silicon Nitride Grinding Damage," presented at the SAE International Congress and Symposium, SAE Paper 930163, SAE, Warrendale, PA, 1993.
22. J. D. Sullivan and P. H. Lauzon, "Shape Profiles of Cracks Formed Under a Vickers Pyramid Indenter," J. Mater. Sci. Let., 5 247-248 (1986).

APPENDIX A

The Newman-Raju stress intensity shape factors for semicircles and semiellipses in beams loaded in bending.[16] All dimensions are illustrated in Figure A1.

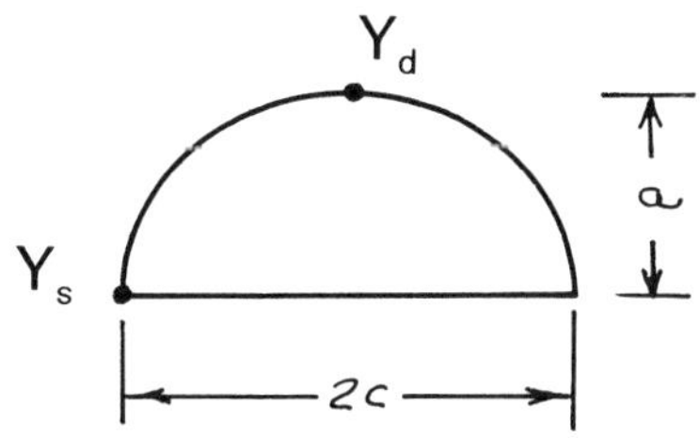

For the deepest point of the flaw:

$$Y_d = (\sqrt{\pi}\, M\, H_2) / \sqrt{Q}$$

where

$Q = 1. + 1.464(a/c)^{1.65}$

$$M = \{1.13 - 0.09(a/c)\} + \{-0.54 + 0.89\cdot[0.2 + (a/c)]^{-1}\}(a/W)^2 + \{0.5 - [0.65 + (a/c)]^{-1} + 14.\cdot(1-a/c)^{24}\}(a/W)^4$$

$$H_2 = 1. - [1.22 + 0.12(a/c)](a/W) + [0.55 - 1.05\,(a/c)^{.75} + 0.47(a/c)^{1.5}](a/W)^2$$

For the point at the surface:

$$Y_s = (\sqrt{\pi}\, M\, S\, H_1) / \sqrt{Q}$$

where

$H_1 = 1. - [0.34 + 0.11(a/c)](a/W)$ and $S = [1.1 + 0.35(a/W)^2]\sqrt{a/c}$

Use the larger value of Y, and compute fracture toughness from equation 1 in the main text.

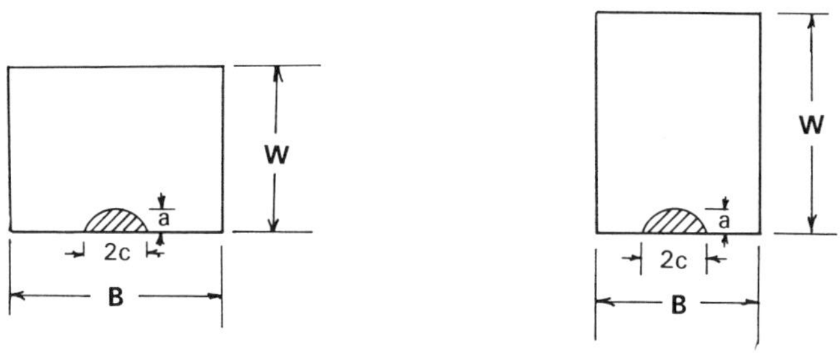

SPECIMEN CROSS SECTION

Figure A1 The cross section of the specimen showing the precrack. The precrack may be in either the wide or narrow specimen face.

APPENDIX B

Procedures to aid precrack detection.

1. Tilt the specimen 1/2° during precracking. Indent into the same location on every specimen (e.g. the exact middle along the length). This will help assess whether fracture actually did occur from the precrack.

2. Remove material by polishing or grinding very carefully when removing the residual stress after indentation. Minimize the amount of new surface damage introduced.

3. After fracture, inspect the fracture surfaces with a stereo optical microscope or other optical microscope if available.
   (Stereo binocular microscopes have superior depth of field and a greater working distance, both of which are very advantageous.)
   a. Use low angle (grazing) incident lighting.
   b. Rotate the specimen to view the fracture surface with different illumination directions.
   c. Select the best fracture half for photography or SEM examination. (Save the other half for later examination if needed.)
   d. Look for color or contrast changes.
   e. Look for halos or ridges.
   f. Use a light gold/palladium coating if necessary due to specimen translucency (e.g., aluminas) but only do this after other inspections since it will obscure some optical color or contrast changes.
   g. Photograph and measure the precrack if possible.

4. Inspect with the SEM if needed.
   a. Sputter coat the surface with gold at a low-incidence angle (shadow coating). Do not use too thick a coating.
   b. Use the mirror to help locate the general region of the precrack. Photograph the origin area.
   c. Use combinations of photos (e.g., one at 100X and one at 200-500X depending on the crack size). The higher magnification photo, which is used for the precrack size measurement, should be normal to the detector.
   d. Try a strong tilt (specimen to detector) at the lower magnification to help highlight the precrack if necessary.
   e. Take stereo photograph pairs of the precrack at the higher magnification. One photo must be with the specimen normal to

the detector (as in 4c above), the other should have the specimen tilted ~15° to the detector.

e. Try the backscattering mode if necessary.

f. Examine the other specimen fracture surface if necessary.

5. Look for: halos, ridges or rims, contrast or color changes, fracture mode changes, or microhackle direction changes.

6. Do not force an interpretation. Cracks may be truncated, uneven and wavy (3-dimensional) or have surface machining damage, or the specimen may have fractured from a natural flaw. Try another specimen or look at the opposite fracture half. Examine fractographs from different specimens, and try to discern feature patterns on the fracture surfaces. Success with one specimen may prompt successful reexamination of earlier specimens thought to be not interpretable.

APPENDIX C

Complications in characterizing SCF Precracks.

Polishing or Machining Damage

This can occur if the polishing or machining to remove the indent is done too aggressively. Specimens with this damage can be repolished to remove the surface damage. If it is necessary to interpret such precracks, then approximate the semiellipse shape as if the surface damage is not present. If the maximum Y factor is at the surface, the datum is to be rejected.

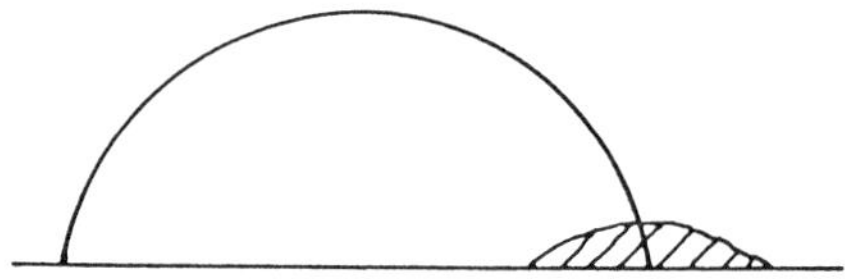

Corner Pop-in

During the fracture test, the precrack reaches critical fracture condition at Point A first. A small crack extends to B. Final fracture starts at Point C. It is recommended that the original ellipse be used to compute fracture toughness, but if the extension at points A-B is excessive, the datum is to be rejected. Polish specimens a deeper amount to force the Y_{max} to be at the deepest point, Y_d.

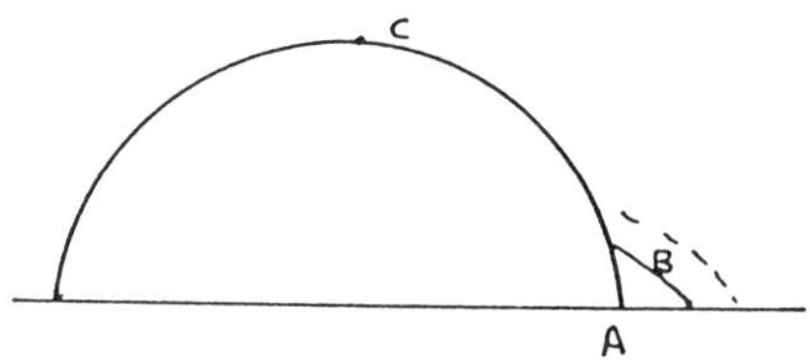

Poorly Defined Crack at the Surface

This can occur in instances where the precrack and the final fracture crack are on the same plane. (The 1/2° tilt may not have been adequate.) Alternatively, a limited depth of field in the optical microscope may hamper focussing the entire precrack. Estimate or approximate the semiellipse shape as best as possible, but if more than 50% of the precrack is not visible, the datum is to be rejected.

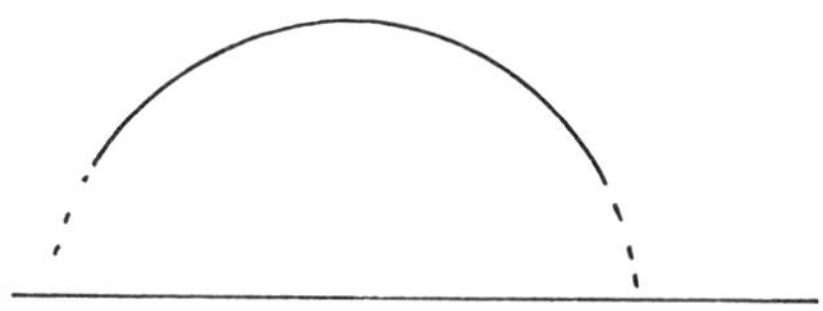

Stable Crack Extension

The crack may extend stably prior to fast fracture, either due to a rising R-curve behavior, or environmentally-assisted crack growth. This can either be an interference (which crack should be measured) or a useful tool to study the stable crack growth phenomena. Definitive interpretation of such stable crack extension markings on a fracture surface is extremely difficult.

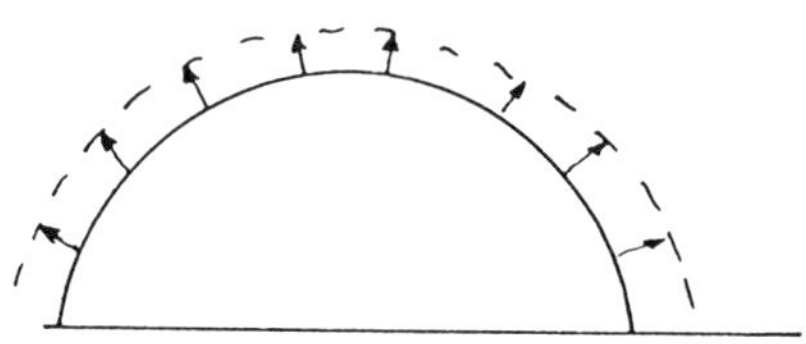

Precrack Truncation

The final crack is on a different plane and only intersects a portion of the precrack. This can occur if the precrack is not perpendicular to the maximum stress in the specimen, and fracture commences from one point on the precrack periphery, but then truncates the remainder of the precrack. In these cases, the datum is to be rejected.

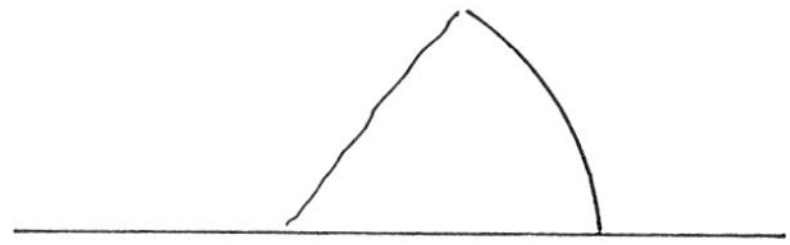

Precrack Segmentation

Precrack is actually made of three segments. The precrack is not flat and has a 3-dimensional aspect shape. It is "rippled" or "corrugated" as shown in the figure. The interference may be from lateral or Hertzian cracks associated with the original indent, or may be due to non-uniform density in the ceramic. (This problem is common in some sintered ceramics.) If the waviness or corrugation is excessive, the datum is to be rejected.

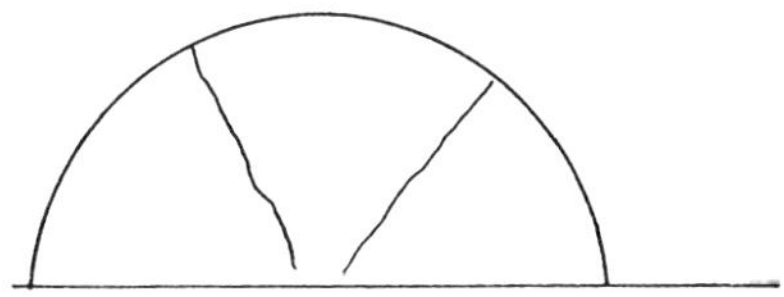

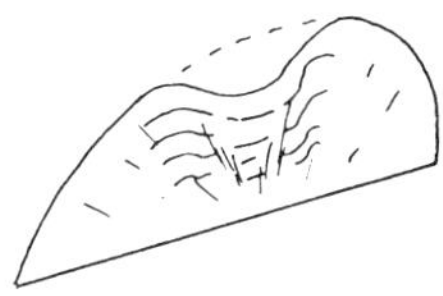

APPENDIX D

A twisted precrack is not always successful in aiding precrack detection since the precrack may be truncated during specimen fracture.

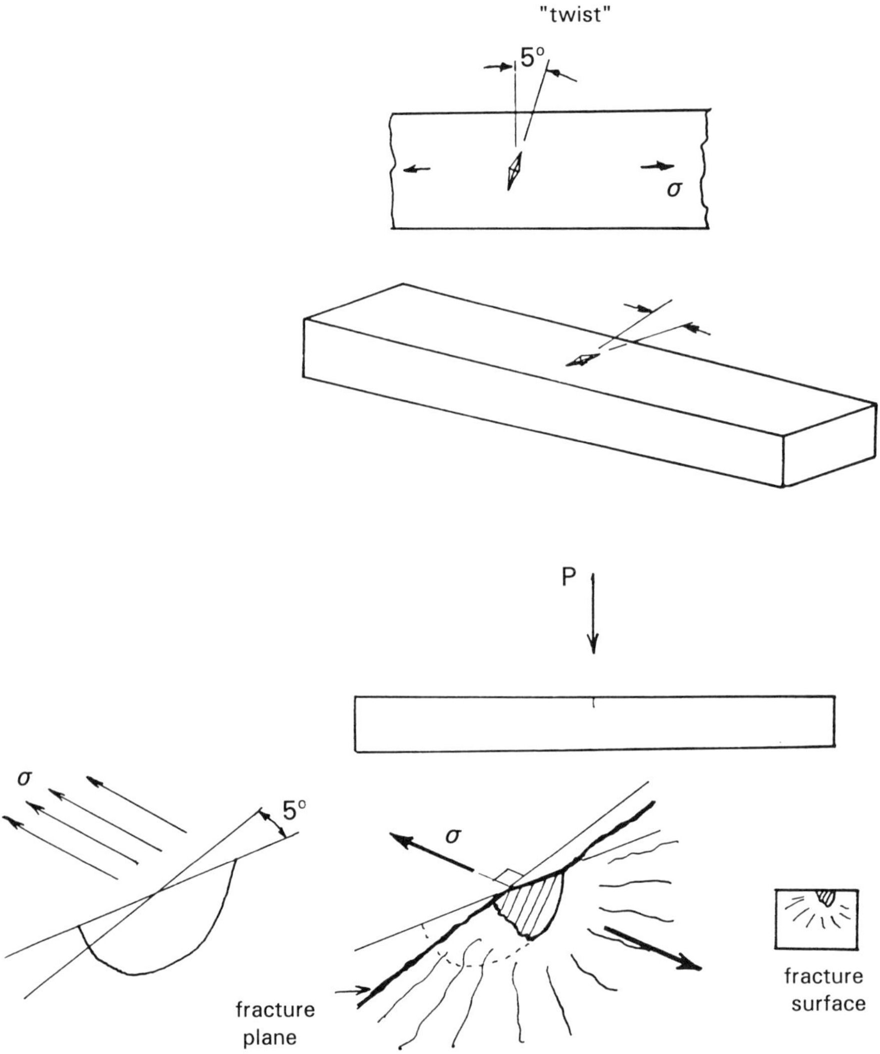

ANALYSIS OF MACHINING DAMAGE IN ENGINEERING CERAMICS BY FRACTURE MECHANICS, FRACTOGRAPHY AND X-RAY DIFFRACTION

Thomas Hollstein, Wulf Pfeiffer, Michael Rombach, Bärbel Thielicke
Fraunhofer-Institut für Werkstoffmechanik
Wöhlerstr. 11, D-79108 Freiburg

INTRODUCTION

The cost for final machining covers a significant percentage of the whole cost of a ceramic component. This is due to the difficult machining of the high performance ceramics. The high values of hardness and wear resistance, which are desired in many applications, hinder the process of machining. Only a few machining procedures are applicable to engineering ceramics e.g. grinding, polishing or ultrasonic lapping, and the rate of material removal is considerably lower than for metals. In addition crack generation in the surface regions during machining is easily possible due to the brittleness of the ceramics.

The material removal during grinding, which is the most important machining procedure of engineering ceramics, takes place mainly by brittle fracture processes but also by ductile material removal (see Fig. 1). The complex stress conditions in the work piece below or in the vicinity of the grinding grits lead to a variability of cracks and crack systems like median cracks, lateral cracks or radial cracks, which extend in general ≤ 50 µm (see Fig. 2) and which lead to the strength anisotropy of ground ceramics, if certain grinding parameters are used e.g. [1, 2].

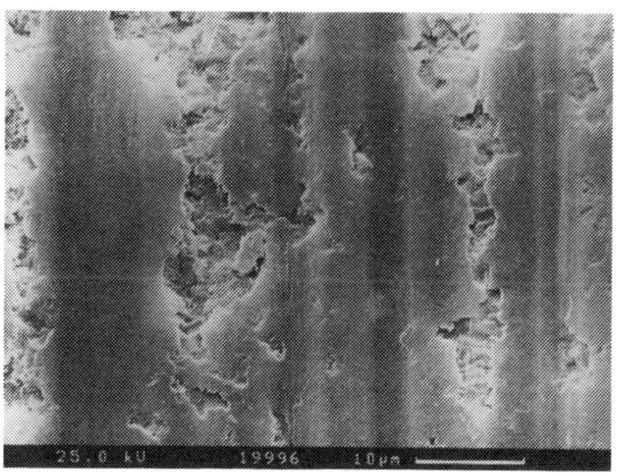

Fig. 1: Brittle and ductile material removal due to grinding

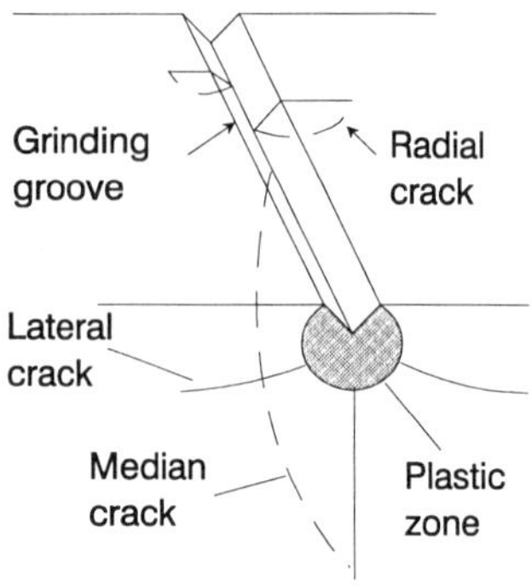

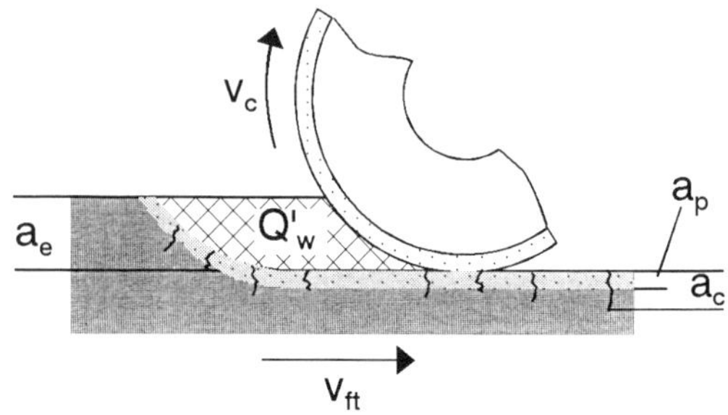

Fig. 2: Sketch showing the different crack types and the plastic zone generated by one single grinding grit (left) and the machining damage, the plastically deformed zone generated by grinding and the machining parameters (right); a_e= infeed of grinding wheel, v_{ft}= tangential feed rate, v_c= grinding wheel speed, $Q'_w = a_e \times v_{ft}$ = specific material removal rate, a_p= plastically deformed zone, a_c= damaged zone with micro strains and micro cracks, respectively.

In the areas of the high compressive and shear stresses plastically deformed regions develop with depths of up to 30 µm. Fig. 2 contains a rough sketch thereof and of the main parameters describing the specific grinding procedure. The tangential motion of the grinding grits, together with this increased plasticity, enables the ductile material removal.

It has been shown by several authors e.g. [3], that cracks or crack systems with longitudinal, lateral and/or radial cracks after grinding can be found, and this is also shown in Figs. 3 and 4 for the Si_3N_4 investigated here. Fig. 3 (left) shows a ceramographic cut through a bend bar, that has been ground transversally: The crack system has developed below a grinding groove, and one can imagine, that the cracks during their development overlap those already generated. Fig. 3 (right) shows the fracture surface, where the median crack, which has been developed to an edge crack, acts as the fracture causing defect. For a longitudinally ground specimen the occurrence of lateral and/or radial cracks, which cannot be separated here, is shown in Fig. 4.

The quantification of machining damage on the basis of one characteristic, such as a roughness parameter, sometimes leads to a wrong assessment of strength properties. As an example Fig. 5 shows the SEM micrographs of two differently ground Si_3N_4 specimens. The unexpected result, that the rougher surface corresponds to the higher strength, is a consequence of both the machining induced residual stresses and subsurface damage, which may influence the strength significantly.

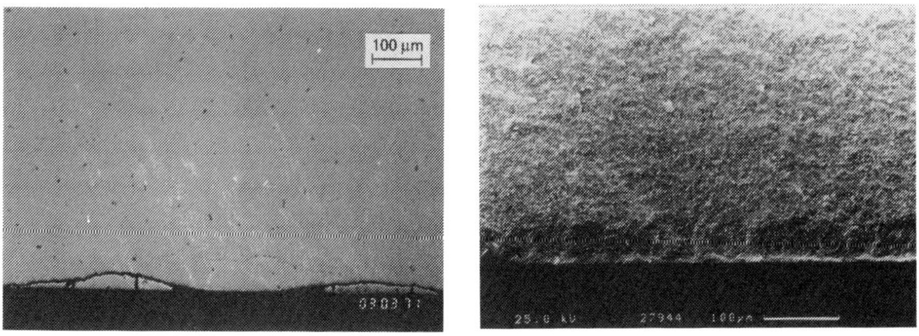

Fig. 3: Median or lateral cracks after transverse grinding of Si_3N_4 bend bars; left: microstructural cut (45°, in grinding direction), right: fracture surface with the fracture causing median crack system forming an edge crack (specimen 21.88)

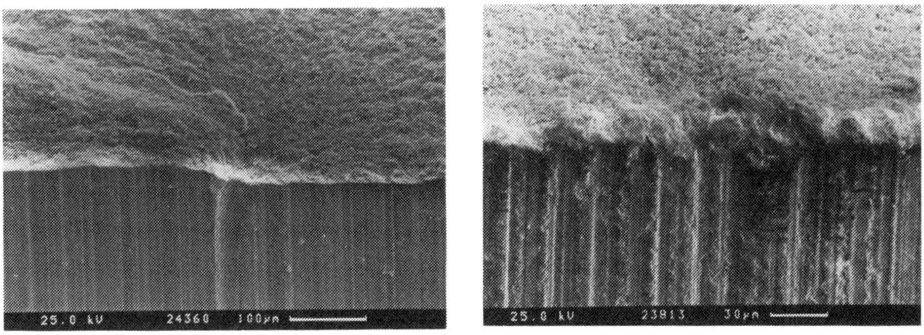

Fig. 4: Fracture surfaces with radial and/or lateral cracks in longitudinally ground Si_3N_4 bend bars. Left: The fracture causing defect can be described by a half-penny shaped surface crack (specimen 13.11). Right: Different cracks have developed into half elliptical surface cracks (specimen 13.14).

X-ray diffraction methods offer a possibility for an accurate assessment of the machining quality of high performance ceramics [1, 3]. These methods quantify machining induced damage by determining the amount of machining induced residual stresses and micro-plastic deformation of the surface layer within the relevant depths of up to 30 µm. In order to analyse the near-surface conditions after grinding the samples are characterised here by conventional roughness measurements, strength tests, X-ray diffraction technique, fractography and fracture mechanics. Good overviews on problems on machining of ceramics are given e. g. in [4-7].

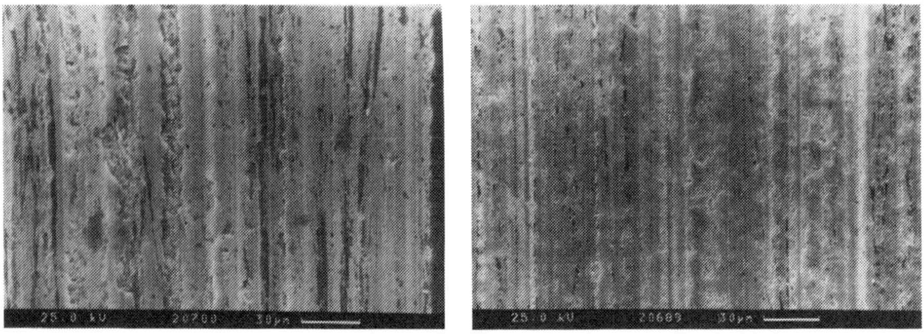

Fig. 5: SEM micro-graphs of ground silicon nitride. Left: conventional grinding (a_e = 0.15 mm), average roughness R_a = 0.42 µm. Right: creep-feed grinding (a_e =3.5 mm), average roughness R_a = 0.90 µm

EXPERIMENTAL PROCEDURES

Materials and Machining

Si_3N_4: The material investigated was a commercial sintered silicon nitride (SN-N3204 from Cremer Forschungs-Institut GmbH, CFI). Four point bend specimens (45 mm × 4 mm × 3 mm) were ground in longitudinal and transverse directions, respectively. A resin-bonded diamond wheel with 126 µm grit size was used, and up-cut surface grinding was performed at different infeeds (0.15 mm $\leq a_e \leq$ 3.5 mm), specific material removal rates (5 mm^3/(mms) $\leq Q'_w \leq$ 25 mm^3/(mms)) and grinding speeds (v_c = 30 and 60 m/s). As coolant Petrofer Isocut R118 grinding oil was used as a coolant and was supplied by a single high pressure jet [8].

Al_2O_3: Concentric ring specimens (50 mm × 50 mm × 3 mm) and round button-head tensile specimens (see Fig. 6) were ground from a 99,7% pure commercial Al_2O_3 (M513 from FRIATEC AG). The concentric ring plates were ground with a resin-bonded diamond wheel with 151 µm grit size) using up-cut grinding with infeeds a_e between 30 and 500 µm, tangential feed rates v_{ft} between 10 and 167 mm/s, and grinding wheel speeds of v_c = 25 and 50 m/s. Some concentric ring specimens were also lapped using grain sizes between 16 and 100 µm, different disc speeds and different normal forces. The tensile specimens were circumferentially ground with grinding wheels with 76 µm and 151 µm grit sizes, whose axes were parallel (grinding grooves perpendicular to the tensile direction) or perpendicular (grinding grooves parallel to the tensile direction) to the axes of the specimens. The infeeds varied between 15 and 25 µm and the tangential feed rates between 0.08 and 20 mm/s, and the grinding wheel speeds varied between 20 and 35 m/s.

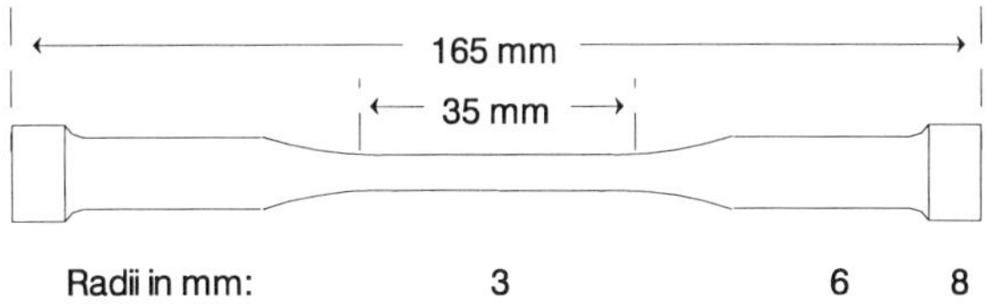

Fig. 6: Button head tensile specimen

Roughness and Strength Testing

The roughness measurements were conducted using a surface profilometer with a stylus of 5 μm radius tip. Average roughness values R_a and average peak-to-valley heights $R_{z,DIN}$ were determined transverse to the grinding direction.

The strengths of the machined surface layers were determined using 4-point bending tests following the DIN-standard [9]. The button-head tensile specimens were tested in a rig with Instron-type super grips. The characteristic strengths σ_0 and the Weibull modulus m were evaluated from the fracture stresses of up to 20 samples with identically machined surfaces. The 90%-confidence ranges of σ_0 and m were calculated using the Monte-Carlo simulation. Due to the small number of samples within several series, the uncertainties of calculated Weibull moduli are sometimes large and differences between the various grinding conditions are therefore not discussed.

X-Ray-Diffraction Analysis

The machining induced near-surface damage and residual stresses were assessed by X-ray diffraction analyses. Due to the small penetration depth of the X-rays used in diffraction experiments, machining induced residual stresses can be directly determined by X-ray stress analysis. The amount of damage is evaluated in an indirect way by measurement of the amount of micro-plastic deformation (dislocation density). In [10, 11] it was shown, that the micro-mechanical loading conditions of the material during machining and therefore the degree of damage of the machined surface can be quantified by the amount of micro-plastic deformation. The X-ray diffraction methods inspect relatively large areas. This takes the statistical meaning of 'damage' into account and avoids a time-consuming search for microscopic small cracks. In most cases the weakening effect of damage dominates compared to the strengthening effect of compressive residual stresses [10, 11]. Thus, from the amount of the micro-plastic deformation the strength of the surface layer can often be assessed. Average values of micro-plastic deformation and residual stresses were determined over the effective penetration depth of X-rays (8 μm). In some cases, also the depth distributions within the first 30 μm were evaluated. Details of the methods used are described in [10].

Fractography

After strength testing the fracture origins and fracture causing defects were analysed fractographically. The classification of the fracture causing defects was done in analogy to MIL HDBK 790 [12]. Fig. 7 as well as Figs. 3 and 4 show some typical fracture origins with the fracture causing defects. The cracks in Fig. 7 can be described by an half-elliptical surface crack (11.11), by an internal penny shaped crack (16.04) and by a half-penny shaped surface crack (13.08). Those in Figs. 3 and 4 by an edge crack (or surface scratch due to machining, 21.88), by a half-penny shaped surface crack (13.11) and by a half-elliptical surface crack (13.14).

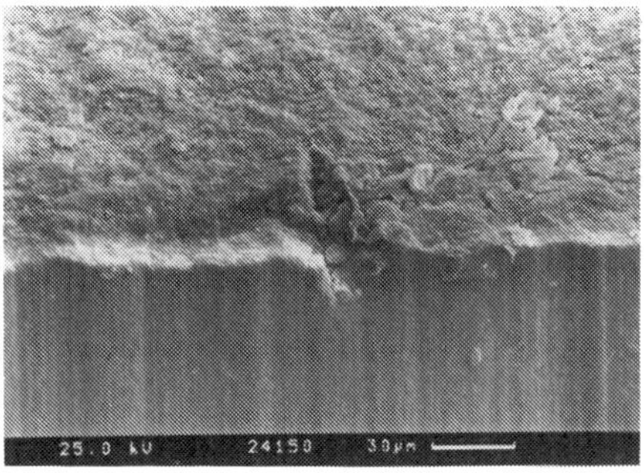

specimen 11.11
material defect (inclusion)

specimen 16.04
material defect (pores)

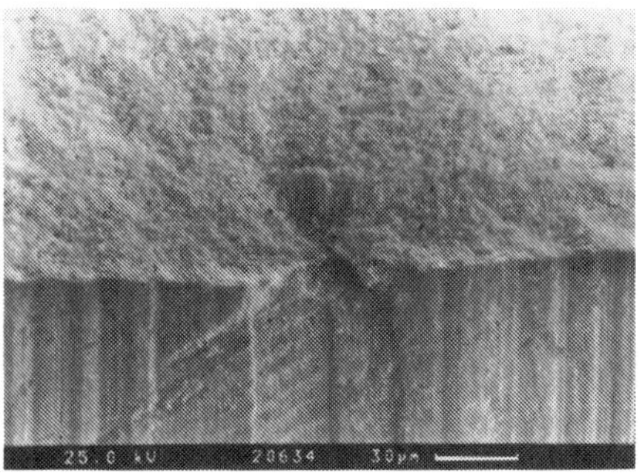

specimen 13.08
machining defect

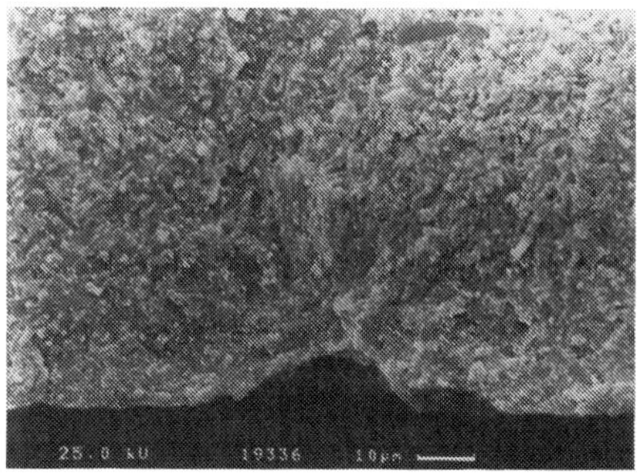

specimen 13.08
fracture surface (see left Fig.)

Fig. 7: Typical fracture origins and fracture causing defects (longitudinally ground)

FRACTURE MECHANICAL APPROACH

Treatment of Fractographically Determined Flaws

One important topic of the fractography is the correlation of the size and shape of the fracture causing defects with the measured strength and the fracture resistance of the material. The fracture behaviour of the ceramics can be described with the fractographic results by linear elastic fracture mechanics (LEFM) quantitatively using some basic fracture mechanics formulae for different defect types. The fractography shows (see e.g. the photos of fracture origins in this paper) that a lot of different shapes of fracture causing defects exist. For practical use one has to limit the number to some selected easily treatable crack geometries. The following crack geometries are considered here:

- edge crack (surface scratch)
- internal penny-shaped crack
- internal elliptical crack
- half-penny shaped surface crack
- half-elliptical surface crack.

Because the crack size is always small compared to the size of the specimen, the surrounding region can be regarded as infinite. The orientation of the main crack plane is always chosen perpendicular to the tensile stress. Since the size of the crack is always small compared to the specimen height, the stress variation over the crack can be neglected and the stress can be assumed to be constant and equal to the fracture tensile stress σ_c. All formulae are based on the Stress Intensity Factors Handbook of Murakami [13]. In the following, the formulae for the stress intensity factor K_I are compiled.

Edge crack: The K_I value for an edge crack of the length 'a' in a semi-infinite plate subjected to constant tensile stress σ, see Fig. 8, is given by

$$K_I = 1.12\ \sigma\ \sqrt{\pi\ a} \tag{1}$$

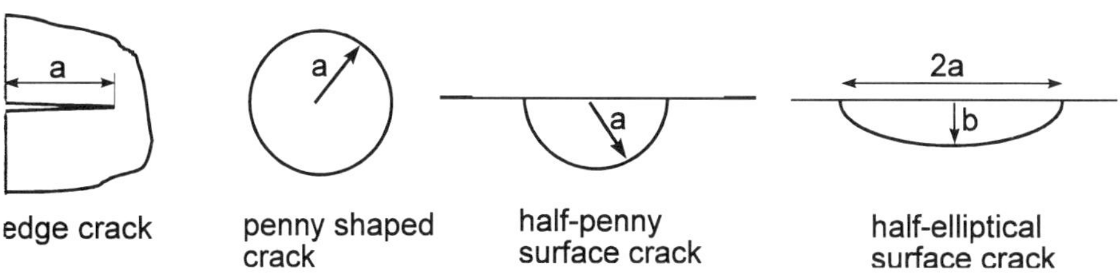

Fig. 8: Crack geometries

Internal penny-shaped crack: The K_I value for a penny-shaped crack of radius 'a' in an infinite plate, subjected to a constant tensile stress σ, see Fig. 8, is given by

$$K_I = \frac{2}{\pi} \sigma \sqrt{\pi a} \tag{2}$$

Fracture causing penny shaped defects can be located near the specimen surface. The free surface effects on K_I for the internal penny and elliptically shaped crack are not considered.

Internal elliptical crack: For an elliptical crack, K_I is not constant along the crack front. For the interpretation of fracture causing defects it is sufficient to supply formulae for the areas of maximum K_I values, at the end points of the half axes a and b, see Fig. 9.

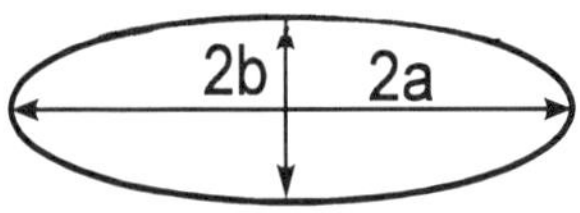

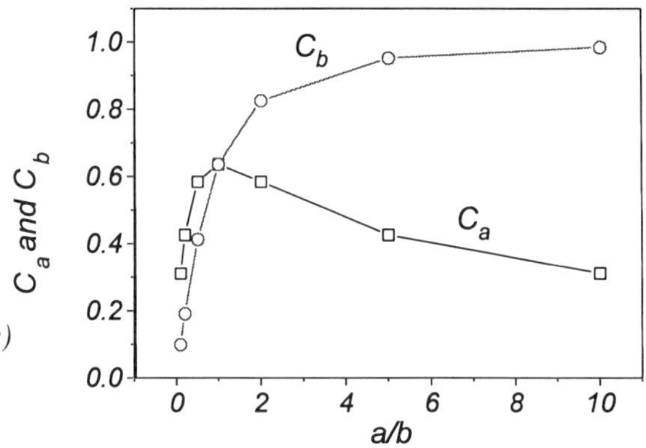

Fig. 9: Elliptical crack (above) and coefficients C_a and C_b as a function of a/b (right)

For an internal elliptical crack with half axes a and b in an infinite plate subjected to a constant tensile stress σ the K_I values at the endpoints of the half axes a ($K_{I,a}$) and b ($K_{I,b}$) are given by

$$K_{I,a} = C_a \sigma \sqrt{\pi b} \quad \text{and} \quad K_{I,b} = C_b \sigma \sqrt{\pi b} \tag{3}$$

C_a and C_b as a function of the ratio a/b are shown in Fig. 9.

Half-penny shaped surface crack: For a half-penny shaped surface crack K_I varies also along the crack front. Here it is sufficient to supply formulae for the extreme K_I values, which are reached at the deepest point (minimum) and at the surface (maximum). K_I for a half-penny shaped surface crack with radius a under constant tensile stress σ is given by

$$K_I = C \sigma \sqrt{\pi a} \tag{4}$$

with C = 0.77 for the surface and C = 0.65 for the deepest point.

Half-elliptical surface crack: For the half-elliptical surface crack, see Fig. 8, the extreme values of K_I are calculated for the deepest point and the surface. The calculation is carried out in two steps: At first K_I is calculated for the corresponding elliptical crack. The influence of the free surface is then considered by multiplying the K_I values at the surface by a factor of 1.21 and the K_I value at the deepest point by 1.02. These factors are the ratios between the K_I value for an internal circular and a half-penny shaped surface crack at the corresponding points. Whether the maximum of K_I is reached at the surface or at the deepest point for this crack geometry depends on the ratio of a/b.

The fracture mechanics approach used here includes the already mentioned approximations with respect to free surface effects and near surface influences. But one has to keep in mind, that the most extensive approximations in a fracture mechanical description of fracture causing flaws are made in the assignment of randomly shaped natural defects to well defined theoretical flaws. The aim of a combination of fractography and fracture mechanics, as it is done here, is not a calculation of stress intensity factors and crack resistance values within an accuracy of a few percent. Rather the aim is a better interpretation and understanding of the fracture events on a microscopic scale.

Surface Defects and Maching Damage under Bend and Residual Stress Loads

As will be shown later, the machining induced compressive stresses may influence the fracture behaviour of e.g. bend bars and may lead to incorrect flaw sizes, if they are not taken into account. In order to get a better correlation between defect size and machining a fracture mechanics approach is used, which considers both the externally applied stress and the residual stresses for a calculation of the size of the so called "substitute crack" for a given material/machining system.

Stress situation: The stress situation and the fracture mechanical method are explained for a bend bar. The method can easily be transfered and applied to any other testing configuration. The stresses $\sigma_{xx}(y)$ in a specimen are divided into two parts (see Fig. 10): The elastic stress field due to bending $\sigma_b(y)$ and the residual stress field $\sigma_r(y)$ due to machining. The variation of σ_b is given by $\sigma_b(y) = \sigma_0(W\text{-}2y)/W$, the variation of $\sigma_r(y)$ is measured by X-ray diffraction and fitted by analytical functions. The total stress σ_{res} is given by the linear superposition of σ_b and σ_r:

$$\sigma_{tot}(y) = \sigma_b(y) + \sigma_r(y) \tag{5}$$

Stress intensity factor: The basic flaw used here is the edge crack (surface scratch) in a semi-infinite body as shown in Fig. 8. Because the typical size of the flaws will be smaller than 100 microns and the bending bar is about 3 millimetres high, the

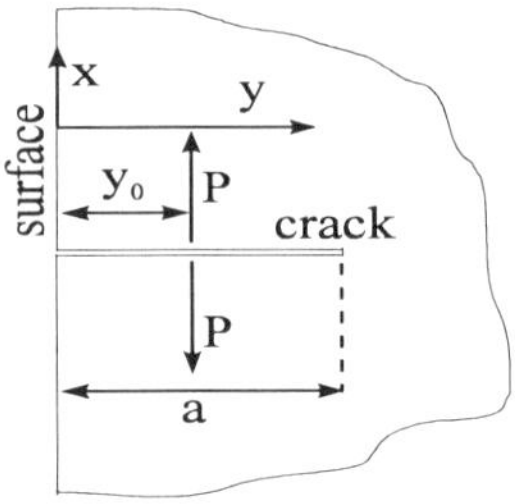

Fig. 10: Surface scratch in a semi-infinite body subjected to a normal load P

flaw can be regarded as being in a semi infinite plate. To calculate the stress intensity factor for any stress distribution $\sigma_{tot}(y)$, an approximate Green´s function given by [14] is used. The stress intensity factor K_I for a surface scratch in a semi-infinite disc under a normal point load P at y_0 is

$$K_I = \frac{2}{\sqrt{\pi a}} P \frac{1}{\sqrt{1 - (y_0/a)^2}} F(y_0/a) \tag{6}$$

where F is given by

$$F(y_0/a) = 1.3 - 0.3\,(y_0/a)^{5/4} \tag{7}$$

The following integration must be carried out to calculate the stress intensity factor $K_{I,tot}$ for a stress distribution $\sigma_{tot}(y)$:

$$K_{I\,tot} = \frac{2}{\sqrt{\pi a}} \int_0^a \frac{\sigma_{tot}(y)}{\sqrt{1 - (y/a)^2}} F(y/a)\, dy \tag{8}$$

Because the surface flaws are much smaller than the specimen height, the variation of the bending stress σ_b in the y-direction can be neglected, and so is σ_b = const. = σ_0 (σ_0 = characteristic bending strength). But the variation of the residual stress σ_r in y-direction is strong and has to be considered. The total stress is then given by

$$\sigma_{tot}(y) = \sigma_0 + \sigma_r(y) \tag{9}$$

Fracture criterion and calculation of defect size: Fracture occurs when $K_I \geq K_{Ic}$. The bending stress for fracture is taken as σ_0 from bending experiments, the residual stress $\sigma_r(y)$ is determined as described above by X-ray diffraction and K_{Ic} is determined in a four point bending test with precracked specimen. Together with Eqs. (5) and (8) this leads to a fracture criterion of the form

$$0 \leq K_{Ic} - \frac{2}{\sqrt{\pi a_c^*}} \int_0^{a_c^*} \frac{\sigma_0 + \sigma_r(y)}{\sqrt{1 - (y/a_c^*)^2}} F(y/a_c^*)\, dy \tag{10}$$

The crack size a_c^* is the only unknown parameter in this equation and it is solved numerically. The crack size a_c^* can be regarded as the typical lower bound estimation of the crack size ($a_c^* \leq a_{c,natural}$) at the surface of the bending bar in the considered material/machining system. Because the residual stress resulting from machining is usually compressive in the near-surface region ($\sigma_r(y) < 0$), a calculation without considering the residual stress would underestimate the crack size.

Variation in crack geometry: So far an edge crack was used to model the surface defects in the material. One reason for choosing this crack geometry is, that the stress intensity factor can be calculated much easier than for a half penny surface crack for example. For a half-penny surface crack the situation gets more complicated for two reasons: At first there are no Green´s functions available for calculating K_I for any given stress situation and secondly the stress intensity factor varies along the crack front. But sometimes the machining cracks might better be described by half penny surface cracks than by surface scratches. For this case the following approximation is used. The stress intensity factor K_I^{hpsc} for a half penny surface crack under constant load σ is given by:

$$K_I^{hpsc} = \sigma \sqrt{\pi a}\, G(\theta) \tag{11}$$

where G is given by

$$G(\theta) = \frac{2}{\pi}\,(1.211 - 0.186\,\sqrt{\sin(\theta)}) \tag{12}$$

The maximum of K_I is reached at the surface ($\theta = 0°$) and the minimum at the deepest point ($\theta = 90°$) of the crack front. The stress intensity factor of a surface scratch (edge crack) K_I^{ec} at constant load σ is given by Eq. (1). The stress intensity factors for a half penny surface crack K_{Itot}^{hpsc} loaded by $\sigma_{tot}(y)$ are determined by multiplying the stress intensity factor K_{Itot} for a surface scratch loaded by σ_{tot} with the correction factors C_0 (K_I at surface) an C_{90} (K_I at deepest point):

$$C_0 = K_I^{hpsc}(\theta = 0°)\ /\ K_I^{ec} \approx 0.687 \tag{13}$$

$$C_{90} = K_I^{hpsc}(\theta = 90°)\ /\ K_I^{ec} \approx 0.58 \tag{14}$$

This method of calculating the stress intensity factors of half penny shaped surface crack is not a straight forward, self consistent fracture mechanics approach, but it is sufficient in this case. The most simplifying assumption in this model is, that random shaped defects are simulated by well defined cracks. The most important topic for calculating the stress intensity factors is, that the strong stress gradient of the residual stress has to be considered because the compressive stresses nearly vanish within the first one hundred microns.

RESULTS AND DISCUSSION

Longitudinal Ground Si_3N_4

The strength distributions of differently longitudinally ground 4-point bend bars are shown in Fig. 11 [7]. Several specimens were investigated after fracture with respect to the fracture causing defects. The datapoints where failure was due to maching defects are indicated by 'M'. Since most other fracture causing defects were production related volume flaws, they are not indicated additionally. That the most machining defects are found with series indicated by the open triangles, can not be explained yet. Because this series exhibits an extreme bimodal distribution, where the machining defects belong to the lower part, one can assume, that during the grinding process some problems arose. The series with the distribution with highest fracture stresses has only material defects as fracture origins. The Weibull diagrams show a significant dependence on the grinding parameters in spite of the fact, that for all series mainly material flaws are responsible for failure. This may be explained by the machining induced micro cracking (as a consequence of the micro-plastic deformation due to the machining), which is not accessible by fractography and which influences the failure in addition to the material flaws by an interaction with them, and which will be understandable, if the depth distribution of the micro strains are considered (see later in this chapter).

The characteristic strengths σ_0 and Weibull moduli m are compiled in the first row of Fig. 12 together with the residual stresses (second row), the roughness values (third row) and the micro-plastic deformation (last row). Shown are in the first column the influence of the specific material removal rate Q'_w for a constant grinding wheel speed of v_c = 30 m/s and a constant infeed (depth of cut) a_e = 1.5 mm, and in the second column the influence of the infeed a_e for grinding at a constant specific material removal rate of 15 $mm^3/(mms)$ and v_c = 30 m/s. The influence of the wheel speed v_c at a constant specific material removal rate of 15 $mm^3/(mms)$ is shown by comparing the second and third column.

Increasing the specific material removal rate only by increasing the tangential feed rate (see Fig. 12) increases the depth of cut at the individual grits and consequently the roughness and the damage. The determined values of R_a and R_z are in agreement with the expected effect of the grinding parameters. Significant compressive residual stresses are created transverse to the grinding direction (TG), whereas parallel to the grinding direction (GD) almost no residual stresses are present.

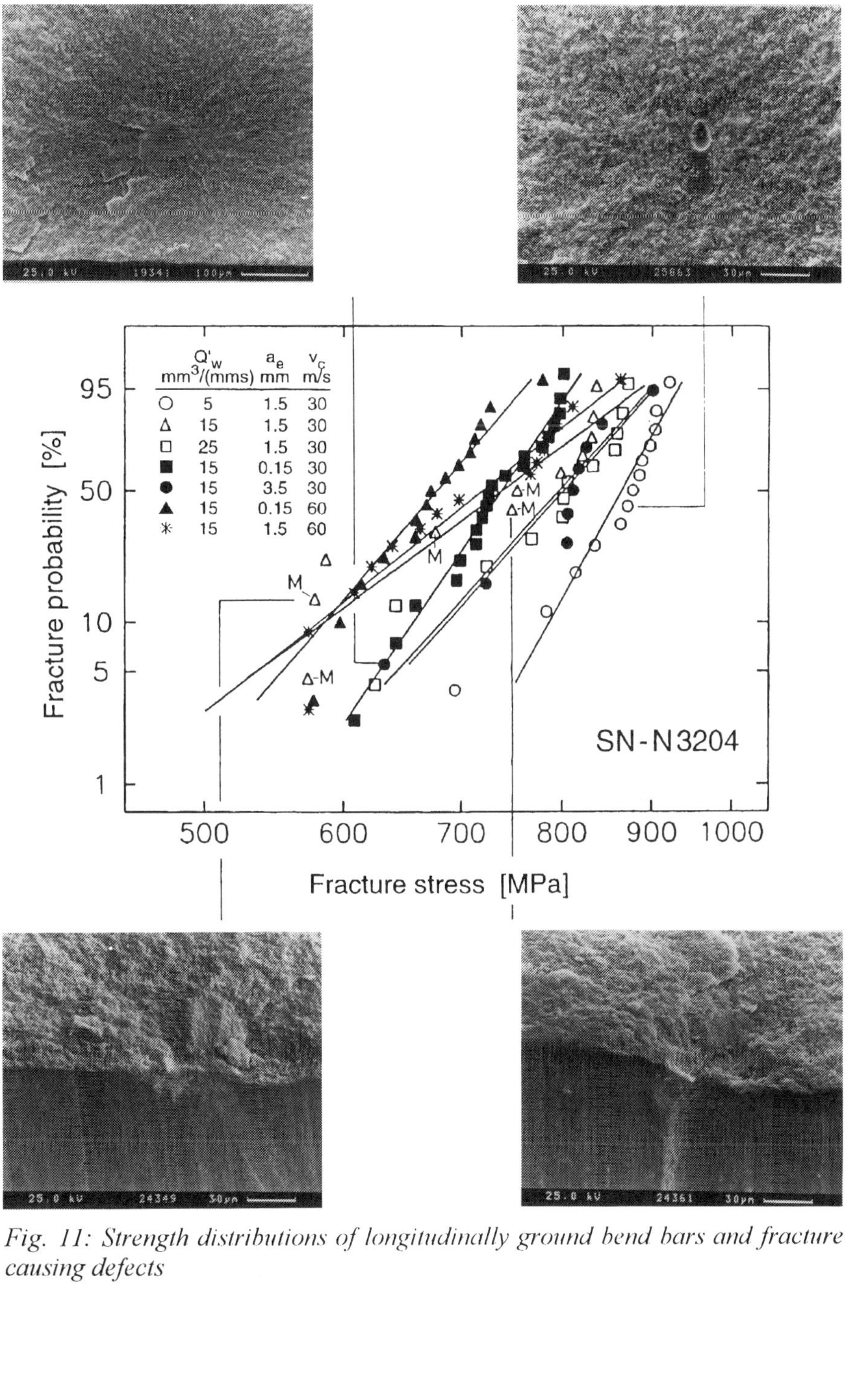

Fig. 11: Strength distributions of longitudinally ground bend bars and fracture causing defects

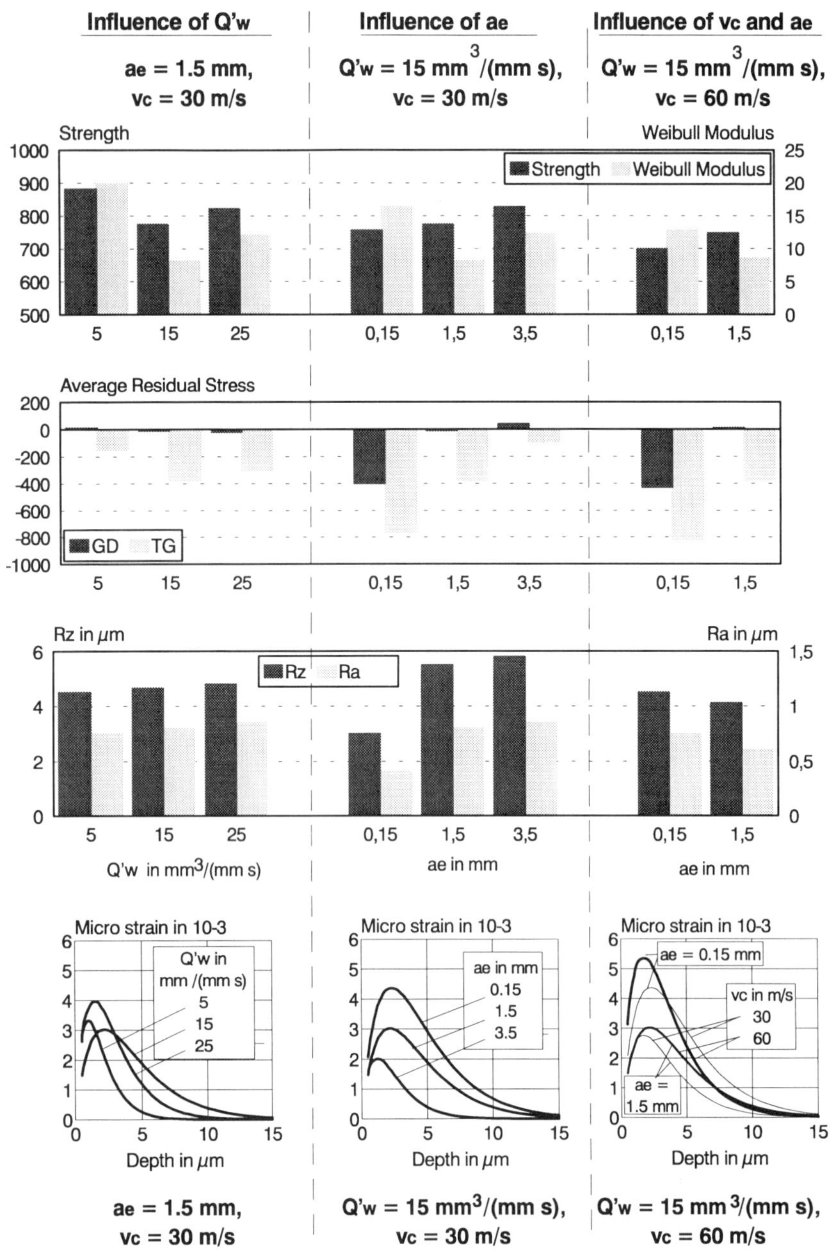

Fig. 12: Correlations between strength, residual stress, roughness and micro strain (damage) and grinding parameters for longitudinally ground bend bars

The highest near-surface influence is evaluated for the medium specific material removal rate of 15 $mm^3/(mms)$: both, compressive residual stresses and depth of micro-plastic deformation are highest. Only at the very near surface the high material removal rate induced the highest deformation. The depth of the micro-plastic deformation (≈damage) corresponds to the strength behaviour: The deeper the damage zones reach, the lower are the strength data.

Changing the grinding conditions for a constant specific material removal rate of 15 $mm^3/(mms)$ from conventional conditions (a_e = 0.15 mm) to creep feed conditions (a_e = 3.5 mm) (see second column in Fig 12), the depth of cut at the individual grits (and by this roughness and damage) is commonly assumed to decrease. The values determined for R_a and R_z indicate, however, the opposite effect, both values are increased for creep feed grinding. The SEM micrographs (see Fig. 5) show large amounts of macro-plastic deformation at the conventional ground surface. This visual impression suggests, that the conventional grinding operation has smeared material into the grinding grooves reducing the roughness of the surface. This is confirmed by the micro-strain and residual stress states. Large compressive residual stresses are created especially transverse to the grinding direction for low infeeds.

The strong dependency of residual stresses on the grinding direction is in agreement with the assumption of a significant contribution of macroscopic plastic deformation in the surface region after conventional grinding. The depth distributions of micro strain show the machining induced micro-plastic deformation of the near surface regions. Both, the amount and depth of micro-plastic deformation are increased from creep feed grinding conditions to conventional grinding conditions. The increase of compressive residual stresses and micro-plastic deformations for conventional grinding conditions, bear out the expected increase of the grinding forces at the individual grits, which corresponds to the decrease of the strength values. Thus, relative high damage and low strength values can be expected in case of conventional grinding, although material defects are responsible for fracture.

As shown in the third column, the increase of the speed of the grinding wheel v_c from 30 m/s to 60 m/s results in a small decrease in strength for both infeeds (0.15 mm and 1.5 mm), if the corresponding values are compared. The residual stresses remain unchanged, but the roughness values R_z an R_a change in a contradictory fashion, these increase for a_e = 0.15 mm and decrease for 1.5 mm. Whereas for the infeed of 1.5 mm the strength reduction corresponds to the increase of damage, the situation for the infeed of 0.15 mm is unclear, because the curves of the micro strains cross each other.

The strength distributions of 4-point bend bars, which are ground transversally with different parameters keeping Q'_w = 15 mm^3/(mm s) constant, are shown in Fig. 13. The low strength values and the deep scratch like machining defects demonstrate the different fracture behaviour of these specimens. As often obtained [1] and as indicated from the strong direction dependency of the residual stresses (see Fig. 12, second row) the bending strength of transversally ground samples is substantially lower than the strength of longitudinally ground samples. There is no clear tendency of the influence of the infeed of strength, but there may be some beneficial effect on strength, if high speeds of the grinding wheel are used.

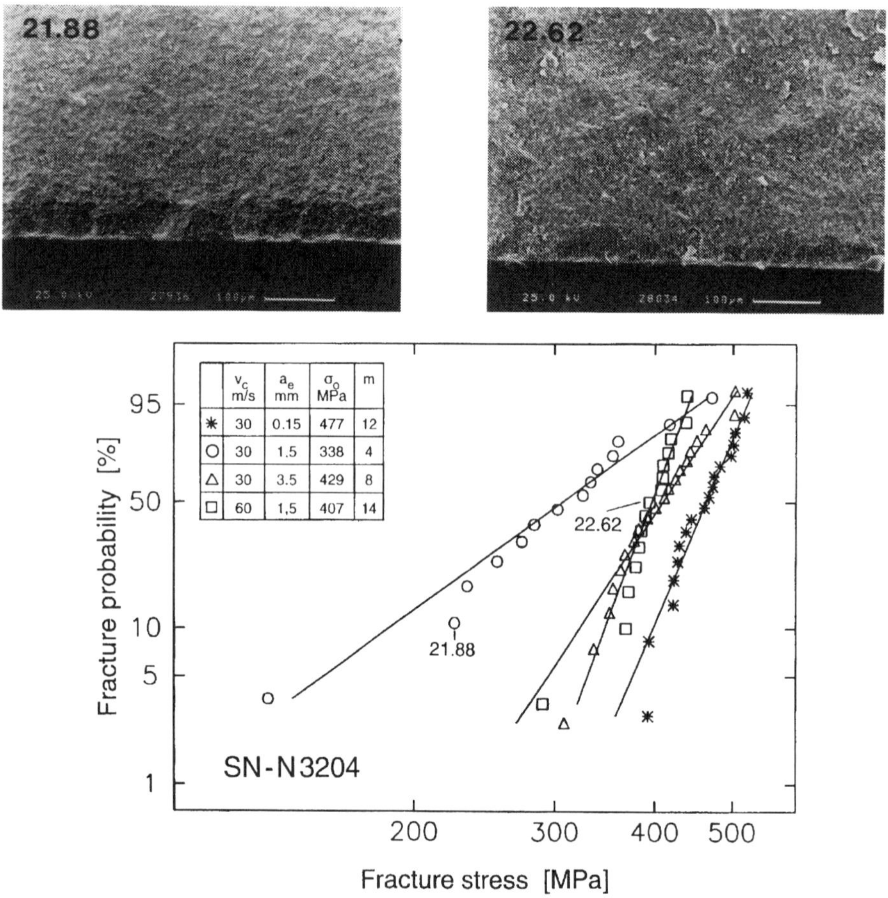

Fig. 13: Fracture stress distribution of transversally ground bend bars at constant Q'_w *= 15* mm^3*/(mms)*

If one compares for $Q'_w = 15\ mm^3/(mms)$ the variation of infeed from 0.15 to 3.5 mm the effect of strength increasing compressive residual stresses is negligible in the case of the longitudinally ground samples and only the damage is responsible for the strength level (see Fig. 12). In case of the transversally ground samples (Fig. 13) the opposite effects of the beneficial compressive residual stresses, which should have been of significant influence here, and the damage may be the reason for the unexpected low strength values of creep feed ground samples.

For both, longitudinal and transversal grinding unexpected strength values occur for the series ground with $a_e = 1.5$ mm and $Q'_w = 15\ mm^3/(mms)$, see Fig. 11 and 13. Since the residual stresses and damage values can explain these values, they may be attributed to some machining problems like e.g. grinding machine resonance.

Fracture Toughness Evaluation, Silicon Nitride

In Tab. I the results of a lot of fractographically investigated specimens are compiled. They are evaluated using Eqs. (1) to (4) according to the fracture mechanical model described above. If one assumes, that fracture begins at the location of the highest stress intensity along the crack front, one can evaluate for all specimens of Tab. I a mean value of the critical stress intensity factors for these small natural and machining defects, which is $K^*_{Ic} = 3.9\ MPa\sqrt{m}$ (standard deviation: $0.75\ MPa\sqrt{m}$). The star represents natural or maching defects. The scatter of the values including a wide range of different shapes and types of crack is relatively small. The last two high values in Tab.I around $5\ MPa\sqrt{m}$ of the transversally ground specimens may be explained by the high machining induced compressive residual stresses of 780 MPa, which have to be compensated through the bending stresses. The influence of the residual stress values of the longitudinally ground specimens is not significant for the evaluated toughness values here, because the residual stresses parallel to the grinding direction (= tension direction) are much lower and the steep residual stress gradient leads to negligible values above about 8 μm. If other formulae for the calculation of the stress intensity factor, which consider the stress gradients due to bending, were used (see e.g. [15, 16]), the fracture toughness values would differ only slighly, and would lie certainly within the standard deviation.

The critical stress intensity factor of the natural and machining defects is surprisingly lower than the material parameter $K_{Ic} = 5.0\ MPa\sqrt{m}$ determined using bend bars with sharp cracks introduced by the bridge method (saw cut ≈0.8 mm, total crack length ≈2.2 mm, a/W≈0.5. This may be explained by different reasons. Firstly, an R-curve effect can yield critical stress intensity factors, which may lie for a Si_3N_4 as investigated here about 1 to 2 $MPa\sqrt{m}$ higher than K_{Ic}.

Table I. Data for evaluation of critical stress intensity factors, Silicon Nitride

spec. no.[1)]	flaw type[2)]	σ_c MPa	σ_r[3)] MPa	a µm	b µm	K_{Ia} MPa$\sqrt{m}$	K_{Ib} MPa$\sqrt{m}$	defect type[4)]	see Fig.
11.05	iec	694	+10	28	23	4.0		G	
11.06	iec	897	+10	11	6	3.1		G	
11.10	iec	871	+10	28	9	3.8		G	11
11.11	hesc	814	+10	52	8	4.8	1.7	I	7
11.12	iec	885	+10	21	5	3.1		P	
11.14	iec	903	+10	9	5	2.8		G	
12.03	iec	872	-20	5	54	3.5		G	
13.01	hpsc	578	-10					M	11
13.03	iec	832	-10	19	7	3.4		G	
13.08	hpsc	677	-10	15		3.3		M	7
13.08	hesc	677	-10	27	11	2.6	3.4	M	7
13.10	iec	831	-10	10	4	3.8		G	
13.11	hpsc	749	-10	20		4.6		M	4
13.14	hesc	572	-10	110	29	3.3	5.9	M	4
14.01	psc	464	+40	70		4.4		G	11
14.04	iec	582	+40	27	12	2.9		G	
15.05	iec	729	-400	22	16	3.6		G	
15.15	hesc	608	-400	52	40	5.2	4.8	M(?)	
16.04	psc	779	-430	13		3.1		P	7
16.10	iec	597	-430	50	42	4.6		G(I)	
21.10	ec	326	-340	33		3.7		M	
21.33	ec	333	-340	29		3.5		M	
21.88	ec	225	-340	61		3.5		M	3
22.31	ec	290	-350	43		3.8		M	
22.32	ec	407	-350	18		3.5		M	
22.62	ec	394	-350	32		4.4		M	
22.92	ec	438	-375	23		4.2		M	
23.03	ec	308	-90	31		3.4		M	
23.11	ec	411	-90	21		3.7		M	
24.11	ec	392	-780	56		5.8		M	
24.66	ec	467	-780	33		5.3		M	

1) first number '1': longitudinally ground, first number '2': transversally ground
2) hesc: half-elliptical surface crack, iec: internal elliptical crack, hpsc: half penny shaped surface crack, ec: edge crack, psc: penny shaped crack
3) residual stresses in the surface layer, acting in tensile direction, weighted mean values over a penetration depth of 8 µm
4) P: pore, G: grain, I: inclusion, M: maching defect

Secondly, the natural and machining defects as measured on the fracture surfaces here are underestimated due to an exclusion of some process zone of cracked or damaged material around the crack, which is not visible on the SEM pictures. Thirdly, linear elastic fracture mechanics as applied above is not applicable, because the microstructural parameters (grain size, aspect ratio of grains, grain size distribution) do not allow for a continuum mechanical description, see e.g. [17]. In addition, one should remember, that the influence of machining on the strength for the longitudinally ground specimens can be explained by the micro-plastic deformation, but that fracture occurs overwhelming at material defects.

Fracture Mechanical Evaluation of Cracks, Silicon Nitride

To clarify (especially for the transversally ground specimens) the influence of the strength relevant effects of the grinding induced residual stresses and damage, they were quantified by the fracture mechanically calculated flaw sizes a_c^* using Eq. (10). In Fig. 14, left, the depths of such 'flaws' a_c^* are shown, that are calculated with the effect of residual stresses included. These 'flaws' represent the actual machining induced damage. From these calculations it can be concluded, that grinding with the highest infeed (a_e = 3.5 mm) induces only shallow flaws parallel to grinding direction, whereas deeper flaws are induced with smaller infeeds. The grinding flaws and their depths, which can be assigned as surface scratches or edge cracks and which act as fracture causing defects in the strengths experiments, are found also fractographically (see Figs. 3 and 13 and Tab. I). They correspond well to the evaluated cracks see Fig. 14.

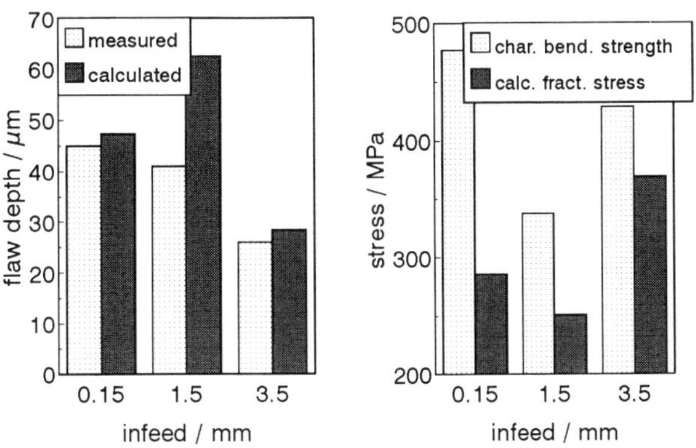

Fig. 14: Depths of fracture causing calculated and measured flaws of transversally ground samples (left), and comparison of measured characteristic bending strengths with calculated fracture stresses (right)

The difference between the measured characteristic bending strengths σ_0 and the calculated fracture stresses σ_{calc} using $K_{Ic} = 3.9$ MPa$\sqrt{m}$

$$\sigma_{calc} = K_{Ic} / (1.12 \sqrt{\pi a_c^*}) \tag{15}$$

(Fig. 14, right) can be regarded as the contribution of the compressive residual stresses to the strengths of the specimens. The largest contribution of machining induced residual stresses is evaluated for the conventionally ground samples ($a_e = 0.15$ mm). Due to the high compressive residual stresses these samples exhibit the best transverse bending strength although relative large flaws were generated. The grinding conditions with medium and large infeeds cause deeper flaws or less strength increasing residual stresses, respectively, leading to lower bending strengths than the conventionally ground samples.

Circumferential Grinding of Al_2O_3

Tensile strength: The influence of cylindrical grinding on the tensile strength of round bars of Al_2O_3 is shown in Fig 15. The 'smoothest' machining with the lowest feed rate of v_{ft}= 0.08 mm/s and with grinding grooves parallel to the tensile axis yields lowest surface damage and consequently highest strength values, and the fracture causing defects are throughout volume flaws ('V' in Fig. 15). For the rougher grinding conditions with higher feed rates and grinding grooves transverse to the tensile direction mainly machining defects 'M' could be determined as fracture origins (see Fig. 16) and the strength values are up to nearly one third lower.

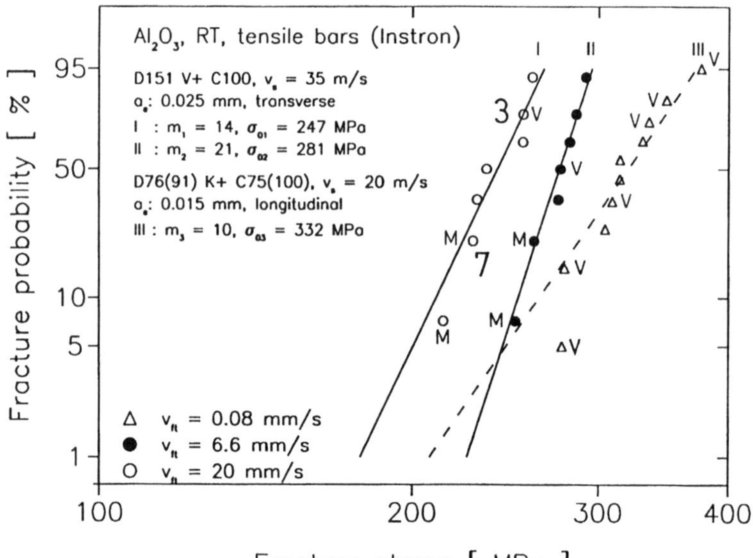

Fig. 15: Strength distribution of differenly ground round tensile bars of Al_2O_3; SEM photos of specimens 3 and 7 are shown in Figs. 16 and 17

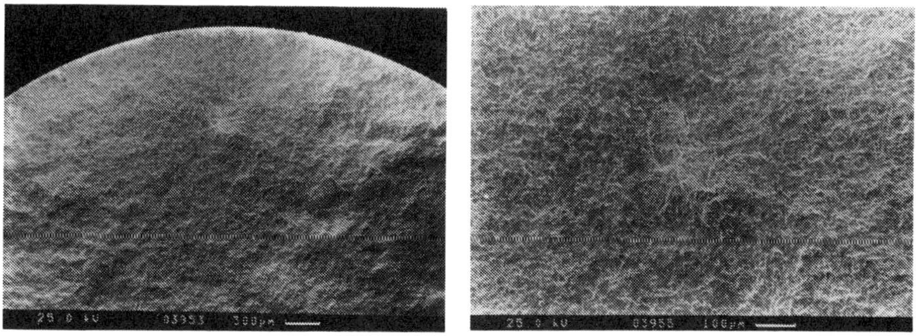

Fig. 16: Fracture causing volume defect (agglomerate) of a tensile specimen

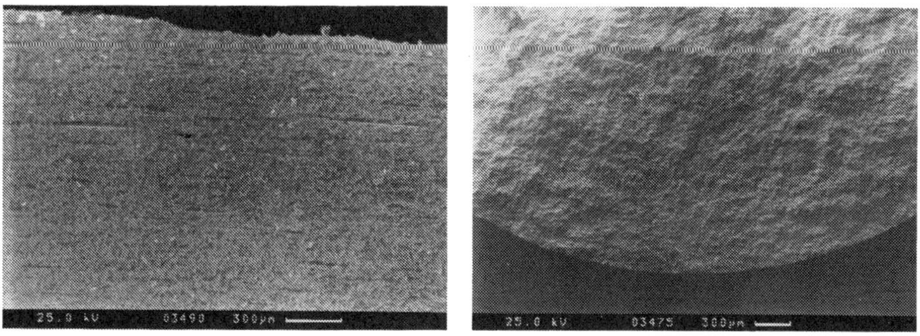

Fig. 17: Fracture causing grinding groove

Maching induced micro strain: The determination of the distribution of microscopic strains as shown in Fig. 12 is very time consuming. Therefore a more easily determinable mean value of the machining induced micro strains of the surface layer was used here for differently ground and lapped Al_2O_3 plates [10]. These mean micro strains are correlated in Fig. 18 with the flaw depths, which are calculated using Eqs. (9) and (10). As for the Si_3N_4 it can be seen, that conventional grinding leads to higher and creep-feed grinding to lower damage and to lower and higher strength values, respectively (see Fig. 19). Also for Al_2O_3 as one would expect, lapping with coarse grains (F150, grain size 100 µm) leads to higher damage and lower strengths, whereas lapping with fine grains (F500, grain size 16 µm) leads to lower damage and higher strengths.

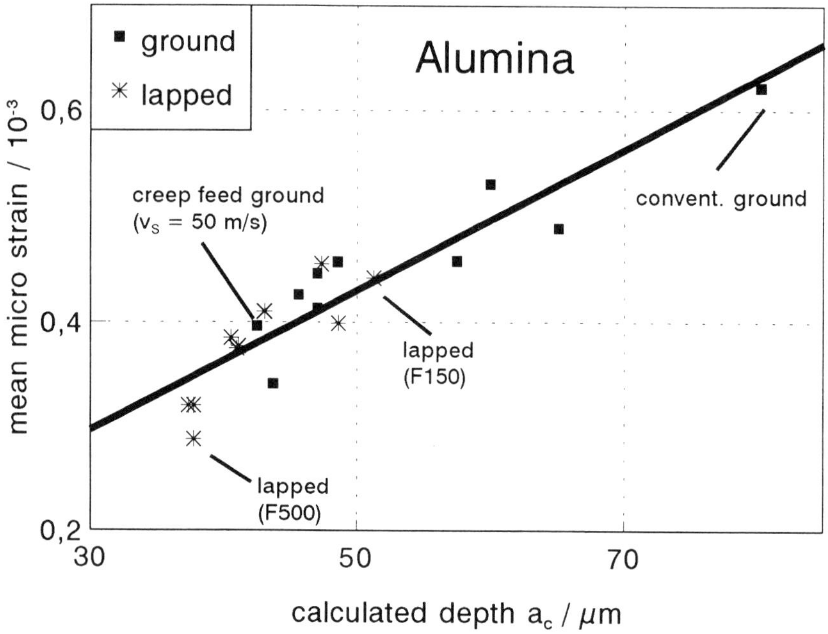

Fig. 18: Micro strains and fracture mechanically evaluated flaw sizes

Effect of residual stresses: With the calculated flaw sizes a_c (taking the residual stresses into account) and with the appropriate $K_{Ic} = 2.4$ MPa$\sqrt{m}$ for this Al_2O_3 the critical fracture stress values σ_c were calculated. Because of their theoretical relation σ_c is plotted versus $1/\sqrt{a_c}$. These values are compared in Fig. 19 with the experimentally determined characteristic bending strength σ_0. The plot shows, that for this Al_2O_3 the contribution of machining induced compressive stresses to the bending strength is small and that the bending strength is dominated by the machining induced damage.

Prediction of strength: It is possible to predict the surface strength of machined components (here: of the tensile specimens), if the residual stresses and the micro strains, from which the crack sizes can be evaluated, are measured [10]. For the tensile specimens these values are compiled in Tab. II. The predicted and the measured strength values, which are plotted also in Fig. 19, are in very good agreement. That the strength values of the tensile specimens are lower than those of the double ring specimens, may be a consequence of the volume effect. A consideration of this (the volume of the tensile specimen is 13 times larger than that of the double ring specimens) would lead to about 11% higher fracture stresses, and the prediction of strength would then even become better.

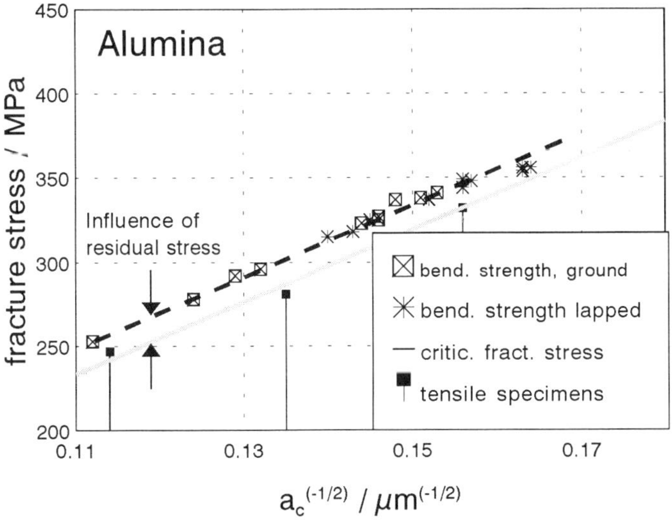

Fig. 19: Strength - damage diagrams showing the effect of machining induced compressive residual stresses on the strength of ground and lapped Al_2O_3 (bend strengths determined with double ring specimens)

Table II: Data for the strength evaluation of the tensile specimens

specimen series[1)]	residual stress MPa	micro strain[2)]	calculated flaw size[2)] µm	predicted strength MPa[3)]	characteristic strength MPa
1	-30	3.7×10^{-4}	41	350	332
2	-60	4.6×10^{-4}	54	303	281
3	-110	6.1×10^{-4}	70	257	247

1) see Fig. 15; 2) see Fig. 18; 3) see Fig. 19

SUMMARY

The strength controlling surface effects of differently machined ceramics have been assessed using experimentally determined bending strengths, residual stresses, depths of micro-plastic deformation, calculated flaw depth, calculated effective fracture stresses and fractographically determined fracture causing defects and using linear elastic fracture mechanics. Quantitative information of machining induced damage can be evaluated from the deformation depth, determined by advanced non-destructive X-ray diffraction methods. These methods additionally allow the determination of machining induced residual stresses and to decide whether damage is the strength controlling effect or whether the strength increasing effect of residual stresses has to be taken into account. Using the results of bend tests, a quantitative assessment of the effect of damage and residual stresses on strength can be evaluated by fracture mechanics based calculations. The depths of the evaluated cracks determined accordingly correspond to those found fractographically.

Furthermore with measured values of machining induced residual stresses and micro strains (damage) and with the materials fracture toughness one can predict the surface strength of a machined component, as it was shown for differently ground round Al_2O_3 tensile bars. The investigations show, that on basis of calculated crack sizes and calculated fracture stresses, 'unexpected' influences of machining parameters on strength can be clarified, if the effect of residual stresses is taken into acount. Thus, effective guidelines for optimized machining parameters can be evaluated. Fracture toughness values calculated for the Si_3N_4 from the fractographically found fracture causing material and machining defects lie below that value determined with sharp cracks. This may be attributed to an R-curve effect, or to a damage or cracked process zone that are not detectable by fractography or to limits of linear elastic fracture mechanics.

REFERENCES

[1] R. W. Rice, The effect of grinding direction on the strength of ceramics, in: The science of ceramic machining and surface finishing II, Nat. Bureau of Standards Spec. Publ. 348, U.S. Governm. Print. Office, Washington D.C. (1972) 365-376

[2] C. A. Andersson, R. J. Bratton, Effect of surface finish on the strength of hot pressed silicon nitride, The science of ceramic machining and surface finishing II, Nat. Bureau of Standards Spec. Publ. 562, U.S. Governm. Print. Office, Washington D.C. (1979) 463-476

[3] R. W. Rice, J. J. Mecholsky, The nature of strength controlling machining flaws in ceramics, in: The science of ceramic machining and surface finishing II, Nat. Bureau of Standards Spec. Publ. 562, U.S. Governm. Print. Office, Washington D.C. (1979) 351-378

[4] B. J. Hockey and R. W. Rice (Eds.), The science of ceramic machining and surface finishing II, Eds. , Nat. Bureau of Standards Spec. Publ. 562, U.S. Governm. Print. Office, Washington D.C. (1979)

[5] S. Jahanmir (Ed.), Machining of advanced materials, Proceedings of the International Conference on Machining of Adsvanced Materials, Nat. Inst. of Standards and Technologie, Spec. Publ. 847, U.S. Governm. Print. Office, Washington D.C. (1993)
[6] G. Spur, Keramikbearbeitung, Carl Hanser Verlag, München, Wien (1989)
[7] G. Warnecke, T. Hollstein, W. König, G. Spur, H.K. Tönshoff (Hrsg), Schleifen von Hochleistungskeramik (Werkstoff, Anwendung, Bearbeitung, Qualität), Verlag TÜV Rheinland, Köln, 1994, ISBN 3-8249-0236-2
[8] G. Warnecke, J. Wimmer, M. Zapp, FBK, Universität Kaiserslautern, persönliche Mitteilungen (1994)
[9] DIN 51 110, Teil 1: Prüfung von keramischen Hochleistungswerkstoffen. 4-Punkt-Biegeversuch bei Raumtemperatur, Deutsches Institut für Normung, Berlin (1990)
[10] Pfeiffer, W., Hollstein T., Damage determination and strength prediction of machined ceramics by X-ray diffraction techniques. In: Machining of advanced materials, Proceedings of the International Conference on Machining of Adsvanced Materials, Nat. Inst. of Standards and Technology, Spec. Publ. 847, U.S. Governm. Print. Office, Washington D.C. (1993) 235-245
[11] Pfeiffer, W., Hollstein, T., Sommer, E., Strength properties of surface-machined components of structural ceramics. Fracture Mechanics, 25th Vol., ASTM STP 1220, Amer. Soc. for Testing and Materials, Philadelphia, USA, 1994
[12] MIL-HDBK-790, Fractography and characterization of fracture origins in advanced structural ceramics, U.S. Department of Defense, July 1992
[13] Y. Murakami, Stress intensity factors handbook, Pergamon Press (1986)
[14] G. Sih, in: The Stress Analysis of Cracks Handbook, H. Tada, P. Paris, G. Irwin (Eds.), Paris Productions Inc., 1973, 8.3
[15] R. C. Shah, A. S. Kobayashi, Stress intensity factor for an elliptical crack approaching the surface of a plate in bending, Stress analysis and growth of cracks, Proceedings of the 1971 Nat. Symp. on Fracture Mechanics, Part I, ASTM STP 513 1972, 3-21
[16] Y. Murakami, Stress intensity factors handbook, Pergamon Press (1986) 822
[17] S. Usami, H. Kimoto, I. Takahashi, S. Shida, Strength of ceramic materials containing small flaws, Engineering Fracture Mechanics, 23, 1986, 745-761

ACKNOWLEDGEMENT

Grinding of the Si_3N_4 samples was performed by J. Wimmer and M. Zapp, University of Kaiserslautern (FBK), Germany. The Si_3N_4 was supplied by the Cremer Forschungsinstitut GmbH, Rödental, Germany, and the Al_2O_3 by FRIATEC AG, Mannheim, Germany. Part of the investigations were sponsored by the Bundesministerium für Forschung und Technologie (BMFT).

FRACTOGRAPHY USED WITH LIFETIME PREDICTION TESTS ON COMMERCIAL GRADES OF ALUMINA AND SILICON CARBIDE

Jakob Kübler, Jarmila Woodtli, Karl Berroth
Swiss Federal Laboratories for Materials Testing and Research (EMPA)
Überlandstrasse 129
8600 Dübendorf, Switzerland

ABSTRACT

Fractographic investigations were conducted on alumina and α-SSiC bend bars from a round robin test comparing a static and a dynamic lifetime test method. The investigations revealed that not all fracture origins found on α-SSiC specimens had contact with the ambient medium and therefore the dynamic lifetime tests did not measure the materials' behaviour under stress in water.

In addition to the conventional SEM fractography an alternate technique, using an auto focus scanning laser profilometer was applied for characterising the fractures. This tool is very promising for fractography because of the ability to handle topographic data by imaging procedures and fractal analysis.

INTRODUCTION

Subcritical crack growth can occur in brittle materials at stress levels below that required to cause instantaneous failure. This effect may be caused by the test environment or by intrinsic crack propagation behaviour of the material. The determination of crack growth parameters can be done with different methods. Therefore a static and dynamic test method were compared in a DKG (German Ceramic Society) round robin test using eight different ceramic materials. The static test method states that the scatter in the lifetime of a sample depends up on the critical flaw size distribution and therefore the scatter in lifetime also correlates to the initial strength scatter. Due to this statistical correlation it is possible to evaluate the lifetime parameters. The dynamic test method states that the scatter in the lifetimes of samples depends on their loading history. Using different loading rates will lead to different stress levels. Therefore the lifetime parameters of a material can be found by varying the loading rate and measuring the obtained stress level. For both test methods, it is necessary to ensure that only relevant flaws limit the measured strengths and lifetimes.

Fractography was not required for the round robin test. The Swiss Federal Laboratories for Materials Testing and Research (EMPA) as a participant decided to investigate some fractured specimens of the commercial grades of alumina and silicon carbide. Therefore this paper is not focused on the comparison of the test methods, but it will demonstrate how fractographic findings can help to clarify the fracture mechanism and thus help to understand measured subcritical crack growth parameters. Additionally it will give information about potential for material optimisation by detection of specific flaws.

MATERIAL, TESTING, AND RESULTS

A commercial grade of each alumina and silicon carbide were tested and are the subject of this paper. Table I. gives some of the relevant material properties taken from the suppliers data sheets additionally figures 1 and 2 show the microstructures.

Table I. Material properties (supplier data)

Material	Density [g/cm³]	Strength [MPa]	Weibull modulus	K_{IC} [MPa √m]
Al_2O_3 - 998	3.86	350 [1) ; 342 [1,3)	21.0 [1,3)	3.5
α-SSiC	>3.06	400 [1) ; 394 [1,3)	>10 ; 10.5 [1,3)	3.0 [2)

1) 4-point bending 2) Chevron notch 3) measured in round robin

From the two materials, about 500 bend bars with a size of 3 x 4 x 45 mm, were prepared by grinding in accordance to standard ENV 843-1. The bend bars were used afterwards by EMPA and its partners (Forschungszentrum Karlsruhe, Karlsruhe, Germany; Bundesanstalt für Materialforschung und -prüfung, Berlin, Germany; BASF, Ludwigshafen, Germany) in the round robin test. First the inert strength in air at room temperature was measured with the four point bending test in accordance with ENV 843-1 on at least 20 specimens from each material. With those strength values, the Weibull-parameters were calculated with the maximum likelihood method. Figure 3 gives the Weibull-plots and table I the characteristic strength and the Weibull-modulus of the two materials. The table shows that the characteristic strength of the Al_2O_3 - 998 is in the range given by the supplier but the Weibull-modulus is unexpected high for this type of material.

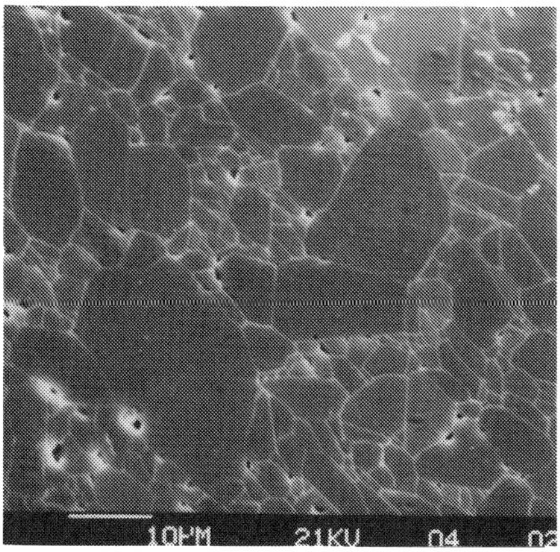

Figure 1
Microstructure of Al_2O_3 - 998

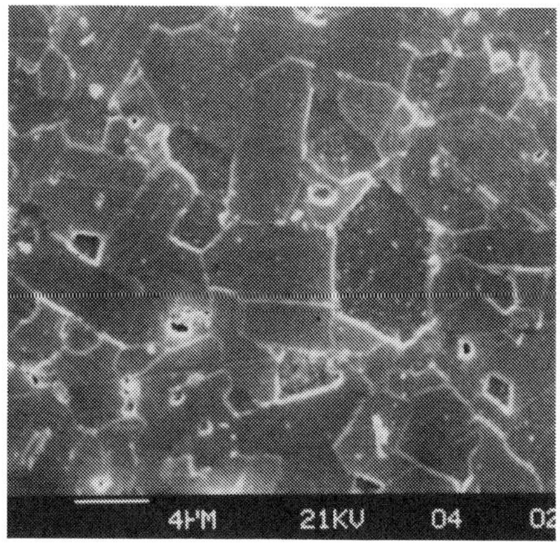

Figure 2
Microstructure of α-SSiC

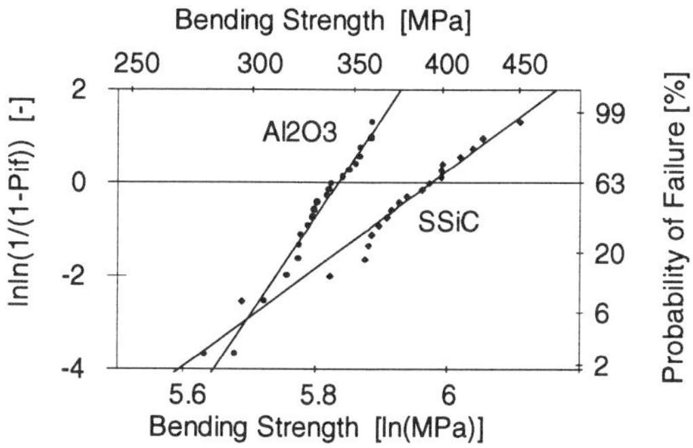

Figure 3
Weibull-plot

After the measurement of the inert strength σ_c, the strength was measured at different stress rates varying by at least three orders of magnitude using 10 specimens per rate. Also the lifetimes of 20 specimens from each material at at least two stress levels was measured. The dynamic and the static tests were conducted in demineralised water (= 100 % relative humidity) at room temperature. The principles of the dynamic and static lifetime test method together with the calculation of the subcritical crack growth parameter A and n (A and n are material constants) using the relation

$$\frac{d\upsilon}{dt} = A \cdot K^n \qquad (1)$$

are described in ENV 843 [1] respectively by Fett and Munz in [2]. Figure 4 reveals that the dynamic lifetime test method showed no significant stress rate dependence for the α-SSiC due to the high scatter (negative slope !). The stress rate dependence of the Al_2O_3 - 99 measured with the dynamic lifetime test method (n = 62) is higher than measured with the static lifetime test method in this round robin (n = 51) and the rate measured for example by Fett at al. [3] for an identical material (n = 48). Figure 5 shows the normalised stress diagram for the Al_2O_3 - 998.

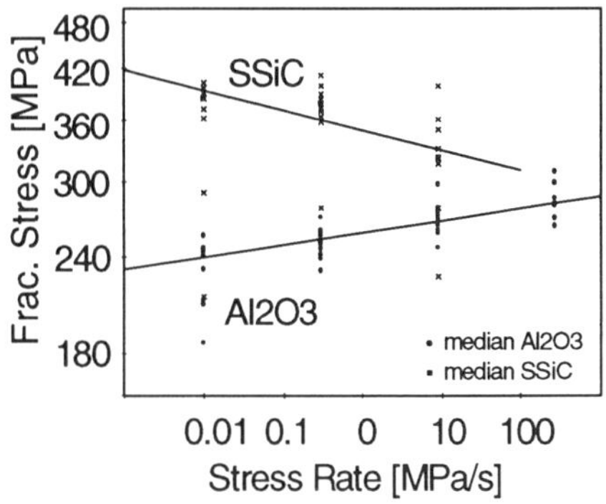

Figure 4
Stress rate vs. strength diagram for Al_2O_3 - 998 and α-SSiC (dynamic test diagram)

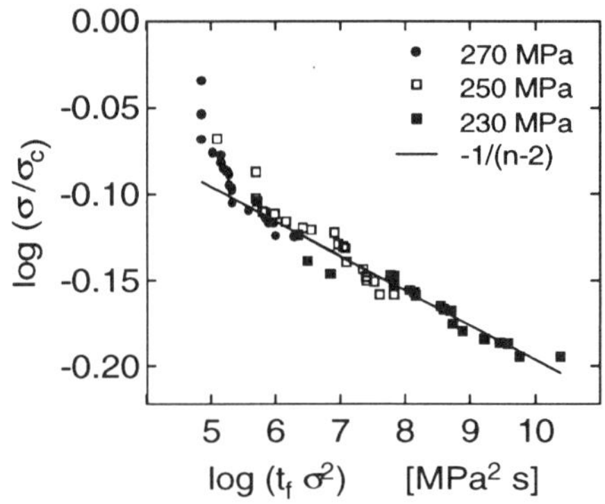

Figure 5
Normalised stress diagram calculated from the inert strengths and lifetimes of Al_2O_3 - 998

FRACTOGRAPHY AND DISCUSSION

Because the two test methods found different subcritical crack growth rates for the alumina and no crack rate for the α-SSiC, EMPA conducted fractographic investigation at some of the 500 specimens tested. Using fractography, it is possible to find fracture origins and also to check if the flaw types (for example pores or cracks) and location (volume or surface) are the same for both test methods. Especially the relevancy for final failure can be obtained. Even though this investigation was recommended by Fett and Munz [2], not many papers about lifetime prediction mention fractography.

Inert strength tests in air

The fractographic results showed that the initial strength of both commercially available materials are primarily controlled by large volume defects. The figures 6 and 7 illustrate that the defects are related to the raw materials and the process. In both materials granules or fragments of granules shells can be seen quite frequently. This indicates, that such granules, usually produced by spray drying are probably not completely destroyed by uniaxial or cold isostatic pressing. The presence of hard granules are attributed to hardening of organic binder during the spray drying process. Also hardened shell fractions of hollow spheres, which often occur in spray dried granules, are observed to be the fracture origins. The defects are randomly distributed in the material and therefore they also occur at the ground surface of the test bars.

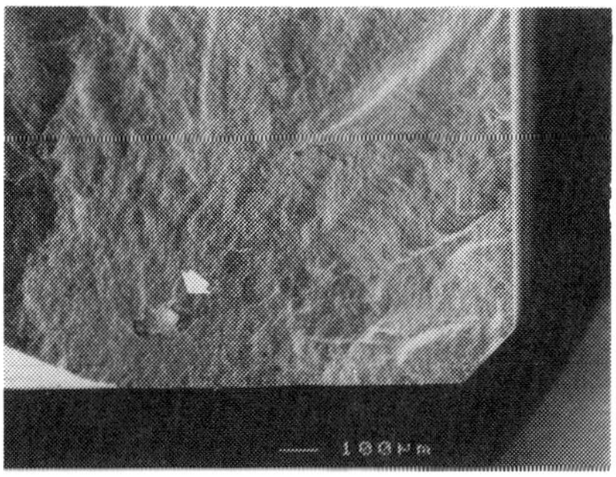

Figure 6
Typical volume flaw near the surface in Al_2O_3 - 998 used in the inert strength tests (292 MPa).

Figure 7
Typical volume flaw in α-SSiC used in the inert strength tests (315 MPa).

With
$$K_{IC} = \sigma \cdot Y \cdot \sqrt{c} \qquad (1)$$

where σ is the failure stress, Y the shape factor for a penny shaped flaw in the interior (Y = 1.13), and c the radius of the failure-initiating flaw the fracture toughness K_{IC} was calculated in accordance to MIL-HDBK-790 [4]. For the specimens shown in figure 6 and 7 a fracture toughness of about 3.5 to 4.0 MPa √m for the alumina and of about 3.0 to 3.5 MPa √m for the α-SSiC were calculated. The K_{IC}-values correspond well with those given in table I. Therefore it can be assumed that a failure-initiating flaw is not open towards the surface as long as the flaw is not touching the specimen surface. In such cases, the media has no direct access to the flaw and can not influence crack growth by chemical reactions.

Dynamic strength tests in water

Figure 4 demonstrates the low (negative !) stress rate dependency of the α-SSiC and on the other hand the high stress rate dependency of the alumina. Since these specimens were contaminated with masking tape, only reduced fractographic information could be gained. After thermally removing the glue from an α-SSiC specimen (stress rate 0.01 MPa/s) a small crack system open to the surface was found (figure 8).

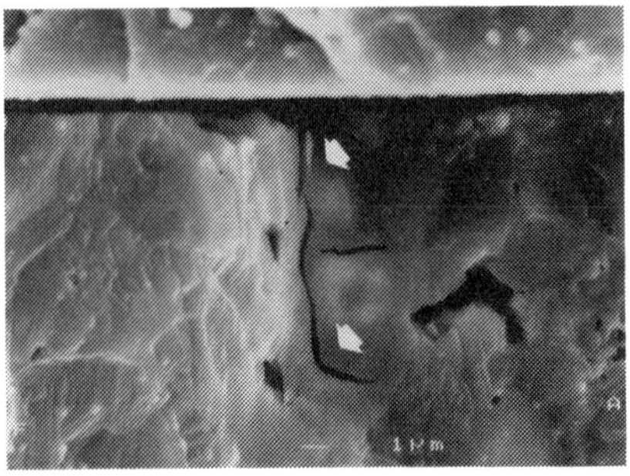

Figure 8
Fracture origin on α-SSiC after thermal removing of the masking tape contamination (stress rate 0.01 MPa/s, 390 MPa).

A fracture surface of an other α-SSiC specimen tested at a stress rate of 0.3 MPa/s that failed at a stress of 373 MPa is shown in figures 9 and 10. The defect, a large agglomerate of granules is located 120 μm away from the tensile surface. The relatively large distance to the tensile surface indicates that it had probably no contact with the water. This is the reason why this specimen should not be used for the calculation of the subcritical crack growth parameters. Consequently the graph for α-SSiC in figure 4 is questionable and does not represent the material's behaviour under stress in water. Moreover, this incidental finding demonstrates that all fractures should be evaluated fractographically, to get the right interpretation of measured values.

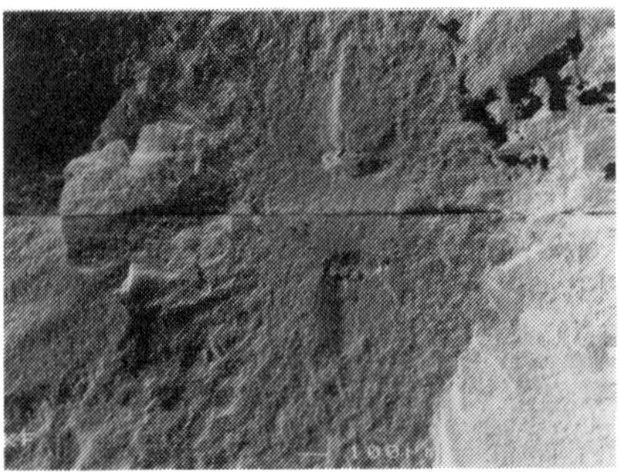

Figure 9
Fracture surface of an α-SSiC specimen broken at 373 MPa (stress rate 0.3 MPa/s).

Figure 10
Detail from figure 9. Agglomerate of granules, which lead to the failure of the specimen.

<u>Static strength tests in water</u>

For the static lifetime tests a time limit of 168 h was specified for the round robin.

Al_2O_3 - 998: Fractography on alumina specimens taken from static lifetime tests in water was conducted for specimens with the highest (figure 11) and lowest lifetimes. Only flaws touching or located near the tensile surface of the bend bars

initiated the fractures. In the specimen with the highest lifetime, the initial crack shape (crack shape before static stress test in water) and the final crack shape (crack shape after subcritical crack growth; just before final rupture) could be detected. Figure 11 shows the fracture surface with the marked crack shapes. For the verification of the crack shapes, the fracture toughness was estimated. For the considered final crack shape together with the stress at fracture (230 MPa) a fracture toughness of 3.6 MPa $\sqrt{m}$ was calculated. Assuming the specimen with the longest lifetime had been used in the inert strength tests, it could be expected that this specimen could carry the highest stress (about 360 MPa). This fracture stress, together with the initial crack shape will give a K_{IC} value of 3.9 MPa $\sqrt{m}$. Those two values correspond well with the 3.5 MPa $\sqrt{m}$ given in table I. Consequently, the area between the marked lines in figure 11 corresponds to the zone of subcritical crack growth controlled by hydrolysis.

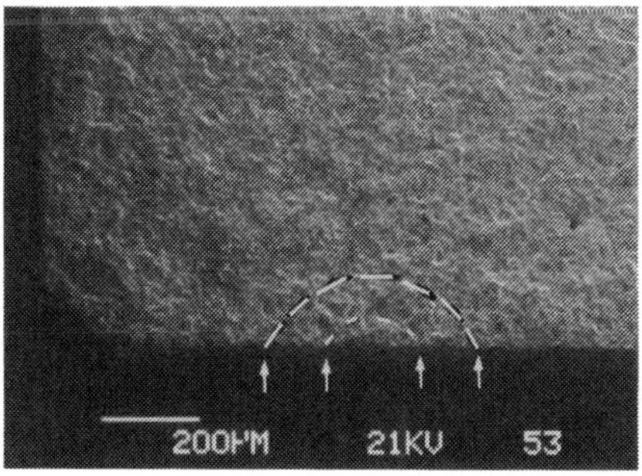

Figure 11
Initial (dashed line) and final (points) crack shape on an alumina specimen (static stress 230 MPa, water, lifetime 128 h; photo tilted 30°).

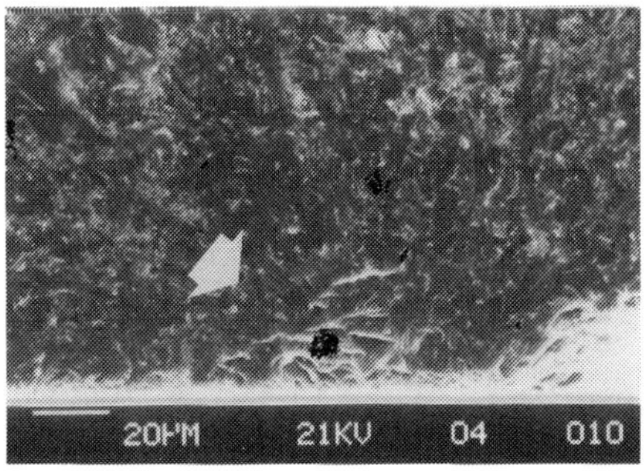

Figure 12
Origin (arrow) of delayed fracture on an α-SSiC specimen (static stress 353 MPa, water, lifetime 6 min).

α *-SSiC*: Except one specimen all other α-SSiC specimen broke immediately or survived. The only specimen with a delayed failure fractured after 6 minutes (353 MPa). As shown by BAM with results from step tests[5] also the lifetime of α-SSiC is effected by hydrolysis [6]. The examination of the specimen with the 6 minutes lifetime showed that the fracture origin was located on a side of the specimen about 200 μm away from the tensile surface. In figure 12 the crack origin is marked. Neither the initial nor the final crack shape could be clearly recognised in the overview SEM-records and due to the short lifetime no hydrolysis products could be found. Therefore the evidence of slow crack growth in this specimen is not clear. On the microscopical scale it was observed, that the area close to the flaw showed mostly a transcrystalline topography and the final fracture ran predominantly intercrystalline. This observation needs further investigation because usually the crack mechanism is described vice versa.

SCANNING LASER PROFILOMETER

In addition to the conventional SEM fractography an alternate technique, using an auto focus scanning laser profilometer (SLP) was applied for characterising the fractures. This tool normally is used for the evaluation and quantification of the surface roughness and topography. The excellent resolution (µm range in the x- and y-direction; nm range in the z-direction) combined with the possibility to scan uncoated surfaces in air as well as the digital processing of the data, make this tool very promising for fractographic tasks. Figure 13 taken with the SLP shows the same rupture surface area as figure 7. Based on this simple application, the possibilities of this technique can be demonstrated.

Figure 13
Fracture origin of in inert condition broken α-SSiC recorded with the SLP (area ~ 1.0 x 1.3 mm). The same specimen is shown in figure 7.

Figure 14
Fracture origin of an Al_2O_3 - 998 specimen broken during the static lifetime tests recorded with the SLP (area ~ 1.0 x 1.4 mm). The same specimen is shown in figure 11.

The potential of this technique occurs in figure 14 which shows the fracture origin of an Al_2O_3 - 998 specimen (same specimen as shown in figure 11) broken during the static lifetime tests. The crack and crack propagation direction can be seen distinctly in the figure. On the low magnification SEM picture these features were barely visible.

CONCLUSION

Mechanical tests, especially lifetime tests, always should be followed by a fractographic investigation. Fractography combined with the calculation of the fracture toughness clarified the fracture mechanism and helped to understand the results. It revealed that not all fracture origins found on α-SSiC dynamic lifetime test specimens had contact with the ambient medium. Therefore those specimens should not be used to calculate the subcritical crack growth parameters and the dynamic test diagram (figure 4) does not show the materials' behaviour under stress in water.

In the presented cases, it could be determined that the fracture origins were mainly flaws derived from an improper processing. So with a limited effort, the spot where to start material optimisation most effectively became rather obvious. For the α-SSiC the supplier has already optimised the spray drying process.

The auto focus scanning laser profilometer which we used for the first time is a very promising tool for fractographic tasks because of the ability to handle fracture topographic data by imaging procedures and fractal analysis.

ACKNOWLEDGEMENT

The authors thank

- Dr. W. Rieger, Metoxit AG, CH-Thayngen for the preparation of the Alumina bend bars,
- Drs. E. Rudolph and Ch. Ullner, BAM, D-Berlin, Dr. K. Keller, BASF, D-Ludwigshafen, and Dr T. Fett KfK, D-Karlsruhe for providing their test results and the fragments of the specimen,
- Dr. M. Dupont for the auto focus scanning laser profilometer measurements, and
- Mr. R. Bächtold for performing the mechanical tests.

REFERENCES

[1] ENV 843: Determination of subcritical crack growth parameters from constant stressing rate flexural strength tests, Third draft, Document CEN/TC: 184 /WG3 /N163, 1992

[2] Fett T. and Munz D., "Methods of Determining Crack Growth Data by Static Lifetime Tests with Natural and Artificial Cracks", Journal of Testing and Evaluation, JTEVA, Vol. 19, No. 6, Nov. 1991, pp. 461-466

[3] Fett T., Hartlieb W., Keller K. and Munz D, "Subcritial crack growth in high-grade alumina for container applications", Technical Report 87-09, Nationale

Genossenschaft für die Lagerung radioaktiver Abfälle, Baden / Schweiz, Sept. 1987
[4] Military Handbook "Fractography and Characterisation of Fracture Origins in Advanced Structural Ceramics", U.S.A., Department of Defence, MIL-HDBK-790, 1 July 1992
[5] Ullner Ch. and Gehrke E., "Effect of Step Size in Incremental Loading Tests", Communications of the American Ceramic Society, J. Am. Ceram. Soc., 69, Nov. 1986, pp. 267-268
[6] Ullner Ch., Beitrag zum 5. Gemeinschaftsversuch des DGM/DKG-AK Festigkeit / Lebensdauer (2. Teil - Ergebnisse), Unpublished Paper BAM-Berlin 1.22/AH, March 1995

THE ORIGIN OF STRENGTH LIMITING DEFECTS IN A TOUGHENED SiC (HEXOLOY SX-SiC)

G. V. Srinivasan, J. Gibson, and S. K. Lau; The Carborundum Company, Niagara Falls, NY; and A. A. Wereszczak and M. K. Ferber; Oak Ridge National Laboratory, Oak Ridge, TN

ABSTRACT

HEXOLOY® SX-SiC, sintered with yttriun, and aluminum containing compounds has higher toughness and strength than α-SiC, sintered with boron and carbon (HEXOLOY SA-SiC). The room temperature fracture toughness, and strength in uniaxial tension and flexure were determined. The strength limiting defects were identified. The composition of the strength limiting defects were analyzed by elemental mapping with Scanning Auger Electron Microscopy and Secondary Ion Mass Spectrometry. The origination of such defects are discussed. The fracture mirror parameter was determined from analysis of tensile specimens fracture surfaces. The experimentally determined fracture mirror parameter is compared with the predicted value for SX-SiC and α-SiC. The local stress state at the vicinity of the defects are discussed.

INTRODUCTION

Pressureless sintered α-SiC (HEXOLOY® SA) has excellent oxidation, corrosion, and wear resistance. Powder availability, price competitiveness, and net shape forming capability make this SiC material an attractive candidate for many structural applications. However, relatively lower strength (primarily due to its low fracture toughness) still represents limitations for some applications involving higher stresses. Various approaches have been considered to enhance fracture toughness and strength. The key approaches are based on the modification of microstructure and fracture characteristics of the material by the incorporation of appropriate sintering additives. Using these approaches, Carborundum company developed a SiC material, designated HEXOLOY SX-SiC, which is sintered with yttrium and aluminum containing compounds.

HEXOLOY SX-SiC had been demonstrated to possess 50 to 100% higher toughness and strength than HEXOLOY SA SiC [1]. A comprehensive study of the thermo-mechanical behavior and microstructural characteristics was conducted, and process-structure-property relationships were established under a program sponsored by Department of Energy Ceramic Technology Development [2,3]. This article will discuss the strength limiting defects at room temperature, defect forming mechanism, the state of local stresses due to the defect, the fractographic technique used, and implications for applications. The failure mechanism and defect types at high temperatures are discussed elsewhere [4].

EXPERIMENTAL PLAN

A 20 lb batch of commercial SiC powder was mixed with the appropriate amounts of yttrium and aluminum compound sintering additives and spray dried into soft, flowable agglomerates. These agglomerates were later compacted into solid rods and plates by isostatic pressing. The rods and plates were pressureless sintered to about 96% theoretical density, and post-heat treated at elevated temperature (post-treatment temperature is about 50 to 100°C less than the sintering temperature) and pressure to >99% theoretical density. The densified rods were machined into button-head-type tensile specimens and the plates were machined into bars of size 3x4x48 mm (MIL-STD-1942 "B") for flexure testing. The flexural strength was measured in a our point bend fixture with 40 mm and 20 mm outer and inner spans respectively. The fracture toughness (K_{IC}) was determined using Chevron notched specimens in flexure. The tensile strength tests were conducted at the High Temperature Materials Laboratory at Oak Ridge National Laboratory. Fractographic investigation was carried out using optical microscopy, Scanning Electron Microscope (SEM), Scanning Auger Electron Microscope (SAM), and Secondary Ion Mass Spectrometer (SIMS).

RESULTS

<u>Microstructure:</u>

The typical microstructure from a polished surface of SX-SiC is shown in Figure 1. The brighter regions in the back scattered SEM image are deduced to be yttrium aluminate(s). X-ray diffraction revealed YAG as the major second phase and YAP as the minor phase. An etched microstructure is shown in Figure 2. The SiC grains are equiaxed with a narrow grain-size distribution. The average grain size is estimated to be 1.5 µm.

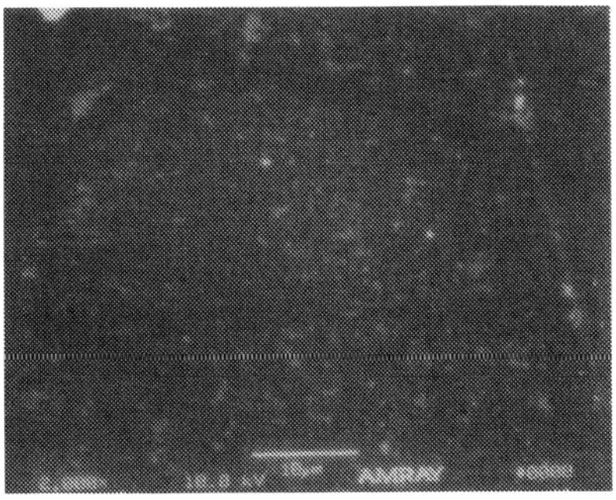

Figure 1. Microstructure from a polished surface

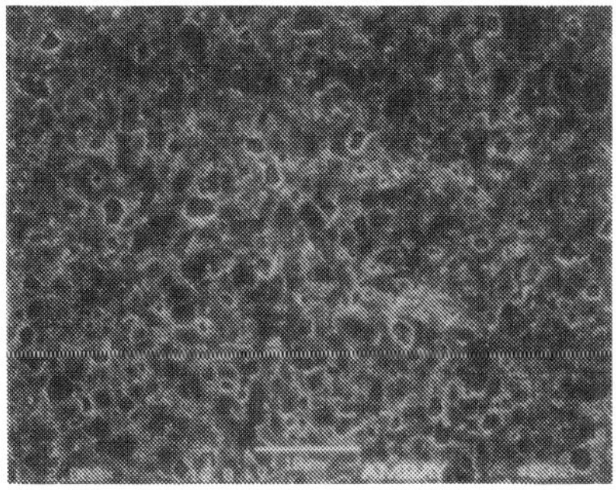

Figure 2. Microstructure from etched surface

Room Temperature Fracture Toughness and Strength:

The average room temperature fracture toughness determined from chevron notched specimens was 4.2 MPa√m. The average flexural strength (from 20 specimens) determined at room temperature was 780 MPa with a Weibull modulus of 5.3. The average tensile strength (from 20 specimens) determined at room temperature was 455 MPa with a Weibull modulus of 3.0.

The Strength Limiting Defects:

Optical fractography was performed on all the fractured flexural and tensile specimens. Selected specimens were analyzed with SEM, SAM, and SIMS to determine the nature and composition of the strength limiting defects. Figures 3a and b show the same volumetric defect from a specimen tested in uni-axial tension at room temperature as seen in an optical microscope and SEM. The same defect is shown in Figures 4a and b as observed in a Scanning Auger Electron Microscope. The tensile strength of the specimen was 589 MPa. All the tensile specimens failed from similar volumetric defects, and the strength was an inverse function of the defect size.

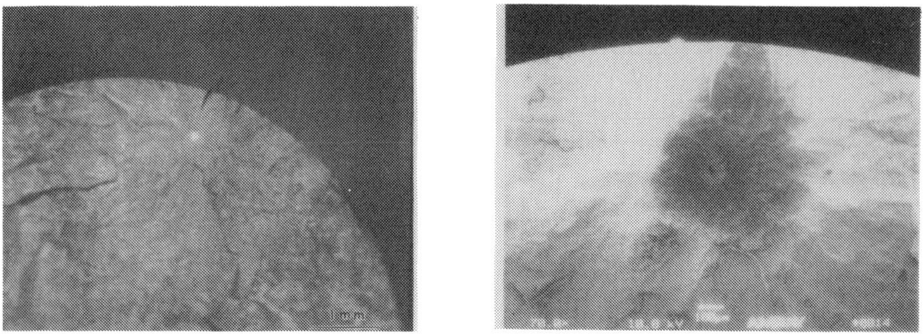

a b

Figure 3. The fracture origin as observed in (a) optical microscope and (b) SEM. Tensile strength = 589 MPa.

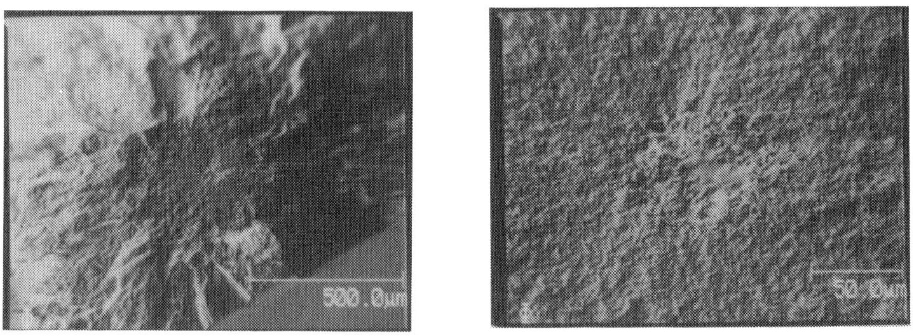

Figure 4. The same strength limiting defect as in Figure 3 as observed in a Scanning Auger Electron Microscope

The compositions of the defects were analyzed by elemental mapping with Scanning Auger Electron Microscope. The elements (silicon, carbon, aluminum, and oxygen) were mapped around the defect and are shown in Figures 5a-d. SIMS was found to be better for the yttrium mapping than Auger, hence, SIMS elemental maps of silicon, carbon, aluminum, and yttrium of the same defect were obtained. These maps are shown in Figures 6a-d. The presence of yttrium can be seen very clearly. Since an oxygen beam was used as the probe, map of oxygen was not obtained. From these analysis it appears that the defect consists of elemental silicon surrounded by compounds containing yttrium, aluminum, and oxygen.

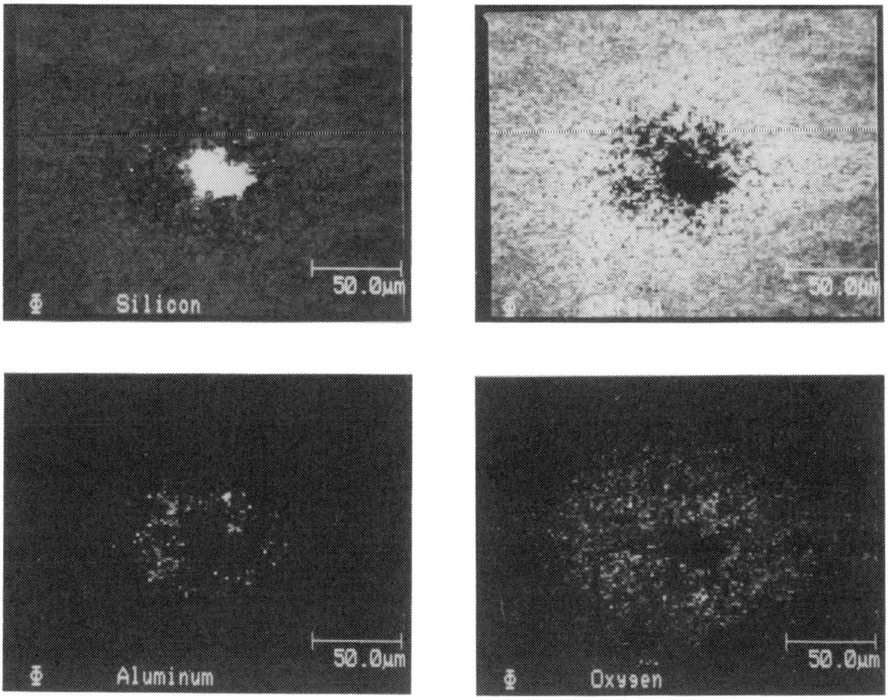

Figure 5. Elemental mapping of the defect region as in Figure 4.

Auger spectra peak shape indicates that the silicon is in the elemental state. Moreover no other element was present in the pool of silicon as shown in Figures 5 and 6. The state of aluminum (whether elemental or compound) could not be clearly ascertained because of its low signal level. The presence of elemental silicon could also be detected by the metallic luster when observed under an optical stereo microscope, as shown in Figure 3a. The strength limiting defects from all the fractured tensile specimens had this luster of elemental silicon and hence it was concluded that the strength limiting defects were the same in all the tensile specimens.

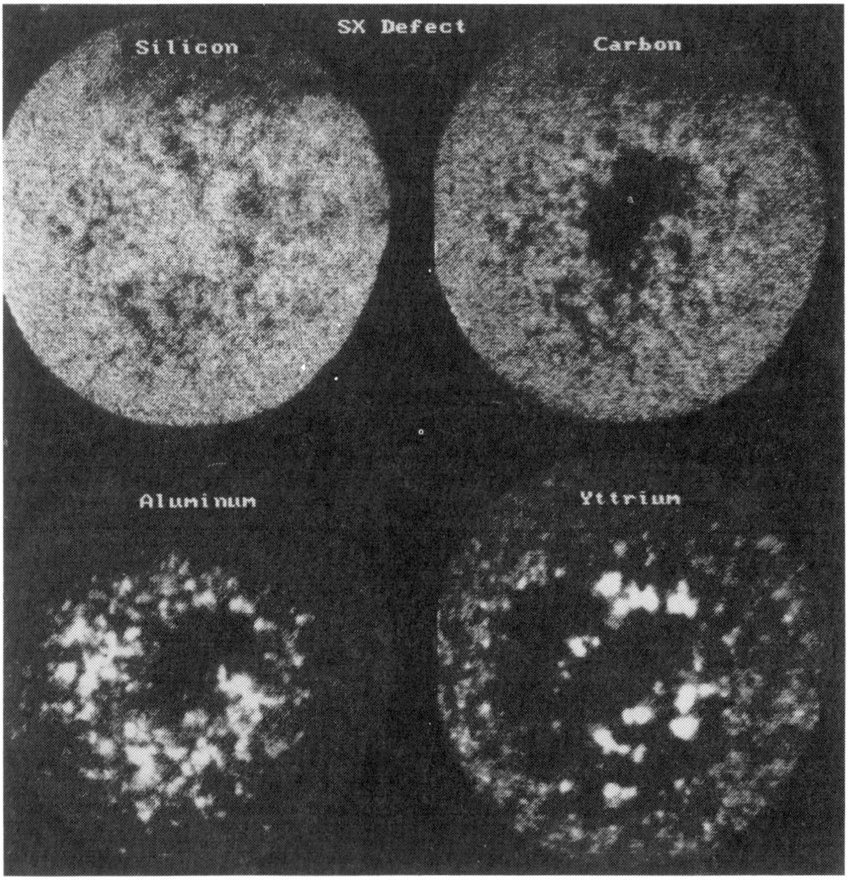

Figure 6. SIMS elemental mapping of the defect shown in Figures 3 and 4.

Seventy percent of the flexural specimens tested failed due to volumetric defects and the rest failed from surface defects characteristic of machining damage. A typical volumetric defect observed the in flexural specimen is shown in Figures 7a and b (SEM back scattered image). The flexural strength of this specimen is 907 MPa. Elemental mapping of the defect is shown in Figures 8a-d. From the mapping, it is clear that the defect consists of elemental silicon, and compound(s) containing aluminum, oxygen, and possibly yttrium. The defect observed in the flexure bar appears to be the same as that observed in the tensile specimens.

However, the size of the defect is much smaller because of the smaller effective volume.

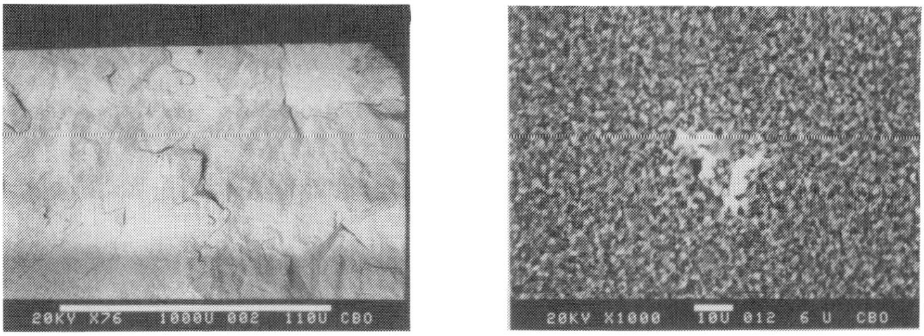

(a) (b)

Figure 7. Fracture origin in a flexure bar. Flexure strength = 907 MPa. (a) low magnification, (b) high magnification

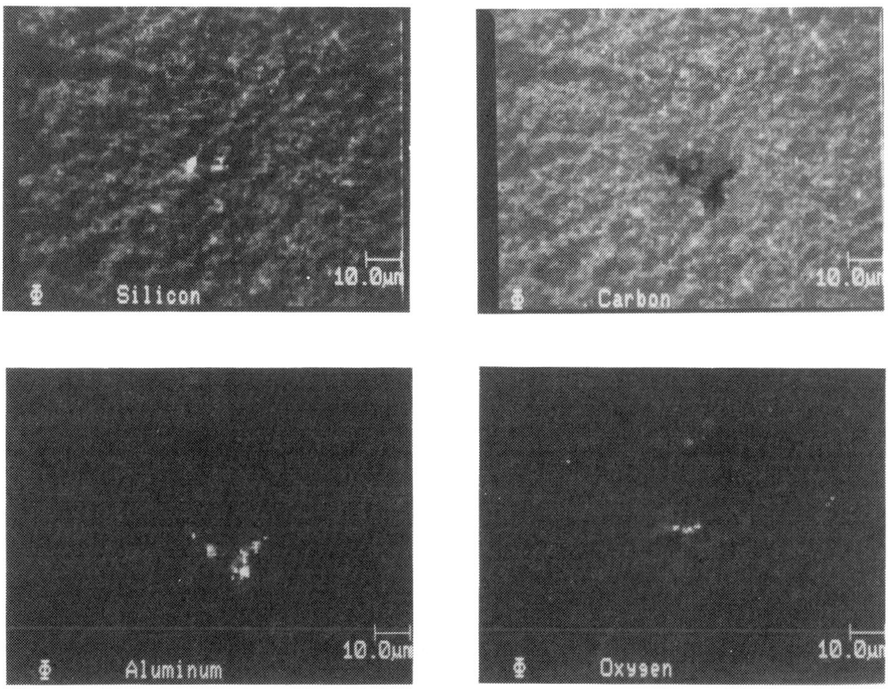

Figure 8. Elemental mapping of the defect shown in Figure 7.

Fracture Geometry Observations

The outer fracture mirror (where the crack branches) radius, r_B, was measured on all of the fracture surfaces of the tensile specimens. A plot of fracture stress (σ_f) vs. $(r_B)^{-1/2}$ was generated and is shown in Figure 9. From the linear fit, of the measured data, the slope of the line was determined to be 7.0 MPa√m.

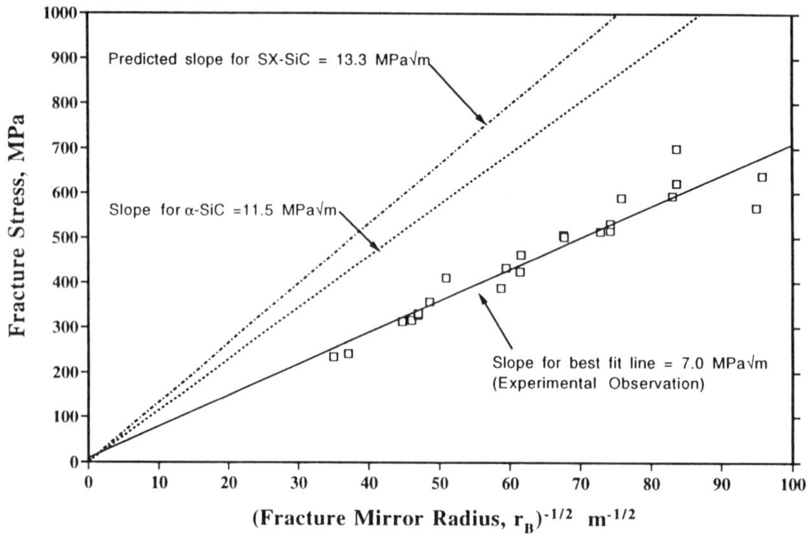

Figure 9. Fracture stress vs. fracture mirror radius.

The basic equation of the stress intensity criterion for simple tension is given as

$$K_B = Y \sigma_f r_B^{1/2} \quad \text{-------------- (1)}$$

where K_B is the stress intensity factor at crack branching and Y is a crack shape factor. The above equation 1 can be rewritten as

$$\sigma_f = A_B r_B^{-1/2} \text{ where } A_B = \frac{K_B}{Y} \quad \text{-------- (2)}$$

where A_B is termed the fracture mirror parameter. The slope of the solid line in Figure 9 (7 MPa√m), is the fracture mirror parameter for HEXOLOY SX-SiC. Figure 9 also represents plots of the theoretical line predicted for SX-SiC and the line reported for α-SiC. The significance of these lines is discussed later in this article.

DISCUSSION

The Origin of Strength Limiting Defects:

From the elemental mapping profiles of the defect it is clear that elemental silicon is present. No other element could be detected with silicon. The elemental silicon is surrounded by a ring containing aluminum, yttrium, and oxygen. It is not clear whether aluminum and or yttrium are present in the elemental or oxidized state. Aluminum may exist as elemental aluminum or as Al_2O_3 or yttrium aluminate (YAG or YAP). The starting powder did not contain elemental silicon to explain the presence of silicon in all the test specimens. Moreover, this elemental silicon is surrounded by a ring of second phase compounds or elements which are added as sintering aids. This configuration and composition suggest that elemental silicon is probably formed by chemical reaction.

Elemental silicon formation can be explained by the possible reaction of yttrium aluminate with SiC as given by

$$xY_2O_3.yAl_2O_3 + SiC \text{ ------> } xY_2O_{3-z}.yAl_2O_{3-z} + Si + CO \quad \text{-----(3)}$$

It is also possible that the yttrium aluminate may become deficient in oxygen at sintering temperatures (>1800°C) in the inert (Argon) environment oxidizing SiC to form silicon and carbon monoxide. At higher sintering temperatures, reaction 3 may proceed to yield elemental silicon and aluminum

$$xY_2O_3.yAl_2O_3 + SiC \text{ ------> } Y_2O_3 + Al + Si + CO \quad \text{------- (4)}$$

Presence of elemental aluminum and yttrium oxide were observed by X-ray diffraction when this material was sintered above 2000°C [3]. Similar reactions have been reported for the system $SiC\text{-}Al_2O_3\text{-}Y_2O_3$ by Omori and Takei [5]. It is possible that the strength limiting defects are formed by a chemical reaction of second phase material with SiC at sintering temperatures.

Stress State around the Defect:

The fracture mirror parameter (A_B) for SX-SiC was found to be 7 MPa√m (see Figure 9). In Figure 9, this fracture mirror parameter is compared to a single phase α-SiC sintered with boron and carbon additions. The stress intensity factor (K_B) at crack branching for α-SiC has been reported to be 13 MPa√m [6]. An estimate of the constant Y (crack shape factor) is needed to obtain an equivalent fracture mirror parameter for α-SiC. The defects seen in SX-SiC are far smaller than the diameter of the test specimen and hence the size effect can be ignored. From the shape of the defect, it can be approximated to a penny shaped crack for which Y=1.13 [7]. Then the fracture mirror parameter for α-SiC is 11.5. This is the slope shown for α-SiC in Figure 9.

SX-SiC is 50% tougher than α-SiC. The chevron notch fracture toughness for SX-SiC has been determined to be 4.2 MPa√m, while the chevron notch toughness for sintered α-SiC (HEXOLOY SA) has been determined to be 2.8 MPa√m [8]. The hardness of both the materials are the same. Hence the stress intensity factor at crack branching for SX-SiC is expected to be higher. The brittleness dependence of crack branching is given by Kirchner and Conway as [6]

$$K_B = MH + \sqrt{2}\,K_{IC} \quad \text{----------- (5)}$$

where M is a regression constant determined to be 2.9×10^{-4} √m, and H is the hardness of the material. From equation 5, K_B for SX-SiC is predicted to be 15 MPa√m and the fracture mirror parameter is predicted to be 13.3 MPa√m (for Y=1.13). A straight line with a slope of 13.3 is also shown in Figure 9.

The experimentally observed tensile strength appears to be lower than the predicted fracture stress for SX-SiC for a particular defect size or fracture mirror radius as shown in Figure 9. This implies the presence of local residual stress around the defect. The origination of the local residual stress can be explained from the presence of elemental silicon as the defect. silicon, like ice, undergoes a volume expansion upon solidification from melt. Based on the defect formation mechanism discussed above, elemental silicon is formed during sintering at temperatures > 1800°C. Silicon freezes at about 1420°C. When the silicon freezes, the surrounding SiC is already dense and rigid. This creates a large tensile hoop stress around the silicon and puts it into compression. The local residual stress enhances the stress concentration around the defect and the ceramic fails at lower than predicted applied tensile stress.

CONCLUSIONS

The use of elemental mapping of the strength limiting defect was very helpful in identifying and understanding the origination of such defects. The possible reaction of yttrium aluminate (the second phase material) with SiC during sintcring leads to the formation of pool of elemental silicon surrounded by other reaction products containing yttrium, aluminum, and oxygen. Residual stresses are introduced when the silicon freezes. The presence of local residual stresses are also deduced from the fracture mirror parameter measurements. The enhanced fracture toughness of SX-SiC is thus offset by the presence of local residual stresses resulting in lower than expected strength enhancement.

ACKNOWLEDGMENT

Research sponsored by the U. S. Department of Energy, Assistant Secretary for Energy Efficiency and Renewable Energy, Office of Transportation Technologies, as part of the Ceramic Technology Project of the Materials Development Program, under contract DE-AC05-84OR21400 with Martin Marietta Energy Systems, Inc.

REFERENCES

1. K. Y. Chia and S. K. Lau, "High Toughness SiC", Ceram. Eng. & Sci. Proc., 12[9-10], 1845-61 (1991).
2. G. V. Srinivasn, K. Y. Chia, S. K. Lau, R. S. Storm, M. K. Ferber, and M. G. Jenkins, "Mechanical Properties Evaluation of Generation I SX-SiC", Proc. of the 37th ASME/International Gas Turbine Institute Conference, Cologne, Germany, June 1992. ASME Publication # 92-GT-410.
3. G. V. Srinivasan, S. K. Lau, and R. S. Storm, "High Temperature HEXOLOY SX-SiC", Final Report, ORNL/DOE Publn. # ORNL/Sub/91-SG341/1, Sep. 1994.
4. G. V. Srinivasan, S. K. Lau, M. G. Jenkins, and M. K. Ferber, "Slow Crack Growth and Creep in a Toughened SiC (HEXOLOY SX-SiC)", publication in preparation.
5. M. Omori and H. Takei, "Preparation of Pressureless Sintered SiC-Y2O3-Al2O3", J. Mater. Sci., 23, 3744-49 (1988).

6. H. P. Kirchner and J. C. Conway, "Fracture Mechanics of Crack Branching in Ceramics", in Fractography of Glasses and Ceramics 1986, Advances in Ceramics, Vol. 22, p 187-214, Amer. Ceram. Soc., Westerville, OH 1988.

7. Military Handbook on Fractography and Characterization of Fracture Origins in Advanced Structural Ceramics, MIL HDBK 790, (1992).

8. G. V. Srinivasan, Carborundum Company, unpublished work, 1991.

® HEXOLOY is a registered trademark of Carborundum Company.

SPATIAL AND TEMPORAL PROBES OF FRACTURE, WEAR, AND DEFORMATION

J. T. DICKINSON, S. C. LANGFORD, SUMIO NAKAHARA, L. SCUDIERO, K. W. HIPPS, MYOUNG-WON KIM, AND NAM-SEOK PARK

Physics Department and Materials Science Program, Washington State University, Pullman, WA 99164-2814

ABSTRACT

Novel spatial and temporal probes of fracture, wear, and deformation provide important information on mechanisms for energy dissipation associated with failure and the micromechanics of crack growth. We present recent studies of a number of spatial and temporal probes that focus on revealing fracture, deformation, and wear mechanisms. These include applications of scanning microscopies, the imaging of photoluminescence, fracto-emission, chemisorptive electron emission, and transient current measurements.

Keywords: Atomic force microscopy, surface force microscopy, deformation, fracture, abrasion, wear, interfacial failure, adhesion, corrosive tribology, photoluminescence, fracto-emission, triboluminescence, charge separation, inorganic crystals, polymers, metals.

INTRODUCTION

In this paper we review recent studies of the use of scanning microscopy to probe wear mechanisms on a nanometer size scale, the application of particle emission to examining dissipative processes during tribological loading and fracture, and applications of transient current measurements to probe in a time resolved fashion highly complex phenomena associated with interfacial failure. The goal of this work is to reveal details of fracture, wear, and deformation mechanisms, in particular, aspects which relate to dissipation and damage processes.

Applications of Scanning Microscopy

Since the development of scanning tunneling microscopy (STM) and scanning force microscopy (SFM), many studies of topography on the nanometer scale have been performed. Fracture surfaces of ceramics,[1-4] polymers,[5] metals,[6] and silicate glasses[7] have been studied in some detail. On extremely flat surfaces of crystalline materials, atomic resolution has been obtained with SFM. STM of the metal phase exposed by interfacial failure of metal-ceramic composites has provided valuable information on energy dissipation during crack growth.[8] A few studies have employed SFM to probe topological changes induced by wear, e.g., on video tapes.[9] With instruments like the SFM, some materials can actually be made to wear during scanning, e.g., a soft, layered, inorganic compound,[10] and a polymer.[11]

$NaNO_3$ Studies

Here, we discuss a modification of the SFM which allows observations of wear and other surface modifications due to tribological loading (the SFM tip rubbing on the surface) in the presence of a chemically reactive environment, i.e., corrosive wear. Corrosive wear involves mechanical interactions between two surfaces in the presence of a corrosive chemical agent. The combination of tribological loading and chemical attack can enhance the rate of material removal. We first examine a model material exposed to dynamic frictional loading in the presence of a "corrosive agent".[12] The surface of interest is cleaved, single crystal $NaNO_3$, a hygroscopic salt with the rhombohedral calcite structure. The corrosive agent is water vapor. The hydrophilic nature of the salt surface favors the formation of a water layer which responds rapidly to changes in ambient humidity. $NaNO_3$ itself is quite soft (Moh hardness, 1.5;[13] Vickers hardness, 200 MPa at 50 g load[14]), and thus is vulnerable to tribological wear.

SFM is a valuable surface probe of nm-scale surface features in a variety of environments, including ambient air and many liquids. SFM has been applied to the dissolution of soluble materials in aqueous solutions[15,16] and long-term surface changes due to ambient humidity.[17,18] Adsorbed water can play an important role in SFM of soft materials in ambient air, where the force applied by the tip can be distributed over a relatively large area of adsorbed material; tip asperities presumably penetrate this layer and allow the underlying substrate to modulate the applied force.[19,20] For example, atomic resolution has been obtained on calcite immersed in a saturated $CaCO_3$ solution.[21] Our goal here is to examine and explain the changes in the surface topography (using the SFM) due to exposure of the surface to mechanical loading (with the SFM tip) in the presence of water vapor.

Figure 1 shows the experimental arrangement. SFM images were obtained with a Digital Instruments Nanoscope III SFM equipped with commercial Si_3N_4 tips with nominal radii of curvature of 40 nm. Observations were performed in both contact and tapping modes. To control the ambient humidity, the SFM apparatus was enclosed with a polyethylene "tent." Depending on the desired humidity, dry or humid air was introduced to the SFM through plastic tubing. Dry air was produced by passing compressed air through a copper coil immersed in liquid nitrogen and warming it back to room temperature by passing it through a second copper coil immersed in room temperature water. Humid air was introduced through a separate tube; high humidities were produced by bubbling compressed air through room temperature water. The air flow from the two air supplies was mixed and adjusted to maintain the desired humidity. The relative humidity (RH) in the SFM chamber was monitored with a solid state humidity sensor and was varied between 10 and 65%. The temperature of the air in the tent was 21±1 °C. Single crystals of $NaNO_3$ were grown from saturated solutions in water over time periods of a few weeks and then stored under dry air conditions. The crystals were cleaved in air and placed immediately in the tent, initially with dry air flow.

Survey scans in both modes (contact and tapping) over numerous $11\bar{2}0$ cleavage surfaces reveal large numbers of steps, most of which are about 3.1 Å high and correspond to single interatomic layers; however, at least 30% of the steps are about 6.2 Å high. [The unit cell of the $NaNO_3$ structure includes two, nonequivalent layers of Na^+ ions and NO_3^- ions, so that these "bilayer" steps are one unit cell high.] Occasionally, ~9 Å steps were also observed.

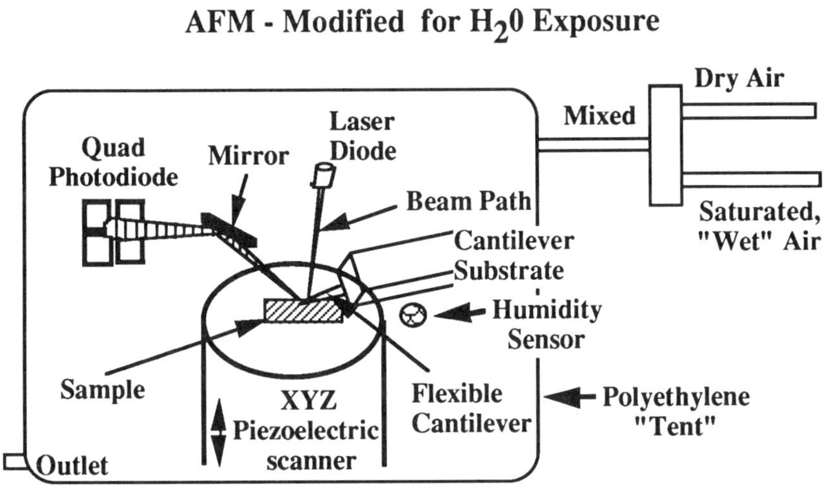

FIG. 1. Diagram of apparatus for SFM under controlled humidity conditions.

Using the SFM in the tapping mode,[22] large areas scans (> 1 $\mu m^2 \times 1$ μm^2) generally produce no observable tip-induced effects and steps and terraces produced by cleavage were stable over periods of hours at RH ≤ 60%. Thus, dissolution of steps was not apparent under these humidity conditions. Under contact mode scanning with normal forces, F_N, of 1-2 nN, consecutive scans in dry air (RH ~ 10%) showed no evidence of wear, including at cleavage steps.

Contact mode scanning + humidity ($NaNO_3$)

In contrast, cleavage steps are quite vulnerable to erosion by the combined effects of high humidity and contact mode scanning. Figure 2(a) shows two cleavage steps imaged at a F_N of 1.7 nN, a scan rate of 5 Hz, and a RH of 20%. These steps are 6.1 Å high—corresponding to two $NaNO_3$ monolayers. This region of the surface was scanned several times at this humidity at various scan sizes and showed no detectable changes as a function of time. After increasing the RH in the SFM chamber to 45%, a sequence of scans was begun, starting with Fig. 2(b). Note that the surface developed a streaky appearance. These streaks were highly irreproducible from scan to scan, indicating that they do not represent permanent changes in the surface topography. This scan also shows the onset of dissociation of the 6 Å steps into pairs of 3 Å steps.

Subsequent scans made at 90 s intervals are shown in Figs. 2(c) and 2(d). In Fig. 2(c) (RH ≈ 50%), both of the 6 Å steps have dissociated into two well-separated 3 Å steps. The "dissociation" of bilayer steps was frequently observed during repeated scanning when the RH reached about 50%. Newly formed terraces appear in the next scan [Fig. 2(d)], along with a series of monolayer trenches parallel to the direction of tip motion. As noted above, images taken in the tapping mode at even higher relative humidities show no step motion. Thus the step splitting and large-scale surface modification in Fig. 2 are due to the combined effects of humidity and scanning: both tribological loading and water vapor are required.

Figure 2(d) also shows monolayer deposits along the upper terrace adjacent to the scanned region. Large area scans taken immediately afterward show no comparable deposits on the lower terrace. This asymmetry reflects a net transport of material "up" the steps, consistent with the enhanced frictional forces applied by the SFM tip as it moves up and over a step. Lateral force measurements confirm that the frictional forces applied by the tip are considerably higher when the tip moves up a step than when it moves down. Similar variations in lateral force have been observed on alkali halide surfaces.[23] Frictional forces therefore play a key role in tribological wear under these conditions.

Relative to cleavage steps, deposits produced during periods of decreasing humidity are much more vulnerable to corrosive wear. In essence, the results shown here represent the interaction of two submicron asperities (a deposit and the tip) in a hostile chemical environment. Approximately 5-nm high deposits (~16 monolayers high) in Fig. 3(a) were produced by simply raising the RH to about 60% and then lowering the RH to less than 15% over the course of half an hour. Regions continuously scanned during the humidity cycle showed no evidence of these deposits, presumably due to a sweeping effect. Three $2 \times 2\ \mu m^2$ scans at a RH of ~55% were performed over the area marked with white lines which partially intersected deposit A and completely intersected deposits B and C. Fig. 3(b) is the $5 \times 5\ \mu m^2$ scan following the smaller scans, showing the remains

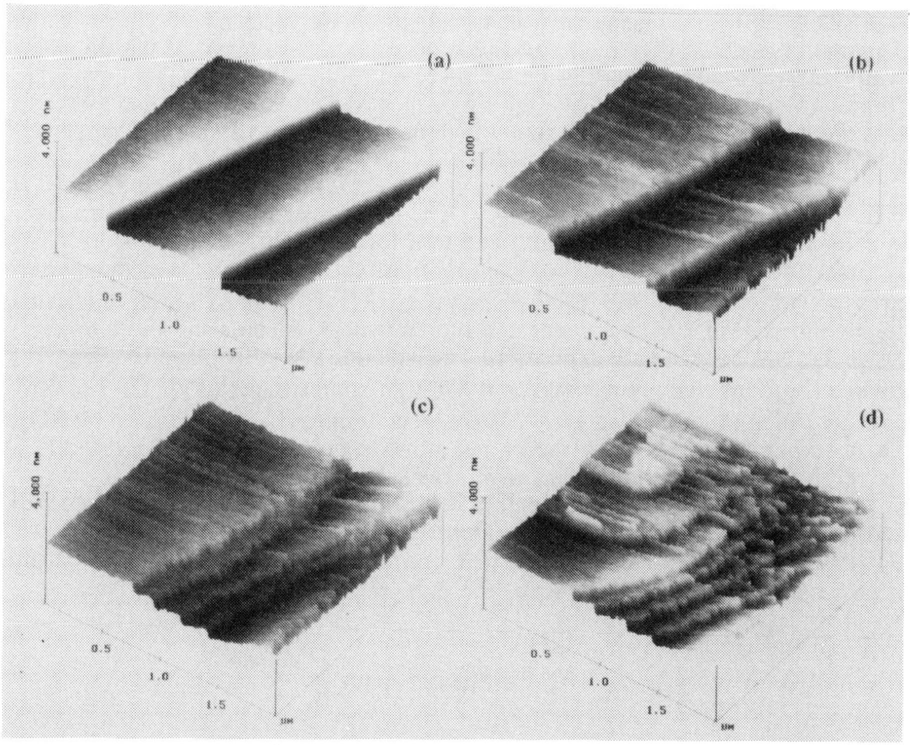

FIG. 2. A series of 2 × 2 μm^2 scans in the contact mode at a scan rate of 5 Hz (a) taken at a relative humidity of about 20%; (b) taken about 6 minutes after turning on wet air supply at a relative humidity of 45%; (Only the central 1 × 1 μm^2 portion of this scan has been previously imaged at high humidity.) (c) a second scan of the same area taken immediately after (b); and (d) a third scan taken immediately after (c).

of A (marked A'), the complete removal of B and C, and no change in the deposits in the regions outside of the small area.

The image during the first $2 \times 2\ \mu m^2$ scan, shown in Fig. 3(c) is revealing. It shows that the tip did not ride over the majority of the deposits, but instead went *down* approximately 3 Å. Subsequent scans of these regions often show depressions ~3 Å deep directly under the original deposit which quickly (within one scan) refill and disappear. Thus, an important mechanical aspect in the removal of material may be a plowing effect, i.e., a "catching" of the tip. This would clearly require asperities on the tip oriented and shaped so that downward forces would be produced when the tip pushes on the deposit. It also indicates that the deposit, which is most likely hydrated, may have mechanical properties which differ from the underlying crystal.

At the humidities employed in this work, ionic crystals such as $NaNO_3$ clearly develop an adsorbed fluid layer. The thickness and structure of this layer is a function of ambient humidity, and to a lesser extent, sample history. For instance, NaCl at 23 °C adsorbs 2-3 monolayers of water at a water partial pressure of 6 torr;[24] which corresponds to RH ≈ 28% in air at the same temperature. Most of this water can be removed by decreasing the humidity; however, NaCl previously exposed to water retains a submonolayer coverage of hydroxyl groups that are removed only at high temperatures (200 °C)[24] or very low pressures (10^{-6} torr water).[25] We expect comparable behavior on $NaNO_3$, i.e., a few monolayers of adsorbed water at modest relative humidities (30-50%), decreasing to low (submonolayer) coverages at low humidities.

The high stresses developed at the SFM tip (due to its relatively small radius) have great potential for altering the surface. Assuming elastic deformations only, the average compressive stress, σ_{avg}, in a soft material beneath a hard, spherical indentor is given by the Hertz relation,[26]

$$\sigma_{avg} = \frac{2}{3\pi}\left(\frac{12\, F_N\, E^2}{R^2\,(1-\nu^2)^2}\right)^{1/3} , \qquad (1)$$

where F_N is the contact or normal force, and R is the radius of curvature of the indentor. E and ν are the Young's modulus and Poisson ratio of the soft phase, respectively. At a modest contact force and tip radius of 1 nN and 50 nm, respectively, and typical moduli for $NaNO_3$ (E = 58 GPa, ν = 0.20), $\sigma_{avg} \approx 400$ MPa. (The corresponding projected area of contact is about 2 nm in diameter.) When applied by μm-sized probe, this stress is more than sufficient to deform the material. For instance, a Vickers indentor supporting a load of 0.5 N will typically produce an indent ~70 μm across the diagonal,[14] corresponding to $\sigma_{avg} \approx$ 200 MPa. We attribute the apparent lack of permanent deformation during contact mode scanning at modest RH to the small size of the SFM tip relative to typical slip bands (1 μm), which hinders deformation.[27] Deformation of ionic material is also retarded at high strain rates,[28] e.g., those associated with rapid motion of the tip across the surface.

It is only under the combined stimuli of contact mode scanning and RH between 45 and 50% that we see this severe surface wear. We argue that sub-

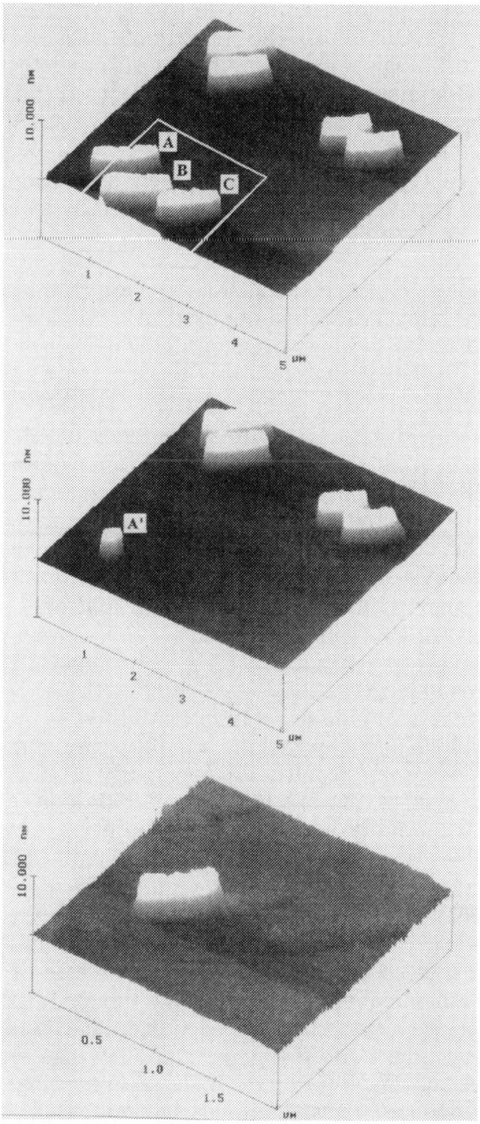

FIG. 3. (a) Image of cleaved $NaNO_3$ after a dry-to-wet-to-dry cycle, showing localized deposits. Immediately after this scan, three $2 \times 2\ \mu m^2$ scans were made in the area outlined in white. (b) A scan of the area imaged in (a) performed after the three $2 \times 2\ \mu m^2$ scans. (c) The first $2 \times 2\ \mu m^2$ scan, which removed most of the deposits. Scanning was performed at a contact force of 1.7 nN, a scan rate of 3 Hz, and a RH of 60%.

saturation ion content, in conjunction with tribological stimulation, promotes this rapid wear along cleavage steps. The stability of these cleavage steps at high RH in the absence of tribological stimulation suggests nondefective cleavage steps do not provide a large number of ions to the thin adsorbed fluid layer at ambient RH. Other potential ion sources include highly defective structures (e.g., adions not imaged by SFM) and the atmosphere itself (e.g., CO_2 incorporated into the water layer as CO_3^{2-} or HCO_3^-). As noted above, the lateral, frictional forces applied by the tip at cleavage steps facilitate wear, as indicated by the preferential transfer of material "up" the steps (high lateral forces), as opposed to "down" the steps (low lateral forces).

At room temperature, a substantial fraction of the cleavage steps (on the order of 30%) are bilayer steps, as opposed to monolayer steps. These bilayer steps would tend to be produced during cleavage when the crack encounters dislocations. In the alkali halides, this process produces mainly monolayer steps.[29,30] Although the cleavage and plasticity in $NaNO_3$ involve similar glide planes and directions, the $NaNO_3$ unit cell is twice the size of the alkali halides (due to two distinct rotations of the NO_3^- ion) and promote the production of bilayer steps. Although both monolayer and bilayer steps are stable under humid conditions alone or tribological stimulation alone, the massive material transport accompanying the combination of stress and water vapor generally leaves monolayer steps only. In the presence of a water film, the energy cost of two monolayer steps appears to be very similar to that of a single bilayer step,[31] which would allow for step splitting.

"Mogul" Features in $NaNO_3$

Repeated, small area scanning at relative humidities of 30-45% often produced sets of parallel ridges and depressions which we refer to as "moguls." (Like moguls on a ski slope, these features are apparently formed by a progressive wear process where material is alternately removed and deposited along the surface.) Depending on the humidity and scanning conditions, a well developed mogul pattern could be developed in a 1×1 μm^2 scan area in as few as ten scans. Figure 4 shows the development of a mogul structure under nearly optimal conditions (40% relative humidity, contact force 4 nN). After one or two scans, the scanned region is covered by 5-6 roughly parallel ridges, shown in Fig. 4(a). The vertical distance between the top of the ridges and the bottom of the troughs averages about 6 nm, as if a monolayer of material were scooped from the depressions and deposited as ridges. Cross sections of this image confirm that the surface modification is volume conserving, i.e., all the material removed from the depressions can be accounted for by the material in the ridges.

During subsequent scans, the ridges and depressions typically coalescence, becoming higher and deeper, and fewer in number. After four additional scans (eight minutes), the 5-6 ridges in Fig. 4(a) have coalesced into 4 ridges [Fig. 4(b)]. Careful examination of these images and others indicate a net motion of material across the surface, here from right to left. This motion leaves a deep depression along one end of the scan (here along the right). The average vertical displacement of the tops of the ridges relative to the bottoms of the troughs in Fig. 4(b) is about 1 nm. Four more scans has reduced the number of ridges to three, as

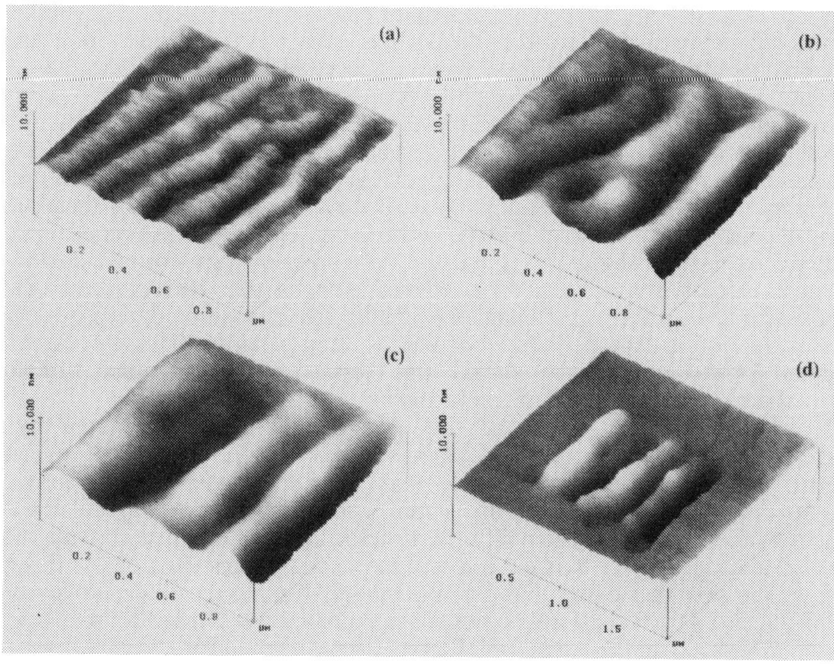

FIG. 4. The development of a “mogul” pattern during by repeated scanning of a $1 \times 1\ \mu m^2$ area in the contact mode at a relative humidity of 40%: (a) second scan; (b) fourth scan; and (c) sixth scan. (d) A $2 \times 2\ \mu m^2$ scan of the same area taken immediately after (c). [Scan rate is 5 Hz for (a), 2 Hz for (b)-(c). Contact force 3.9 nN.

shown in Fig. 4(c), and has increased the average vertical ridge-to-rough displacement to about 1.6 nm. A large area scan of the resulting mogul structure appears in Fig. 4(d). Again, the structure as a whole appears to be volume conserving: the volume of the ridges is equal to the volume of material removed from the depressions.

Mogul formation during small area scans requires a fairly narrow range of RH, but occurs over a wide range of scan rates and contact forces. The number of ridges and depressions formed in the scanned area was independent of scan rate and contact force, providing these were in acceptable ranges. This suggests that the mobility or frictional behavior of the near-surface material is critical for mogul formation. We attribute initial ridge/depression formation to a stick-slip like instability, due to the accumulation of abraded material on the SFM tip. When the accumulated mass reaches a critical value, it is deposited in place by the SFM tip, either by riding over the deposited mass (if the material is piled up ahead of the tip) or pulling away from the deposited material (if the accumulated material adheres to the trailing edge of the tip). Depressions are formed during the removal process, and ridges are formed during deposition. Ridges formed during one pass of the tip across the image would stimulate deposition during subsequent passes, reinforcing the periodicity of the resulting patterns. This process would account for the apparently small effect of the scan rate on the number of ridges formed in the early stages of mogul formation.

Subsequent evolution of the moguls would be dominated by the preferential removal of material along one edge of each ridge and redeposition as the tip passes over the top of the ridge, i.e., the ridges appear to drift across the scanned region as they grow. This is similar to the evolution of moguls on a ski slope, where snow is preferentially removed from the slope just above a mogul (where the edges of the skis cut into the snow) and is deposited on the downhill side as the skis pass over the top of the mogul. Unlike the skier, however, the SFM tip passes back and forth across the surface in a symmetric fashion.

We attribute the existence of a preferred direction for material transfer to the combined effects of anisotropic resistance to wear and the variation in the lateral force as the tip passes over a ridge. As noted above, the variation in lateral force as the tip passes over an obstacle favors material transfer as the tip mounts a step. Although the tip must rise up over each ridge from both directions, we expect that the basic anisotropy of step growth in calcite-like minerals (where steps of a given sign tend to grow faster in supersaturated solutions) will favor growth in one direction over the other. This establishes a preferred direction for material transfer, allowing a net transfer of material along one scan direction.

Indentation of $NaNO_3$ - Nucleation of Deposition

$NaNO_3$ is a soft material and deforms readily when the SFM tip is brought down on the surface at a constant dz/dt without scanning up to an F_{max} ~20 nN. At RH $\lesssim$ 30%, this indent is readily observed by scanning in the contact mode loading. A small bump is produced next to the indent due to the associated plastic deformation. At RH > 40%, contact mode images after indentation show no indent—either because an indent was not formed or because it disappeared before or during the imaging process. A dramatic transition occurs when the humidity is raised further to about 55%. Although images taken immediately after a force calibration do not show an indent, they do reveal a growing conical structure. An image of a typical structure is shown in Fig. 5. The nearly circular step-islands which form the edges of the conical structure have a petal-like appearance which we designate as an indentation "rose." Indentation roses grow faster as the relative humidity is increased from 55% to 60%. The rose in Fig. 5 grew over the course of about an hour.

The most likely mechanism for the formation of rose-like structures appears to be spiral step growth in the vicinity of a pair of screw dislocations of opposite sign. These dislocation pairs would very likely correspond to the ends of dislocation loops "punched" into the bulk during indentation. The two ends of such a loop would be no farther apart than the diameter of the depression formed by the SFM tip. Since this distance is less than the diameter of the SFM tip, imaging the step joining the ends of the dislocation pair would be difficult unless a second, much sharper tip was used to image the indented region. Each revolution of the growth "spirals" radiating from the two dislocations would produce a new terrace—in a fashion similar to the operation of Frank-Reed sources, which nucleate dislocation loops in metals.

We therefore conclude that tribological stresses applied to $NaNO_3$ cleavage surfaces by contact mode scanning can seriously alter the surface in the presence of sorbed water. Corrosive wear on nanometer size scales can involve several forms of material transport, including step erosion and splitting, terraceplowing, and redeposition. Over a significant range of operating

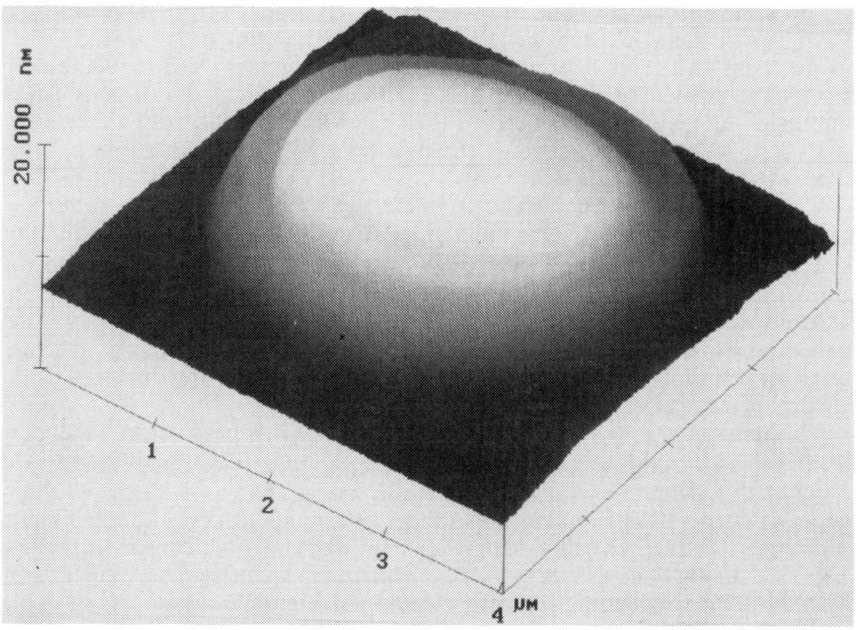

FIG. 5. A contact mode image of an indent formed by the tip and the resulting redeposition that occurred.

conditions and humidities, this material transport can produce quasiperiodic arrays of depressions and ridges. Over the small range of normal forces studied here, both scanning-induced stresses *and* the presence of water vapor are essential to induce wear. The vulnerable structures are initially the cleavage steps, followed by eventual plowing of the ledges. Scanning the tip over nanometer scale deposits sitting on terraces involves the interaction of two sub-micron asperities in a hostile environment (water), where again corrosive wear was observed. This system and presumably other solid-solvent combinations can serve as model systems for corrosive wear which can be studied in this fashion.

Corrosive Wear in Calcite

$CaCO_3$ (Icelandic spar) has the same crystal structure as $NaNO_3$ but is considerably harder (Moh hardness 3; Vicker's hardness, 1.3 GPa at 80 g load[14]) and considerably less soluble in water (by a factor of ~10^5). To observe effects with the SFM, we found it necessary to perform our studies "under water" as opposed to humid air. For this purpose we used a solution cell (Digital Instruments, CA) of approximately 20 μL in volume, sealed with a soft o-ring and filled with a saturated $CaCO_3$ solution adjusted to a pH of ~9. This buffer solution slowed the dissolution of the crystal in the absence of abrasion to low rates (step velocities < 1 Å/s). Further, we found that much higher normal forces were required to induce wear in calcite (consistent with its hardness). As we describe below, under these conditions, dissolution plays a dominant role in the wear generated by scanning at these higher normal forces.

Cleavage surfaces of optical grade $CaCO_3$ scanned in ambient atmosphere (RH ~30%) is quite resistant to tip-induced effects. Figure 6 shows an SFM image taken immediately after three small area scans (marked by white box) in the central portion of the figure at a high contact force of 270 nN. The step running along the central portion of Fig. 6 has not been noticeably affected by the small area scans. Thus, scanning in air at relatively low humidity results in no wear features at this contact force.

When a cleavage surface is immersed in 60 μM $CaCO_3$ solution, the surface features initially evolve rapidly due to dissolution. This initial dissolution is consistent with the observations of Hillner et al.,[32] who report that supersaturations of about 170% (100 μM) are required for a neutral dissolution/deposition environment in the presence of steps and pits. In the present work, the 60 μM $CaCO_3$ solution was allowed to remain on the surface (no flow) until it reached equilibrium with the underlying material and the velocity of steps along the etch pits was < 1 Å/s.

Spontaneous etch pit formation and growth has been observed under the SFM[32,33] and using optical microscopy measurements on much larger pits.[34] Liang et al. have proposed a dissolution mechanism based on Hirth and Pound's model of evaporation from crystal surfaces, where nucleation of double kinks on step edges produces weakly bound ions (at the kink) which are subsequently desorbed. Liang et al. use this model to explain the significant asymmetry seen in dissolution rate depending on step orientation. Figure 7 shows a portion of the calcite cleavage surface showing two parallel steps, one

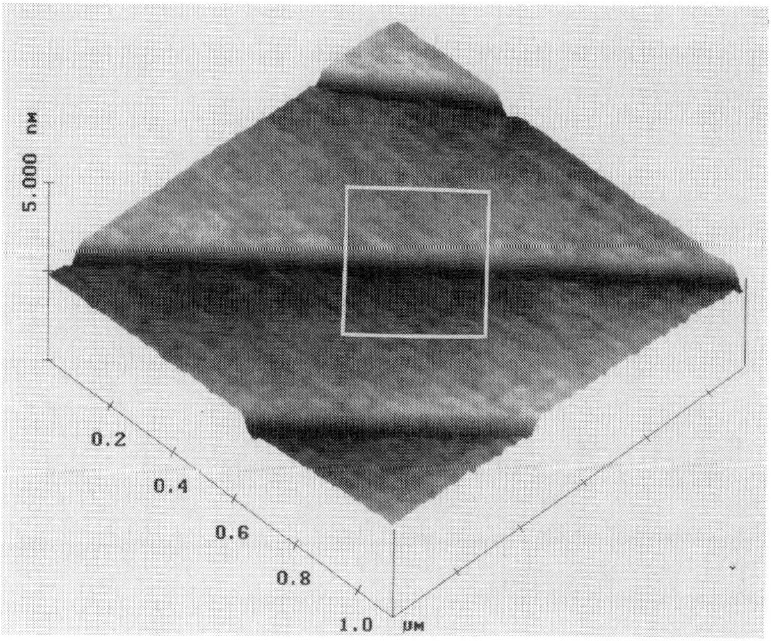

FIG. 6. SFM image ($1 \times 1\ \mu m^2$) of a $CaCO_3$ cleavage surface imaged in ambient air, taken immediately after three small area scans along the central step. The location of the small area scans is outlined in white.

"fast" and the other "slow". As seen, the angles of inclination of the lattice along these steps are significantly different (78° and 102°, respectively). The more open step (102° inclination) propagates more rapidly; slower motion occurs for the 78° step. Liang et al. attribute these different growth rates to differences in the activation energy for double kink nucleation in the two different step orientations.

The effect of scanning under these solution conditions is shown in Fig. 8(a). Three small area scans (marked by white box) at a contact force of about 270 nN have been made in the central portion of Fig. 8(a) immediately prior to taking the image. These small area scans overlap the edge of a monolayer deep etch pit which corresponds to the "fast" growing direction. The jog in the central region of the region corresponds approximately to the area of the small area scans. Note first that the terraces above and below the step are unharmed—wear does not occur over the entire small area scan. Second, the wear pattern appears to be due to *dissolution* which has been enhanced in the region covered by the small area scan and in fact has extended slightly beyond the extent of the small area scan. Repeated large area scans of similar regions confirm that under these conditions, the step velocity associated with dissolution alone (without small area scanning) is less than 1 Å/s. In terms of a wear "rate", we observe here an approximateremoval of ~30 nm per scan (12 Å/s). This enhancement is consistent

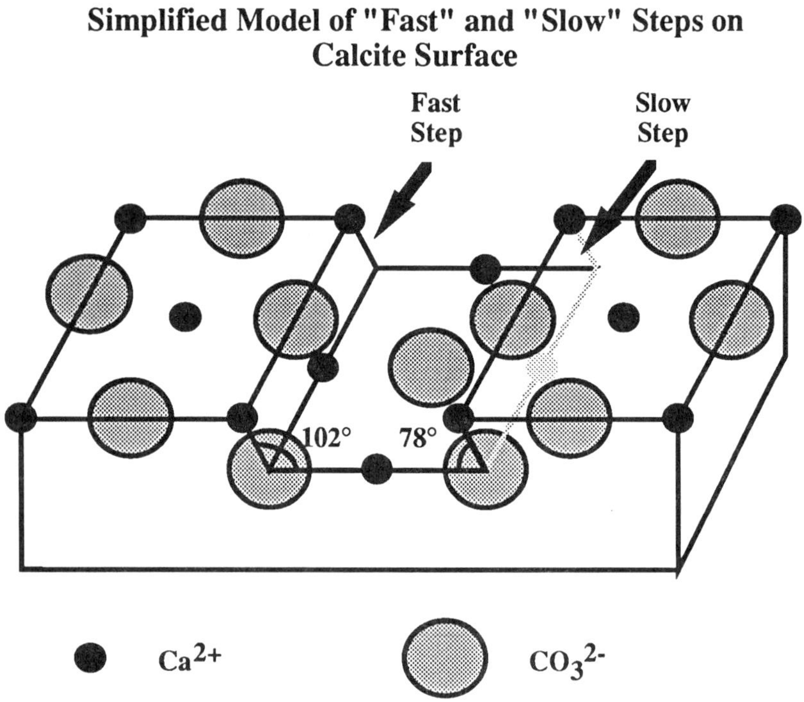

FIG. 7. Model of "slow" and "fast" dissolving steps on the calcite surface.

with the stress-induced nucleation of double kinks by small area, high contact force scanning. Some of these kinks would propagate outside the small area scan, allowing for limited dissolution outside the scanned area.

Similar experiments in which the direction of scanning was varied ±45° with respect to the direction in Fig. 8 (normal to the fast step) showed no change in the step velocity as a function of scan direction. This suggests that the frictional forces applied by the tip to the step (which depend strongly on scan direction) do not significantly affect the dissolution process.

Significantly, the wear produced by scanning under these conditions does not continue after small area scanning is stopped. Step advance on either side of the small area scans continues until the edge of the pit is again straight, leaving no sign of the previously enhanced step growth. The wear, which we interpret as enhanced dissolution, does not appear to be due to a permanent or persistent change in the surface induced by small area scanning, but to a transient effect during or immediately after scanning itself.

The wear rate or step velocity enhancement (the velocity in the scanned region less the velocity outside the scanned region) is, however, a strong function of the normal force applied to the tip. (See Fig. 9.) The velocity enhancements are shown for the two non-equivalent steps produced during $CaCO_3$ dissolution,

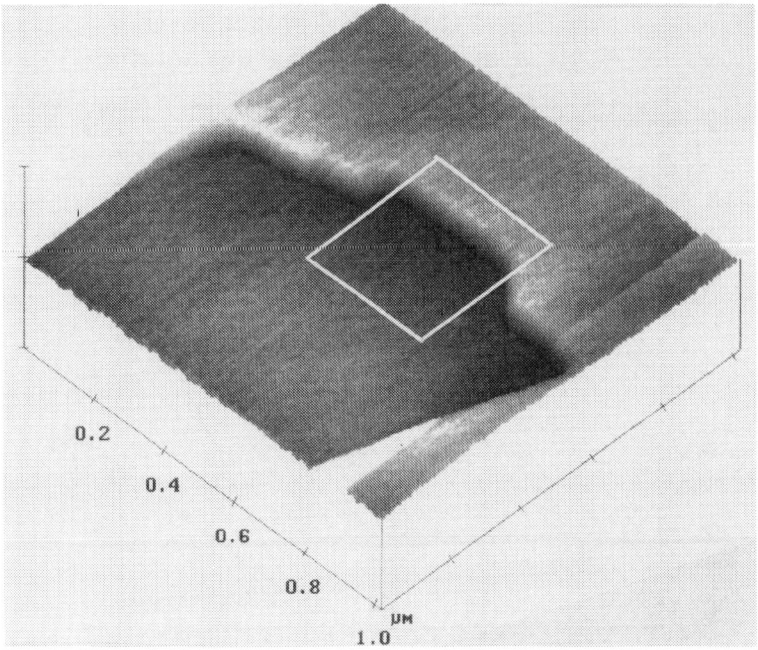

FIG. 8. SFM images of a "fast" step taken immediately after three small-area scans, where the direction of scanning is oriented normal to the cleavage step. The location of the small area scans is outlined in white.

the "fast" and "slow" steps mentioned above. Velocity enhancements for both step orientations require a minimum normal force of about 160 nN under these solution conditions and scan sizes. At higher contact forces, the step velocity increases roughly linearly with contact force. This strong dependence of step velocity on normal force rules out enhanced material transport (due to mixing of the solution by the SFM tip) as the mechanism for enhanced dissolution. Mixing effects would be quite insensitive to the normal force applied by the tip.

The average stress applied by the SFM tip at the threshold contact force for dissolution enhancement is quite sufficient to deform calcite. Assuming elastic deformations only, the average compressive stress, σ_{avg}, in a soft material beneath a hard, spherical indentor is given by Eq. 1.[26] At the threshold contact force for the onset of enhanced dissolution (~160 nN), a tip radius of 50 nm, and average moduli for $CaCO_3$ (E = 81 GPa, ν = 0.21), $\sigma_{avg} \approx$ 3.7 GPa. (The corresponding projected area of contact is about 13 nm in diameter.) A similar stress applied to µm-sized probes would significantly deform the material. For instance, a Vickers indentor supporting a load of 0.8 N will typically produce an indent ~35 µm across the diagonal,[14] corresponding to $\sigma_{avg} \approx$ 1.3 GPa.

High dislocation and twin densities have long been known to enhance the rate of dissolution of ionic and metallic materials, including calcite.[35] Under

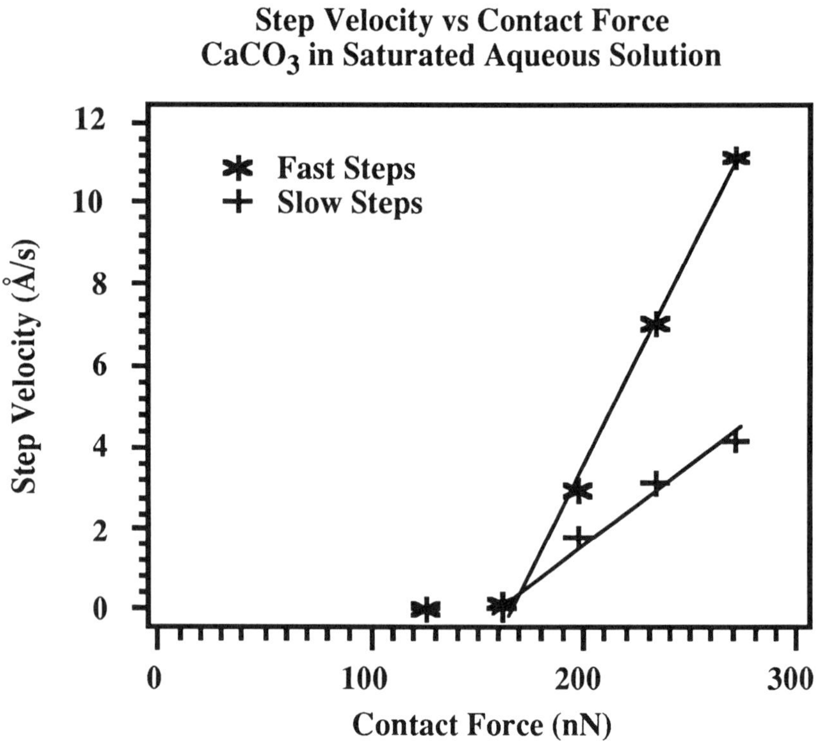

FIG. 9. Step velocity as a function of applied normal force for fast and slow steps.

normal conditions, this involves the formation of spiral-like steps centered on screw dislocations. These spirals provide for very high step densities. In this case the rate of dissolution is enhanced by increasing the step density, with no increase in step velocity—in contrast with observations in this work. We note that transient dislocation emission (e.g., by "punching in" dislocation loops) along the step would provide high densities double kinks, and thus account for the observed enhancement in step velocity. These dislocations (or equivalent twin structures) would have to be stable over time scales sufficient for kink propagation away from the nucleation site. After initial kink propagation, subsequent relaxation would produce additional kinks and further enhance dissolution. We expect that virtually all the deformation produced in a scanning situation relaxes rapidly, on time scales of 10's of seconds or less. Any persistent deformation structures should enhance step velocities long after scanning is terminated, which is not observed.

For $0 < F_N < 270$ nN, repeated scanning on broad flat terraces had no effect on the appearance of the $CaCO_3$ surface. However, repeated scanning with extremely small scans (10 × 10 nm^2) and high contact forces (270 nN) could nucleate dislocation pits at arbitrary positions along terraces. Dissolution pits initiated by a sequence of three 10 × 10 nm^2 scans at 270 nN are shown in Fig. 10. Once the pit is nucleated, growth continues in a fashion similar to that of "self-nucleated" pits. The nucleation of a pit on a flat terrace requires significantly more energy than double kink nucleation along a step—and therefore correspondingly stronger mechanical stimuli are required. Significantly, indents formed by bringing the SFM tip directly down onto to the surface did not generally nucleate dissolution pits. (SFM "indentation" did nucleate pits if the tip was within about 500 nm from a step edge, as discussed below.)

Effect of indentation (Calcite)
Observable indents can be made in many soft materials by bringing the SFM tip down vertically onto the surface, without scanning. This process is part of a typical force calibration procedure in most SFM equipment. With the tips and normal forces employed in this work, this procedure did not modify the appearance of the surface in subsequent images unless the force calibration was performed within 500 nm of a surface step. SFM indentation on flat surfaces more than 500 nm away from the nearest dissolution pit left no mark visible in

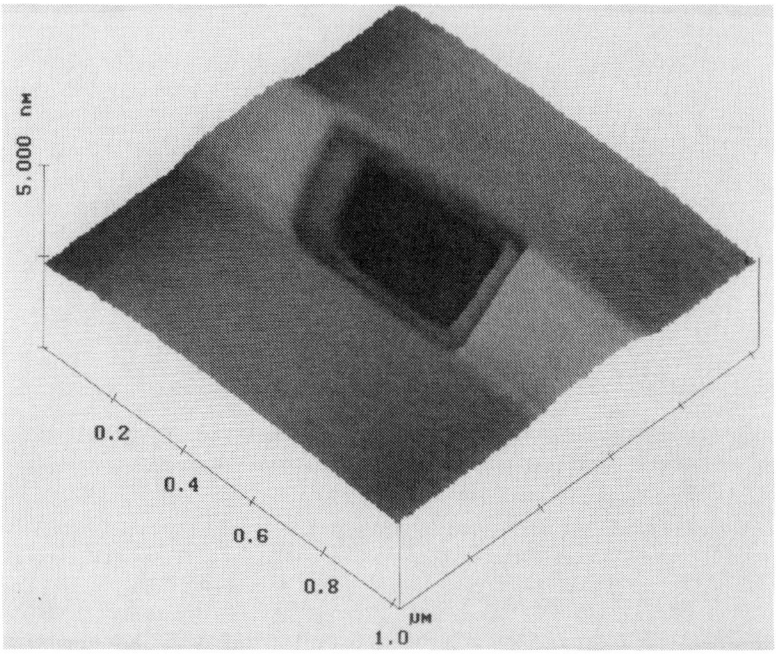

FIG. 10. Dissolution pit nucleated in a broad terrace by three 10 × 10 nm^2 scans at a contact force of 270 nN.

subsequent scans. (It is possible that a transient indentation was produced that disappeared prior to the next scan, which typically required 30-60 s to perform after the indentation procedure.) However, when the same procedure is performed near a previously existing dissolution pit, a readily visible pit is observed in subsequent scans.

Figure 11 shows a dissolution pit where two SFM indents have been made: one on either side of the pit. The indent on the extreme right edge of the scan was made immediately prior to the scan of Fig. 11 and is about 190 nm from the "slow step" along the right side of the dissolution pit; this indentation has produced a small pit, one monolayer deep and 50 nm on a side. This indent has not affected the propagation of the nearby slow step. A similar indent was made several minutes earlier about 200 nm from the left side of the dissolution pit (a "fast step") and has produced a much larger pit. (The elongated shape of the pit at the indentation site is unusual, but not unique.) Further, the edge of the dissolution pit nearest the indent (a fast step) has advanced toward the indent. In general, dissolution along fast steps (but not slow steps) can be strongly enhanced by nearby indents. Again, we emphasize that indents father than 500 nm away from the edge of a pit leave no visible mark under these conditions.

We attribute the growth of dissolution pits at indentation sites to the production of extensive, long-lived deformation structures. The steps surrounding the hole produced by the SFM tip would typically be highly "kinked" and allow

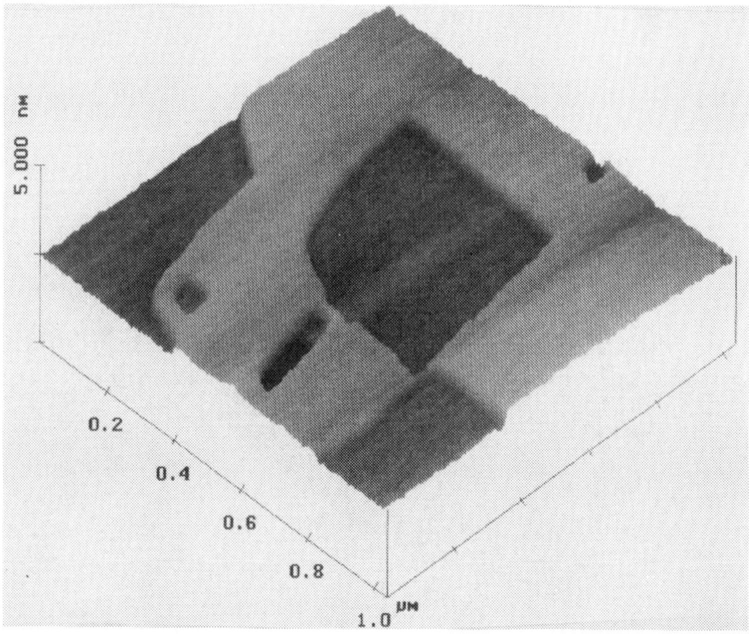

FIG. 11. SFM image of a large dissolution pit with indents positioned approximately 190 nm from the right edge (slow step) and 190 nm from the left edge (slow step).

for rapid initial growth. Subsequent pit growth would proceed in the same manner as spontaneously nucleated dissolution pits discussed above. The enhancement of dissolution along nearby fast steps suggests that the residual strain due to indentation is strong enough at a distance of 200 nm to enhance dissolution. Enhanced dissolution of nearby steps is generally confined to rather narrow strips oriented along particular crystallographic directions with respect to the indentation (parallel to the pit edges). This strongly suggests that indentation has produced slip bands or narrow twins. The lack of pits at indentation sites far from surface steps suggests that the associated deformation either does not occur or that it relaxes quickly after the tip is withdrawn. We propose that the nearby step serves as a stress concentrator or weak point in the lattice structure at which deformation is nucleated.

Scanning conduction microscopy—a probe of thin film wear

The SFM is a useful tool for examining the consequences of tribological wear, particularly on inherently flat surfaces such as single crystals. However, on moderately rough surfaces, changes in topography due to tribological loading are often difficult to interpret in terms of wear processes. We have examined the wear of an important class of coatings, namely fluorocarbon thin films deposited on stainless steel, using a simple modification of standard SFM techniques. Here we examine polytetrafluoroethylene (PTFE) thin films on stainless steel substrates. Metal-coated silicon nitride tips are used to probe current flow between the tip and the conducting substrate. Simultaneous topography and conduction images are acquired. During wear (performed outside of the SFM), localized thinning of the film and exposure of bare metal are easily and unambiguously detected by this method on size scales less than 50 nm. We describe the technique[36] and present results on fluorocarbon thin films.

Film topography was characterized with the Nanoscope III SFM operated in the contact mode. SFM scans were acquired at scan sizes ranging from 200×200 nm^2 to 100×100 µm^2 and at tip velocities from 0.2 to 200 µm/s. A simple modification of the SFM allowed simultaneous, spatially resolved measurements of film conductivity. A schematic of the experimental arrangement is shown in Fig. 12. Scanning conduction microscopy (SCM) measurements were made using commercial triangular Si_3N_4 SFM tips (115 µm from tip to base)[37] that were sputter coated with 300 Å of Au or Ag at room temperature. Both sides of the cantilever were coated to minimize cantilever deflection due to stresses in the metal coating. The SFM was operated in the contact mode, using a nearly constant compressive force, typically 30-60 nN. Assuming a tip radius of 50 nm, the pressure applied by the tip is about 20 MPa, well below the yield stress of most materials. Forces were determined from the measured displacements and quoted force constants for the commercial cantilever beams used in this experiment. The sample was mounted on double stick tape to insulate it fromground. Electrical contact to the conductive cantilever was made through the substrate. The circuit was completed by a 30-µm wire spot welded to the conducting cantilever clamp.

Current measurements were performed with a Keithley 602 electrometer whose output was amplified with a Tektronix 3A9 differential amplifier. The

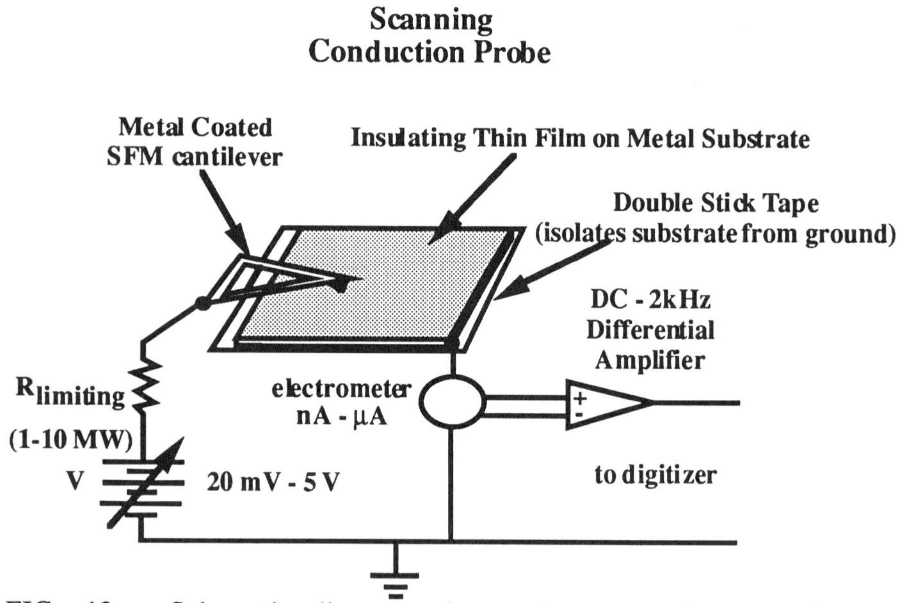

FIG. 12. Schematic diagram of scanning conduction experimental arrangement.

amplifier output was digitized in phase with the acquisition of the SFM image, providing a SCM image that corresponded to the topographical image acquired simultaneously. A bias voltage (typically, 1 volt) was applied to the sample through a limiting resistor ($R_{limiting}$, typically 1-10 MΩ) to prevent current-induced damage to the metal coating of the SFM tip when metal-to-metal contact was made. When the SFM tip is over a region where the polymer film is thick, the effective tip-to-substrate resistance is high ($>10^{12}$ Ω) and little current flows through the circuit. When the tip crosses a bare region of the substrate, the tip-to-substrate resistance falls to low values ($< 10^5$ Ω) and the resistance of the circuit is approximately equal to that of the limiting resistor. Regions of intermediate conductivity are also evident. A typical SCM image is a 3-dimensional plot of the current magnitude vs position.

The useful lifetimes of the conducting tips were limited by wear of the metal coating and ranged from 10-30 scans on the polymer. High currents (> µA) associated with low limiting resistances or high bias voltages, however, dramatically shortened tip lifetimes. Both positive and negative bias voltages produced nearly equal currents with no apparent nonlinearities, suggesting ohmic conduction. Little effort was made to electrically shield the sample; pickup or perhaps ground loops generated small but detectable modulations in the current signal which are visible in the images presented here.

Samples of polytetrafluoroethylene (PTFE) coated stainless steel were prepared in our laboratory by laser ablation deposition.[38-40] PTFE films, ~200 nm thick, were produced by pulsed laser deposition using a KrF excimer laser. Stainless steel exposed by wear processes would be oxide coated. Film abrasion was performed by manually drawing a cotton swab dampened with acetone across the surface. The applied normal force was approximately 2 N over a nominal contact area of 3 mm^2. Although not well controlled, the wear produced by successive linear strokes was indeed cumulative and served as a simple means to demonstrate the technique.

Figures 13(a)-13(e) show 10 × 10 μm^2 SCM images of a PTFE film on stainless steel prior to and following varying amounts of abrasion. The scans were acquired at a tip velocity of 20 μm/s with a 100-mV bias voltage applied through a 10 MΩ limiting resistance; a full-scale reading on the electrometer corresponds to 10 nA. As shown in Fig. 13(a), the nonabraded films are basically insulating. Sampling several regions of the film, we typically observed one or two submicron-sized regions of enhanced conductivity [similar to the small region in the foreground of Fig. 13(a)] in approximately 20% of the scans. These features were reproducible upon repeated scans. The large area of no current indicates that there is no evidence of noise due to discharges that might be generated by contact electrification.[41-43] Spatial resolution, as determined by the smallest patches of conductivity on otherwise non-conducting films, is as small as 10 nm. We suspect that this resolution is significantly influenced by the structure of the conducting patch (steep changes in conductivity) and the morphology of the conducting coating at the tip (i.e., perhaps only a limited portion of the tip is participating). One would expect that small modulations in conductivity would result in much lower resolution.

Figure 13(b) shows a typical SCM scan after a small amount of abrasion (5 strokes of the swab). Some conduction through the film is evident in the SCM scan. All of the conduction zones shown show currents well below the value for tip-to-bare metal contact (10 nA). The total area of the conduction zones (regions exhibiting currents above the noise level) corresponds to 0.6% of the scanned area.

Further abrasion (a total of 8 strokes) increases the area of the conduction zones, as shown in Fig. 13(c). The total area of conduction zones is now 2% of the scanned area. A few of the conduction zones display full scale current readings (10 nA) but most display intermediate values. All of the conduction zones in Fig. 13(c) are a micron across or less, with the majority being much smaller.

Additional abrasion (a total of 16 strokes) further increases the fractional area of the conduction zones, as shown in Fig. 13(d); the conduction zones now occupy approximately 6% of the surface. Most of the conduction zones still display less than full scale current readings.

Wiping the surface with the cotton swab 50 times produces large conduction zones, as shown in Fig. 13(e). Much of the surface now displays full-scale current readings. About 30% of a random sample of 10 × 10 μm^2 scans on this surface showed large, highly conducting patches. The remainder of the scans displayed large numbers of small, highly conducting regions over most of the scan, similar to the upper third of Fig. 13(e).

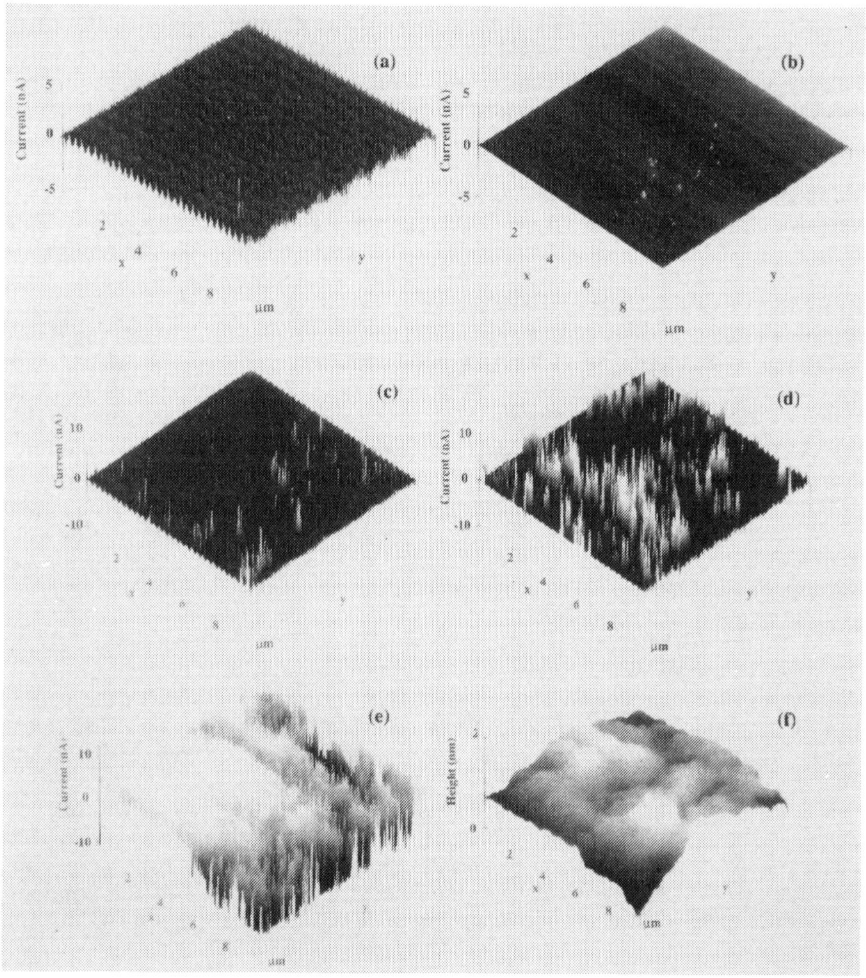

FIG. 13. (a) Scanning Conduction Microscope (SCM) scan of a non-abraded PTFE coated stainless steel sample; (b) SCM scan of the same PTFE film after the light abrasion (5 strokes) (c) SCM scan after a total of 8 strokes; (d) SCM scan after a total of 16 strokes; (e) SCM scan after a total of 50 strokes; (f) SFM topographic scan corresponding to (e).

SFM topographs acquired simultaneously with these particular SCM scans generally showed little correlation with the regions of conduction. The surface relief actually dropped progressively with abrasion, perhaps due to the removal of particulates (a common problem with thin films grown by pulsed laser deposition). In Fig. 13(f) we show the SFM topograph corresponding to the SCM scan of Fig. 13(e); the highly conducting zone appears to be in a region which tends to be elevated.

Because our abrasion was not well controlled, quantitative analysis of the data is not particularly meaningful. Nevertheless, the fraction of the scanned area occupied by conduction zones is a clean, monotonically increasing function of the "extent of abrasion" (number of strokes). The conducting area fraction increased rapidly over the first several strokes, and subsequently displayed a slower increase proportional to the square of the number of strokes. An improved abrasion apparatus under construction will allow for controlled wear under more technologically interesting conditions.

The influence of the bias voltage on conduction was examined over a range of 10-1000 mV. Basically, the images remain invariant when the current scales are normalized. Over this limited range of tip bias, the measured current is a linear function of bias. Higher biases allowed for the detection of intermediate conducting regions that were in the noise at lower bias voltages. Typical resistances of the tip + film in these intermediate conductivity regions were 10-30 MΩ. Conduction occurred with both plus and minus bias voltages.

The current at a fixed position can be examined as the tip approaches and leaves the surface. Figure 14 shows simultaneous force (a) and current (b) measurements as the tip approaches a region of intermediate conductivity on an abraded PTFE film (275 mV tip bias; full scale current corresponds to 100 nA). Note first that the onset of measurable current (labeled **A**) occurs after contact when the force has reached ~15 nN; second, the current continues to rise as the force increases (**A** to **B**), but then stops rising at **B**, at a force of ~50 nN, even though the force continues to increase. Force (c) and current (d) measurements made as the tip is withdrawn are shown in Figs. 14(c) and 14(d). The current does not start to drop until **C**, when the force has dropped to ~25 nN, and decreases continuously until the force is very nearly zero (**D**). Note that adhesion produces an attractive force between the tip and surface as tip is withdrawn further; nevertheless, no current is observed after **D** even though there is obviously a connection between the tip and surface up to the instant of detachment. This physical connection could be due to direct polymer-tip adhesion, perhaps dominated by electrostatic forces.

On high current patches ("metal-to-metal"), slightly higher forces were often required to produce measurable currents, with onsets at ~20 nN; maximum currents were reached at forces of ~65 nN. During tip retraction the current begins to drop at ~60 nN and reaches zero at ~55 nN. Even higher adhesive forces were observed in the final stages of tip withdrawal from the high conductivity regions, again implying continued physical contact *without* measurable current. This higher adhesion may be due to an intervening water layer (which would have poor conductivity), consistent with the hydrophilic nature of an oxidized metal surface.

SFM Force and Current vs Time for PTFE Film on Stainless Steel During Contact and Detachment—Intermediate Current Region

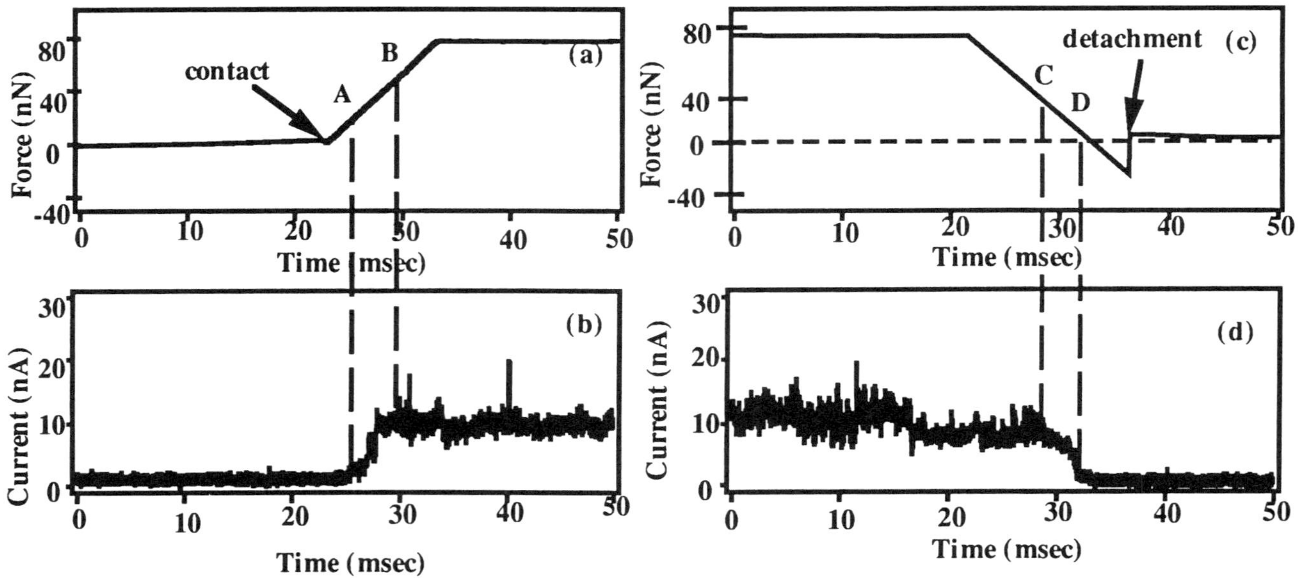

FIG. 14. Simultaneous force and current measurements during tip approach and contact [(a) force, (b) current] and withdrawal and detachment [(c) force and (d) current] on a region of intermediate conduction.

With slight differences due to the hysteresis during make vs break, the force dependence of the current is nearly the same for these two regions, which display different levels of conduction. However, the current fluctuations were noticeably higher for the intermediate currents. We suggest that this is due to the conduction through thinned polymer, generally involving a hopping mechanism, which would tend to be "noisier".[44] Conduction from the tip to the stainless steel may also be complex, but definitely displays lower amplitude fluctuations.

Numerous efforts are currently under way to produce stable conducting cantilevers which would facilitate SCM studies, e.g., by Thomson and Moreland.[45] Coatings employing higher cohesive energy metals such as W and Rh may provide more durable, longer-lasting tips, with greater tolerance for higher contact forces and higher currents. Doped diamond tips could be very robust and allow a variety of material systems to be studied. In principle, diamond tips would allow for the study of nm-scale wear of very hard coatings on metals and semiconductors. Significant improvement in electronic methods (e.g., lock-in techniques) to enhance signal-to-noise would allow a larger dynamic range of film thicknesses and conductivities to be probed by this method.

Applications of Photoluminescence and Fracto-Emission

Photoluminescence imaging of mechanically produced defects in MgO

Deformed inorganic crystals often exhibit detectable increases in point defect densities. MgO is an example which displays a strong blue photoluminescence when illuminated by (defocused) 248 nm excimer laser radiation. This photoluminescence is readily imaged in situ (during laser exposure) using microscope optics. We have produced a number of photoluminescence images of indented, cleaved, abraded, and laser-damaged surfaces and relate the observed features to the expected deformation patterns. Photoluminescence imaging has potential for the identification of flaws in materials as well as for fundamental studies of deformation, fracture, and wear. Resolutions of up to 10 μm are currently possible with room for significant improvement.

We employ single crystal MgO, an ionic metal oxide with the rock salt structure. Deformation of MgO produces high densities of point defects, including F- and F^+-centers.[46,47] Collectively, these defects absorb strongly at 248 nm [48,49] and produce a blue photoluminescence (PL) which peaks at about 410 nm when excited at 248 nm.[50] In previous work, we have shown that very similar PL is produced by laser damaged material, and that this damage was largely due to *mechanical deformation* in the heated region.[51,52]

A schematic of the experimental arrangement appears in Fig. 15.[53] The PL was excited by exposing the sample in air to 10^3-10^6 20-ns pulses of low fluence (200 mJ/cm^2) radiation from a Lambda Physik EMG203 excimer laser operating at 248 nm (KrF) or at 308 nm (XeCl). The laser beam was incident on the sample at a 30°-45° angle to the surface normal in a darkened room. The sample was positioned in the focal plane of a Zeiss stereomicroscope equipped with 50 mm and 100 mm focal length objectives. Images were obtained with a 35 mm camera using high speed (ASA 3200) black and white or color film. Optical micrographs of the laser-irradiated regions were also taken in white light to compare the surface features with the luminescence features. Scanning electron microscope (SEM) micrographs of the irradiated regions were often taken for additional comparisons.

Time resolved spectra were obtained with a Thermo Jarrell Ash Monospec-18 spectrometer (300 lines/mm grating) equipped with an EG&G Model 1421 gated, intensified, detector array, sensitive to 200-830 nm light. Spectral data was acquired with an EG&G OMA-III system.

The MgO samples were cleaved from single crystal MgO of nominal 99.99% purity obtained from W. & C. Spicer, Ltd., and Tateho Chemical Industries, Ltd. Some cleaved material was indented (diamond-tipped Vickers indentor with 2 kg load applied for 5 s), or scribed with a diamond prior to PL imaging.

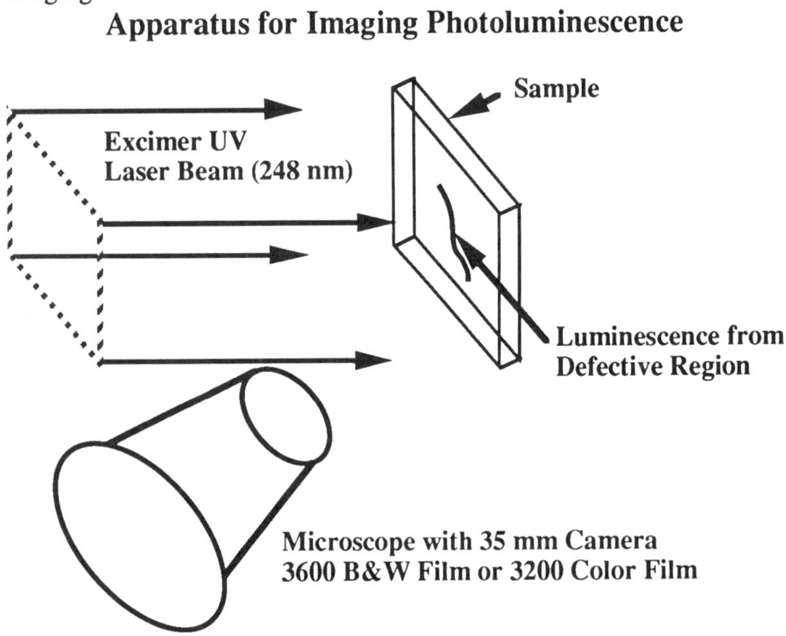

FIG. 15. Apparatus for photographing magnified images of PL from mechanically treated MgO surfaces.

Photoluminescence Spectrum

All mechanically treated surfaces of MgO have been found to produce PL. Figure 16 shows PL spectra from cleaved, diamond paste polished, and abraded MgO crystals taken during identical exposures to low fluence radiation at 308 nm.

As the extent (including depth) of damage increases, the integrated area under the PL peak increases. (This damage also increases the amount of light scattered out of the laser beam into the spectrometer; thus the height of the 308 nm peak increases as well.) The spectra are consistent with the PL spectra from deformed MgO previously reported by Chen et al.,[40] who attributed these emissions to clusters of F^0 and F^+ centers.

Plastic deformation in MgO and other ionic crystals with the rocksalt structure involves the growth of a relatively small number of dislocations to form extremely long, convoluted structures (dislocation "multiplication").[54] Jogs areproduced along dislocations at points where they intersect and "pass through" other dislocations. Crystallographic constraints require that jogs along dislocations with screw components produce vacancies (or interstitials) as they move through the crystal;[55,56] thus dislocation motion can produce dense trails of vacancy defects. (These crystallographic constraints do not apply to pure

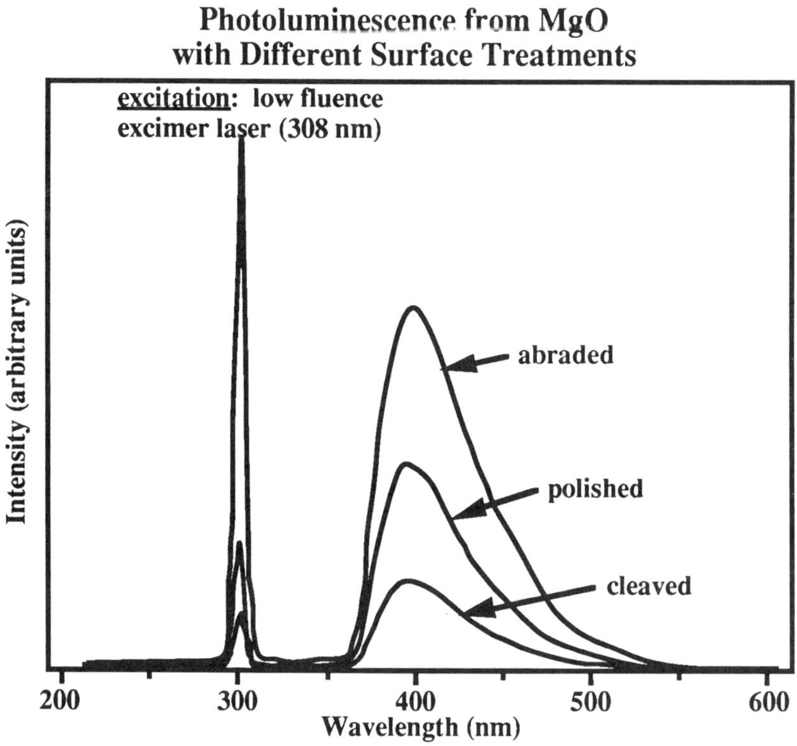

FIG. 16. PL spectra from cleaved, polished, and abraded MgO crystals at an excitation wavelength of 308 nm. The peak at 308 nm is due to scattering from the crystal into the spectrometer.

edge dislocations; thus jog dragging along edge dislocations does not in general produce point defects.) The high energy required to produce interstitials in MgO assures that deformation produces primarily vacancies. Very high defect densities may be produced. Dislocation growth in deformed LiF, which has a structure similar to that of MgO, has been shown to yield vacancy concentrations on the order of 10^{18} cm^{-3}.[57] The cyclic stresses produced by repetitive laser irradiation provide an ideal environment for defect formation, promoting dislocation growth to replenish the absorption centers removed by annealing, evaporation, and fragmentation.

PL from Indented surfaces

Indenting MgO produces a distinctive pattern of deformation associated with a bright luminescence in the central portion of the indent. Matching white-light and luminescence images of a pair of indents on an MgO(100) surface appear in Fig. 17. In addition to the bright central luminescence, up to three bright lines emanate from each edge of the indent along the <110> directions. Figure 17(c) shows the expected orientation of the cracks and slip bands produced by indentation,[58] along with the approximate location of the observed luminescent lines. Dislocation motion in MgO is largely restricted to the <110>{110} slip systems, producing high densities of luminescent point defects along these planes. The bright central region in the PL image is significantly larger than the visible portion of the indents, reflecting the large size of the deformed region compared to the size of the indent itself. Rotating the indentor 45° produces a very similar pattern of deformation and luminescence, indicating a strong dependence on the orientation of the favored slip planes and only a weak dependence on the orientation of the indentor.

The bright lines along the <110> directions correspond to cracks and slip bands which intersect the surface at right angles, parallel to the [001] viewing direction. The high defect densities along the line of sight produce unusually bright, linear structures. Other slip bands intersect the surface at 45°, at an oblique angle to the viewing direction, and produce a diffuse luminescence surrounding the indent.

The white light photographs show (110) cracks intersecting the sample surface along all four <110> directions, as well as ring-like cracks around the indents. (The indentor load employed in this work was sufficient to produce subsurface cracking.) The {110} planes are secondary cleavage planes in MgO; fracture along these planes is usually initiated by stress concentrations at intersecting slip bands. (Mobile dislocations on intersecting slip bands react to form sessile dislocations; subsequent dislocation pileup behind these sessile dislocations produces very high, localized stresses.) Thus these cracks grow through highly deformed material, and actually relieve some of the deformation-related stresses. This may account for the weakness of the luminescence along the cracks relative to the slip bands on either side.

The intersection of slip bands with MgO surfaces has been imaged with high resolution by cathodoluminescence (CL) imaging.[59] Interestingly, CL images show bright lines along the $\{110\}_{45}$ slip bands (intersecting the surface at

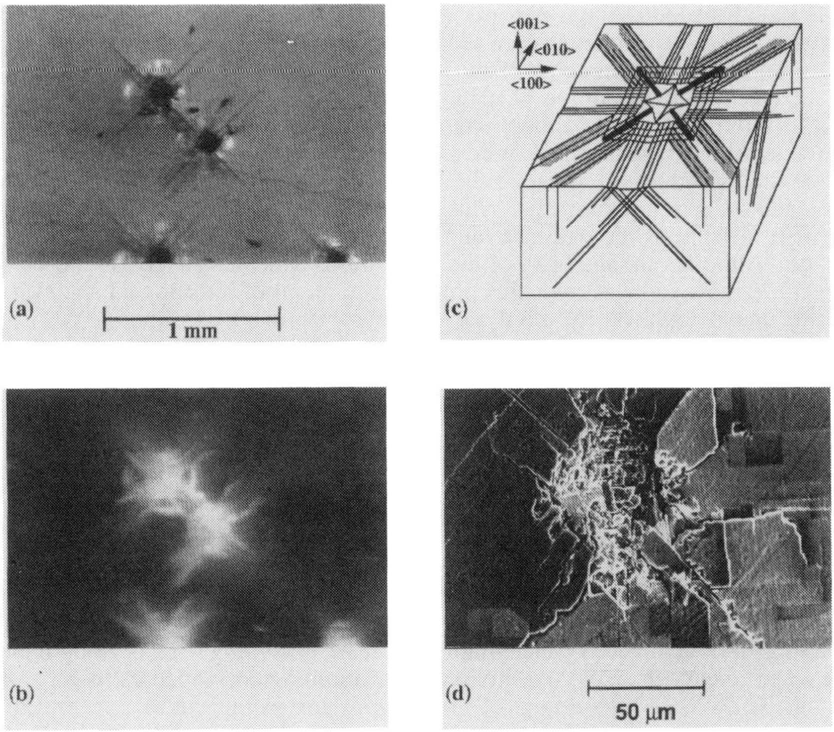

FIG. 17. (a) White light and (b) luminescence images of two indents on a single crystal MgO surface. (c) A schematic diagram of the expected pattern of deformation around these indents. The heavy dark lines indicate the {110} cleavage cracks around the indent and the light heavy lines mark the $\{110\}_{90}$ slip planes responsible for several of the observed linear luminescent features. (d) An SEM micrograph of an indent at an MgO fracture surface after exposure to high fluence laser radiation.

45° angles); in contrast, the PL image of Fig. 17(b) shows especially bright lines along the $\{110\}_{90}$ slip bands (intersecting the surface at 90° angles). Dislocations in the $\{110\}_{45}$ slip bands intersect the surface as screw dislocations, while dislocations in $\{110\}_{90}$ slip bands intersect the surface as edge dislocations. As noted above, the motion of pure edge dislocations does not generate vacancy defects (by the jog dragging mechanism). Thus, dislocation motion along the $\{110\}_{90}$ slip bands should not produce high defect densities in the near surface region. Near-surface defects are required for cathodoluminescence imaging due to the short penetration depth of the electron beam, so that one would not expect to see bright lines along the $\{110\}_{90}$ slip bands in cathodoluminescence images. In the bulk, however, dislocations on both sets of slip bands display mixed character and can generate PL producing point defects during dislocation motion. Since the uv light used to excite the luminescence penetrates deep into the bulk, both sets of slip bands in the bulk yield PL. As noted above, only when the viewing direction lies within a slip plane will this PL appear as a line in the image. Although the intensity of the PL features depend strongly on the viewing direction, most damage processes produce slip on all of the allowed slip systems, so that one is not likely to "miss" a damaged region in a PL image.

Deformed material has been shown to be far more vulnerable to laser damage than undeformed material. Figure 17(d) shows an SEM micrograph of an indented region after exposure to high laser fluences of 248 nm irradiation. Interestingly, the most vulnerable regions are along the edges of the indent, rather than the center. This may be due to the constraint provided by the indentor, the greatest deformation (and highest point defect densities) may actually lie adjacent to the indentation, rather than underneath it. Residual tensile stresses around the edge of the indentor may also play an important role.

PL from Cleaved surfaces

Cleavage is the gentlest way to create a free surface. Nevertheless, PL imaging shows enhanced concentrations of laser-absorbing defects along many of the larger cleavage steps. An image of the luminescence appears in Fig. 18(b), alongside an SEM micrograph of the same region in Fig. 18(a). Several of the cleavage steps in the SEM micrograph are clearly imaged by their PL. An especially bright luminescence is observed in a triangular region in the lower portion of Fig. 18(b). The pattern of cleavage steps surrounding this triangular region suggests that the triangular region is undercut and thus would have been severely deformed prior to ultimate failure. The PL image also shows several bright spots which correspond to particles lying on the surface in the SEM image. These spots are MgO fragments produced during cleavage (what we have called "ejecta"[60,61] which the intense PL reveals to be highly deformed.

Cleavage steps on semi-brittle materials are generated by cracks running in parallel (but not coplanar) crystallographic planes. Swain, Lawn, and Burns have analyzed the process of step formation in semi-brittle materials,[62] finding that adjacent but non-coplanar cracks tend to overlap prior to ultimate failure. The deformation process in the material between overlapping cracks is illustrated in Fig. 19, where: (i) overlapping cracks propagate, (ii) shear stresses in the intervening material result in substantial deformation and dislocation motion, and

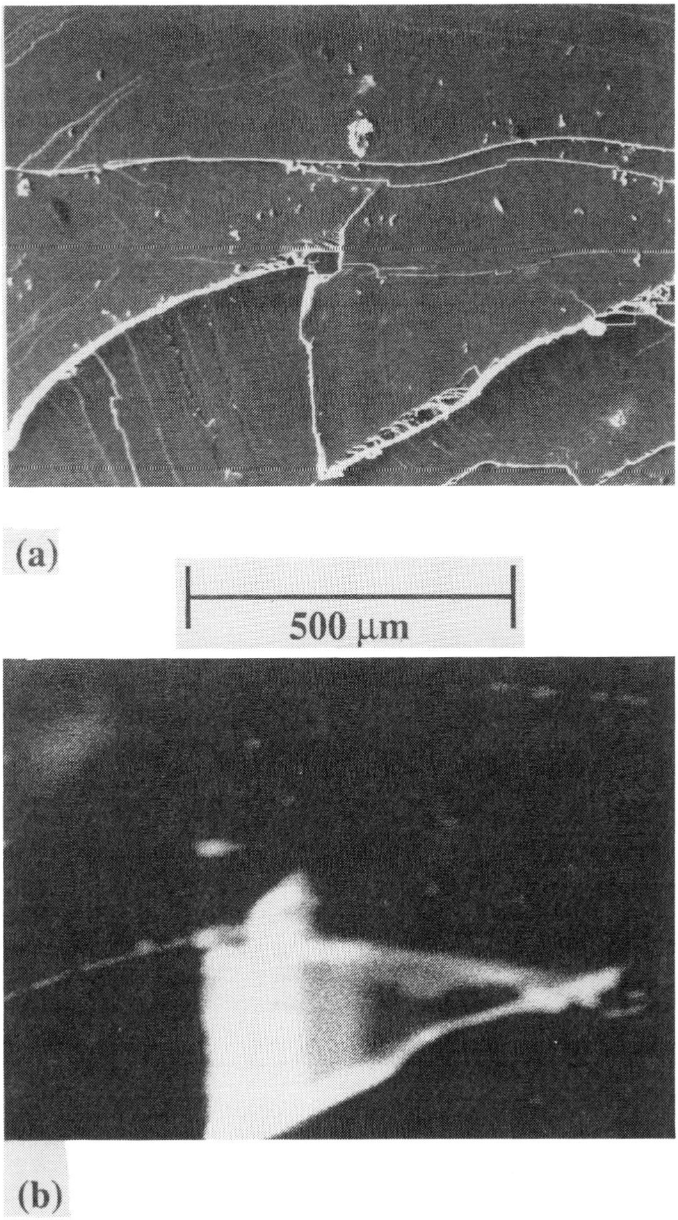

FIG. 18. (a) SEM and (b) PL images of a cleaved MgO surface.

(iii) failure of the adjoining ribbon of material leaves a undercut ledge containing high dislocation densities. High dislocation densities along many cleavage steps are readily observed by dislocation etch techniques in MgO[63] and LiF[62] (also a semi-brittle ionic crystal with the rock salt structure).

The amount of deformation along any given cleavage step is a strong function of its size. Cleavage steps which form under conditions of minimal overlap of crack fingers experience relatively little deformation and thus yield little luminescence. Similarly, the thick ribbon of material between large cleavage steps is often too rigid to deform significantly, so that large cleavage steps often show weak or insignificant luminescence. These effects account for the cleavage

Cleavage Step Deformation

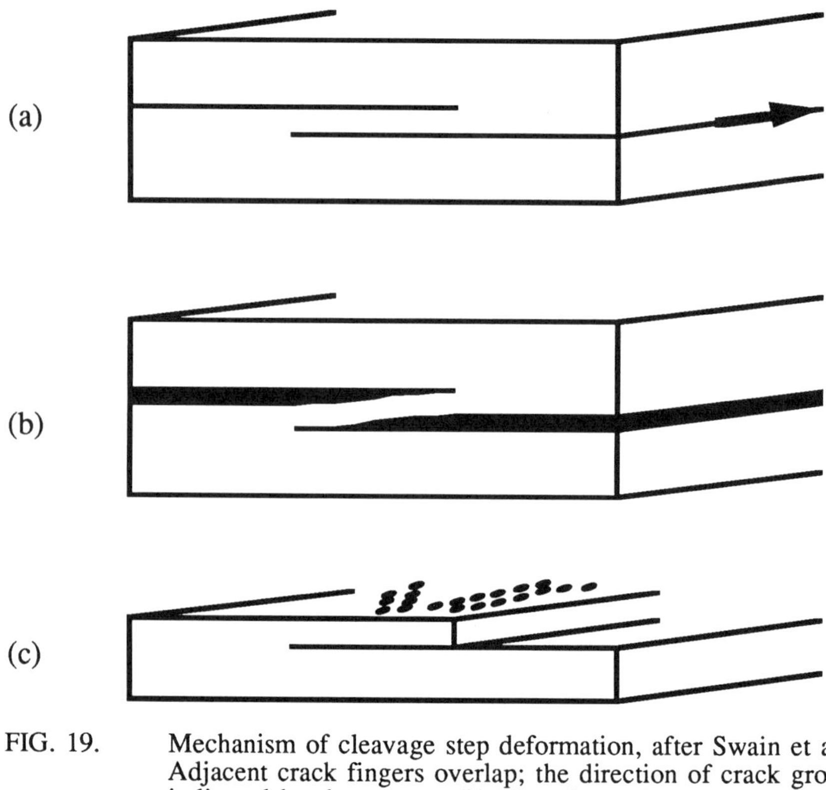

FIG. 19. Mechanism of cleavage step deformation, after Swain et al. (a) Adjacent crack fingers overlap; the direction of crack growth is indicated by the arrow. (b) The sliver of material between the crack fingers is deformed strongly in shear prior to failure. (c) A schematic diagram of the pattern of pits produced by a chemical dislocation etchant applied to the fracture surface of (b).

steps visible in the SEM image of Fig. 18(a) which display little, if any, luminescence in Fig. 18(b). Deformed, undercut material along cleavage steps can interact strongly with laser radiation at higher fluences; this material is often highly fractured after exposure to a single pulse of 248 nm radiation at fluences higher than 1 J/cm^2.[63] (This is significantly lower than the threshold fluence for steady-state laser etching of cleaved MgO, which is about 3-4 J/cm^2.) The ability of luminescence imaging to distinguish between highly deformed and relatively undeformed regions may prove useful in inspecting optical materials for optically important defects.

PL from Abraded Surfaces

Distinctive patterns of luminescence are also produced by abrasion. Typical white light and luminescence images of a surface damaged by drawing a diamond point from left to right appear in Fig. 20. The luminescence image clearly shows the characteristic triangular pattern of deformation due to scribing,[64] even in regions where the white-light image is obscured by coherent reflections from a pre-existing, subsurface crack. (The thickness of thc undercut material may have prevented significant deformation during cleavage, would account for the insignificant luminescence from the undercut material.)

The photoluminescence from mechanically damaged regions of single crystal MgO can be imaged with conventional stereomicroscope optics with resolutions of 10 μm. Higher resolutions could undoubtedly be obtained with improved optics. Further improvements in sensitivity and resolution could be attained by imaging the PL with CCD arrays. In principle, confocal microscopy (very small depth of field) could produce luminescence images of "sections" through the interior of the sample, providing three dimensional images of the deformation zone. The images of scratches indicate that wear tracks formed in a ceramic during tribological loading could be readily imaged and monitored during a wear test to quantify the extent of plastic deformation during the test.

Although this study employed MgO, similar phenomena are expected in other semi-brittle optical materials. For instance, F-centers in LiF produce a green luminescence, and we expect that the complex defect centers produced by deformation can be excited to produce luminescence in a similar region of the spectrum.

Dynamic Measurements of electron and photon emission during abrasion

When a stylus made of a hard material is drawn across a softer substrate, the substrate can respond by deforming elastically, plastically, or catastrophically by crack growth. The onset of substrate fracture increases substrate wear dramatically and often terminates the useful life of the substrate in tribological applications. Friction measurements usually show a marked change at the onset of substrate fracture, but typically yield little insight into the evolution of the surface prior to failure. To understand the processes leading up to failure, other probes of the wear process are usually necessary, e.g., microscopic examination of the wear track. However, these observations are difficult to make in real time during the course of a wear experiment. In the case of MgO, deformation and

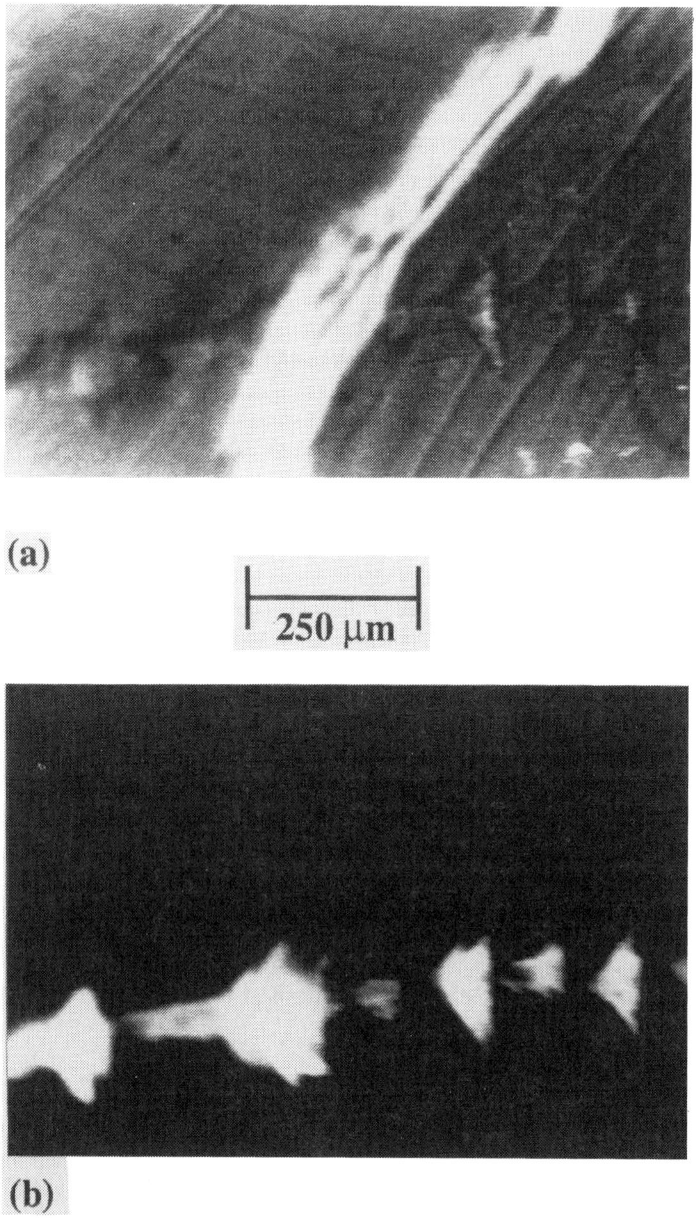

FIG. 20. (a) White light and (b) luminescence images of a scribed MgO surface. A diamond was drawn from left to right along a <100> direction.

fracture produce readily measurable photon emission (phE)[65,66] and electron emission (EE) which, like the coefficient of friction, can be measured in a time-resolved fashion during the wear process. These emissions form two components of so-called "fracto-emission",[67] (reviewed most recently in Refs. 68 and 69). We show here that these emissions provide complementary information on the evolution of MgO substrates during wear *in a time resolved fashion*, and shed light on the sequence of events which ultimately lead to ultimate failure.

MgO is a model material for wear studies,[70-74] largely because of its simple crystal structure and the availability of single crystal material. MgO also displays limited plasticity and significant work hardening at stresses well below those required for fracture. This property greatly simplifies the study of the deformation processes which ultimately lead to fracture. Although analogous processes are important in more brittle ceramics (e.g., TiC), the small difference between the yield and fracture stresses make it difficult to study the role of deformation in promoting the onset of failure.[75] In this section, we show that phE stimulated by deformation processes in MgO provides time resolved information on the course of plastic deformation during failure—with at least ms resolution. In contrast, the exposure of freshly formed MgO surfaces to the outside world from fracture is both necessary and sufficient to produce EE, and thus monitors the onset and evolution of localized catastrophic failure. Consequently, the two signals together provide significant and complementary probes of tribologically induced deformation and fracture.

A diagram of the apparatus for simultaneous phE, EE, and fracture measurements is shown in Fig. 21. As electron multipliers require vacuum conditions, the experiments were performed in a vacuum system maintained at pressures below 1×10^{-4} Pa. [We point out that photon emission measurements do not require vacuum and allow for a great deal more flexibility in the experimental arrangement.] Electron emission was detected with a Channeltron electron multiplier (CEM), Galileo Electro-optics Model 4039 mounted near the substrate. The front cone of the CEM was biased at +1000 V to attract and efficiently detect electrons. Photon emission was detected with a EMI Gencom 9924QB photomultiplier tube with a fused silica window; this tube is sensitive to 180-600 nm photons, with peak sensitivity in the blue and near UV region of the spectrum. The photomultiplier tube viewed the sample from below, through the transparent MgO substrate and poly(methyl methacrylate) (PMMA) sample mount. To provide high sensitivity, the electron multiplier and the photomultiplier were operated in the pulse-counting regime; the outputs were amplified, discriminated, and counted with standard nuclear physics instrumentation.

Single crystal MgO substrates ($2 \times 2 \times 0.3$ cm^2) were cleaved from a block from Tateho Chemical Industries, Ltd. The resulting surfaces were characterized by high densities of cleavage steps. To provide smoother surfaces for testing, some samples were polished to a 0.25-μm finish with diamond paste. Some polished samples were annealed at 1400 °C for two hours and rapidly cooled (10 °C/min) to provide smooth material with a lower yield stress. Abrasion was performed with a diamond scribe (nominal radius of curvature about 100 μm) mounted in a pivoted cantilever which allowed for the application of a controlled load. By moving the sample mount at a constant velocity, abrasion could be performed without moving point of contact relative to the detectors. This arrangement allowed for a high, constant photon collection efficiency, and a reasonable electron collection efficiency throughout the experiment. All of the time-resolved measurements described below involved single passes of the scribe over fresh, previously untested portions of the substrate.

The frictional force was determined by measuring the strain at the end of the aluminum rod with foil strain gauges with a sensitivity of 190 μV/N lateral force.

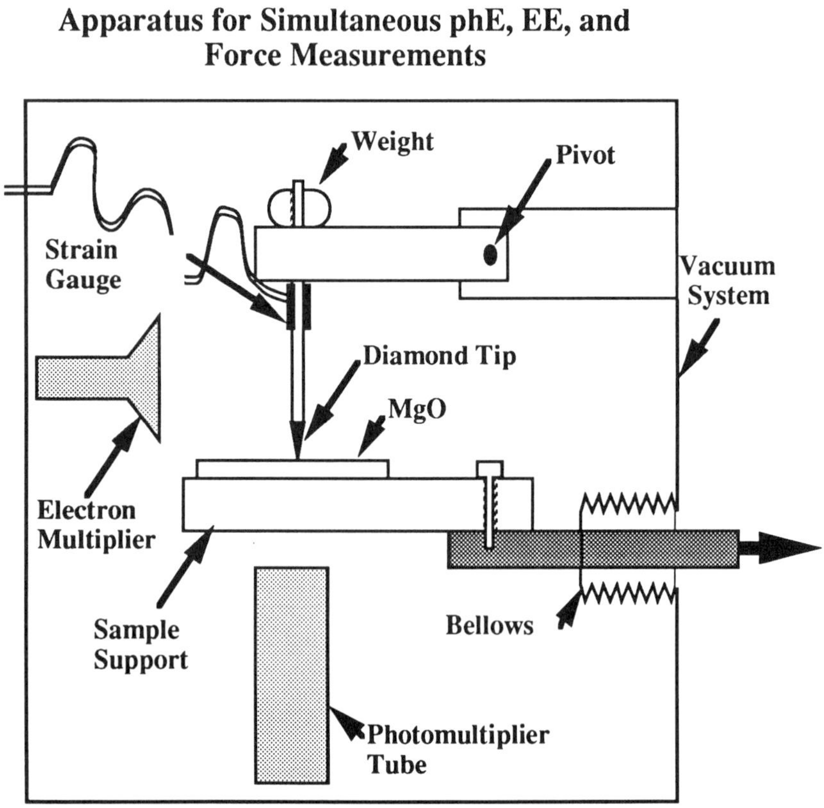

FIG. 21. Diagram of the apparatus employed for simultaneous phE, EE, and frictional force measurements in vacuum.

Nakayama and Hashimoto have measured the photon and charged-particle emission during the abrasion of Si_3N_4, Al_2O_3, and ZrO_2 with a diamond stylus in a variety of atmospheres. They attributed the observed emissions to gaseous discharges accompanying contact charging and dielectric breakdown of the ambient gas.[76,77] This does not appear to be the case for MgO. A phE spectrum generated during abrasion of single crystal MgO in air with a diamond stylus is shown in Fig. 22. The spectrum is characterized by three emission bands centered at about 710 nm, 520 nm, and 420 nm, respectively, and a broad shoulder into the UV. The absence of N_2 spectral lines is inconsistent with gaseous discharge in air. (Numerous studies in our laboratory of triboluminescence created by interfacial failure[41,78] show strong N_2 spectral lines due to breakdown caused by intense charge separation.) Thus we conclude that phE due to diamond on MgO is *not* due to electrostatic breakdown of gases; rather, phE is produced by the electronic excitation of defects in the solid phase. Furthermore, the same

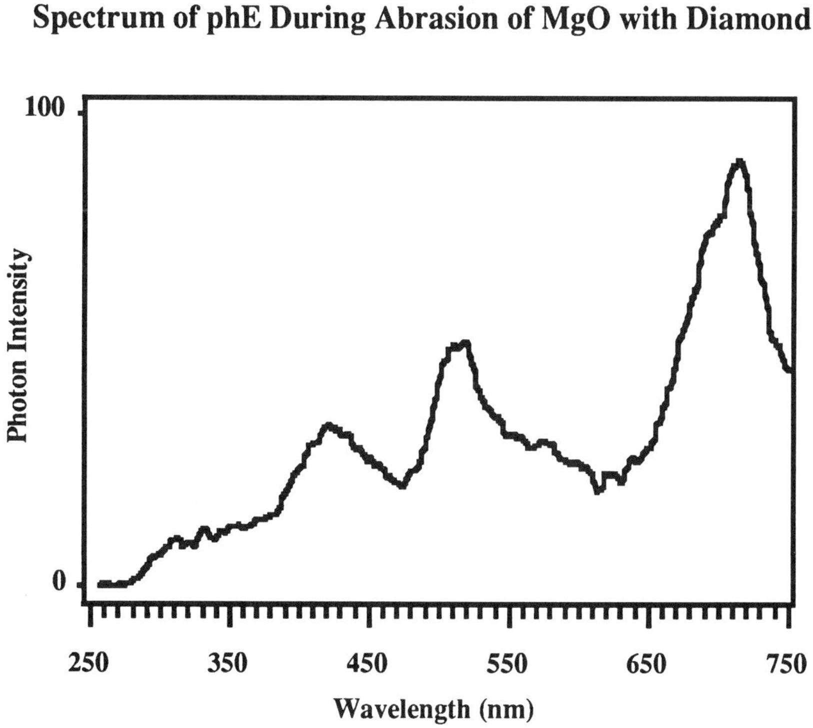

FIG. 22. phE spectrum produced by repeated passes of a diamond stylus over an MgO substrate.

mechanism should hold for experiments performed in vacuum. (The air plays no role.)

The major features of the phE spectrum during repeated abrasion can be readily interpreted in terms of previously observed excitations in MgO. The red, 710-nm band is characteristic of impurity Cr^{3+} (the Cr^{3+} R line),[79] with sidebands at longer and shorter wavelengths due to vibrational structure. The green, 520-nm band has been previously assigned to the decay of excited F-centers[80] (oxygen vacancies containing two electrons). Significantly, emission in this region has not been observed in the photoluminescence (PL) or cathodoluminescence (CL) spectra of previously deformed MgO. The blue, 420-nm band corresponds to the broad peak seen in the deformation-induced PL spectrum described above. (See Fig. 16.)

The short wavelength tail in the PL spectrum is especially intriguing as it requires the production of rather energetic excitations (> 4 eV). MgO CL spectra often show a peak in the 235-250 nm region,[81-84] which is attributed to electron tunneling from electron traps (e.g., F and F^+ centers) to magnesium vacancies (V- and V^--centers). Fe^{3+} impurities in the MgO employed in this work absorb relatively strongly at wavelengths below 260 nm and may account for the lack of a definite peak at shorter wavelengths.

Emission mechanisms

The mechanism for deformation-related phE clearly involves dislocation motion. We know that high densities of point defects are produced by deformation in MgO and the alkali halides via a jog dragging mechanism. Jogs formed at the intersection of moving dislocations impede (but do not necessarily stop) further dislocation motion. Crystallographic constraints require that a trail of vacancy pairs (or interstitial pairs) be produced when a screw dislocation "drags" a jog through the lattice. The relatively high energy required for interstitial formation strongly favors the formation of vacancy defects. Thus dislocation motion in the presence of high dislocation densities can produce high densities of vacancy defects.

Williams and Turner suggest that charge exchange between point defects and moving dislocations can excite point defects.[65] The best explanation for the characteristic 520-nm and 420-nm emissions are produced when conduction band electrons are captured *by* F^{2+}- and F^+-centers to form excited F^+- and F-centers, respectively. The unfilled centers (e.g., F^{2+}-) could be the result of newly created oxygen vacancies or the result of an ionization process (creating either F^{2+}- and F^+). There is some evidence that the conduction band along an edge dislocation is 1-2 eV below the normal conduction band.[85] If the F-center and F^+-center energies lie high in the band gap, as suggested by Gibson et al.,[86] moving dislocations could steal electrons *from* F-type centers. Refilling and subsequent radiation at 520-nm and 420-nm by retrapping of conduction band electrons at nearby oxygen vacancies or vacancy clusters would then produce these emissions. An additional possibility for free carrier production is through F or F^+ center annihilation by moving edge dislocations. Locally, an edge dislocation looks like a row of interstitials, and individual ions along the dislocation would be vulnerable to "recombination" with appropriate vacancies. The annihilation of oxygen vacancies containing trapped electrons would tend to promote these electrons to the conduction band.

Both phE and EE are observed when MgO is fractured in vacuum. One component of the phE is derived from the dislocation motion mechanism just described, and both phE and EE result from the recombination of charge carriers at appropriate traps.[66] In the present work, the majority of the phE is from deformation due to the high volume of material deformed by the stylus relative to the area of newly fractured material. Fracture-related phE may also be especially weak under wear conditions if the associated crack velocities are relatively low. phE and EE intensities during fracture tend to be strong functions of crack velocity.[87]

Simultaneous EE and phE Measurements

The yield stress of MgO, and consequently the wear behavior, can be significantly lowered by annealing treatments. At modest load and low stylus velocities, deformation and fracture are competing mechanisms for energy dissipation. Thus lowering the yield stress enhances plastic deformation and reduces the amount of fracture. Emission measurements were made on polished (not annealed) and polished-annealed MgO as a function of stylus velocity. The total number of phE and EE counts recorded during the production of a 1.5-cm wear track with an applied normal force of ~0.1 N are shown in Fig. 23. (Note log-log scales.) At stylus velocities below about 0.3 cm/s, the emissions from polished and polished-annealed samples both increase rapidly with increasing stylus velocity. In this range of velocities, the annealing treatment lowers the EE intensities an order of magnitude and raises the phE intensities almost a factor of two. This is consistent with our extensive observations that phE is principally due to plastic deformation processes, while EE is associated with fracture. One can perform indentation experiments on MgO showing that deformation alone produces *phE only*, whereas fracture produces both phE and EE. A significant portion of the phE generated by fracture is in fact due to dislocation motion accompanying fracture.

Typical phE, EE, and frictional force measurements during the abrasion of cleaved MgO with a diamond point are shown in Fig. 24. The normal force was 0.1 N and the stylus velocity relative to the substrate was 0.25 cm/s. We note that the background phE and EE signals (in the absence of abrasion) are very low (< 1 per second for EE; ~100 per second for phE); thus, virtually all the emission shown results from the abrasion process. Transient phE and EE signals were often observed near the onset and completion of the abrasion due to transients in the frictional force associated with starting and stopping the motion. Both the phE and EE signals display marked fluctuations during abrasion, principally in the form of numerous spikes. The EE signal also displays a slowly varying component; after each EE spike, the intensity does not immediately return to low levels, but decays with a characteristic time of ~100 ms. This is consistent with EE decays observed previously from fracture of MgO.[66] Due to the convolution of this decay from numerous spikes, the overall envelope of the EE is quasi-continuous. In contrast, the phE intensity after a phE spike returns to near-zero intensities on ms-time scales; thus, the great majority of the phE counts occur in bursts, with little "steady state" emission.

Measurements are readily made during abrasion at much faster stylus velocities. phE and EE produced at a stylus velocity of about 4 cm/s is shown in Fig. 25. About 100 ms were required for the stylus to come up to speed. (The emissions during acceleration are strongly nonlinear functions of tip velocity—especially for the EE—consistent with the velocity dependence of the emissions

Total phE & EE Counts during 1.5-cm Stylus Travel from Cleaved, Polished, and Polished-Annealed MgO

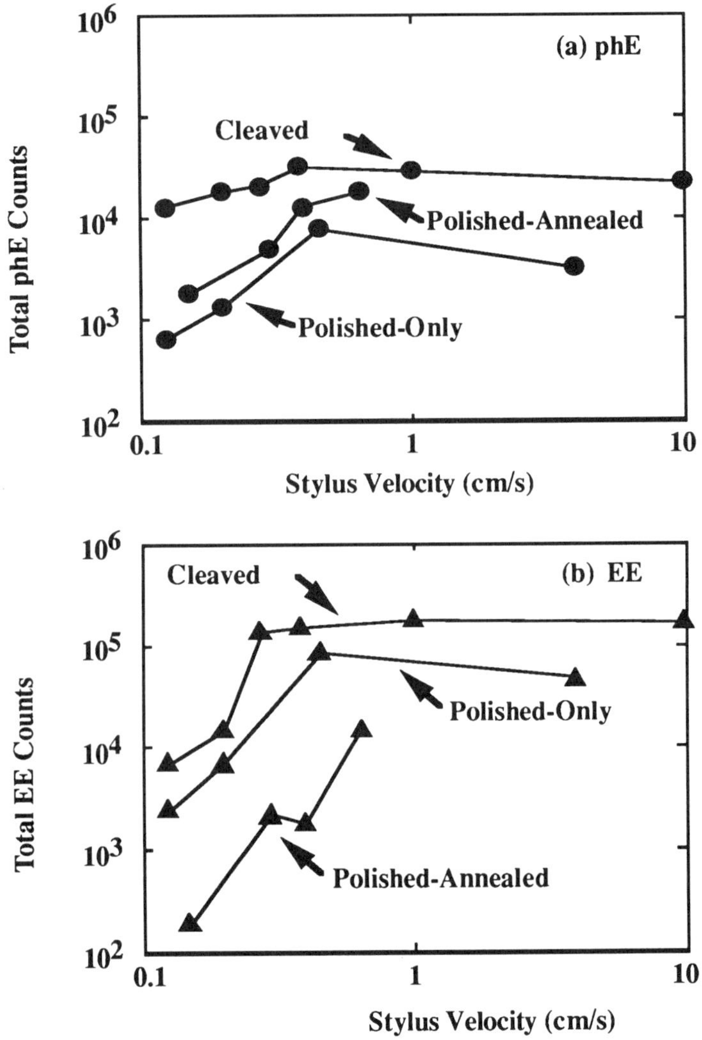

FIG. 23. Total number of phE and EE counts detected during the formation of a 1.5 cm wear track as a function of stylus velocity for cleaved, polished, and polished-annealed MgO. In each case, the stylus was loaded with a 10 g mass.

shown in Fig. 23.) During constant velocity stylus motion, all three signals show roughly periodic fluctuations with a period of about 50-ms, reflecting stick-sliplike behavior. We note that the natural period of the stylus mount was ~800 μs—much too short to account for these fluctuations.

The relationship between the three signals in Fig. 25 is complex. However, the overall behavior of the phE correlates with the frictional force quite well; i.e., the strong minima in the frictional force coincide with the minima in the phE. (These minima are marked by the dark vertical lines in Fig. 25.) Likewise, high frictional forces are accompanied by intense phE. We interpret this behavior in terms of the expected correlation between regions of high plastic deformation (thus high phE), significant plowing, and high frictional forces. The slip-like events are marked by rapid drops in the lateral force, presumably due to less plowing and therefore less deformation and less phE.

The EE fluctuations are similar to the phE fluctuations, except that the rise in EE intensity near the onset of a stick-like event lags the rise in phE and friction force by typically 5-10 ms. After a clear minimum in the EE (correlating with minima in the phE and frictional force), the rise in EE does not begin until the phE and lateral force have plateaued and steady-state plowing has commenced. Thus, at least at these faster stylus velocities, the more intense parts of the EE coincide with regions of extensive plastic deformation. MgO lacks the five independent slip systems required to accommodate arbitrary deformations, so that regions of extensive deformation are expected to display significant amounts of fracture. Optical microscopy of this wear track confirm the presence of large numbers of microcracks oriented perpendicular to the direction of stylus motion along potions of the wear track.

Wear Process

At modest loads, the passage of the stylus over a fresh portion of the substrate will nucleate slip bands and cause existing slip bands to grow. Fracture is attributed to the interaction of mobile dislocations forming immobile dislocation structures. These serve as defects (stress concentrators) that raise the effective yield stress and ultimately nucleate crack growth.[76]

These strong dislocation interactions are of substantial practical importance because of their role in concentrating stress and nucleating fracture. This process is likely responsible for the ultimate failure of many hard ceramics in wear applications. These dislocation interactions rarely produce progressive changes in the frictional force which might reflect the evolution of the surface toward catastrophic instability. However, the combined phE and EE emissions in this work show clear indications of fracture activity associated with high levels of deformation (marked by peaks or plateaus in the phE) prior to fracture (marked by peaks in the EE *following* the phE)—consistent with this scenario. Thus, phE and/or EE emissions can serve as time resolved probes of the evolution of the surface toward ultimate failure, including loading conditions where many passes of the stylus over the wear track occur.

The emissions and wear behavior are strong function of surface condition, load, and stylus velocity. In particular, phE is seen to be sensitive to deformation

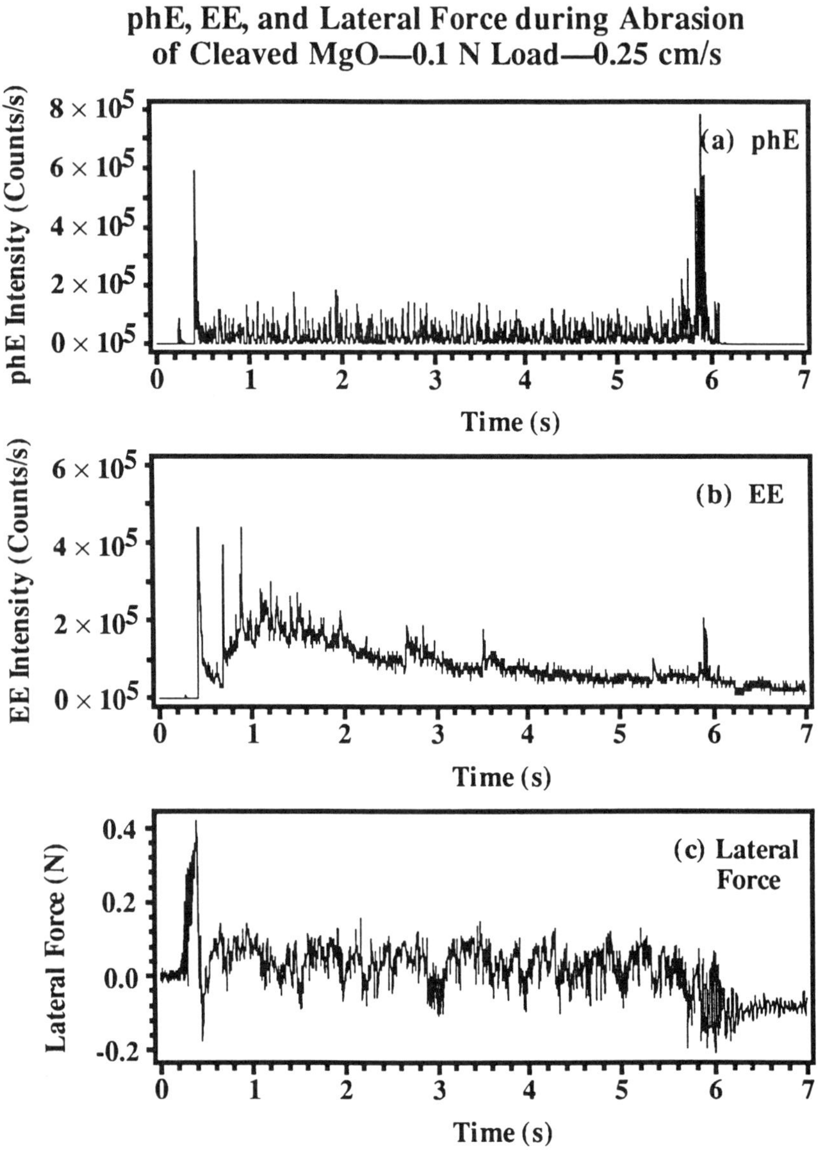

FIG. 24. Simultaneous measurements of (a) phE, (b) EE, and (c) lateral force produced by drawing a diamond stylus across a cleaved MgO surface at a velocity of 0.25 cm/s. A 0.1 N normal force was applied to the stylus.

processes in the bulk, corroborated by the increased phE intensities from annealed MgO. Similar increases in phE are observed during indentation of annealed vs unannealed MgO. Conversely, the surface-sensitive nature of EE, its especially strong dependence on stylus velocity, and the depressed EE intensities from annealed material are all consistent with EE due to localized fracture. During indentation of MgO, prior to any fracture, we observe phE only; no EE observed. The relative intensities of these signals can therefore be used to follow the progress and extent of plastic deformation and fracture during wear on ms-time scales.

Although this study was conducted in vacuum under ideal conditions for phE and EE detection, we point out that phE in particular can be detected with comparable efficiency in air, even in the presence of lubricating films. Opaque substrates are readily viewed from above with a slight loss of sensitivity due to solid angle. Fiber optics can be used to enhance photon collection efficiencies. Although complicated by the presence of electrical breakdown events, charged particle emissions (detected as total positive or negative current) accompanying wear have been made in gaseous[88] or liquid[89] environments. Measurements suggest that charge measurements accompanying wear of brittle substrates in gaseous atmospheres may reflect fracture behavior in a manner similar to the EE in this work.[88]

Concerning the present state of time resolution, we are confident that in many ceramic systems we can go to even faster time scales due to faster relaxation (decay) times and therefore even higher stylus speeds. Finally, phE and EE signals should provide similar information on the evolution of hard ceramic surfaces during cyclic wear and thus help elucidate the sequence of events responsible for the onset of excessive substrate wear in more practical tribological applications.

Chemisorptive electron emission as a probe of plastic deformation in metals during interfacial failure.

During chemisorption of highly reactive gases such as oxygen and halogens on clean metal surfaces, a non-adiabatic process can occur which results in the emission of electrons. This process is called chemisorptive electron emission (CSE). In previous work,[90] we showed that CSE is a sensitive measure of plasticity during tensile deformation of a number of metals. In principle, the method is similar to the use of photoelectron emission as a probe of tensile deformation[91] and fatigue damage[92] in metals such as Al and Mg. Steps produced by the intersection of dislocations with the surface increase the surface area of the low work function metallic phase and thus increase the photoelectron emission intensities. The chief advantage of CSE is its uniform sensitivity to deformation in a variety of sample geometries. Gas molecules tend to be adsorbed uniformly on complex surfaces, whereas in photoelectron measurements, portions of the surface are often shadowed from the UV source. Further, the background CSE signals from undeformed material are often quite low due to the presence of a thin oxide layer. CSE requires adsorption on fresh metal, whereas uv photons and photoelectrons can penetrate sub-μm oxides to

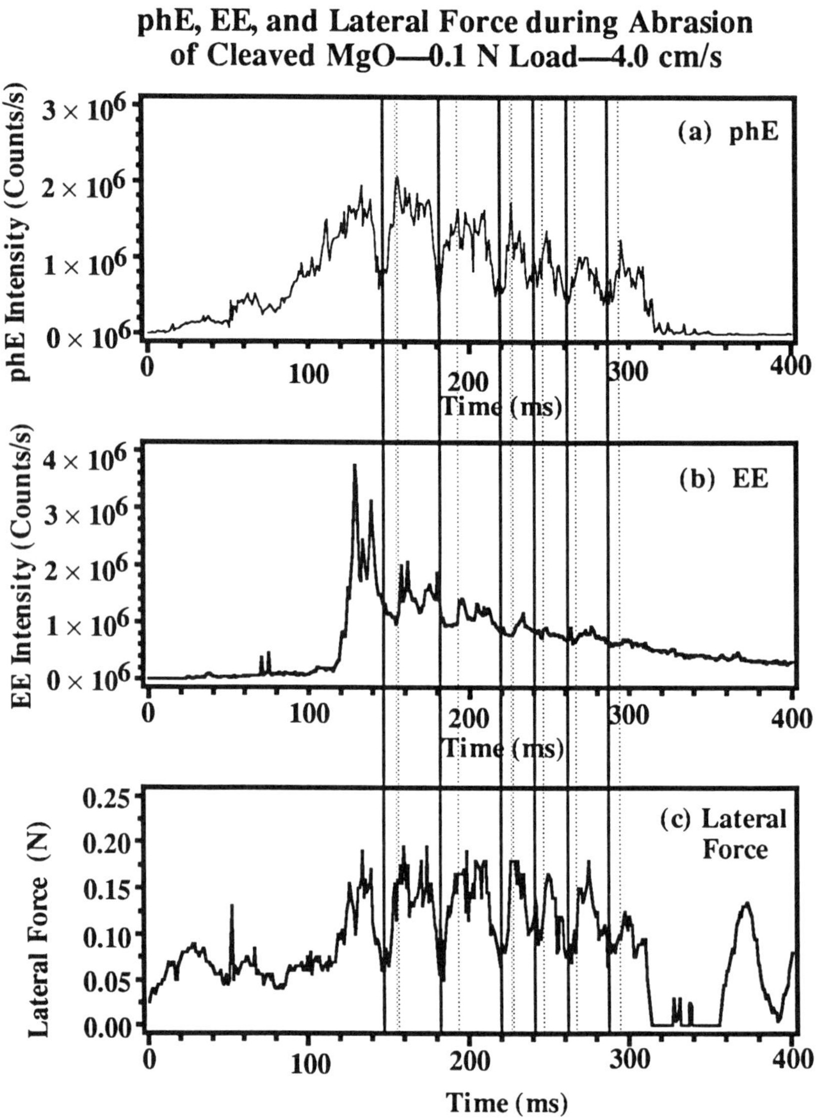

FIG. 25. Simultaneous measurements of (a) phE, (b) EE, and (c) lateral force produced by passing a diamond stylus across a cleaved MgO surface at a velocity of 4 cm/s. A 0.1 N normal force was applied to the stylus. The dark vertical lines mark minima in the phE and strain signals, which correspond to the slip portion of the stick-slip-like cycle; the light vertical lines mark the rise in EE after each slip event.

produce intense signals even in the absence of deformation. Thus, CSE is a superior probe of the increase in surface area accompanying plastic deformation.

Here we show how this method can be used to probe the extent of plasticity occurring during interfacial failure at a metal-glass interface.[93] In metal-ceramic composites, the plastic deformation associated with void formation in the metallic phase is often the dominant dissipative process during interfacial failure. This deformation is a strong function of the tortuosity or roughness of the interface,[94,95] the thickness of the metallic phase,[96,97] and the presence of interphase materials.[98] When competing mechanisms of energy dissipation operate simultaneously, the contribution of deformation to the measured toughness must be estimated by careful fractographic examination and numerical modeling. More direct methods are desirable, and can be especially useful when ductile deformation of the metallic phase and brittle fracture at or near the interface occur simultaneously.[99] CSE thus provides an alternative probe of the ductile phase.

The experiments here were conducted in vacuum with a base pressure of less than 1×10^{-7} Pa. A schematic of the experimental arrangement is shown in Fig. 26(a). Oxygen was used as the reactive gas and was leaked into the system through a clean manifold and leak valve, raising the pressure in the vacuum system to 4×10^{-4} Pa. Electron emission was detected with a Channeltron electron multiplier (CEM). The front cone of the CEM was biased at +1000 V for efficient electron detection. Within about 2 s after the completion of peel, the two sides of the sample were drawn apart, allowing for efficient detection of electrons emitted from the Mg surface.

Measurements of CSE intensity as a function of Mg surface area (by exposing a freshly evaporated Mg film to oxygen) yielded 3×10^8 electrons/cm^2, which corresponds to a yield of 1×10^{-7} electrons/Mg surface site. SFM scans of the Mg film surface confirmed that its surface roughness was negligible.

The Mg/glass interfaces were formed by evaporation in vacuum. After cleaning, the slides were mounted in an ion-pumped deposition system with a base pressure of $< 1 \times 10^{-5}$ Pa. The film thickness was monitored by a quartz microbalance positioned near the glass slide. Small angle x-ray scattering measurements on a typical 1–µm film indicated a high degree of orientation in the polycrystalline film, with the c-axis of the grains aligned normal to the surface. A similar preferential orientation has been observed in Mg films deposited by "cluster-beam" techniques onto room temperature glass substrates.[100]

The sample geometry employed is shown in Fig. 26(b). A triangular piece of Mg-coated glass was epoxied between two other glass slides to which the load was applied. Sandwiching the metal film between brittle substrates minimized the effect of residual stresses on crack growth.[99] Failure initiated at the tip of the triangle and proceeded along the Mg/glass interface, the Mg remaining on the epoxy. After initial "pop-in" of a small starter crack, steady crack growth followed at roughly constant load. Microscope observations indicated that the starter crack represented only a small fraction (< 1%) of the total surface area.

The total energy expended during crack growth (here designated the peel energy) was determined from the total area of the constant load portion of the load vs displacement curve. We note that the energy so derived is not simply related to the critical strain energy release rate at the onset of crack growth (except possibly in very brittle materials), but represents an average energy for surface formation during *continuous* crack growth. As the great majority of the surface is formed during continuous crack growth, the peel energy is a more appropriate measure of deformation for comparison with emission measurements.

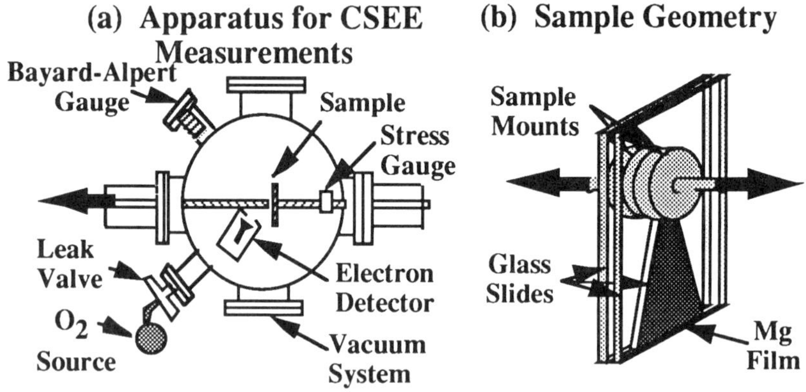

FIG. 26. Diagram of the (a) apparatus employed in the CSE measurements and (b) geometry of the peel sample.

Figure 27 shows the CSE emission intensities (log) vs time from a high peel energy surface (a 1-μm film with a peel energy of about 16 J/m^2) and a low peel energy surface (a 0.5-μm film with a peel energy of 3.2 J/m^2). Both samples yielded intense emissions during the peel event itself, saturating the electron counting electronics. The emission from the 0.5-μm film drops rapidly after peel and continues to drop for the duration of data acquisition. Although the emission from the 1-μm film drops two orders of magnitude after peel, it subsequently returns to high levels before beginning a long, slow decay. Furthermore, the emission from the 1-μm film depends strongly on the partial pressure of O_2, falling sharply when the O_2 pressure is reduced and rising when the O_2 pressure is restored. A strong pressure dependence is characteristic of CSE. In contrast, the emissions from the 0.5 μm-film are not significantly affected by changing the O_2 partial pressure, implying very small CSE intensities. Thus peeling the 0.5-μm film has exposed very little metallic Mg, consistent with failure through a brittle interphase. In contrast, the emissions from the 1-μm film correspond to a metallic Mg surface coverage of about 16% of the nominal surface area. This is still a relatively small fraction of the nominal surface area, consistent with localized patches of deformed material in conjunction with failure through a brittle interphase. This apparently small amount of deformation has a dramatic effect on the peel energy.

Electron emission during and immediately after crack growth has been investigated previously,[101,102] and occurs with or without O_2; the fracture-

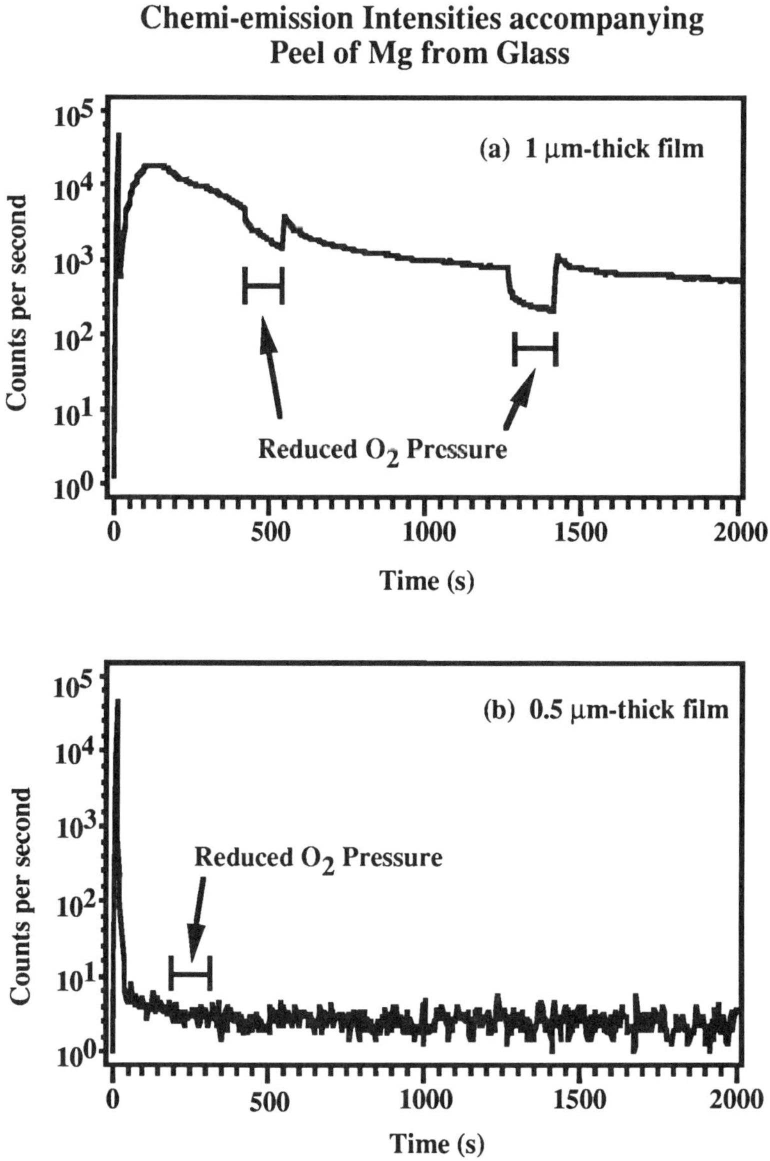

FIG. 27 Typical CSE signals following the peel of Mg films from soda-lime glass: (a) a 1-µm film and (b) a 0.5-µm film. The partial pressure of O_2 was lowered from 4×10^{-4} Pa to about 7×10^{-6} Pa twice in (a) and once in (b).

induced emission alone is weakly correlated with peel energy. For instance, both samples in Fig. 27 show intense emissions during peel. To reduce the influence of these prompt emissions on sample-to-sample comparisons, the emissions during the first fifty seconds after peel were not included in the analysis. Good results were obtained when the electron counts were integrated from 50 s to 1200 s after the onset of peeling. The resulting CSE intensities are plotted in Fig. 28 as a function of peel energy. The peel energies of the 1-μm films (determined from the load curves) range from about 8 to about 16 J/cm^2 and are strongly correlated with the CSE intensities. The low peel energy point corresponding to the 0.5-μm film of Fig. 27(b) is also included in Fig. 28, along with a least squares fit to the data corresponding to the 1-μm films (omitting the 0.5-μm film point). The least squares fit to the 1-μm film data crosses the x-axis near the point corresponding to the 0.5-μm film, which yielded extremely weak emissions. Thus the peel energy of the 0.5-μm film appears to represent the "zero plastic deformation" limit of failure in this system.

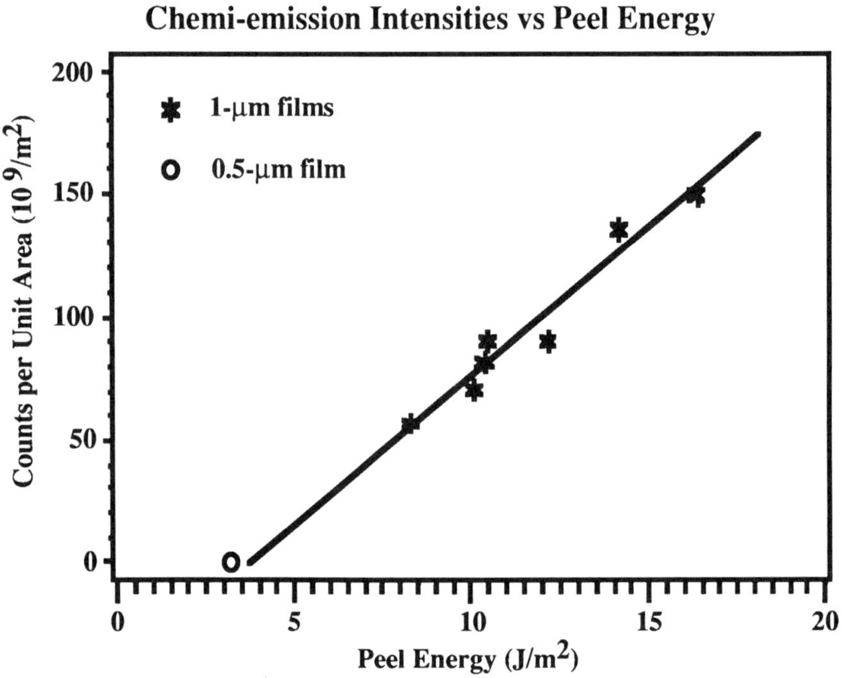

FIG. 28. A plot of CSE intensities vs peel energy, where the CSE intensities include the total number of electrons detected beginning 50 s after the peel event and ending 1200 s later. The line represents the least squares fit to the data from the 1-μm films only.

The peel energy of the 0.5-μm film, about 3 J/m^2, is well below metallic fracture energies, and is significantly lower than the fracture energy of the soda lime glass substrate (about 9 J/m^2). Magnesium oxides [e.g., MgO(100)[103]] have fracture energies similar to that of the 0.5-μm film and would readily form in the early stages of deposition at the pressures employed in this work. The presence of a brittle interphase can have dramatic effects on the fracture of metal-ceramic composites.[96] We attribute the rather large sample-to-sample variation in peel energies of the 1-μm films to variations in the metal microstructure and interphase composition. Both phases are strongly affected by the deposition rate, which varied somewhat from sample to sample. The good correlation between peel energy and CSE intensities suggest that the sample-to-sample variations in peel energy are due to real differences in material behavior, rather than measurement uncertainties.

Figure 29 shows typical SFM images of the Mg surfaces which yielded the emissions of Fig. 27. All of the high peel energy, high emission surfaces showed large numbers of submicron-sized voids along the surface, similar to those in Fig. 29(a), confirming the presence of ductile deformation. No similar voids were observed on the low peel energy surface, as shown in Fig. 29(b). The voids in the high peel energy samples were typically about 10 nm deep, with relatively flat bottoms and angular sides. The angular sides contrast markedly with the smooth ellipsoidal profiles of ductile voids in FCC metals. The formation of smooth ellipsoidal features in Mg (HCP structure) is hindered (relative to the FCC metals) by the absence of the five equivalent slip planes required to accommodate arbitrary deformations.[104]

SFM images of the mating glass surfaces show no evidence for voids or protrusions comparable to the voids on the high peel energy Mg surfaces. Therefore we attribute void formation in the Mg to nucleation and ductile growth, similar to the formation of ductile dimples in thicker metallic films.[105]

Chemisorptive electron emission occurs immediately prior to the collision of an electronegative atom or molecule (e.g., O, O_2, X, X_2—where X = halogen) with an appropriate metal surface.[106-108] As the gas atom approaches the surface, the energy of its lowest unfilled electron state (whose energy defines the electron affinity of the molecule, E_A) drops. If E_A falls below the Fermi level of the metal, E_F, an electron can tunnel from the metal to the atom to produce a negative ion. In the rare event that the difference between E_A and E_F is greater than the metal work function ϕ, tunneling leaves behind a hole in the valance band deep enough to stimulate electron emission. Although the emission probability per incident molecule is small (typically $\leq 10^{-6}$), the number of surface sites is very large (typically 10^{15} cm^{-2}). With our nearly single-electron sensitivity, CSE is readily observable for reactive metals such as Mg, Ti, Zr, Al, and alkali metals exposed to gases such as oxygen and the halogens.

Assuming a simple geometry for the deformation process, the relation between the creation of surface area and the work of deformation can be made more quantitative. An edge dislocation of length l oriented parallel to the surface and moving into the material will produce a step of length l and height b, where b is the magnitude of the Burgers vector. The corresponding change in surface area is $\Delta A = b \cdot l$, and the work of performed as this dislocation moves a distance d into the material is on the order of $\sigma_y \cdot b \cdot d \cdot l$, where σ_y is the stress required to move

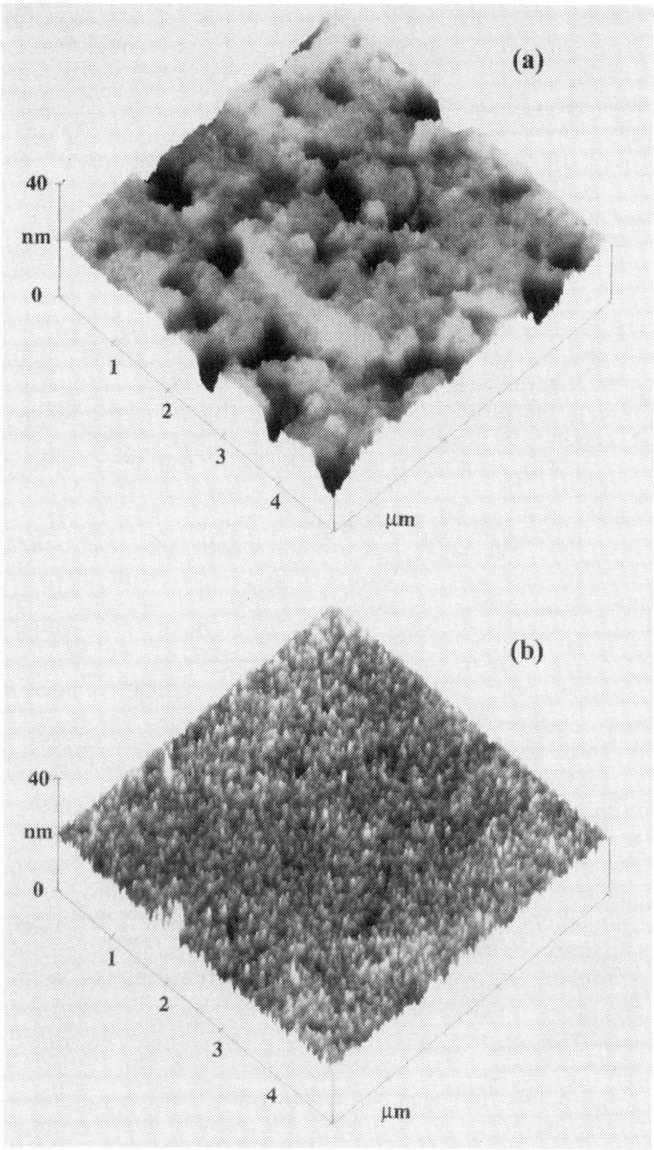

FIG. 29. SFM images of peeled Mg surfaces: (a) the 1-μm film yielding the intense CSE shown in Fig. 27(a), and (b) the 0.5-μm film yielding the very weak CSE shown in Fig. 27(b).

the dislocation. The work done per unit surface area, $W_{plastic}/A$, can then be expressed as

$$\frac{W_{plastic}}{A} \approx \sigma_y \cdot d \cdot \frac{\Delta A}{A} \quad , \tag{2}$$

where $\Delta A/A$ is the ratio between the surface area produced by deformation and the nominal area of the peeled surface. This line of reasoning equates $\Delta A/A$ with the effective strain at failure, ε_o, where one expects $W_{plastic}/A \sim \sigma_y \cdot d \cdot \varepsilon_o$.

Quantitative CSE measurements can be used to infer $\Delta A/A$; thus the only two unknowns in Eq. 2 are σ_y and d. When the distance between ductile voids is much greater than one tenth the film thickness (as in this work), it is generally adequate to assume that the film is uniformly deformed throughout its thickness.[98] However, the effective yield stress is somewhat uncertain. Cold rolled Mg sheet yields substantially at a stress of 100 MPa. With $\sigma_y \approx 100$ MPa and $d \sim 1$ µm, the maximum observed value of $\Delta A/A \approx 0.16$ corresponds to a predicted work of deformation $W_{plastic}/A \approx \sigma_y \cdot d \cdot \Delta A/A \approx 16$ J/m^2. The corresponding experimental peel energy was about 16 J/m^2, which corresponds to a work of plastic deformation $W_{plastic}/A \approx 13$ J/m^2. In view of the uncertainties, better agreement is not expected. The surface areas inferred from CSE measurements are clearly of the correct order of magnitude to account for the observed enhancements in peel energy due to plastic deformation.

CSE signals are sensitive to small amounts of deformation. The work of deformation corresponding to a CSE signal ten times the "zero deformation" limit (from the 0.5-µm film point in Fig. 28) is less than 0.1 J/m^2. In smaller samples, with correspondingly small background count rates, a fresh Mg surface of area $\Delta A = 10^{-2}$ mm^2 would yield readily detectable 10^3-10^4 counts.

Roughness measurements based on SFM observations were also made for potential correlation with peel energy measurements. These measurements displayed significant "magnification effects,"[109] in that the measured roughness was a strong function of the scan size. Quantitatively, the measured roughness was poorly correlated with peel energy at the smallest scan sizes attempted (500 × 500 nm^2) and only moderately correlated with peel energy at larger scan sizes (up to 50 × 50 µm^2). The poor correlation with peel energy is apparently due to the wide range of length scales required for proper characterization—from atomic steps (associated with the Burgers vectors of individual dislocations) to the dimensions of the ductile voids (some hundreds of nm in diameter). Atomic resolution has not yet been demonstrated by either SFM or STM of surfaces as rough as the Mg surfaces in this work. Novel acquisition methods, such as large-area scans with extremely high digitization densities, may help solve this problem. Due to its high lateral resolution, scanning tunneling microscopy (STM) can also be a valuable tool. For instance, Warner and Bonnell[110] have used STM to examine the metal side of a fractured ceramic-metal interface and noted correlations between the observed topography and fracture energy. Nevertheless, the resolution of STM can be limited by surface roughness and the presence of insulating interphases and oxides formed after failure.

Applying these techniques to interfaces involving other non-ferrous metals would be interesting. For instance, FCC metals like Al deform more homogeneously (due to local strain hardening) and yield smoother (although often larger) surface features. In contrast, the HCP metals, such as Mg, Ti, and Zr, tend to deform more inhomogeneously (due to local strain softening) and produce rougher surfaces. CSE measurements on these materials would yield reproducible surface area values despite the contrasting morphologies.

Dynamic measurements of transient currents generated during interfacial failure

In this last section we present briefly methods still under development for investigating dynamically the micromechanical aspects of fracture at interfaces. It involves the detection of small currents generated by the separation of two dissimilar materials when fracture is occurring along an interface. This has been applied to the fracture of an embedded rod in an insulating matrix,[111] and to the peeling of pressure sensitive adhesives from a metal substrate.[41,42] The extremely nonlinear, viscous behavior of pressure sensitive adhesives results in considerable energy dissipation during peeling over a wide range of peel speeds and substrates. In particular, the formation and elongation of adhesive fibrils during adhesive detachment is generally the dominant dissipative process during peeling, provided that these fibrils do not fail prematurely (cohesively or adhesively). The current fluctuations during peel change dramatically with peel speed and can be analyzed to extract information on the dynamics of the peel process. Numerical analysis of these signals shows evidence of deterministic chaos.[42]

There are several geometries possible for detection of these transient signals. A diagram of the experimental geometry is shown in Fig. 30. Figure 31 shows typical current measurements in air at three peel speeds. The magnitude of the average current (between negative spikes) increases with peel speed. Under these conditions a strong correlation between load and low-frequency current fluctuations are observed, where the highest currents are observed during periods of falling load. This is consistent with predictions that the current is proportional to the rate of detachment. When the tape is drawn at a constant speed, transient increases in the rate of detachment decrease the applied load and increase the current. On a finer scale, localized, rapid detachment results in a contribution to current at higher frequencies. The sum of these events produces the observed current.

The current fluctuations are readily analyzed by a variety of numerical techniques. For instance, the power spectra of the current signals at three different peel speeds are shown in Fig. 32. These spectra were computed by two distinct methods, once from the Fourier transform of the current signals and once by a maximum entropy technique.[112] Maximum entropy methods typically render broad band features in a smoother fashion, but are also capable of representing isolated, sharp features. We have interpreted these spectra in terms of the various size and time scales involved in the fibril formation and detachment process. Obviously, strong contributions are seen in the 40-120 Hz range and change with peel speed.

The maximum entropy "spectral estimates" follow the average behavior of the power spectra computed from the Fourier transforms quite well. At the lower two peel speeds, the general shape of the spectral estimates are similar to the

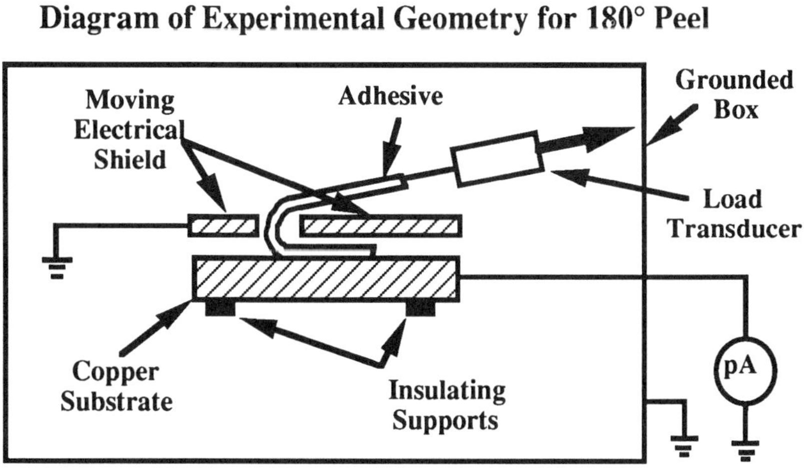

FIG. 30. A typical arrangement for measuring transient currents generated by peeling dielectrics (e.g., pressure sensitive adhesives) from conducting substrates.

spectra of an underdamped harmonic oscillator with a resonant frequency of 50-60 Hz. As noted below, this comparison can be made quantitative. It is important to note that the features in the 50-60 Hz range are at frequencies too low to be attributed to individual discharge events (which produce the negative current spikes) and are too broad to be attributed to electrical noise (which produces distinctive, but much smaller features). Thus they reflect a mechanical resonance along the peel front.

As the peel speed is increased to 6.0 mm/s, the power spectra of the current fluctuations change dramatically. A broad peak centered at zero frequency appears, suggesting a highly overdamped mechanical response, although a shoulder at about 75 Hz suggests a weak underdamped response remains. The power spectrum at 6 mm/s is well described by two harmonic oscillators in parallel, one overdamped and one underdamped. This behavior may reflect an inhomogeneity in the mechanical response. For instance, void formation and interfacial failure *ahead of* the nominal peel front could yield current fluctuations with an entirely different (highly damped) character than interfacial failure along the peel front itself. We have presented simple

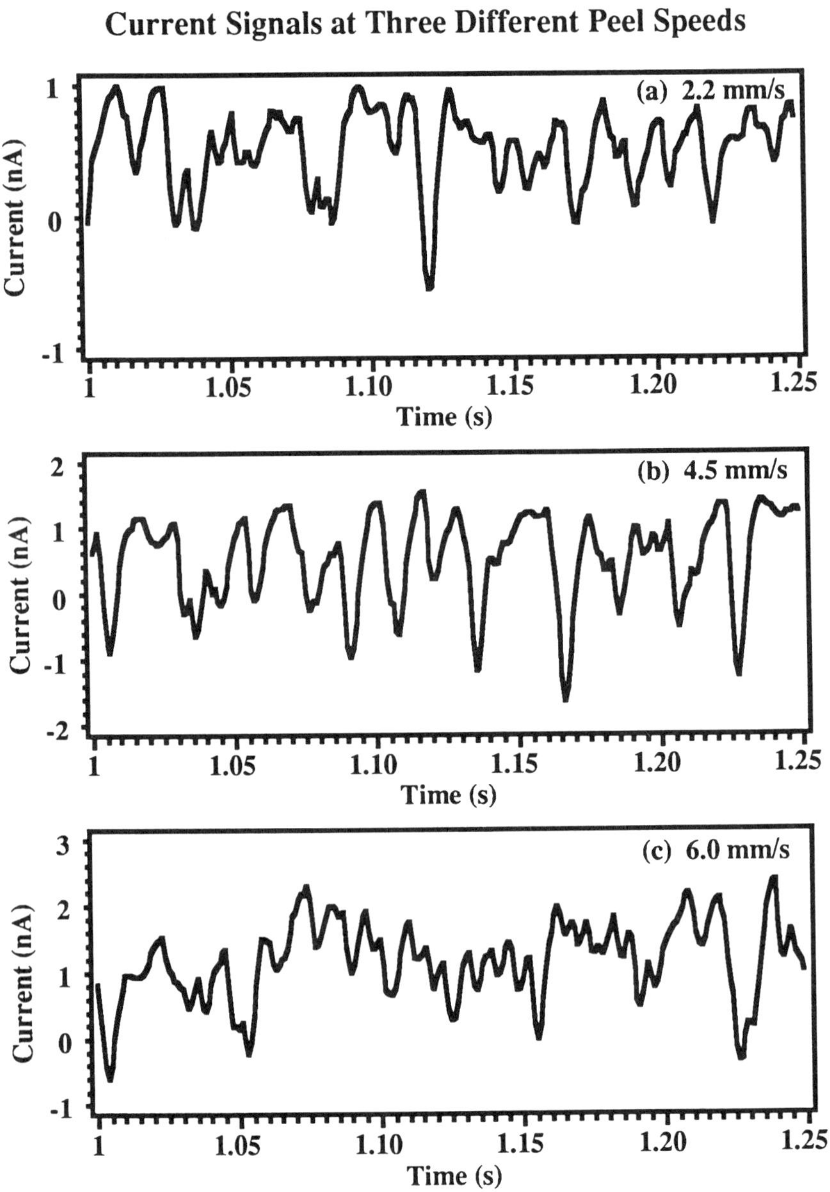

FIG. 31. Current observed at peel speeds of (a) 2.2 mm/s, (b) 4.5 mm/s, and (c) 6 mm/s.

mechanical models involving damped nonlinear oscillators that can generate these maximum entropy spectra fairly well.[113]

Chaotic systems typically display broad band noise similar to that observed here.[114] The definitive test for chaos is an extreme sensitivity to initial conditions, which can be quantified in terms of Lyapunov exponents. Chaotic systems have at least one positive Lyapunov exponent. Some of our data sets with very high signal to noise ratios are now yielding positive Lyapunov exponents. In addition, we are able to numerically test the data for fractalbehavior (the search for a "strange attractor").[115] Although fractal character does not mathematically require chaos, few fractal, nonchaotic systems have been identified. This analysis confirms that the current signals are indeed fractal ($2.5 < D < 4.5$), consistent with a chaotic origin. These dimension measurements also exclude the possibility that the corresponding current fluctuations (reflecting mechanical and/or electrical perturbations along the peel front) are random. Random signals are inherently infinite dimensional,[116] so that dimension estimates on random signals fail to converge altogether.

Perhaps of more interest to the ceramics community, we are now extending these methods to measurements of currents generated by peeling metals from glass, attempting to resolve the currents on μs time scales. One important implication on these current measurements is that the magnitude of the current can be related to a *minimum* charge density that existed at the interface before separation (reneutralization of tunneling of charge in the cracktip reduces the charge separation). This yields the minimum contribution of contact electrification (through Coulomb forces) to the work of adhesion. To date, our measurements show this to be non-trivial for a number of metal-glass combinations.

Conclusions

Current signals accompanying interfacial failure between conducting and nonconducting phases can provide time-resolved information on the course of failure. In the case of pressure sensitive adhesives, visual observations of adhesive detachment reveal a highly branched structure of adhesive "fingers" in a constant state of flux—branching, merging, extending, and retracting. However, these visual observations are very difficult to quantify in a time resolved fashion. Current measurements provide a time-resolved picture of the collective behavior of this very complex dynamical system. Numerical analysis of the current signal yields direct information on the mechanical response of the peel front, and indicates that the peel front is perturbed in a nonrandom, probably chaotic manner. Our present efforts are focused on improving the sensitivity and frequency range of the electrical measurements, was well as improving observational techniques for characterizing the events occurring in the peel zone.

In ceramic metal composites involving reactive metals, CSE measurements have considerable potential as a probe of deformation in material and process evaluation. The ductility of thin metal films may be a key to understanding deformation 10-100 nm ahead of the crack tip, where the local

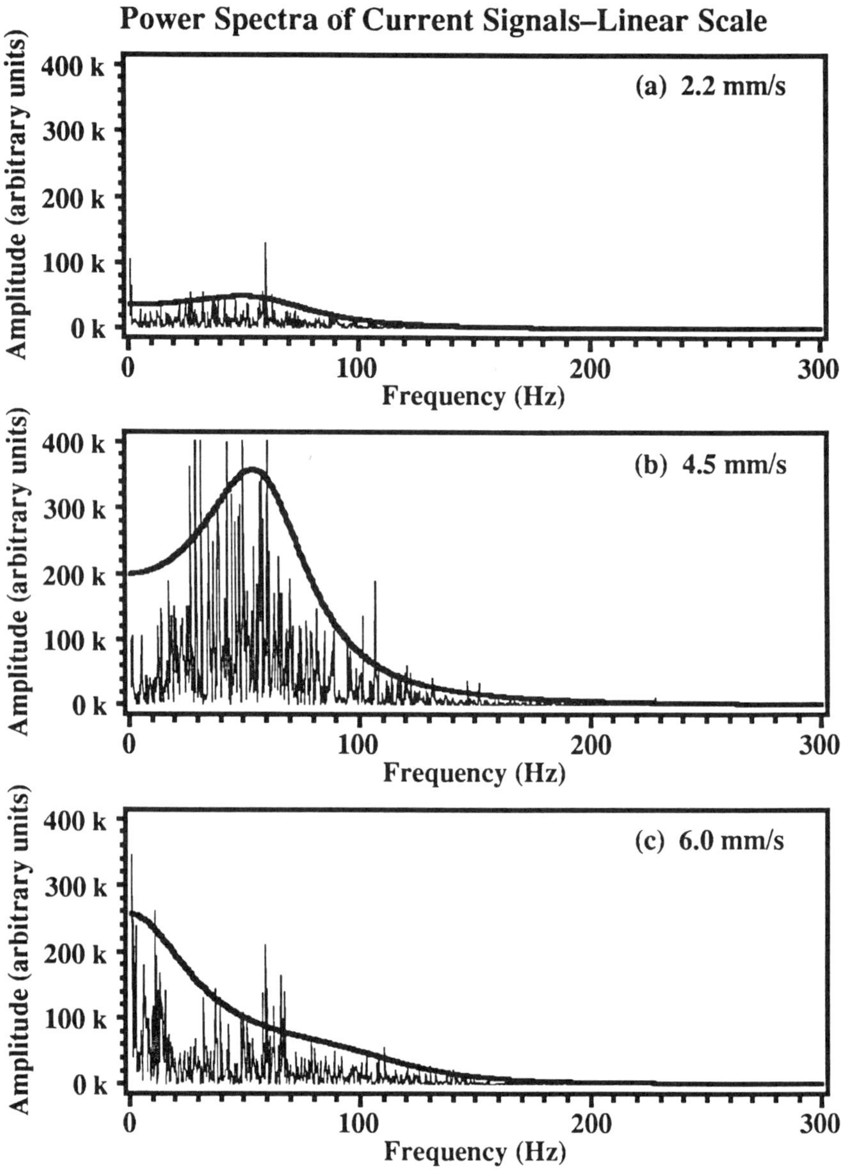

FIG. 32. Power spectra of the current signals of Fig. 31 accompanying peel at (a) 2.2 mm/s, (b) 4.5 mm/s, and (c) 6 mm/s. The vertical bars show the power spectra computed from the Fourier transforms of the current signals, while the smooth curves show power spectra estimates computed by the maximum entropy method.

stresses are not well understood.[96] The sensitivity of CSE measurement to deformation in the metallic phase is particularly valuable in systems where the contribution of competing fracture processes (e.g., fracture through a brittle interphase or fracture of brittle reinforcements) is variable or not known. We expect that this technique can be readily extended to thicker metallic structures (possibly commercial metal-ceramic composites) and to other reactive metals, including Ti, Zr, and Al, to provide quantitative measures of the integrated work of plastic deformation over entire fracture surfaces.

Current and particle emission processes ultimately depend on atomic-level events, which allows for potentially very high spatial and temporal resolution. Departures from *adiabatic* separation of atomic planes or molecular bonds are reflected in the topographical, electrical, and chemical states of the resulting fracture surface. Thus these signals typically reflect energy-consuming processes of great interest for their potential to limit the rate of crack growth. The details provided by these and similar studies complement recent developments in fracture mechanics, where atomic scale models of fracture are being extended to still larger length scales. Microscopic, dynamic descriptions of crack growth and related processes may require new syntheses of mathematical formalisms, where fractal geometry and chaos play key roles.

Acknowledgments

We wish to thank John Hirth, Richard Webb, Kah Siek, Mary Dawes, and Les Jensen, Washington State University, for valuable discussions and assistance in the laboratory. This work was supported by the National Science Foundation Surface Engineering and Tribology Program under Grant CMS-9414405, a National Science Foundation Instrumentation Grant DMR-9205197, the Department of Energy, BES-Chemical Sciences DE-FG06-92ER14252, the Murdock Charitable Trust, the Japan Private School Promotion Foundation, and the Yonam Foundation.

References

1. M. D. Antonik and R. J. Lad, “Faceting, reconstruction, and defect microstructure at ceramic surfaces revealed by atomic force microscopy,” J. Vac. Sci. Technol. A 10(4), 669-673 (1992).
2. S. C. Langford, L. C. Jensen, Ma Zhenyi, and J. T. Dickinson, "Scanning tunneling microscopy analysis of MgO fracture surfaces, J. Vac. Sci. Technol. A 8(4), 3470-3478 (1990).
3. A. L. Shluger, R. Mark Wilson, and R. T. Williams, “Theoretical and experimental investigation of force imaging at the atomic scale on alkali halide crystals,” Phys. Rev. B 49(7), 4915-4930 (1994).
4. F. Ohnesorge and G. Binnig, “True atomic resolution by atomic force microscopy through repulsive and attractive forces,” Sci. 260, 1451-1456 (1993).

5. D. M. Kulawansa, S. C. Langford, and J. T. Dickinson, "Scanning tunneling microscope observations of polymer fracture surfaces," J. Mater. Res. 7(5), 1292-1302 (1992).
6. D. M. Kulawansa, J. T. Dickinson, and S. C. Langford, and Y. Watanabe, "Scanning tunneling microscope observations of metallic glass fracture surfaces," J. Mater. Research 8(10), 2543-2553 (1993).
7. D. M. Kulawansa, L. C. Jensen, S. C. Langford, and J. T. Dickinson, and Y. Watanabe, "Scanning tunneling microscope observations of the mirror region of silicate glass fracture surfaces," J. Mater. Res. 9(2), 476-485 (1994).
8. C. P. Warner and D. A. Bonnell, in *Interface Dynamics and Growth*, edited by K. S. Liang, M. P. Anderson, R. F. Bruinsma, and G. Scoles, (Materials Research Society, Pittsburgh, PA, 1992), 393-398. [Mater. Res. Soc. Symp. Proc. 237].
9. Bharat Bhushan and Vilas N. Koinkar, "Macro and microtribological studies of CrO_2 video tapes," Wear 180, 9-16 (1995).
10. T. Thundat, B. C. Sales, B. C. Chakoumakos, L. A. Boatner, D. P. Allision, and R. J. Warmack, "Atomic layer-by-layer surface removal by force microscopy," Surf. Sci. Lett. 293, L863-L869 (1993).
11. J. A. Lin and W. N. Unertl, "Nanoscale deformations of polyimide with a force microscope," J. Adhes. Sci. Technol. 8(8), 913-27 (1994).
12. S. Nakahara, S. C. Langford, and J. T. Dickinson, "Surface force microscope observations of corrosive tribological wear on single crystal $NaNO_3$ exposed to moist air," submitted to Tribology Letters.
13. W. L. Roberts, G. R. Rapp, Jr., and J. Weber, *Encyclopedia of Minerals* (Van Nostrand Reinhold, New York, 1974), pp. 567-568.
14. P. Anuradha and I. V. K. Bhagavan Raju, "Study of plastic flow in $CaCO_3$ and $NaNO_3$ crystals under concentrated loads," Phys. Status Solidi (a) 95(1), 113-119 (1986).
15. P. E. Hillner, A. J. Gratz, S. Manne, and P. K. Hansma, "Atomic scale imaging of calcite growth and dissolution in real time," Geology 20, 359-362 (1992).
16. P. E. Hillner, S. Manne, A. J. Gratz, and P. K. Hansma, "AFM images of dissolution and growth on a calcite crystal," Ultramicroscopy 42-44, 1387-1393 (1992).
17. A. Sato and Y. Tsukamoto, "Observation of changes in surface structure of sputtered glasses by AFM upon water adsorption," J. Ceram. Soc. Jpn. 101, 400-404 (1993).
18. Yoshihisa Watanabe, Yoshikazu Nakamura, J. T. Dickinson and S. C. Langford, "Changes in air-exposed fracture surfaces of silicate glasses observed by atomic force microscopy," J. Non-Cryst. Solids. 177, 9-25 (1994).
19. R. T. Williams, R. M. Wilson, and Hanli Liu, "Defects in alkali halides: Atomic force microscopy, cluster spectroscopy, and ultrafast time resolution," Nucl. Instrum. Meth. Phys. Res. B 65(1/4), 473-483 (1992).
20. A. L. Shluger, R. M. Wilson, and R. T. Williams, "Theoretical and experimental investigation of force imaging at the atomic scale on alkali halide crystals," Phys. Rev. B 49(7), 4915-4930 (1994).

21. F. Ohnesorge and G. Binnig, "True atomic resolution by atomic force microscopy through repulsive and attractive forces," Sci. 260, 1451-1456 (1993).
22. Y. Martin, C. C. Williams, and H. K. Wickramasinghe, "Atomic force microscope—force mapping and profiling on a sub 100-Å scale," J. Appl. Phys. 61(10), 4723-4729 (1987).
23. L. Howard, R. Luthi, E. Meyer, G. Gerth, H. Haefke, R. Overney, and H.-J. Guntherodt, "Friction force microscopy on clean surfaces of NaCl, NaF, and AgBr," J. Vac. Sci. Technol. B 12(3), 2227-2230 (1994).
24. R. A. Lad, "Adsorption of water on sodium chloride: the effect of prior exposure to hydrogen chloride, carbon dioxide, and water vapor," Surf. Sci. 12, 37-45 (1968).
25. J. Estel, H. Hoinkes, H. Kaarmann, H. Nahr, and H. Wilsch, "On the problem of water adsorption on alkali halide cleavage planes, investigated by secondary ion mass spectroscopy," Surf. Sci. 54, 393-418 (1976).
26. S. P. Timoshenko and J. N. Goodier, *Theory of Elasticity*, 3rd Ed. (McGraw-Hill, New York, 1970).
27. S. J. Bull, T. F. Page, and E. H. Yoffe, "An explanation of the indentation size effect in ceramics," Philos. Mag. Lett. 59(6), 281-288 (1989).
28. W. W. Walker and L. J. Demer, "Effect of loading duration on indentation hardness," Trans. Metall. Soc. AIME 230, 613-614 (1964).
29. J. J. Gilman, "Creation of cleavage steps by dislocation," Trans. Metall. Soc. AIME 212, 310-315 (1958).
30. S. J. Burns and W. W. Webb, "Plastic deformation during cleavage of LiF," Trans. Metall. Soc. AIME 236, 1165-1174 (1966).
31. A. J. Gratz, P. E. Hillner, and P. K. Hansma, "Step dynamics and spiral growth in calcite," Geochim. Cosmochim. Acta 57, 491-495 (1993).
32. P. E. Hillner, A. J. Gratz, S. Manne, and P. K. Hansma, "Atomic Scale imaging of calcite growth and dissolution in real time Geology 20, 359 (1992).
33. Y. Liang, D. R. Baer, and A. S. Lea, "Dissolution of $CaCO_3(10\underline{1}4)$ surface," preprint.
34. Jeanne Paquette and Richard J. Reeder, "New type of compositional zoning in calcite: Insights in to crystal-growth mechanisms," Geology 18(12), 1244-1247 (1990).
35. Ian N. MacInnis and Susan L. Brantley, "The role of dislocations and surface morphology in calcite dissolution," Geochim. Cosmochim. Acta 56, 1113-1126 (1992).
36. J. T. Dickinson, L. C. Jensen, K. H. Siek, and K. W. Hipps, "The use of scanning conduction microscopy to probe abrasion of insulating films," Rev. Sci. Instrum. 66(7), 3802-3806 (1995).
37. Digital Instruments, Inc., Santa Barbara, CA
38. Graciela B. Blanchet and S. Ishmat Shah, "Deposition of polytetrafluoroethylene films by laser ablation," Appl. Phys. Lett. 62(9), 1026-1028 (1993).
39. G. B. Blanchet, C. R. Fincher, Jr., C. L. Jackson, S. I. Shah, and K. H. Gardner, "Laser ablation and the production of polymer films," Science 262, 719-721 (1993).

40. Wenbiao Jiang, M. G. Norton, Lancy Tsung, and J. T. Dickinson, "Pulsed laser deposition of polytetrafluoroethylene," J. Mater. Res. 10(4), 1038-1043 (1995).
41. Sunkyo Lee, L. C. Jensen, S. C. Langford, and J. T. Dickinson, "Electrical transients generated by the peel of a pressure sensitive adhesive from a copper substrate. Part I: Initial Observations," J. Adhesion Sci. Technol. 9, 1 (1995).
42. L. Scudiero, J. T. Dickinson, L. C. Jensen, and S. C. Langford, "Electrical transients generated by the peel of a pressure sensitive adhesive from a copper substrate. Part. II. Analysis of fluctuations," J. Adhesion Sci. Technol. 9, 27 (1995).
43. R. G. Horn and D. T. Smith, "Contact electrification and adhesion between dissimilar materials," Science 256, 362 (1992).
44. J. T. Dickinson, S. C. Langford, and L. C. Jensen, "Recombination on fractal networks: Photon and electron emission following fracture of materials," J. Mater. Res. 8, 2921 (1993).
45. R. E. Thomson and J. Moreland, "Development of highly conductive cantilevers for atomic force microscopy point contact measurements," to be submitted.
46. J. E. Wertz, J. W. Orton, and P. Auzins, "Electron spin resonance studies of radiation effects in inorganic solids," Disc. Faraday Soc. 31, 140-150 (1961).
47. G. K. Walters and T. L. Estle, "Paramagnetic resonance of defects introduced near the surface of solids by mechanical damage," J. Appl. Phys. 32(10), 1854-1859 (1961).
48. W. A. Sibley, J. L. Kolopus, and W. C. Mallard, "A study of the effect of deformation on the ESR, luminescence, and absorption of MgO single crystals," Phys. Stat. Solidi 31(1), 223-231 (1969).
49. T. J. Turner, C. Murphy, and T. Schultheiss, "A study of the effect of deformation on optical absorption of MgO single crystals," Phys. Status Solidi (b) 58(2), 843-857 (1973).
50. Y. Chen, M. M. Abraham, T. J .Turner, and C. M. Nelson, "Luminescence in deformed MgO, CaO, and SrO," Philos. Mag (Series 8) 32(1), 99-112 (1975).
51. R. L. Webb, S. C. Langford, L. C. Jensen, and J. T. Dickinson, "Ablation of single crystal MgO by UV excimer laser radiation," Mater. Res. Soc. Symp. Proc. 236, 21 (1992).
52. R. L. Webb, L. C. Jensen, S. C. Langford, and J. T. Dickinson, "Interactions of wide bandgap single crystals with 248 nm excimer laser radiation: I. MgO," J. Appl. Phys 74(4), 2323-2337 (1993).
53. J. T. Dickinson, L. C. Jensen, and R. L. Webb, "Photoluminescence imaging of mechanically produced defects in MgO," J. Non-Cryst. Solids. 177, 1-8 (1994).
54. W. G. Johnston and J. J. Gilman, "Dislocation multiplication in lithium fluoride crystals," J. Appl. Phys. 31(4), 632-643 (1960).
55. W. T. Read, Jr., *Dislocations in Crystals* (McGraw Hill, New York, 1953) pp. 85-88.

56. J. P. Hirth and J. Lothe, *Theory of Dislocations*, 2nd Ed. (John Wiley, New York, 1982), pp. 588-609.
57. G. A. Andreev and B. I Smirnov, "Formation of point defects by the plastic deformation of LiF crystals," Fiz. Tverd. Tela 10(6), 1693-1698 (1968) [Sov. Phys. Solid State 10(6), 1336-1340 (1968)].
58. R. W. Armstrong and C. CM. Wu, "Lattice misorientation and displaced volume for microhardness indentations in MgO crystals," J. Am. Ceram. Soc. 61(3/4), 102-106 (1978).
59. M. Munawar Chaudhri, "The displacement of material and the formation of $\{110\}_{90}$ cracks around spherical indentations in MgO crystals," Philos. Mag. A 53(4), L55-L63 (1986).
60. E. E. Donaldson, J. T. Dickinson, and S. K. Bhattacharya, "Production and properties of ejecta produced by fracture of materials," J. Adhesion 25, 281 (1988).
61. E. E. Donaldson, M. H. Miles, and J. T. Dickinson, "Electrical charge measurements on ejecta from impact loading of explosive crystals," J. Mater. Sci. 24, 4453 (1989).
62. M. V. Swain, B. R. Lawn, and S. J. Burns, "Cleavage step deformation in brittle solids," J. Mater. Sci. 9, 175-183 (1974).
63. J. T. Dickinson, L. C. Jensen, R. L. Webb, M. L. Dawes, and S. C. Langford, "Interactions of wide band gap single crystals with 248 nm excimer laser radiation. III. The role of cleavage-induced defects in MgO," J. Appl. Phys. 74(6), 3758-3767 (1993).
64. Tadaaki Sugita and Takanori Hasegawa, "Wear damage of MgO single crystals," J. Mater. Sci. 13(7), 1471-1479 (1978).
65. G. P. Williams, Jr. and T. J. Turner, "Triboluminescence in single crystal alkaline earth oxides," Solid State Commun. 29(3), 201-203 (1979).
66. S. C. Langford, J. T. Dickinson, and L. C. Jensen, "Simultaneous measurements of the electron and photon emission accompanying fracture of single-crystal MgO," J. Appl. Phys. 62(4), 1437-1449 (1987).
67. J. T. Dickinson, P. Braunlich, L. Larson, and A. Marceau, "Characteristic emission of negatively charged particles during tensile deformation of oxide-covered aluminum alloys," Appl. Surf. Sci. 1, 515 (1978).
68. J. T. Dickinson, "Fracto-Emission," in *Non-Destructive Testing of Fibre-Reinforced Plastic Composites* - II, J. Summerscales, ed. (Elsevier Applied Science, London, 1990), pp. 429-482.
69. J. T. Dickinson, S. C. Langford, and L. C. Jensen, "Dissipative Processes Accompanying Fracture," in *Fracture Mechanics of Ceramics*, Vol. 10, R. C. Bradt, D. P. H. Hasselman, D. Munz, M. Sakai, and V. Ya. Shevchenko (Plenum, New York, 1992), pp. 1-32.
70. C. A. Brookes and J. B. O'Neill, "Frictional anisotropy in magnesium oxide," in: *Anisotropy in Single-Crystal Refractory Compounds*, F.W. Vahdiek and S.A. Mersol, eds. (Plenum, New York, 1968), pp. 291-305.
71. R. P. Steijn, "Friction and wear of single crystals," Wear 7(1), 48-66 (1964).
72. Tadaaki Sugita and Takanori Hasegawa, "Wear damage of MgO single crystals," J. Mater. Sci. Lett. 13(7), 1471-1479 (1978).
73. Yuji Enomoto, "Deformation by scratches and frictional properties of MgO crystals," Wear 89(1), 19-28 (1983).

74. Kazuo Suzuki and Tadaaki Sugita, "Characteristics of magnesium oxide single crystals polished with vibrational sliding in water," Wear 75(2), 269-283 (1982).
75. C. A. Brookes, M. P. Shaw, and P. E. Turner, "Non-metallic crystals undergoing cumulative work-hardening and wear due to softer lubricated metal sliding surfaces," Proc. Royal Soc. London A 409(1836), 141-159 (1987).
76. Keiji Nakayama and Hiroshi Hashimoto, "Triboemission of charged particles and photons from wearing ceramic surfaces in various gases," Tribology Trans. 35(4), 643-650 (1992).
77. Keiji Nakayama, "Triboemission of electrons, ions, and photons during wear of solids in dry sliding," to appear in Proc. Internat. Symp. on Exoelectron Emission and Applications 1994.
78. Ma Zhenyi, Fan Jiawen, and J. T. Dickinson, "Properties of the Photon Emission Accompanying the Peeling of a Pressure Sensitive Adhesive," J. Adhesion 25, 63 (1988).
79. B. Henderson and J.E. Wertz, "Defects in alkaline earth oxides," Adv. Phys. 17(70), 749-855 (1968).
80. L. A. Kappers, R. L. Kroes, and E. B. Hensley, "F^+ and F' centers in magnesium oxide," Phys. Rev. B 1(10), 4151-4157 (1970).
81. R. T. Williams, J. W. Williams, T. J. Turner, and K. H. Lee, "Kinetics of radiative recombination in magnesium oxide," Phys. Rev. B 20(4), 1687-1699 (1979).
82. J. L. Grant, R. Cooper, P. Zeglinski, and J. F. Boas, "Time resolved luminescence spectroscopy of alkaline earth oxides after pulsed electron beam irradiation. II. Spectra and thresholds of MgO," J. Chem. Phys. 90(2), 807-812 (1989).
83. K. J. Caulfield, R. Cooper, and J. F. Boas, "A temperature effect in the luminescence emission from electron irradiated MgO," J. Chem. Phys. 92(11), 6441-6446 (1990).
84. R. Leszczyński, M. Mieńska-Tomczyk, S. Wysocki, and S. Karolczak, "Luminescence and conductivity induced in polycrystalline MgO by high energy electrons," Radiat. Phys. Chem. 36(2), 117-121 (1990).
85. S. J. Pennycook and L. M. Brown, "Cathodoluminescence at dislocations in divalent oxides," J. Luminescence 18/19, 905-909 (1979).
86. A. Gibson, R. Haydock, and J.P. LaFemina, "The electronic structure of neutral and charged surface vacancy defects in periclase," Appl. Surf. Sci. 72, 285-293 (1993).
87. K. A. Zimmerman, S. C. Langford, J. T. Dickinson, and R. P. Dion, "Electron and photon emission accompanying deformation and fracture of polycarbonate," J. Polym. Sci. Polym. Phys. 31, 1229-1243 (1993).
88. Yuji Enomoto and M. Munawar Chaudhri, "Fracto-emission during fracture of engineering ceramics," J. Am. Ceram. Soc. 76(10), 2583-2587 (1993).
89. Keiji Nakayama, "Triboemission of charged particles from various solids under boundary layer lubrication," Wear 178(1/2), 61-67 (1994).

90. J. T. Dickinson, L. C. Jensen, S. C. Langford, and R. G. Hoagland, "Chemisorptive electron emission as a probe of plastic deformation in reactive metals," J. Mater. Res. 9(5), 1156-1165 (1994).
91. B. Sujak and A. Gieroszyński, "On the kinetics of intensity growth of exoemission of electrons during plastic deformation of aluminum coated with an oxide film," Acta Phys. Polon. 28, 249-266 (1968).
92. W. J. Baxter, "Exoelectron measurements of the rate of development of fatigue," Fatigue Engin. Mater. Struc. 1, 343-350 (1979).
93. Sumio Nakahara, S. C. Langford, and J. T. Dickinson, "Chemisorptive electron emission and atomic force microscopy as probes of plastic deformation during fracture at a metal/glass interface," to appear in J. Mater. Res.
94. A. G. Evans and J. W. Hutchinson, "Effects of non-planarity on the mixed mode fracture resistance of bimaterial interfaces," Acta Metall. 37(3), 909-916 (1989).
95. Z. Chen and J. J. Mecholsky, Jr., "Control of strength and toughness of ceramic/metal laminates using interface design," J. Mater. Res. 8(9), 2362-2369 (1993).
96. A. G. Evans and B. J. Dalgleish, "The fracture resistance of metal-ceramic interfaces," Mater. Sci. Engin. A162, 1-13 (1993).
97. J. Kim, K. S. Kim, and Y. H. Kim, "Mechanical effects in peel adhesion test," J. Adhesion Sci. Technol. 3(3), 174-187 (1989).
98. A. G. Evans, B. J. Dalgleish, M. He, and J. W. Hutchinson, "On crack path selection and the interface fracture energy in bimaterial systems," Acta Metall. 37(12), 3249-3254 (1989).
99. I. E. Reimanis, B. J. Dalgleish, and A. G. Evans, "The fracture resistance of a model metal/ceramic interface," Acta Metall. Mater. 39(12), 3133-3141 (1991).
100. O. F. Hagena, G. Knop, R. Fromknecht, and G. Linker, "Magnesium cluster-beam deposition on glass and on Si(111)," J. Vac. Sci. Technol. A 12(2), 282-288 (1994).
101. L. A. K'Singam, J. T. Dickinson, and L. C. Jensen, "Electron and photon emission accompanying failure of metal/glass interfaces," J. Am. Ceram. Soc. 68, 510 (1985).
102. D. L. Doering, S. C. Langford, J. T. Dickinson, and P. Xiong-Skiba, "Fracto-emission during the interfacial failure of a metal-oxide-semiconductor system: Au-SiO_2-Si," J. Vac. Sci. Technol. A 8, 2401 (1990).
103. J. J. Mecholsky, Jr., S. W. Freimam, and R. W. Rice, "Fracture surface analysis of ceramics," J. Mater. Sci. 11(7), 1310-1319 (1976).
104. Richard W. Hertzberg, *Deformation and Fracture Mechanics of Engineering Materials*, 3rd Ed. (John Wiley, New York, 1989), pp. 81-83.
105. A. G. Evans and B. J. Dalgleish, "The fracture resistance of metal-ceramic interfaces," Acta Metall. Mater. 40, S295-S306 (1992).
106. B. M. McCarroll, "Chemisorptive luminescence," J. Chem. Phys. 50(11), 4758-4765 (1969).
107. B. Kasemo, E. Törnqvist, and L. Walldén, "Metal-gas reactions studies by surface chemiluminescence," Mater. Sci. Engin. 42, 23-29 (1980).

108. R. H. Prince and R. Persaud, "Contribution of broadening effects to chemisorptive emission," Surf. Sci. 207, 207-14 (1988).
109. E. E. Underwood and K. Banerji, "Fractals in fractography," Mater. Sci. Engin. 80, 1-14 (1986).
110. C. P. Warner and D. A. Bonnell, in *Interface Dynamics and Growth*, edited by K. S. Liang, M. P. Anderson, R. F. Bruinsma, and G. Scoles, (Materials Research Society, Pittsburgh, PA, 1992), pp. 393-398.
111. K. A. Zimmerman, S. C. Langford, and J. T. Dickinson, "Electrical transients during interfacial debonding and pullout of a metal rod from an epoxy matrix," J. Appl. Phys. 70(9), 4808-4815 (1991).
112. W. H. Press, B. P. Flannery, S. A. Teukolsky, and W. T. Vetterling, *Numerical Recipes in Pascal: The Art of Scientific Computing* (Cambridge University, Cambridge, 1981), pp. 473-479.
113. J. T. Dickinson, L. C. Jensen, Sunkyo Lee, L. Scudiero, and S. C. Langford, "Fracto-emission and electrical transients due to interfacial failure," J. Adhesion Sci. Technol. 8(11), 1285-1309 (1994).
114. Heinz Georg Schuster, *Deterministic Chaos: An Introduction*, 2nd Ed. (VCH, Weinheim, Germany, 1988), pp. 1-16.
115. P. Grassberger and I. Procaccia, "Measuring the strangeness of strange attractors," Physica 9D, 189-208 (1983).
116. Avraham Ben-Mizrachi and Itamar Procaccia, "Characterization of experimental (noisy) strange attractors," Phys. Rev. A 29(2), 975-977 (1984).

Combination of Fractography and Computed Tomography for the Determination of the Cause of Fracture (Case Study)

Jarmila Woodtli, Karl Berroth, Thomas Lüthi
Swiss Federal Laboratories for Materials Testing and Research, Dübendorf, Switzerland

Abstract

The utilization of fractography to determine crack origins as well as the mechanisms of crack propagation is an established method in failure analysis. In the case where structural parts show crack patterns consisting of multiple branched cracks, the interpretation of the sequence of the cracking is needed. If the cracked part separates into numerous parts during the sampling, the reconstruction of the original shape can be critical. By improper handling, moreover, the key information can be lost due to possible secondary damage. Therefore computed tomography (CT) investigation was found to be a useful technique in tracing the crack paths.

The insulating load switch used for railway tracks failed due to multiple cracks in the porcelain insulator body. A case study to illustrate the applicability of CT prior to the sampling for a fractographical investigation was performed. The evaluation of the digital tomograms showed the approximate location of the crack origins and the sampling could be conducted purposefully. The fractographical investigation of the fragments confirmed the results of CT and revealed the crack origin sites. Using optical and scanning electron microscopy, detailed information about the material used in the insulator assembly was gained and the cause of the failure was detected.

Introduction

Ceramic insulators are widely used in high voltage engineering for the past 100 years and show service times of several decades. But occasionally, earlier failure by damage or catastrophic rupture occurs for individual parts. In order to make sure that, a whole series is not concerned, failure analysis has to be done with such components.

Insulators are made of porcelain type ceramics with an elevated content of alumina. They are produced by wet ballmill grinding of the raw materials such as quartz, fieldspar and kaolin. After partially removing the water by pressure filtration, the plastic material is extruded to cylinder shapes. Subsequent partial drying of the cylinders leads to a state where shaping of the complex insulator geometry can be done by turning. The machined insulators are then fully dried. The glaze is applied by a spray process and after drying of this glaze the insulators are fired at a temperature of about 1250–1300°C. During this firing cycle densification causes a shrinkage of 13–15% and formation of the final ceramic. Shrinkage during the sintering cycle may cause high residual stresses within the ceramic body leading to micro- and macrocracks when the procedure is not performed carefully, following exact process steps and parameters.

After firing, certain extensions of the insulator which are necessary for proper positioning in the furnace during firing are removed by diamond wheel cutting. Finally the metallic fittings have to be attached to the ceramic insulator. This also is a very sensitive processing step demanding extreme care. Several different techniques are successfully applied. In order to minimize stresses caused by the mismatch of coefficient of thermal expansion (CTE) usually elastic materials like lead alloys, tar pitch or resins are introduced as one interlayer of the zone between metal and ceramic.

A final inspection, including visual and ultrasonic checks as well as electrical high tension load is performed to assure that the part is defect free and meets the break-down voltage specifications.

Although there are some standards concerning minimal specifications of mechanical properties and chemical composition, these values cannot guarantee a proper operation whenever the whole production processing is not carefully performed. Nothing is specified concerning allowed residual stresses, inhomogenities, flaws, cracks in both the ceramic body and the fixture of the fittings. But such imperfections can cause failures.

Attachment of the fittings

Because the attachment of the fittings is very complex and may influence the failure, the specific solution must be known exactly.

In the case of the insulators presented here, the cast iron fittings are fixed inside large bores on the top and bottom side of the ceramic body by a Portland cement with quartz additions (Fig. 1).

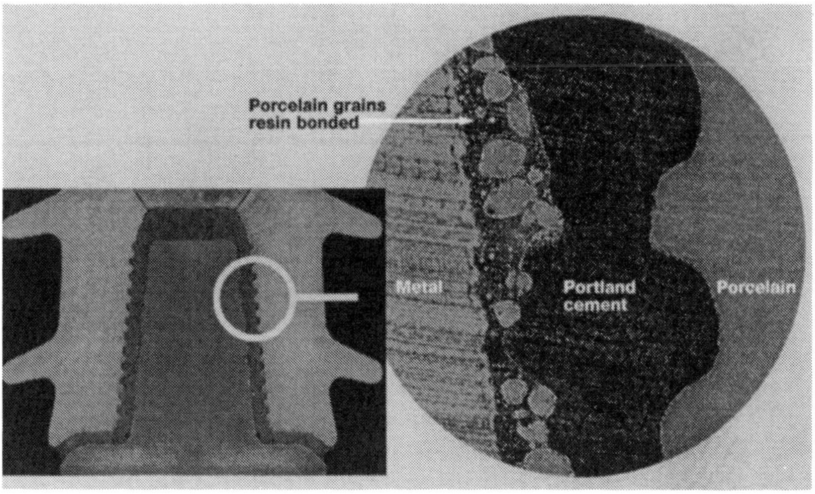

Fig. 1: Longitudinal section through an insulator

The fittings of cast iron are covered with a zinc layer to avoid corrosion. The elastic layer is represented by an epoxy resin. The resin is additionally used to glue quartz or alumina particles on the surface of the metal part to reach a higher roughness and improved bonding to the Portland cement.

Inside the bore of the insulator, circumferential grooves (ribs) are machined for an improved mechanical bonding of porcelain and cement.

The gap between the bore and the zinc and resin-coated metal fitting is filled with Portland cement with quartz additions. The quartz is used to give higher strength and to adjust the CTE at a certain level between cast iron (12 x 10^{-6}/K) and porcelain (5–6 x 10^{-6}/K).

Failure description

After several years of service, sudden transverse rupture separated an insulator in two parts and caused an interruption in electrical power supply for a railway line. Consequently numerous insulators underwent a visual inspection, and longitudinal cracks were found in some of them. These insulators, however, were not yet completely ruptured and still operated properly. The longitudinal cracks, as displayed in Fig. 2, are located in the upper and lower part where the metal fittings are placed.

Fig. 2: Longitudinal crack within the area of metal fitting

Since the transverse fracture spread square to the longitudinal fractures, and moreover, it showed features of fast fracture (Wallners lines), it was assumed, that this was secondary damage. Therefore, the main interest of the investigation was focused on the question of how such longitudinal cracks could be initiated in the insulators.

The cement seams between metal fitting and porcelain additionally show a rather high number of radial cracks, which might be caused by lack of humidity and shrinkage during the drying operation, or by stresses due to temperature changes and a high mismatch in CTE between the metal and cement. In some cases, the metal also was not centered in the bore, which leads to different thickness of the cement layer in different locations.

Failure hypothesis

We assumed that a combination of the following effects may have led to failure:

- stress concentration at inhomogenities, cracks and flaws
- residual stresses in the ceramic material
- stresses caused by temperature changes and the mismatch in CTE between metal and porcelain
- improper mounting of metallic fitting
- additional stresses caused by freezing of water which had penetrated into the cracks in the cement.

Investigations

Nondestructive evaluation

First, to find the surface locations of the cracks and to determine any further actions by means of NDE, the sample was investigated by dye- penetrant testing, and the crack positions were marked. All other investigations were carried out by industrial computed tomography (CT).

Like medical CAT-scanners, industrial CT-systems non destructively produce artificial cuts through an object; different densities lead to different gray levels in the resulting image. In contrast to medical systems, industrial scanners use a wider range of X-ray energy, in this case 450 kV at 1 mA with a focal spot of 1 mm. In medical systems, the necessary rotation is carried out by the X-ray tube and tfe detectorbank, industrial systems move and rotate the object itself, which is much easier to control. Finally, industrial objects normally do not move and are not sensitive to tfe applied X-ray dose, so tfe procedure is less critical relative to the time. In this case, in order to get a high signal-to-noise ratio, one cut took about 5 minutes. Tfe information gained this way was processed to an image by filtered back projection.

Industrial scanners have a wide range of density resolution – for extreme applications up to 20 g/cc, a high spatial resolution down to 0.1 mm, and with 3rd generation CT (rotation only) even for larger objects, long cracks can be detected if they have a crack width of only 20 µm or even less. If a crack is laying directly in tfe fanbeam layer (Fig. 3, crack B), the contrast usually is not sufficient to make it visible as a slightly darker area. In all otfer situations the interception between tfe crack (crack A) and tfe fanbeam layer can be seen as a line or a curve. However, it is not possible to make a general statement about tfe necessary length of this interception (microcracks), as this is strongly dependent on tfe object size, material characteristics, and tfe ability of tfe observer to see slight differences in contrast.

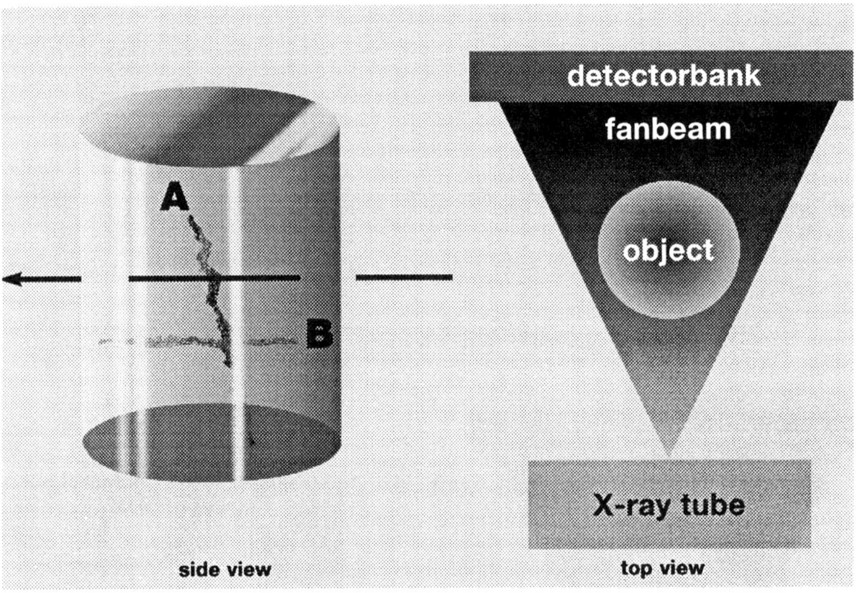

Fig. 3: Schematic showing visible (A) and not visible cracks (B)

Fig. 4 shows the general structure of tfe insulator: a metal core with higfer density, which produces some edge artifacts, surrounded by the Portland cement and the porcelain. Within tfe Portland cement, and especially at tfe interface between cement and porcelain, many flaws can be seen (Fig. 5). Tfe cracks do not always stop at this interface; sometimes the cement itself is broken also.

With additional cuts a situation was found where tfe crack clearly followed the cement/porcelain interface, and finally one possible origin of tfe crack was identified at tfe same interface (Fig. 5).

Tfe nondestructive evaluation resulted in a rough 3D record of tfe crack planes and allowed systematic further investigations.

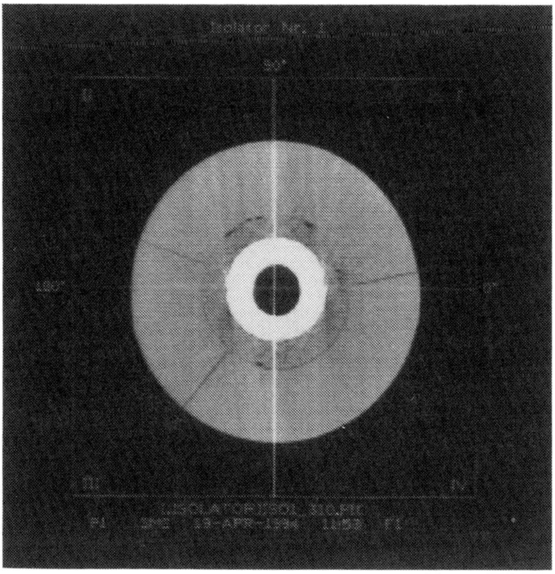

Fig. 4: CT-record. General structure of tfe insulator: white metal core surrounded by Portland cement and porcelain with three longitudinal cracks and a crack along tfe cement/porcelain interface (pixel size 0.17 mm x 0.17 mm)

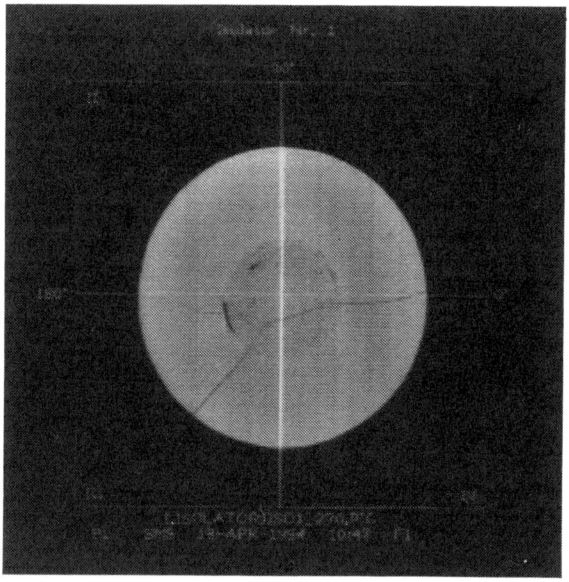

Fig. 5: CT-record. Possible origin of the crack at the cement/porcelain interface

Sampling

Although one completly broken insulator was delivered, for the fractographic investigation, only the precracked insulators, as shown in Fig. 1, were chosen. Based on the CT results, it was obvious that the primary, longitudinal fractures originated inside the insulators, near the boundary between the porcelain and the Portland cement.

The aim of the sampling was to completely open the longitudinal cracks, thus exposing their fracture surfaces for fractographic examination. By opening cracks, care has to be exercised to prevent breaking apart of the bond between the porcelain and cement. For this reason, longitudinal sections parallel to the crack planes (crack surfaces) were cut using a dia-mond cutting wheel. After macroscopic examination was completed, specimens from crack-origin areas were cut for the SEM investigation.

Because of the rotational symmetry of the insulators, precautions were taken for the exact identification and location of the samples and frag-ments. Markings were placed on the specimens to establish a reference point. This helped in the reconstruction and correlation between CT photographs and the fracture area.

Longitudinal sections

Longitudinal sections which were created for opening of the cracks also revealed the internal geometrical structure of the fitting and cement layer. In addition, significant information with regard to the quality of the materials could be gathered. As shown in Fig. 6, the cement layer exhibits multiple irregularities such as pores, second-phase inhomogenities, cavities and porous seams. Locally these flaws are associated with cracks. In addition, Fig. 6 shows that the metal fittings of the fixing arrangements were eccentrically mounted into the porcelain body. Due to this eccentricity, the permissible tolerance could be exceeded which very probably contributed to increasing the internal stresses in service.

The surface of the cast-iron fitting was covered by a Zn layer and was thus sufficiently protected against corrosion. No corrosion products could be found on the sections.

Additionally, a particle-loaded resin layer was applied. This elastic layer varies strongly in thickness and particle concentration.

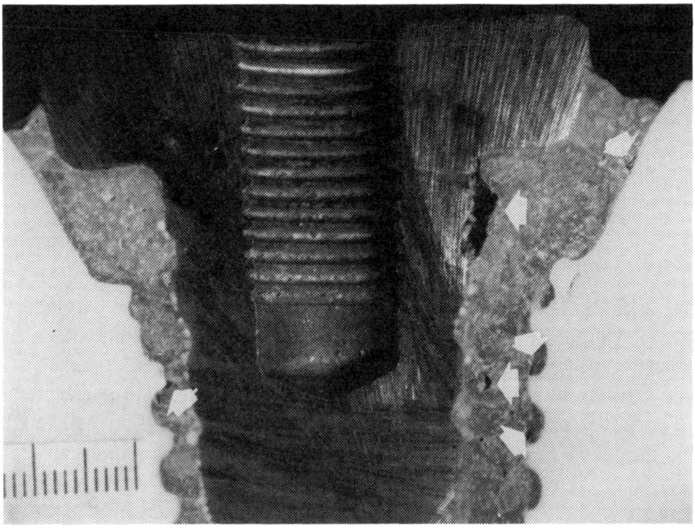

Fig. 6: Eccentricly mounted fitting. Arrows indicate flaws in cement layer

Fractographic investigations

As shown in CT photographs (Fig. 3, 4), the fracture path in the Portland cement spreads mostly in the same plane as in the porcelain material. Whereas the crack propagation direction could be traced in the porcelain due to the distinct fracture markings, the cement fracture surface did not reveal any significant morphology which would allow a determination of the crack propagation direction. In the porcelain, the coarse hackle lines and ridges radiate from, and thus point the way back to, the fracture origins. The numerous crack origins in the porcelain are regularly distributed on the internal phase boundary with the Portland cement (Fig. 7).

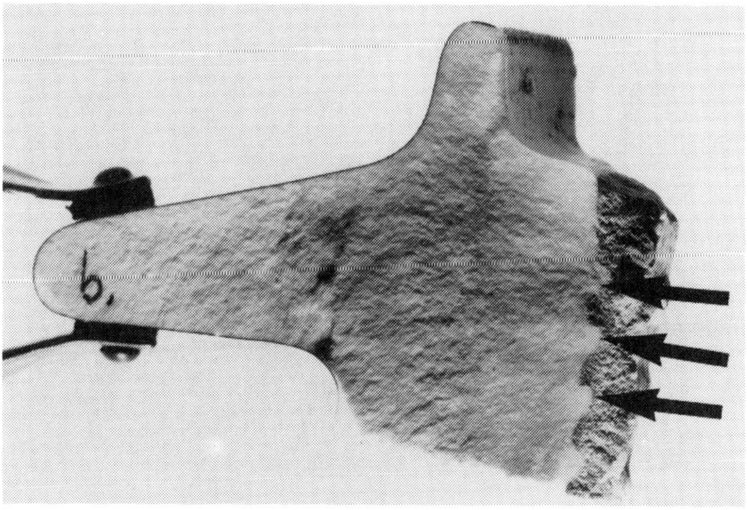

Fig. 7 Fracture origins in porcelain (arrows)

The investigation by SEM revealed a typical fracture topography in the porcelain (Fig. 8) consisting of spherical quartz particles embedded in a mullite matrix. No significant material flaws could be found at the crack origins. Moreover, the origin sites seem to be geometrical continuations of the crack path within the cement. The transitions of the crack propagation from the cement to the porcelain occur on the stress-raising ribbed structures.

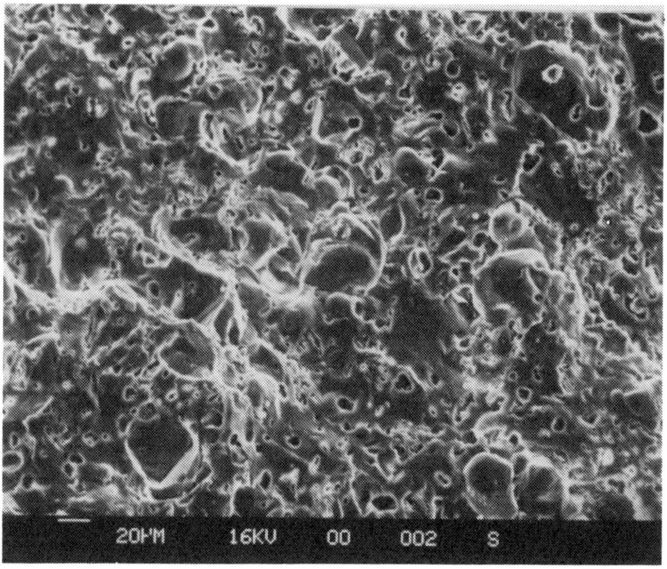

Fig. 8: Detail of one fracture origin area in porcelain

Mechanical testing

In order to check whether the mechanical properties of the material are within the specification of the IEC standard, four-point bending tests were performed, following procedures specified in ENV 843.

The values measured in a total of five series of 30 samples each, showed characteristic strengths of 75 to 84 MPa. The measured strength therefore does not meet the standard, which requires a minimum strengths of 90 MPa. The scatter of the individual values was rather narrow as indicated by Weibul moduli in the range of 15 to 23, but there were some individual samples with strength values as low as 36 MPa.

Also, Young's modulus was found to be only 65 GPa, which is lower than the required 70 GPa.

The density was found to be at the lower end or even slightly below the standard.

Physical properties of ceramic insulators IEC type C-120

	IEC 672-2+3	measured
Bending strength	90 MPa	75-84 MPa
Weibull (m)	-	15-23
Youngs mod.	$\geq$70 GPa	65 GPa
Density	2.3 g/cm^3	2.25 -2.29

The fractographic inspection of the test bars showed no significant microstructural defects and inhomogenities. Only some coarse-grain agglomerates were observed with porous seams and microcracks between the quartz grains and the matrix.

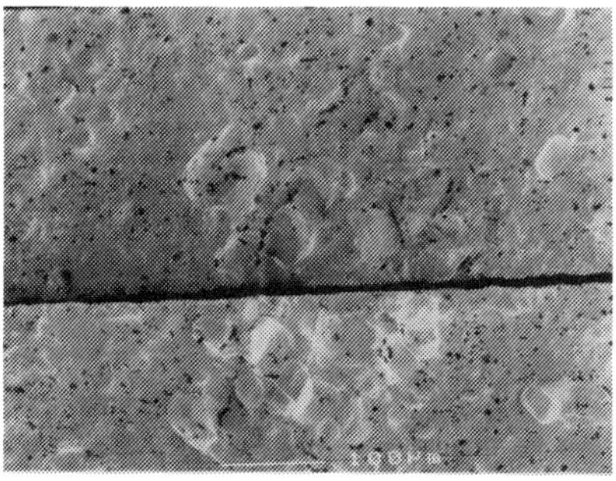

Fig. 9: Facture origin area in test bar

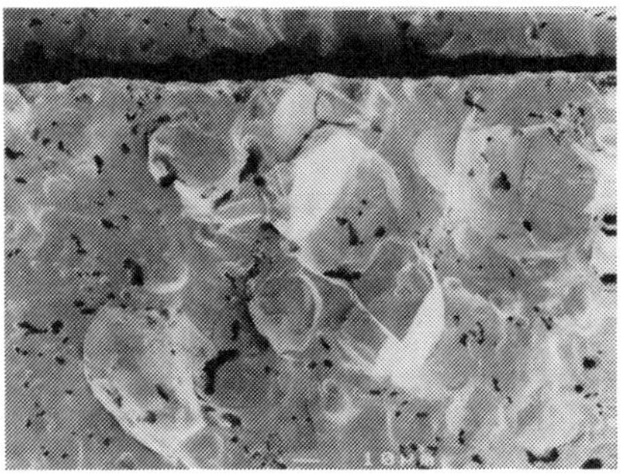

Fig. 10: Detail from fig. 9

Only in the sample with the lowest strength was the fracture origin found to be a crack perpendicular to the sample axis.

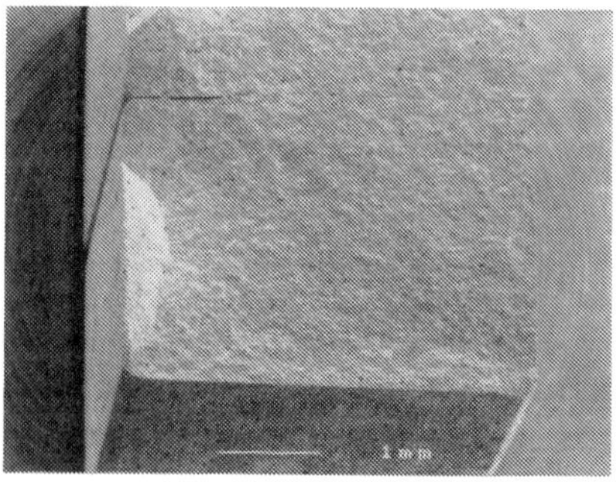

Fig. 11: Fracture of the test bar with the lowest strength showing precracked region

This type of defect became obvious even during preparation of the test samples for the mechanical tests. Cracks in the range of 1 to 3 mm in length where found at the surface of ground test bars by visual checks prior to testing. This damaged samples were not used for the strength determinations. But it must be considered that such cracks randomly present in the insulators can reduce the local strength to very low values, thus allowing crack initiation and growth even with very low initial stress.

Damage mechanism

The nondestructive testing of the delivered insulators indicated that the cracks were initiated inside the insulators. This finding is in good agree-ment with the fractographic results. The distinct crack propagation lines and ridges as well as the evaluation of the roughness of the fracture allow the determination of the crack propagation direction. Moreover, the crack origins in porcelain could be found. They are situated at the phase bound-ary between the cement and porcelain. Since the fractures in the cement as well as in the porcelain spread in the same plane, it must be assumed that the fractures run from one material into the other, probably from the porcelain into the cement.

Cracks and other flaws which were observed in the test bars for the strength measurements must be considered as possible fracture-initiating material defects.

Crack initiation therefore is produced by stresses in the area of the fitting. The stresses are caused by environmental conditions for example temperature change, and they act on intrinsic flaws in the cement layer, such as pores or inhomogenities. Once initiated, the crack propagation is controlled by the mismatch in CTE between materials (metal/cement/porcelain), which leads to enhanced internal tensile stresses. In addition, residual stresses in porcelain, as well as stresses due to eccentrical mounting of the fitting, and service stresses must be considered.

This crack growth can take place over a period of several years, leading to longitudinal cracking in a first state which does not attact proper operation, but which ultimately leads to transverse rupture and catastrophic failure of the insulator.

Summary

The presented case study demonstrates that for failure analyses in complex components, an interdisciplinary attitude using different inspection methods is necessary and effective. The failure mechanism of a high-voltage insulator was evaluated by using CT, fractography, mechanical testing, and analysis of the fabrication processes.

Acknowledgement:

The autors wish to acknowledge A. Flisch for the CT investigation, J. Kübler, and R. Bächtold for conducting the mechanical testing.

GROWTH OF INDENTATION INDUCED FLAWS IN RBSiC AND RBSiC-TiB_2 COMPOSITES UNDER AN APPLIED STRESS

K A Kibble and D Turner
School of Construction, Engineering and Technology
University of Wolverhampton, UK
WV1 1SB

ABSTRACT

The influence of microstructure on the mechanical properties of a fine grain size reaction-bonded silicon carbide (RBSiC) and RBSiC-TiB_2 composites has been evaluated. Incorporation of TiB_2 particles in a 1200# RBSiC, (4 μm average grain size), resulted in a reduction in the flexural strength but an increase in the indentation fracture toughness. This increase in toughness was the result of a smaller induced indentation flaw, because of crack arrest and deflection mechanisms by TiB_2 particles. This gave the composite grades an enhanced damage or flaw tolerance in comparison with fine grain size, (4 μm and less), grades of RBSiC. Indentation crack measurements taken on the fracture face of indentation strength bars also showed that crack growth to failure occurs via an expanding cavity mode that is dependent on indentation load.

INTRODUCTION

Reaction bonded silicon carbide is a well established engineering ceramic. The basic mode of manufacture consists of infiltrating liquid silicon at temperatures around 1600°C into a porous powder compact of fine silicon carbide particles mixed with colloidal graphite. The carbon reacts with the liquid silicon to form new silicon carbide which is deposited epitaxially on the original silicon carbide grits thus bonding the whole compact together.[1] Residual porosity and any space not occupied by silicon carbide, such as fabrication defects, is filled with the liquid silicon. The fired material consists of a matrix of silicon carbide with some 10-15% of interconnected silicon occupying the residual pore space that is necessary for full impregnation of the silicon into the compact. The volume change on firing is negligible thus the material can be made into complex shapes with sufficiently good shape tolerance to require minimal machining of the final dense solid.

In applications involving high contact and structural stresses RBSiC, and similar SiC ceramics are limited because of their relatively low fracture toughness, (3-4 $MPa.m^{1/2}$). In order to improve their toughness additions of second phase reinforcing materials have been made that are either rod-like (whiskers) or particulate in shape.

Extensive research has been carried out on the use of whiskers in reinforcing ceramics to improve their fracture toughness. There are comprehensive reviews in the literature documenting the extent and success of this research.[2-4] The results of whisker reinforcement have been very inconclusive with at times an increase in toughness being reported and at other times a decrease in toughness that is very much dependent on the volumetric extent of whisker addition.[5] The major difficulty is in dispersing the fine sub-micron diameter rods in a ceramic matrix without causing surface damage to the whiskers or processing defects through the agglomeration of the whiskers. Both defects remain a limiting case to the use of whisker reinforced ceramic composites.

Particulate reinforcement provides an attractive alternative to whisker reinforcement in improving the fracture toughness of engineering ceramics. In general they are easier to fabricate although toughening improvements tend to be lower than with successful whisker reinforcement. In addition any processing route that narrows the flaw size distribution, and therefore the fracture strength range, will increase the reliabilty of a component. Reliability can be improved through the addition of second phase additions to form a composite, providing the problems of dispersion can be overcome. Either whisker reinforcement or particulate additions can be utilised. Whisker reinforcement has not been successfully carried out with RBSiC,[6] possibly because of the nature of the fabrication process. Particulate reinforcement may offer an acceptable compromise solution in improving toughness and reliability.

Studies on particulate reinforcement of SiC have been concentrated on the use of either TiC or TiB_2 as the particulate addition. An increase in the fracture toughness of SiC, through the addition of either TiC or TiB_2, of the order of 30-100% has been reported by several workers.[7-11] It has been shown by Higgins and Whitehead[12] that it is possible to incorporate secondary phases within green powder RBSiC compacts and to infiltrate these with pure silicon. Many of the materials tested in this study were found to be incompatible with molten silicon, however, TiB_2 was observed to show a good resistance to the reaction-bonding process.

In assessing the strength and fracture toughness of ceramic materials indentation fracture techniques are particularly suitable. In these techniques a sharp indenter such as a Knoop or Vickers diamond is used to induce surface cracks. These controlled flaws are much closer in size to those encountered in service conditions than the macro-cracks that are a feature of conventional fracture toughness testing. RBSiC can be etched following indentation so that if a mechanical test specimen is broken on the controlled flaw than the flaw shape and size can be clearly identified and measured. Thus it is possible to characterise RBSiC and RBSiC composites, using indentation techniques, where the diamond indenter is used as a mechanical microprobe.

INDENTATION FRACTURE MECHANICS

Sharp contact indentation, (Vickers or Knoop), produces radial cracks on median planes that contain the axis of loading and the symmetry axes of the contact hardness impression, Fig. 1. In addition a secondary, less severe, flaw system develops from the base of the deformation zone in the form of lateral cracks that run, initially, parallel to the specimen surface. If a critical indentation load threshold is exceeded then the lateral cracks can break through to the specimen surface, creating surface damage in the form of chipping. The most important observation concerning the formation of the lateral and radial cracks is that they usually grow to their final lengths as the indenter is unloaded. This feature is attributed to the presence of a residual stress field, generated as a result of incompatibility strains between the deformation zone and the surrounding elastic material.

The complex elastic-plastic field has been shown[13] to consist of separable "elastic" and "residual" terms. The residual component is more than just a correction factor to be taken into consideration in indentation fracture mechanics; rather, this component assumes a dominating role in determining the size of the radial crack.

Two major indentation methods have emerged for the determination of K_c where a residual stress term is incorporated in the analysis. The first technique involves direct measurement of the surface trace of the radial crack, c_o, when post indentation equilibrium has been obtained, and using a simple relation of the type given by Lawn et al[14]:

$$K_c = \chi_r P/c_o^{3/2} = \xi(E/H)^{1/2}(P/c_o^{3/2}) \quad (1)$$

where χ_r incorporates residual stress correction factors that are dependent on indenter geometry, ξ , and the response of the ceramic to resist deformation under a given indentation load, P, (i.e. χ_r is a function of Young's modulus, E, and the hardness, H, of the ceramic under test). The correction factor for indenter geometry, ξ , is usually determined empirically by a calibration against reference ceramics where K_c has been derived using conventional fracture toughness methods.[15] The validity of the technique depends on the radial crack being planar and well defined on the specimen surface. In addition it is assumed that a residual stress damage zone develops underneath the indenter and that it behaves in an ideal manner. Thus the technique can only be used where c_o, the half-crack length, can be measured easily. In two-phase ceramics such as RBSiC this is difficult.

The second method does not depend on the measurement of the half-crack length c_o. In this technique the indent is located in the centre of a strength test specimen. If this is a flexure bar the indent and the associated radial crack are aligned transverse to the flexure stress axis. The net stress intensity is therefore

$$\begin{aligned} K &= K_{(residual)} + K_{(bend)} \\ &= K_r + K_a = \chi_r P/c^{3/2} + \sigma_a(\pi\Omega c)^{1/2} \quad (c \geq c_o) \end{aligned} \qquad (2)$$

Here Ω is a crack geometry factor which includes free-surface effects and ellipticity in the radial/median profile.

For growth under equilibrium conditions, $K = K_c$, Eq. (2) may be rearranged to give the applied stress as a function of crack size,

$$\sigma_a = [K_c/(\pi\Omega c)^{1/2}][1 - (\chi_r P/(K_c c^{3/2}))] \qquad (3)$$

Differentiating Eq. (3) and setting $d\sigma_a/dc = 0$ shows that this function has a maximum at

$$c_m = (4\chi_r P/K_c)^{2/3} = 4^{2/3}c_o \quad \text{(i.e. } c_m = 2.5\ c_o) \qquad (4a)$$
$$\sigma_m = [(27/256)K_c^4/(\chi_r(\pi\Omega)^{3/2})]^{1/3}P^{-1/3} = 3K_c/4(\pi\Omega c_m)^{1/2} \qquad (4b)$$

According to this description,[16] Fig. 2, the indentation crack undergoes a stage of precursor stable growth, from c_o to c_m, in attaining an instability configuration at $\sigma_a = \sigma_m$ which now defines the as-indented strength. It should be noted that the strength no longer depends on the value of c_o, for all $c_o<c_m$. Thus the indentation crack grows, under an applied stress, as an expanding cavity until the critical flaw size, c_m, is attained.

An expression for the toughness has been derived by Chantikul et al,[16] from Eqs. (4a/4b), making use of the definition of χ_r in Eq. (1),

$$K_c = \eta(E/H)^{1/8}(\sigma_f P^{1/3})^{3/4} \tag{5}$$

$$\text{where} \quad \eta = [(256/27)(\pi\Omega^{3/2}\xi]^{1/4} \tag{6}$$

η is a general expression, incorporating residual stress terms, and is usually derived from an empirical calibration against reference ceramics.[16] Under an applied stress the residual stress field drives the radial crack, on an expanding front, to a critical size. It is the indentation load and the materials resistance to deformation that determines this critical size. Therefore, K_c can be determined, using the above indentation-strength expression, from a measurement of the failure strength and a knowledge of the indentation load, Young's modulus, E, and the hardness, H, of the ceramic under test. The flaw size does not enter into the evaluation.

Probably the main source of error in using indentation fracture mechanics expressions to determine K_c is the accuracy of the calibration constants ξ and η. For example Chantikul et al[16] estimated that $\eta_v = 0.59 \pm 0.12$ for a Vickers indenter (a subsequent revision by Lawn[17] gives $\eta_v = 0.62$). Thus use of this value in the indentation-strength expression, (Eq.(5)), leads to an error in determining K_c of around ± 20% of the K_c values derived from conventional fracture toughness test methods. Where possible residual stress correction factors, such as χ_r and η, should be derived for the particular grade or class of material under test rather than rely on general expressions derived from averaging data from several ceramic reference materials. This should lead to greater accuracy in the determination of K_c. Care must also be exercised in the use of the general residual stress constant term, η , in the determination of K_c as the model, (Eq. (5)), assumes that pre-cursor crack extension takes place on an expanding crack front that is semi-circular, (i.e. half-penny shaped), at failure. However, recent work,[18,19] on soda-lime glass has shown that is not always the case. Where departure occurs from the classic half-penny shape of the radial crack, i.e. the flaw is semi-elliptical, then an ellipticity correction factor may have to be made.

Where doubt exists concerning the viability of K_c derived from an indentation-strength analysis then strict comparison with other ceramic materials should not be made. However, comparisons can be made if the controlled flaw strength and its variability with indentation load are compared within a class of materials, provided that it is understood that these are approximations.

EXPERIMENTAL PROCEDURE

For this investigation a 1200# grade (4 μm average grain size) was examined as a standard reference material. Two composites were investigated, where the original 1200# grit SiC was mixed with 15% and 30% by volume of 1-10 μm particle size TiB_2. Analysis by x-ray diffraction gave a final volume % of TiB_2 in the reaction sintered products of 5.3 % and 10.5 % respectively for the two composite grades.* All of the materials were manufactured as extruded 5 mm X 7 mm bar. For the strength results test bars were ground, parallel to the flexure tensile axis, with a 100# wheel and a 0.01 mm depth of cut. The cross section of the final machined bars was 5mm X 3 mm. For indentation strength testing, following grinding, the flexure bars were fine ground, (6 μm diamond suspension), then polished, (6 and 1 μm diamond suspension). A minimum of 60 μm was removed from the surface of the ground flexural test bars in the fine grinding/polishing operations in order to eliminate any influence of grinding induced surface flaws in the subsequent indentation evaluation of fracture toughness. In Fig. 4, a micrograph of the 1200#/10% TiB_2 grade, the TiB_2 particles and SiC matrix are shown.

Fracture strength tests were carried out at ambient temperature using 4 point bend test fixtures. As-ground fracture stress was determined using a fixture with an inner span of 15 mm and an outer span of 45 mm. Smaller length test bars, around 30 mm in length, were assessed in a second fixture with an inner span of 5 mm and an outer span of 20 mm. Further information on these test fixtures can be found elsewhere.[20]

All fracture toughness assessments were made using indentation-strength equation (5), with η assumed to be material independent and set equal to 0.59. Microhardness indentation was carried out with a Knoop diamond to provide controlled flaws. A Vickers indenter creates more lateral flaw damage, in the form of surface chipping, for a given indentation load with RBSiC and so was not adopted. To measure the size of flaw induced by indentation testing selected samples were etched, after indentation, in a boiling saturated solution of NaOH and $K_3Fe(CN)_6$. The fracture surfaces were examined in a Phillips SEM 515 Scanning Electron Microscope (SEM). Indentation strength bars that were not etched were indented with a drop of silicone oil placed on the indentation site to eliminate any possible influence from environmental effects.

---

*Material supplied by AEA Technology Ltd, Risley, UK

As-ground strengths were evaluated using the two parameter Weibull distribution

$$P_f = 1 - \exp-(\sigma_f/\sigma_o)^m \quad (7)$$

where P_f is the probability of failure, σ_f the fracture stress, σ_o a scaling parameter and m the Weibull modulus. Failure probabilities were calculated using the probability estimator

$$P_f = (i - 0.5)/N \quad (8)$$

where i is the rank in ascending order and N the total number of samples.

RESULTS AND DISCUSSION

Fracture strength and toughness

The presence of TiB_2,(5-10 vol % in the final compacts), results in a reduction in the fracture strength. For example 539 MPa for the reference 1200# grade down to 424 MPa for the composite containing 10 vol% TiB_2. However, the indentation strength of the composite grades, in comparison with the reference 1200# grade, show a reversal of this trend. For example, with an indentation load of 19.6 N, the indentation strength is 240 MPa for the composite containing 10 vol % TiB_2, whereas the indentation strength of the equivalent reference 1200# grade is 183 MPa, Table1. This is indicative of a significant toughness increase from the particulate addition. Examination of the indentation flaws revealed that a smaller flaw was induced in the composite grades, Fig. 3, and this could be attributed to crack deflection and pinning, (crack arrest), mechanisms.

The resulting K_c values, calculated using the residual stress calibration factor η in Eq.(5), are best treated with caution. Using this indentation strength approach η will vary according to the grade of RBSiC under test. η and therefore K_r will be influenced by grain size and the distribution of free silicon. Although K_c values are presented in Table 1 for these grades, calculated using $\eta = 0.59$ and Eq. (5), the indentation strength response may give a better indication of toughening mechanisms in these materials.

The SEM investigation of indentation flaw size, Table 2, is revealing in that the indentation flaw is smaller in the 1200#/10% TiB_2 grade, (a = 50 μm, 2c = 162 μm), in comparison with the reference 1200# grade, (a = 62 μm, 2c = 200 μm). A study of the fracture surfaces indicates that the TiB_2 particles are effective in increasing toughness by crack deflection. The particles in some areas, apparently, pinning the approaching crack front, Fig. 5, or, effectively deflecting the indentation crack out of its vertical plane. The relatively modest increase in indentation strength, and therefore toughness over the reference 1200# grade, is consistent with the relatively low residual tensile and hoop stresses around the TiB_2 particles (σ_{mr} = 42 MPa; $\sigma_{m\theta}$ = -21 MPa, respectively) that have been determined from the formula proposed by Evans and Langdon.[21] These stress calculations, (given in Appendix A1), indicate that the residual stress fields generated by TiB_2 particles in RBSiC have only a small effect in crack deflection, it is the "barrier" presence of the hard particles that is most relevant.

With the 1200#/5% v/o TiB_2 grade the lower percentage addition of particulate material results in a similar level of strength reduction, (406 MPa), to that of the 1200#/10% TiB_2 grade, (424 MPa), in comparison with the 1200# reference grade, (539 MPa), Table 1. These results indicate that at the microstructural level it is the TiB_2 particle size that is in part strength controlling and not the volume fraction of particles. It should be noted that all bars were ground parallel to the flexure tensile axis so that the transverse/lateral flaw induced through grinding is not of a size that dominates the fracture behaviour in comparison with the 10 μm maximum observed diameter of the TiB_2 particles.

In terms of the mean indentation strength (19.6 N load), the lower % of TiB_2 in the 1200#/5% TiB_2 grade is clearly less effective in toughening as only a modest increase in indentation strength of 203 MPa is achieved in comparison with 240 MPa for the 1200#/10% TiB_2 grade over the 183 MPa achieved for the reference 1200# grade, Table 1. Clearly the volume percentage of TiB_2 particles is of importance and is indicative of crack deflection being the prime mechanism for increasing toughness. Assuming that fabrication problems can be overcome, with respect to particulate dispersion, than an increase in TiB_2 content above 10% should lead to a further increase in toughness.

The materials containing TiB_2 are also more consistent with respect to their measured strength in comparison with the 1200# reference grade. The variability with respect to strength is given in Table 1, by Weibull modulus. The Weibull modulus estimates should only be compared within this sample group as the sample size is small, (n = 10 to 11). Further work is required to establish if a more uniform distribution of flaws is present in the composite grades and if damage tolerant mechanisms are operating, either at the indentation stage or during subsequent crack propagation to failure under an applied load.

Pre-cursor crack growth under an applied stress

To investigate the analysis given by Chantikul et al[16] and the use of the constant η in Eq.(5), that an expanding cavity develops from the indentation flaw under an applied stress, a pre-load of 87-88% of the unetched, indented failure load was applied to the reference 1200# and 1200#/10% TiB_2 materials. The load was applied in approximately 2 minutes, and then was removed. The test bars were then etched, returned to the test fixture, and tested to failure. From this evaluation, the results of which are given in Table 2, it can be seen that stable pre-cursor crack growth does occur but this appears to be confined to the surface zone. The crack does not appear to extend in the "a" flaw direction, Figs. 6 and 8 for the 1200# grade and Figs. 7 and 9 for the 1200#/10 vol% TiB_2 composite grade, i.e. into the body of the test bar. However, close examination of Figs. 10 and 11, from a different 1200# grade test bar, shows a change in the fracture texture that provides evidence of further crack advance in the "a" flaw direction by an expanding cavity model. This crack extension has developed on further stressing to failure above the pre-load.

Crack growth along the surface of the flexure bars will be favoured as the stress intensity will be high at the crack tips and there will always be some surface damage, even on carefully polished surfaces. However, the crack appears to be restrained in the vertical direction deep into the main body of the test bar.

Examination of the indentation flaw crack face in both 1200# and 1200#/TiB_2 composite grades, Figs. 12 and 5, shows that fracture of the SiC grains is primarily transgranular as imaging in the SEM by secondary electron contrast differences reveal the original SiC grits. Faulting, (either inclusions or small voids), along the epitaxial boundaries of the 1200# SiC grits with the newly deposited SiC is more in evidence, Fig. 12, in comparison with a coarser grain size 600# material, Fig. 13, examined in another study.[20] The significance of this microstructural feature is difficult to quantify.

Influence of an increase in indentation load on 1200# and 1200#/TiB_2 grades

Examination of the flaw size at the higher indentation load of 49 N confirms the findings of the flaw analysis at the lower indentation load of 19.6 N that a larger flaw is induced in the reference 1200# grade in comparison with the 1200#/TiB_2 materials,

Table 2 and Fig. 3. In Fig. 3 the initial indentation flaw size is plotted against the indentation fracture strength for both 19.6 N and 49 N indentation loads. The superiority of the 1200#/10% TiB_2 grade with respect to controlled flaw strength, in comparison with the 1200# control grade, is clearly shown. For the 10% TiB_2 grade a mean flaw size, a_o , of 89 μm, obtained with a 49 N indentation load, resulted in a higher controlled flaw strength of 235 MPa than that found for the 1200# grade of 212-222 MPa where a lighter indentation load of 19.6 N was applied. Fig. 3 also shows that it is not the initial flaw size that dictates the final controlled flaw strength i.e. the response from the residual stress field does not necessarily scale with flaw size. Subsequent crack deflection mechanisms as the crack propagates must also be taken into account.

Well defined semi-elliptical indentation flaws were obtained in the reference 1200# grade, Fig. 14, but not in the composite grades, Fig. 15. Therefore, the flaw dimensions given in Table 2 for the composite grades are the average of the maximum measured values for a_o and c_o and do not give a true representation of flaw ellipticity. The smaller observed flaw in the composite grades can be attributed to a combination of crack deflection around TiB_2 particles and crack arrest by TiB_2, as shown in Figs. 15 and 16.

The presence and development of an expanding cavity, from the indentation flaw, with an applied stress is now confirmed for the reference 1200# grade and an indentation load of 49 N. The size of the cavity at failure, shown in fig. 14 by a change in fracture texture, is approximately twice that of the flaw depth a_o. The results given in Table 2 show that the unetched reference 1200# grade has only approximately 85-90% of the measured strength of the equivalent etched bars, thus substantial stress relief has clearly taken place during etching. It is reasonable to predict therefore that the flaw dimension a_o will increase by a factor of X2.5, (i.e. $c_m = 4^{2/3} c_o$), in line with the analysis of Chantikul et al,[16] if we assume that the critical dimension for flaw propagation is the flaw depth dimension a_o for semi-elliptical flaws rather than the dimension c_o for the half-penny flaw configuration.

The change in texture that marks the final boundary of the expanding cavity, Fig. 14, may be the result of the change in crack growth velocity. Equilibrium crack growth of the expanding cavity from the original indentation flaw is relatively slow in comparison with fast fracture once the critical failure condition a_m (or c_m) is exceeded. It is probable that under fast fracture a more irregular fracture surface develops and this will influence secondary electron contrast.

A comparison of the flaw dimension analysis results for the higher indentation load of 49 N, (where an expanding cavity was observed on the fracture face of the flexure bars), with the analysis for pre-cursor crack growth at the lower indentation load of 19.6 N, (where a pre-load is applied to the flexure bar before flaw etching), is useful. The results indicate, Table 2, that for an indentation load of 19.6 N, pre-cursor crack growth takes place along the surface of the flexure bar in the "c" flaw direction but is constrained in the flaw direction "a", as shown in Figs. 6 to 9. The extent of constraint may be influenced by the initial indentation load. An indentation threshold load may exist below which crack growth to failure only occurs by crack extension along the surface, ("c" flaw direction failure only), and therefore a critical crack length dictates failure. When the critical indentation threshold load is exceeded then the residual stress field is sufficiently strong for an expanding cavity to be developed under an applied stress, Fig. 17. Initial crack growth still takes place along the surface, then at a certain applied stress, that is approaching the failure condition, constraint is lost and the crack expands to the final flaw cavity, (e.g. as shown in Fig. 14).

For the reference 1200# grade this indentation threshold load appears to be in the 20-50 N region, as clear evidence is shown of an expanding cavity at a load of 49 N, Fig. 14, but evidence is not as conclusive at 19.6 N, (compare Fig. 8 with Fig. 11). For the composite grades the indentation threshold load will be greater, on average, than that found for the fine grain size 1200# reference grade. However, the influence of etching in stress relieving the surface of the flexure bars is a complicating factor in these assumptions. Further work is being undertaken in order to establish if an expanding cavity mode of failure occurs, from indentation flaws, in the composite grades.

These results show that an expanding cavity model is suitable for characterising the fracture toughness of fine grain size RBSiC. However, an accurate calibration of the residual stress term χ_r , (and therefore K_r), should be obtained. In addition, the contribution to the fracture toughness of the microstructure should also be determined, i.e. a fracture toughness term, K_m, can be derived that incorporates microstructural factors, such as grain size and shape, presence of a second phase and crack deflection mechanisms.

CONCLUSIONS

1. Incorporation of TiB_2 particles to increase the toughness of a 1200# RBSiC, (4 μm average grain size), results in a reduction in the flexural strength but an increase in the controlled flaw strength and Weibull modulus in comparison with a reference

1200# RBSiC. The decrease in strength is thought to be strongly influenced by the larger observed grain size of the TiB_2 particles, (up to 10 μm diameter), in comparison with the maximum observed grain size, (4-5 μm), of the 1200# grade material.

2. The increase in fracture toughness with incorporation of TiB_2 into 1200# RBSiC is the result of a smaller induced indentation flaw in comparison with the 1200# grade. An increase in the volume fraction of TiB_2 particles also results in an increase in indentation fracture toughness.

3. Indentation crack measurements taken on the fracture face of indentation strength bars show that crack growth to failure occurs via an expanding cavity mode that is dependent on indentation load. Pre-cursor crack growth takes place along the surface of the flexure bars, (the "c" flaw direction), but appears to be constrained in the flaw direction "a" below a critical indentation load threshold. These results show that an expanding cavity model is suitable for characterising the fracture toughness of RBSiC.

REFERENCES

1. Ness J N and Page T F, "Microstructural Evolution in Reaction Bonded Silicon Carbide"; J. of Mater. Sci., **21**, 1986, pp. 1377-1397.

2. Swain M V, "Toughening Mechanisms for Ceramics"; Materials Forum, **13**, 1989, pp. 237-253.

3. Evans A G, "Perspective on the Development of High-Toughness Ceramics"; J. Am. Ceram. Soc., **73** (2), 1990, pp. 187-206.

4. Mecholsky J J, "Engineering Research Needs of Advanced Ceramics and Ceramic-Matrix Composites"; Am. Ceram. Soc. Bull., **68** (2), 1989, pp. 367-375.

5. Tiegs T N and Becher P F, "Sintered Al_2O_3-SiC Whisker Composites"; Ceram. Eng. Sci. Proc., **7**, 1986, pp. 1182-1186.

6. Whitehead A J, PhD dissertation, University of Cambridge, England, 1990.

7. Wei G C and Becher P F, "Improvements in Mechanical Properties in SiC by the Addition of TiC Particles"; J. Am. Ceram. Soc., **67** (8), 1984, pp. 571-574.

8. Lin B-W and Iseki T, "Effect of Thermal Residual Stress on the Mechanical Properties of SiC/TiC Composites"; Br. Ceram. Trans. J., **91** (1), 1992, pp. 1-5.

9. Janney M A, "Mechanical Properties and Oxidation Behaviour of a Hot Pressed SiC-15-vol%-TiB_2 Composite"; Am. Ceram. Soc. Bull., **66** (2), 1987, pp. 322-324.

10. McMurty C H, Boecker W D G, Seshadri S G, Zanghi J S and Garnier J E, "Microstructure and Material Properties of SiC-TiB_2 Particulate Composites"; Am. Ceram. Soc. Bull., **66** (2), 1987, pp. 325-329.

11. Jenkins M G, Salem J A and Seshadri S G, "Fracture Resistance of a TiB_2 Particle/SiC Matrix Composite at Elevated Temperature"; J. of Composite Mater., **23** (1), 1989, pp. 77-91.

12. Higgins I and Whitehead A J, in "Advanced Structural Inorganic Composites", Ed. P. Vincenzini, Pub. Elsevier Science B.V., 1991, pp. 583-592.

13. Marshall D B and Lawn B R, "Residual Stress Effect in Sharp Contact Cracking: Part 1 - Indentation Fracture Mechanics."; J. Mater. Sci., **14**, 1979, pp. 2001-.

14. Lawn B R, Evans A G and Marshall D B, "Elastic/Plastic Indentation Damage in Ceramics: the Median/Radial System"; J. Am. Ceram. Soc., **63** (9-10), 1980, pp. 574-.

15. Anstis G R, Chantikul P, Lawn B R and Marshall D B, "A Critical Evaluation of Indentation Techniques for Measuring Fracture Toughness: I, Direct Crack Measurements"; J. Am. Ceram. Soc., **64** (9), 1981, pp. 533-538.

16. Chantikul P, Anstis G R, Lawn B R and Marshall D B, "A Critical Evaluation of Indentation Techniques for Measuring Fracture Toughness: II, Strength Method"; J. Am. Ceram. Soc., **64** (9), 1981, pp. 539-543.

17. Lawn B R, "Fracture of Brittle Solids - 2nd Edition"; Cambridge University Press, 1993, pp. 253-257.

18. Cook R F and Pharr G M, "Direct Observation and Analysis of Indendation Cracking in Glasses and Ceramics"; J. Am. Ceram. Soc., **73** (4), 1990, pp. 787-817.

19. Smith S M and Scattergood R O, "Crack-Shape Effects for Indentation Fracture

Toughness"; J. Am. Ceram. Soc., **75** (2), 1992, pp. 305-315.

20. Kibble K A, PhD dissertation, University of Wolverhampton, England, 1993.

21. Evans A G and Langdon T G, in "Structural Ceramics"; Prog. Mater. Sci., **21** (3-4), 1976, pp. 171-441.

Table 1: Test data for 1200# RBSiC and 1200#/TiB_2 RBSiC (4 point bend, 15/45 mm span)

	1200#	1200#/10% TiB_2	1200#/5% TiB_2
Mean strength /MPa (As-ground)	539	424	406
Weibull Modulus, m (Sample size)	11.69 (10)	28.49 (11)	19.95 (11)
Indentation strength (19.6 N load) /MPa	183	240	203
Fracture toughness, K_c /MPa.m$^{1/2}$ (Sample size)	2.94 (10)	3.60 (11)	3.17 (10)
Hardness, HK_2 /GPa (19.6 N load)	14.96	15.56	15.13

Table 2: Mean indentation strength σ_f and flaw dimensions (crack depth a, crack length 2c) of 1200# RBSiC and 1200#/10% TiB_2 RBSiC. (4 point bend, 5/20 mm span)

Indentation load,	1200#			1200#/10% TiB_2		
whether etched	σ_f /MPa	a /μm	2c /μm	σ_f /MPa	a /μm	2c /μm
19.6 N, unetched	191			236		
19.6 N, etched	212	62.4	199.6	263	50.4	162.2
19.6 N, pre-load, etched	222	60.2	277.8	287	48.5	238.7
49 N, etched	179	109.6	389.0	235	88.8	360.3

Sample size n = 4-5; pre-load ≈ 87% of indented and unetched failure strength.

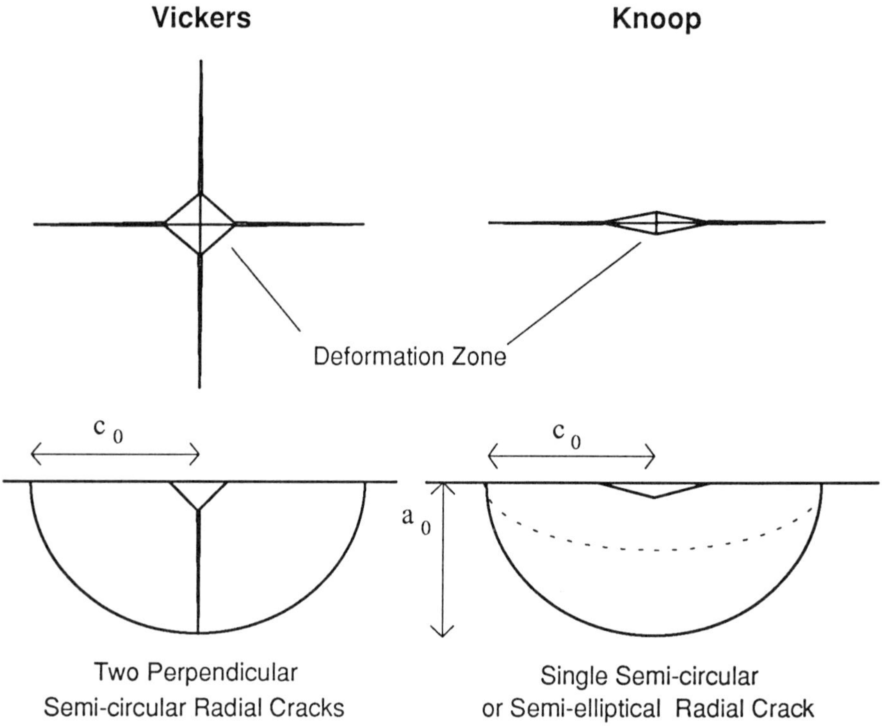

Fig. 1. Radial crack configuration for Vickers and Knoop indentation.

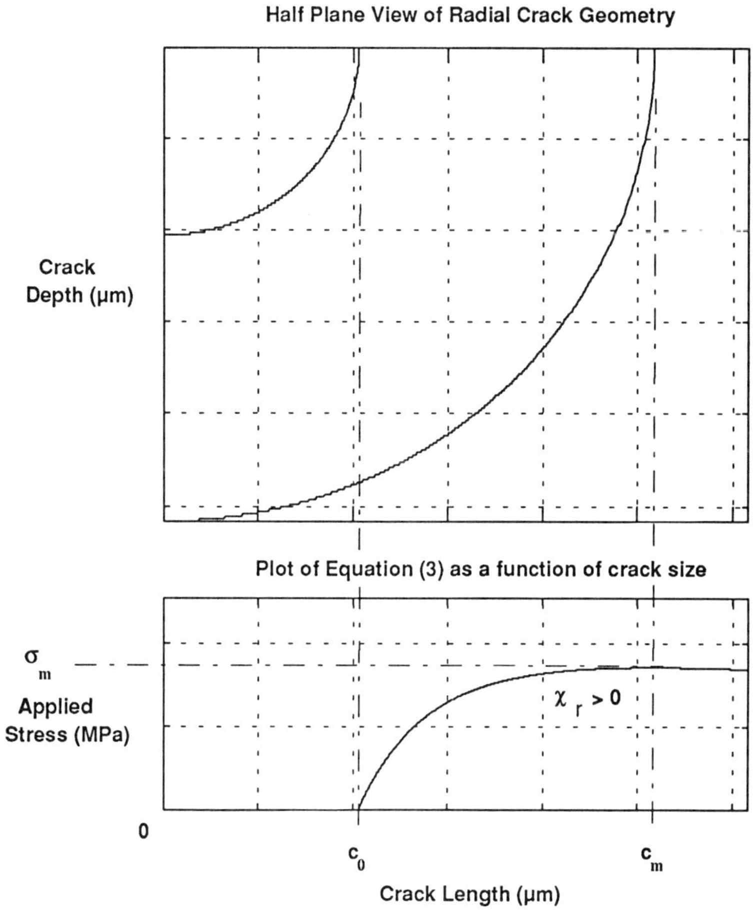

Fig. 2. Schematic model of pre-cursor crack growth from the indentation flaw under an applied stress. According to this model $c_m = 2.5\ c_0$.[16]

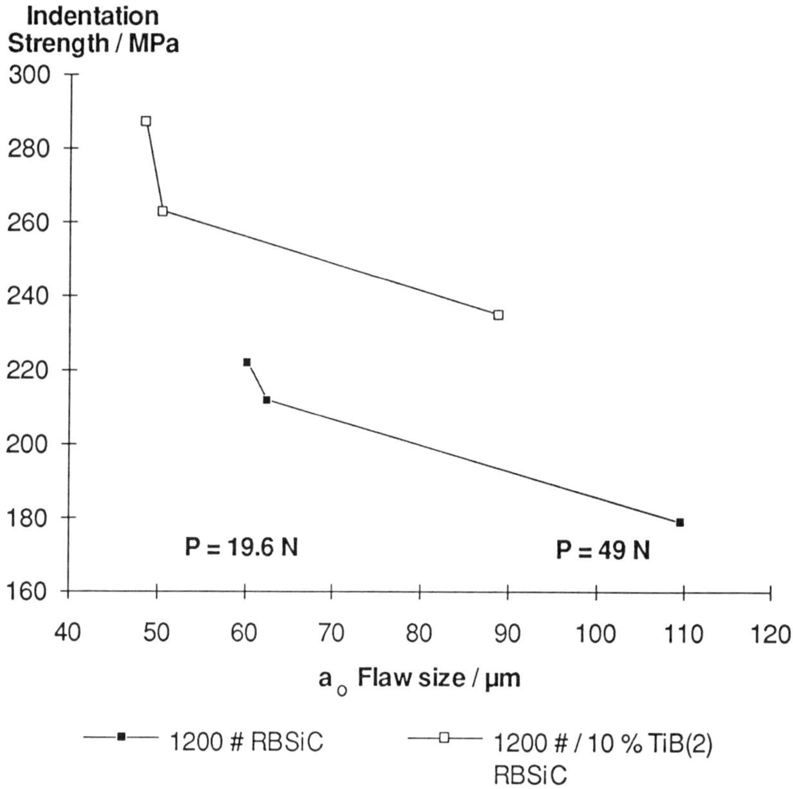

Fig. 3. Initial controlled flaw dimension a_0 v indentation strength.

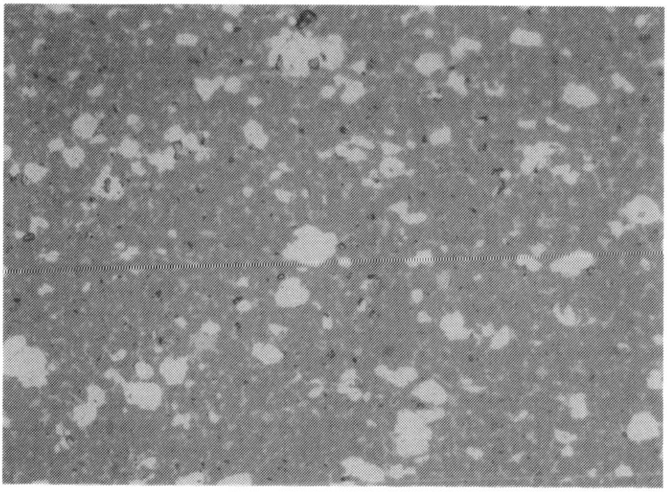

Fig. 4 Light micrograph of 1200#/10% TiB_2 RBSiC showing TiB_2 particles (white) in the SiC matrix. (Field of view 250 μm wide)

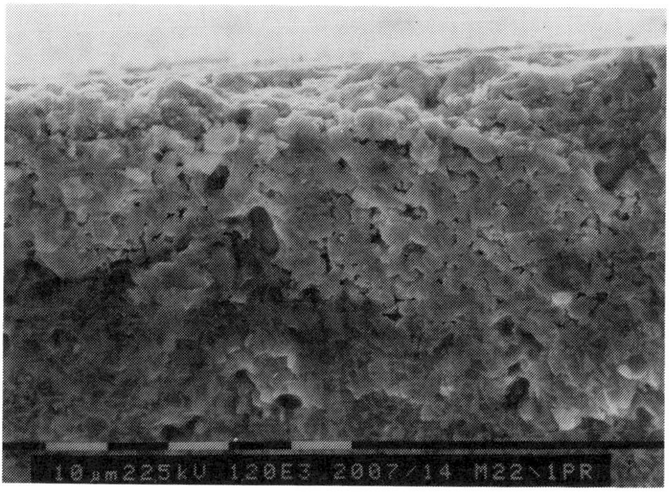

Fig. 5 SEM fractograph of the indentation flaw (19.6 N load) in a 1200#/10% TiB_2 RBSiC, etched, bend test bar, showing crack pinning and deflection by TiB_2 particles.

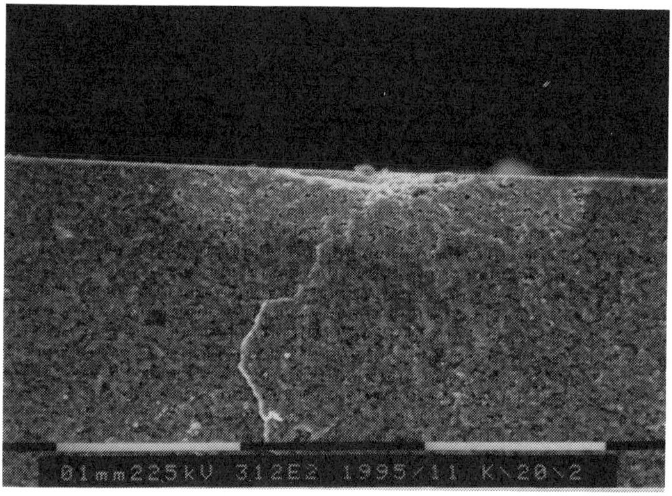

Fig. 6 SEM fractograph of the indentation flaw (19.6 N load) in a 1200# RBSiC, etched, bend test bar showing pre-cursor crack growth under an applied stress.

Fig. 7 SEM fractograph of the indentation flaw (19.6 N load) in a 1200#/10% TiB_2, RBSiC etched, bend test bar. The application of an applied stress results in pre-cursor crack growth along the surface of the bar.

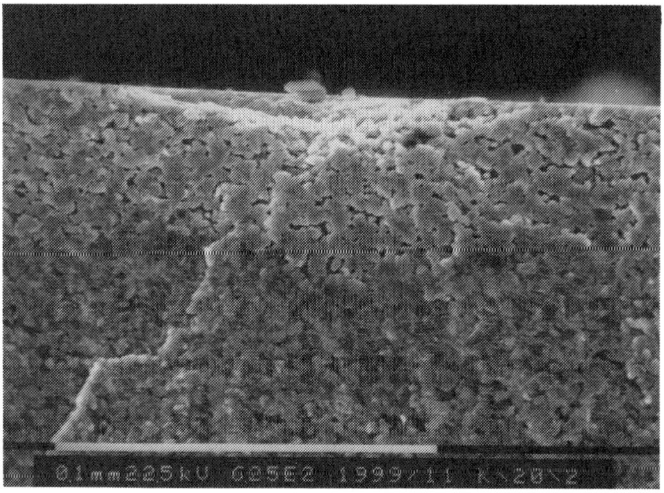

Fig. 8 Further detail of the indentation flaw shown in Fig. 6, from a 1200# RBSiC bend test bar. Pre-cursor crack growth takes place along the surface of the bar but appears to be constrained in the vertical direction i.e. into the body of the test bar.

Fig. 9 Further detail of the indentation flaw shown in Fig. 7, from a 1200#/10% TiB_2 RBSiC bend test bar. The indentation crack is smaller than that observed in the 1200# reference grade, (Fig. 8). A large (10 μm diameter) TiB_2 particle can be seen pinning the base of the indentation crack.

Fig. 10 SEM fractograph of the indentation flaw (19.6 N load) in a 1200# RBSiC. Precursor crack growth can be seen along the surface of the bar but a change in the texture of the fracture face also provides some evidence of crack growth in the vertical direction, (i.e. into the body of the test bar), through an expanding cavity mode.

Fig 11 Further detail of the indentation flaw shown in Fig. 10, from a 1200# bend test bar. The change in the fracture texture provides evidence of further crack advance, under an applied stress, in the "a" flaw direction by an expanding cavity model.

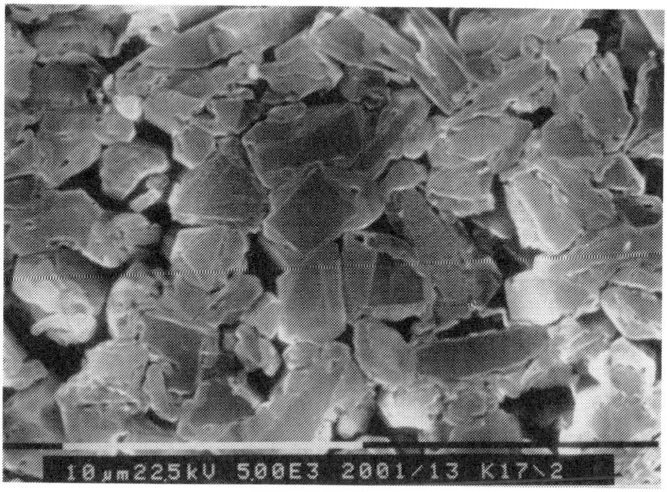

Fig. 12 SEM fractograph of the indentation flaw crack face, shown in Figs. 10 and 11, for the 1200# RBSiC bend test bar. Fracture is clearly transgranular and faults along the epitaxial boundaries of the original grits and the new deposited SiC are much in evidence. The degree of boundary faults is greater than that observed in a 600# grade, Fig. 13.

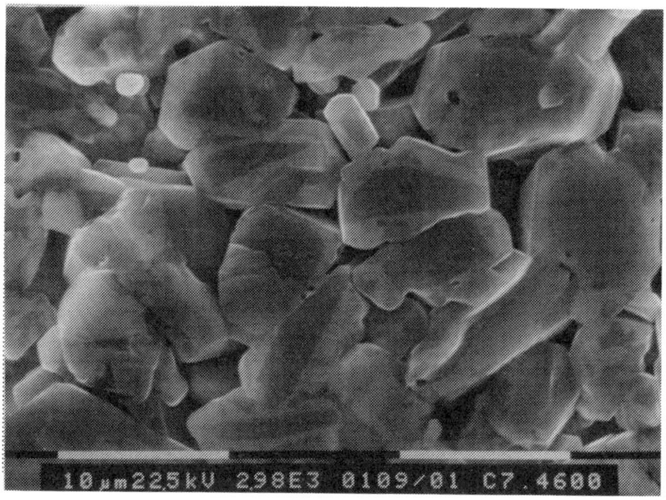

Fig. 13 SEM fractograph of the indentation flaw crack face (19.6 N load) in a 600# RBSiC, etched, bend test bar showing transgranular fracture.

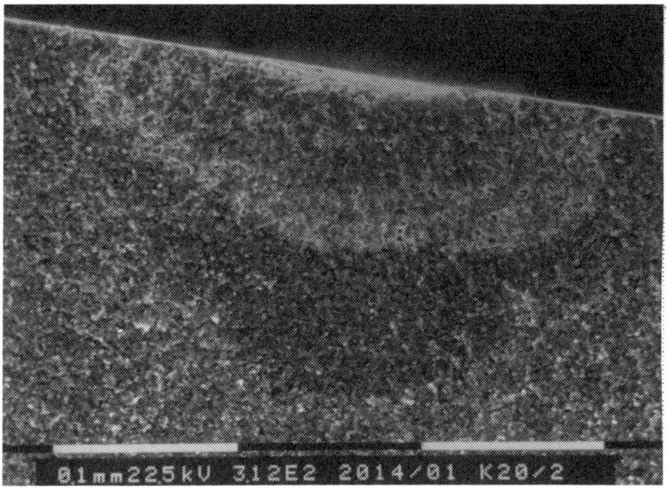

Fig. 14 SEM fractograph of the indentation flaw (49 N load) in a 1200# RBSiC, etched, bend test bar. The change in fracture texture provides clear proof of further crack advance, under an applied stress, by an expanding cavity mode.

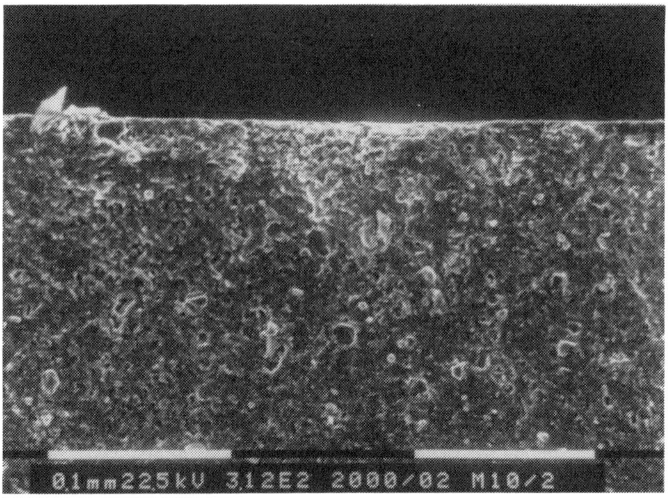

Fig. 15 SEM fractograph of the indentation flaw (49 N load) in a 1200#/10% TiB_2 RBSiC, etched, bend test bar. The indentation crack is much smaller, (and poorly delineated), in comparison with the indentation flaws observed in the 1200# reference grade, for the same indentation load, (e.g. Fig. 14).

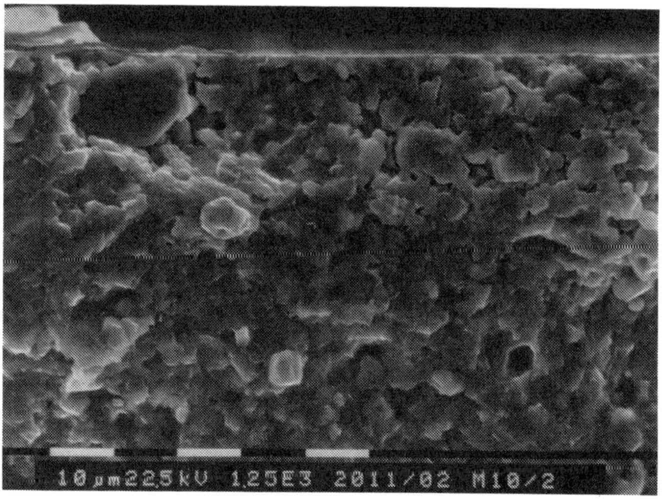

Fig. 16 Further detail of the indentation flaw shown in Fig. 15, from a 1200#/10% TiB_2 RBSiC bend test bar. The TiB_2 particle in the top left corner has either pinned or deflected the advancing surface crack.

APPENDIX A1:

Residual stress field around TiB_2 particles

The residual stress field exerted by differential thermal contraction of a spherical particle in an infinite matrix has been determined using the Evans and Langdon[21] analysis given below.

Assuming the bulk (matrix) controls the contraction, the differential thermal expansion coefficient is given by,

$$\Delta\alpha = \alpha_m - \alpha_p$$

where the subscripts `p` and `m` refer to the particle and matrix respectively

Differential contraction on cooling gives rise to a residual pressure (σ_R), at the particle-matrix boundary, $p_r = R_b$, where, p_r is the particle radius and R_b the boundary distance from the particle centre.

The hydrostatic pressure (σ_R) is estimated by,

$$\sigma_R = \frac{\Delta\alpha\,\Delta T}{\dfrac{1+\nu_m}{2E_m} + \dfrac{1-2\nu_p}{E_p}}$$

where, ν is Poisson's Ratio, E is Elastic Modulus and ΔT is the temperature range over which stresses are not relieved by a diffusive process.

The hydrostatic pressure in the matrix comprises two superimposed stress components, a radial (σ_{mr}) and hoop tangential ($\sigma_{m\theta}$) stress field.

$$\sigma_{mr} = -\sigma_R\left(\frac{p_r}{R_b}\right)^3, \qquad \sigma_{m\theta} = \frac{1}{2}\sigma_R\left(\frac{p_r}{R_b}\right)^3$$

cont..

Inserting the reference values given below for the constants in the above formulae and taking ΔT to be 1000°C then,

$$\sigma_{mr} = 42 \text{ MPa}, \quad \text{and} \quad \sigma_{m\theta} = -21 \text{ MPa}$$

	TiB_2 [10]	SiC [22]
Coefficient of linear expansion, α (RT. to 700°C) [10^{-6}/K]	4.6	4.5
MOE, [GPa]	531	440
Poisson's Ratio, ν	0.25	0.25

REFERENCES

10. McMurty C H, Boecker W D G, Seshadri S G, Zanghi J S and Garnier J E, "Microstructure and Material Properties of SiC -TiB2 Particulate Composites"; Am. Ceram. Soc. Bull., **66** (2), 1987, pp. 325-329.

21. Evans A G and Langdon T G, "Structural Ceramics"; Prog. Mater. Sci., **21** (3-4), 1976, pp. 171-441.

22. Edington J W, Rowcliffe D J and Henshall J L, "Powder Metallurgical Review no. 8. Mechanical Properties of Silicon Nitride and Silicon Carbide: 1"; Powder Metall. Int. **7** (2), pp. 82-96.

FAILURE ANALYSIS OF AN α-SiC TUBE SUBJECTED TO THERMAL CYCLING IN AN OXYGEN-STEAM ATMOSPHERE*

Kristin Breder and James R. Keiser
Metals and Ceramic Division
Oak Ridge National Laboratory
Oak Ridge, TN 37831-6069

ABSTRACT

Failure analysis of a 0.965 m long closed end α-SiC tube which failed during thermal cycling in an oxygen-steam atmosphere was conducted. The tube failed during heating after having survived several cycles. The shattered tube was re-assembled and the fracture origin was determined to be a flaw in the lower part of the tube on the outer surface. The flaw was likely an oxidation pit produced during the previous cycles. The stresses in the tube at the time of failure were a combination of stresses from the internal pressure and from thermal gradients.

INTRODUCTION

Oak Ridge National Laboratory (ORNL) is participating in an industry led effort to develop an advanced high pressure heat exchanger for a convective steam methane reformer. The heat exchanger utilizes ceramic tubes in an effort to raise the temperature and subsequently the effectiveness of the system and thereby reduce the energy consumption. As part of this project two experimental furnaces which can reach temperatures up to 1550°C and accommodate large test tubes have been constructed at ORNL. The purpose of the experimental work has been to expose candidate tubes to temperatures and environments as close to reality as possible, while carefully monitoring the temperature and pressure.

* Research sponsored by the U. S. Department of Energy, Assistant Secretary for Energy Efficiency and Renewable Energy, Office of Industrial Technologies, Industrial Energy Efficiency Division, under contract DE-AC05-84OR21400 with Martin Marietta Energy Systems Inc.

During a test cycle one of the tubes failed catastrophically. The purpose of the present failure analysis was to locate the failure initiation point and from the nature of the initiation point draw a conclusion about the failure event and the reason why the tube failed.

EXPERIMENTAL HISTORY

The test tubes were made of α-SiC from Carborundum (Hexoloy SA) which is a pressureless sintered ceramic with boron as a sintering aid. This material has a room temperature flexure strength of about 460 MPa and this strength level is retained up to 1400°C[1]. The fracture toughness is 3.0 MPa $\sqrt{m}$[2,3] and density is 3.1 g/cm.[4] The material is a good candidate for heat exchangers because of its good thermal properties and relatively high resistance to thermal shock. The tubes were 0.965 m long with an inner diameter of 76.2 mm (3") and an outer diameter of 88.9 mm (3.5"), and had a flange at the open top end and a hemispherical closed bottom end. The tubes were hung in the furnace by fitting the flange into a steel collar. The steel collar also functioned as a lid with metal tubes coming through to the inside of the SiC tube for pressurizing and flush through of gas. About one third of the tube was entirely inside the furnace, one third was surrounded by an insulation lid and one third was above the furnace. The furnace was heated by SiC resistance heaters mounted horizontally in two opposite walls of the furnace.

Before the start of the experiment the tubes were all pressurized individually at room temperature and at 1260°C as a proof test measure. The tube in question was pressurized three times at room temperature up to 2 MPa (300 psi) for a total duration of 254 h, and twice at 1260°C; 0.4 MPa (56 psi) for 5 h and 2 MPa (300 psi) for 18 h. The tubes were then subjected to cycles of heat and steam pressure as described in Table I. Upon heating in the ninth cycle, one of the tubes failed catastrophically and the furnace was automatically shut down. At the time of failure only one tube was in the furnace, and the temperature was at 755°C. The fracture event was sudden and high in energy as evidenced by an abrupt loss of pressure and the subsequent damage seen inside the furnace, see Fig 1.

Table I. Test history of Hexoloy α-SiC tube.

Run No.	Max Pressure MPa	(psi)	Max. Temp. ℃	Time at Temp. h	Heat-up Time+ h
SS-1	1.3	(198)	1177	24	6
SS-2	1.4	(201)	1177	24	8
SS-3	1.4	(200)	1177	48	8
SS-4	1.3	(183)	1177	48	8
SS-5	1.4	(201)	1260	50	10
SS-6	1.5	(211)	1260	50	10
SS-7	1.5	(216)	1177	50	10
SS-8	1.5	(216)	1177	50	10
SS-9*	[1.4	(200)]	[1038]		

* Did not reach operating conditions

+ Cool-down time was in addition to these times

Figure 1. The furnace insulation and heating rods were damaged during the tube failure.

FAILURE ANALYSIS

The hundreds of pieces of the fractured tube were carefully collected from the furnace and the tube reassembled. Most of the tube fragments could be put back together, except for the area right below the flange which had shattered into very small pieces, probably by impact with the steel hardware of the furnace. The reassembled tube is shown in Fig. 2 a) - c). There are several things that can be learned from studying the macro-pattern of the fractured tube. First, the overall reassembled tube indicates that the failure must have initiated in the bottom part. If the initiation had happened in the top, the bottom would not have been so shattered, and further with the lower end having smaller pieces indicating a large stored elastic energy here right before the explosion. Also, the tube pieces retrieved from the furnace were partially found in the side walls, indicating that they had been thrown horizontally rather than vertically down. Second, fracture initiating from internal pressure and also from bending and torsion typically has the fracture origin at a straight line before branching occurs. The angle between the crack branches can give an indication of the stress state, with equibiaxial tension resulting in large angles and torsion resulting in small angles.4 These types of fracture patterns can be seen in several places in the lower end of the tube. Third, several long and wavy cracks can be seen in Fig. 2 c). These are indicative of a slow moving fracture typically resulting from thermal stresses. These are also lower in energy and it is reasonable to assume that these cracks are secondary to the high energy cracks and occurred after the failure event had been initiated. The direction and also to some degree the velocity with which cracks move can be found from studying the Wallner lines which are curved markings on the fragment edges.[4] The crack has moved in the direction of convex curvature, and the Wallner lines can therefore give an indication as to where cracks might have been traveling from, i.e. ideally one could find one place where all cracks had traveled outward from a common origin. Several of the tube fragments had distinct Wallner lines, and these were utilized in order to identify possible fracture origins.

There are a few areas of the tube which have a fracture branching pattern typical of a fracture initiation site. A longer distance before branching starts is an indication of a larger fracture initiating flaw, and hence, a lower stress would be needed to initiate failure. An empirical relationship between the fracture stress and the branching radius has been developed:[5]

$$\sigma\, r^{1/2} = M, \tag{1}$$

where σ is the fracture stress, r is half the length of the unbranched portion of the crack, and M is a material and environment constant. By applying this it is possible to estimate the fracture stress for the initiation of the explosion. Then having an estimate of the stress level it is possible to estimate the flaw size at the initiation point by using the analysis of Newman and Raju:[6]

$$K_{IC} = \frac{F}{\sqrt{Q}} \sigma \sqrt{\pi \ a}, \tag{2}$$

where Q and F are functions of the flaw dimensions and flaw and specimen dimensions respectively, K_{IC} is the fracture toughness and the flaw depth. By assuming a semielliptical flaw in an "infinite" specimen the functions take on the specific values of:[7]

$$Q = \frac{\pi^2}{4} \quad \text{and} \quad F = 1.144$$

This way it is possible to make an evaluation of the flaw size one will be looking for and this will greatly aid the microscopic part of the failure analysis.

Four fragments were studied in the scanning electron microscope (SEM): two from possible initiation points, one from a fragment that had a rough zone close to the inner surface, and one with a "clean" fracture surface. These fragments are marked in Fig. 2 c). X-ray analyses on these surfaces were performed in the SEM.*

Figure 2 a) and b) The re-assembled tube viewed from two different sides.

* Hitachi SEM S 800, with Tracor Northern XRD

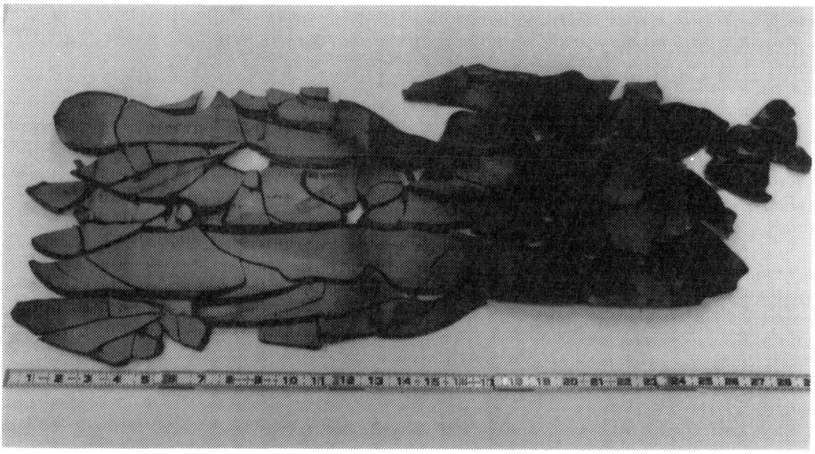

Figure 2 c) The re-assembled tube as opened up to view the pieces.

Fragment 1, a possible initiation point, was located 170 mm from the bottom of the tube and had a branching length of 2r = 32 mm. Using Eq. 1 with r = 16 mm and M = 9.77 MPa $\sqrt{m}$ for sintered SiC[8] results in a calculated stress at failure of 77.2 MPa. Then, using Eq. 2 with the given values of F and Q and K_{IC} = 3.0 MPa $\sqrt{m}$ it is found that the fracture initiating flaw must have had a depth of 0.9 mm. As can be seen from Fig. 3 a) this particular fragment had a relatively large (1.5 mm) area which had microcracks and something that looked like oxidized material. The depth of the area can be estimated from the micrographs to be of the order 800 μm, close to the size calculated above. The X-ray analysis showed that there was oxygen present in the areas which look like they have a smaller and different grain size than the rest of the surface. (Fig. 3 b, and c).

Fragment 2, a "clean" surface shown in Fig. 4 was analyzed for comparison. The X-ray analysis for this showed, as expected, basically SiC.

Fragment 3 had a surface with some irregularities to it, and was picked for that reason. This showed no oxidation, but had several small cracks going parallel to the surface, see Fig. 5. It is difficult to say whether these were produced by the overall fracture event or not, but it is not likely that they were there in the as-received tube. X-ray analysis from this area showed no evidence of oxygen.

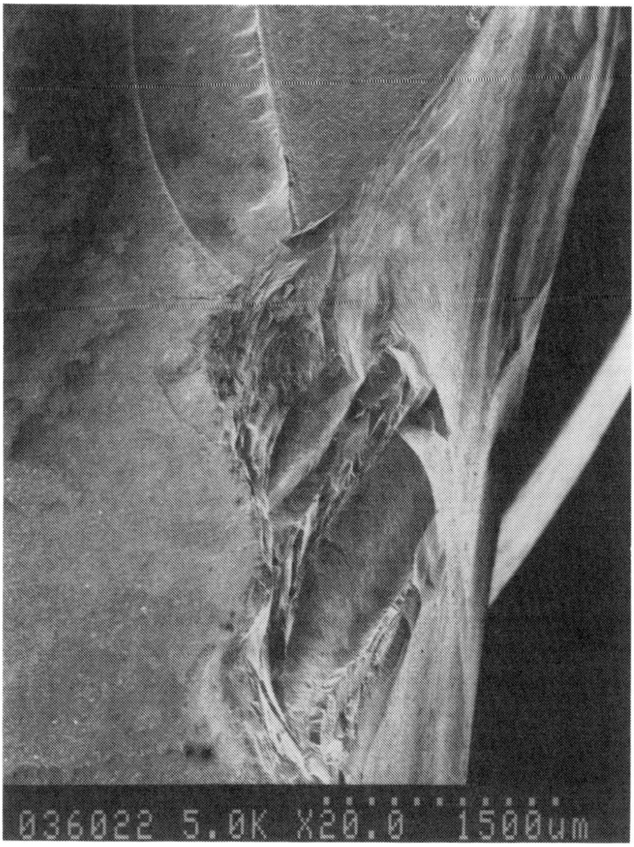

Figure 3. a) Possible fracture origin from fragment #1 in Fig. 2 c).

Figure 3 b) The possible initiation point was a crack of approximately 1.5 by 0.8 mm.

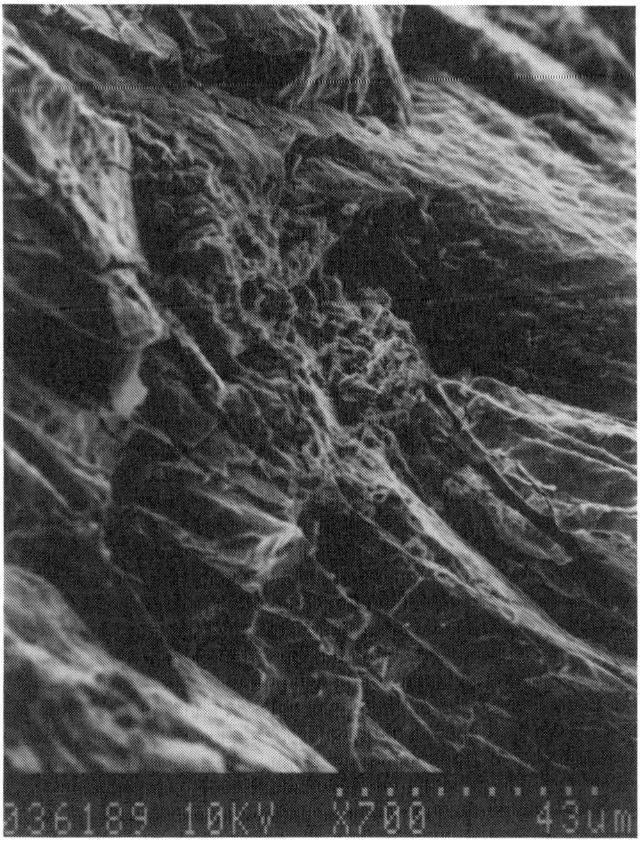

Figure 3 c) Close up of the small grained area in the crack.

Figure 4. A clean fracture surface from fragment #2.

Figure 5. a) Fragment #3 contained numerous small cracks and the surface contained an oxidation layer.

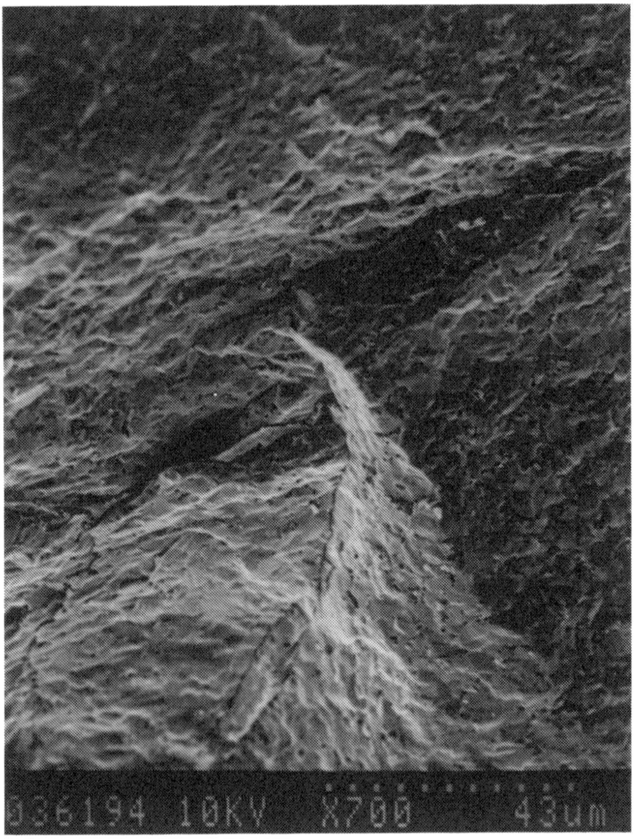

Figure 5. b) A fracture surface from fragment #3 containing numerous small cracks.

Fragment 4 was from another possible initiation point. This had a crack branching length of the same order of magnitude as Fragment 1 and should therefore contain a flaw of the same order of magnitude as described above. As can be seen from Fig. 6 this had a large crevasse - like crack which extended into a closed up crack. This was located somewhat down into the crack but seemed to be connected to the surface. The X-ray analysis showed a presence of oxygen inside this crack. The unusual "bridge" seen in Fig. 6 is an indication that this is a complicated fracture event and that part of this crack existed before the catastrophic failure occurred. The SEM analysis gave, however, no indication that this could have been a pore existing from the beginning, because the grain morphology in a pore is distinctively different from that produced by fracture.

DISCUSSION

The above analysis resulted in the finding of two possible failure initiating flaws. The most likely initiation point is the one located on fragment #1 (Fig. 3), because the flaw located here was situated in the fracture pattern in a way that is typical for initiation points in tubular geometries. Further, the good correlation between the crack branching length and the flaw size indicates this as the initiation point. The calculated stress level of 77 MPa is higher than the stresses set up in the tube due to the internal pressure of 1.4 MPa (200 psi). However, if the stresses due to the temperature differences both along the length of the tube and between the inner and outer surfaces is taken into account, hoop stresses between 25 MPa (closed end) and 100 MPa (open end) may exist in the tube[9] and these stress levels would be high enough to initiate failure at the given circumstances.

The flaw in Fig. 3 is quite large and must have developed over time during the previous eight cycles, or the failure would have occurred at an earlier stage. Sintered α-SiC is not prone to slow crack growth (SCG) damage at temperatures which occurred under the present experiment.[10] However, the nature of the flaw is more like that of an oxidation pit, with oxide interspersed with cracks. It is possible that the cyclic stresses resulting from the thermal cycling may have caused repeated cracking of the SiO_2 scale at preexisting sites like pores and thereby gradually creating a larger crack. This is consistent with the observations by Quinn and Katz[11] who in a stepped temperature stress rupture test saw time dependent failure typical of slow crack growth, but they found no fractographic evidence of SCG. They indicated that oxidation of surface connected porosity might be the reason for the observed behavior. Normally under the conditions described above SiC will have a protective SiO_2 layer produced by passive oxidation. This will protect the ceramic from further oxidation damage. Active oxidation will not occur until the oxygen partial pressure drops below approximately 13 MPa (10^{-2} Torr).[12] The presence of steam in the system will however alter this value, and it is possible that further local active oxidation may be aided by the presence of material impurities. An additional complicating factor is that the oxidation behavior of SiC has been shown to change with the applied stress level.[13] It is not known how the stress level may affect the passive to active oxidation transition.

Figure 6. A possible initiation site from fragment #4.

CONCLUSIONS

The fractographic investigation of the fractured SiC tube indicates that the fracture initiated from a flaw approximately 0.8 mm deep located at the inner tube surface approximately 170 mm from the bottom of the tube. The flaw contained oxide and it is assumed that the flaw developed during the previous eight temperature cycles. The oxidation could have occurred at an already existing pore, and might have been enhanced by the presence of steam and stress. The tube failed at a hoop stress of approximately 77 MPa, a stress which could arise from the combined action of internal pressure and thermal gradients.

REFERENCES

1. D. E. McCullum, N. L. Hecht, L. Chuck and S. M. Goodrich, "Summary of results of the Effects of Environment on Mechanical Behavior of High-Performance Ceramics," *Ceram. Eng. Sci. Proc.*, **12** [9-10] 1886-1913 1991.

2. M. G. Jenkins, A. Gosh, K. W. White, and A. S. Kobayashi, "Fracture Resistance of Sintered SiC Using Chevron Notch," pp. 241-252 in Ceramic Transactions, Vol. 2, edited by J. K. Collie and C. E. Semler, American Ceramic Society, Westerville, OH, 1989.

3. G. V. Srinivasan, Carborundum Co., Personal Communication 1995.

4. V. D. Frechette, Failure Analysis of Brittle Materials, Advances in Ceramics Vol. 28, American Ceramic Society, Westerville, OH 1990.

5. J. J. Mecholsky, A. C. Gonzalez and S. W. Freiman, "Fractographic Analysis of Delayed Failure in Soda-Lime Glass," *J. Am. Ceram. Soc.,* 62 [11-12] 577-580 1979.

6. J. C. Newman and I. S. Raju, "An Empirical Stress Intensity Factor Equation for the Surface Crack," *Eng. Fract. Mech.*, **15** [1-2] 185-192 1981.

7. G. S. Glasemann, K. Jakus and J. E. Ritter Jr., " Strength and Variability of Indented Soda-Lime Glass," *J. Am. Ceram. Soc.*, **70** [6] 441-444 1987.

8. D. L. Shelleman, Personal Communication 1993.

9. D. L. Shelleman, R. E. Tressler and J. C. Conway Jr., "Finite Element Analysis of Heat Exchanger Tubes," Quarterly Report to DOE under Contract No. PS-004416 between DOE and Penn State, August 1993.

10. K. Y. Chia and K. S. Lau, "High Toughness Silicon Carbide," *Ceram. Eng. Sci. Proc.*, **12** [9-10] 1845-61 1991.

11. G. D. Quinn and R. N. Katz, "Time Dependent High-Temperature Strength of Sintered α-SiC," *J. Am. Cer. Soc.,* **63** [1-2] 117-119 1980.

12. E. A. Gulbransen and S. A. Jansson, "The High-Temperature Oxidation, Reduction, and Volatilization Reactions of Silicon and Silicon Carbide," *Oxidation of Metals*, **4** [3] 181-201 1972.

13. T. E. Easler, R. C. Bradt and R. E. Tressler, "Strength Distributions of SiC Ceramics After Oxidation and Oxidation under Load," *J. Am. Cer. Soc.,* **64** [12] 731-734 1981.

FRACTOGRAPHIC INVESTIGATION OF FLAWS IN SINTERED SiC FIBER

G. V. Srinivasan and V. Venkateswaran, The Carborundum Company, P.O. Box 832, Niagara Falls, NY 14302.

INTRODUCTION

The Carborundum Company has developed a continuous process to produce multifilament, polycrystalline SiC fiber tow [1-3]. The process uses a melt spinning approach to make the green fibers wherein the submicron SiC powder is mixed with polymers and the resulting compound is melt spun and extruded through a spinnerette. After extraction, the green fibers are sintered in a continuous sintering line. The properties of this fiber are given in Table 1 [3]. This paper will describe the use of fractography in directing the fiber developmental efforts, specifically on characterization of strength-limiting defects and their origination. Fractographic analysis is also extended to determining strength of fibers at any gauge length and in identifying artifacts associated with the tensile strength testing technique.

Table 1. Properties of SiC Fiber

Diameter	17-20 μm
No. of Fibers/tow	200 - 400
Density	3.15 g/cm^3
Elastic Modulus	400 GPa
Room Temperature Tensile Strength @ 25.4 mm gauge length	1.5 - 1.75 GPa

EXPERIMENTAL PROCEDURE

To test the tensile strength of the fiber, a single fiber was gently separated from the tow, cut to appropriate length, and mounted on a paper tab. It is typical to test a minimum of 30 samples for characterizing the tensile strength. The fiber samples were randomly collected from different sections of the tow. The fiber was glued to the paper with fast drying epoxy glue at the two ends. In this work, two types of tabs differing in stiffness were used. One of them was a regular copy paper, and the other one was a stiffer manila folder paper. To capture the fractured fiber sample without shattering, a thin layer of petroleum jelly (semi-liquid paste) was squeezed on the fiber after mounting the fiber on the paper tab. The effect of petroleum jelly was investigated by conducting tests with and without the petroleum jelly.

The tab with the fiber was then attached to the load cell via air-pressure-activated snap-on grips. The testing setup is schematically shown in Figure 1. The paper tab has a hole punched in the center. The paper is burnt from the sides of the hole to transfer the load to the fiber. A DC motor provides a linear movement of the load train. A personal computer controls the movement of the load train and is also used for data acquisition.

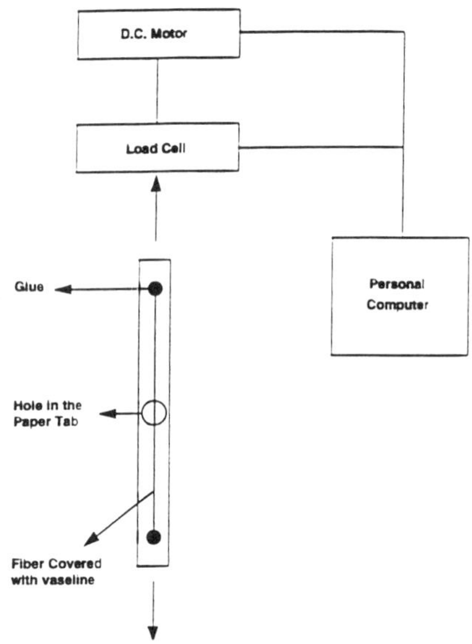

Figure 1. Single-Fiber Tensile Testing Apparatus

After the test, the two pieces of the broken fiber are measured to make sure that they add up to the starting gauge length. If a piece of the fiber is missing through shattering during testing, that sample is removed from the data pool. After ascertaining that no pieces of fiber are missing, the petroleum jelly on the fiber is dissolved in a solvent in an ultrasonic bath. The fracture ends were carefully mounted on the sides of a hexagonal nut with conducting graphite paint for observation in the Scanning Electron Microscope (SEM). Typically, more than 90% of the fractured samples were recovered for fractographic observation. Some samples are lost during the cleaning and transport process. Again, those data points were not included in strength determination. It should be noted that no fractography could have been possible without the use of petroleum jelly, since the fiber sample usually shatters upon fracture.

RESULTS AND DISCUSSION

Tensile strengths of single fiber filaments were determined over gauge lengths of 254 mm (10 inches) and 25.4 mm (1 inch) respectively. The observed average strengths over different gauge lengths, and variations in test procedure are summarized in Table 2. The strength results along with fractographic details are discussed below.

Table 2. Fiber Testing Method and Tensile Strength

Gauge Length	Test Procedure	No. of Tests	Avg. Strength ± SD MPa
254 mm	Paper Tab with Petroleum Jelly (Soft Tab)	36	1262 ± 255
25.4 mm	Paper Tab with Petroleum Jelly (Soft Tab)	78	1565 ± 331
25.4 mm	Stiff Tab (Manila folder paper) No Petroleum Jelly	60	1310 ± 290

Characterization of Types of Defects and Their Origination

Fractured fiber samples from 254 mm gauge length tests and 25.4 mm gauge length tests with the use of petroleum jelly were observed in the SEM for classification of defects. The flaws observed from 254 mm gauge length tests will be discussed first.

The observed flaws were classified into two groups, namely "Processing Flaws" and "Microstrucural Flaws." An example of a "Processing Flaw" would be a porous region or agglomerate; these defects originate during processing of the fiber in the green state. When the failure is not due to an obvious flaw related to green processing, then the resulting microstructure itself is the strength-limiting defect, and it is then referred to as a "Microstructural Flaw." The frequency of these two flaw types with their average strengths is shown in Table 3 for fibers tested over a gauge length of 254 mm.

Table 3. Characterization of Different Types of Flaws
(254 mm Gauge Length)

% Process Defect	42
Mean Strength of Process Defects	1090 MPa
% Microstructural Defect	58
Mean Strength of Microstructure	1427 MPa
Mean Strength of Whole Population	1255 MPa

Processing Flaws:

Figure 2 shows a fracture surface wherein a porous region is the strength-limiting flaw. A porous region in the sintered body suggests lower green density areas in the green fiber. This indicates that some areas are richer in the thermoplastic polymer binder, indicating poor mixing of SiC particles with the polymer. The other type of processing flaw is SiC agglomerates. This indicates the presence of higher green density areas consisting mainly of SiC particles. This is also likely caused by the poor mixing of polymer with SiC particles in the green processing stage. Thus both type of process flaws are related to the same phenomena in the green processing stage. These are common types of defects observed in injection-molded parts where the thermoplastic polymer is mixed with ceramic powder in the green forming stage [4].

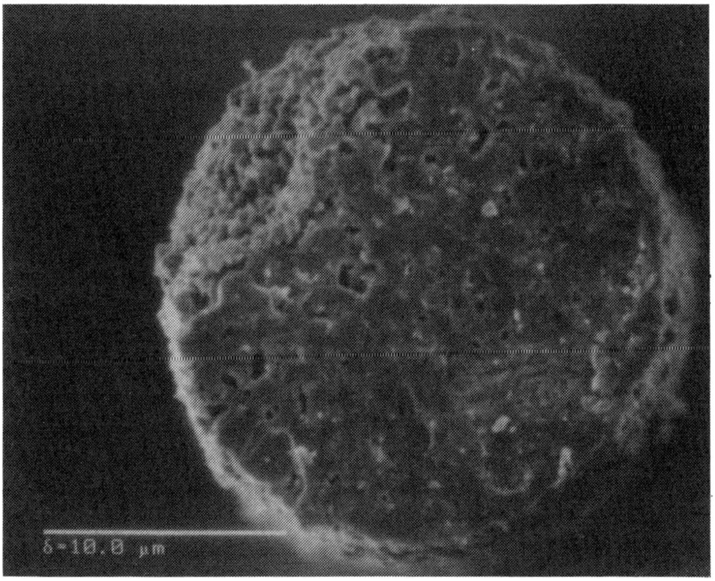

Figure 2. Fracture surface of the fiber showing a porous region as the strength-limiting defect. Failure stress = 1006 MPa.

Determination of K_{IC}:

It is practically impossible to determine K_{IC} directly on small-diameter fiber samples with controlled crack techniques. However, with the help of fractography, it is possible to determine the K_{IC} of this sintered SiC fiber. The porous region in Figure 2 can be considered equivalent to a semi-elliptical crack in a rod of finite diameter and infinite length with parameters shown schematically in Figure 3. The stress intensity factor, K_{IC}, formulation for such a crack is given as

$$K_{IC} = F_I \sigma_f \sqrt{\pi b} \quad \text{-------} \quad (1)$$

where σ_f is the fracture stress, b is a crack-size parameter shown in Figure 3, and F_I is a function of b/r and b/a. The value of F_I is readily obtained from a stress intensity factor handbook [5]. The failure stress was 1006 MPa for the fiber with fracture origin shown in Figure 2. The crack-length parameter from Figure 2 is measured to be

$b = 4\ \mu m$, $b/a = 0.66$, and $b/r = 0.4$

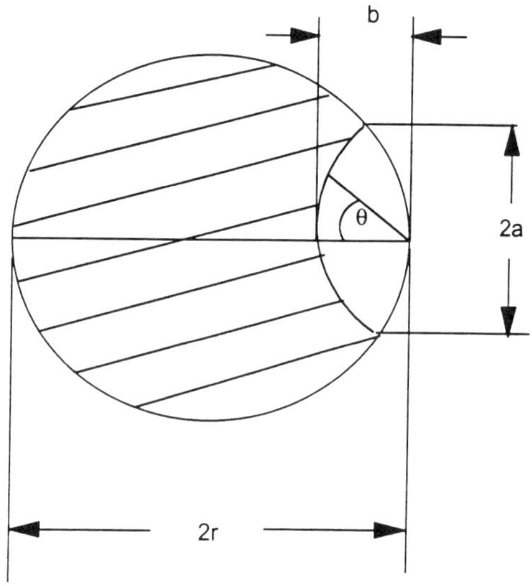

Figure 3. Schematic of a round bar containing a semi-elliptical surface crack under uniform tension.

A value for F_I as 0.751 was obtained from the stress intensity factor handbook for the above crack size and geometry [5]. The calculated K_{IC} is 2.7 MPa$\sqrt{m}$. This value is in good agreement with the K_{IC} value of 2.8 - 3.0 MPa$\sqrt{m}$ obtained from chevron notch and controlled flaw fracture toughness measurements on monolithic, bulk sintered α-SiC (HEXOLOY SA) [6,7]. The sintered SiC fiber is very similar to HEXOLOY SA in composition and microstructure, and hence the toughness of the fiber is also expected to be the same.

Microstructural Defects:

Figure 4 shows a fracture surface where there is no obvious flaw present. In this case the microstructure of the fiber itself acts as the defect. This fiber failed at a stress of 1517 MPa when tested over a 254 mm (10 inches) gauge length. The tiny, isolated, individual pores cannot be the failure origin, since they are too small (~0.1 μm) to be associated with a failure stress of 1517 MPa. The specific element or the constituent in the microstructure that corresponds to the failure origin is not very obvious. Sometimes the failure origin could be located with the use of a fracture mirror pattern. In Figure 4, a clear fracture mirror, as observed in monolithic bulk SiC, is not evident.

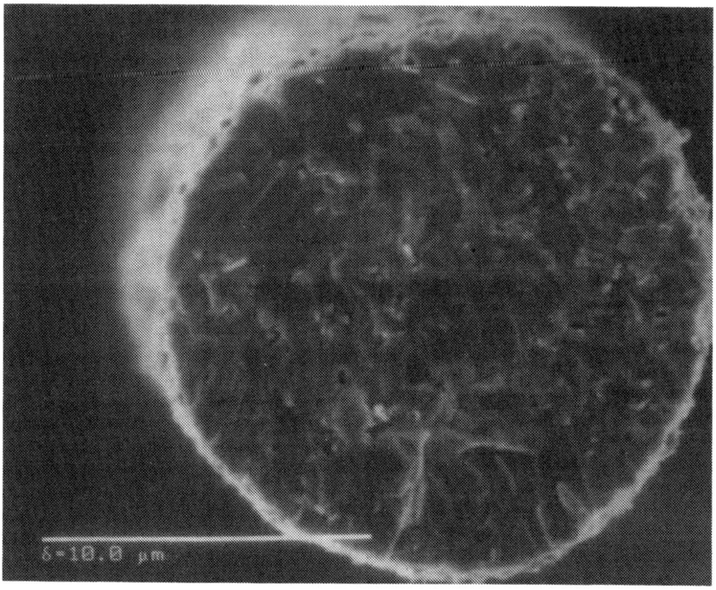

Figure 4. Fracture surface of fiber showing no obvious defect.
Failure stress = 1517 MPa.

The fracture mirror radius can be estimated as a function of failure stress from the stress intensity relation

$$K_B = F_I \sigma_f \sqrt{\pi r_b} \quad \text{-----------------(2)}$$

where K_B is the stress intensity where crack branching occurs and r_B is the fracture mirror radius. K_B is taken as 13 MPa$\sqrt{m}$ from literature [8]. F_I is taken as 0.65, the choice of this value is explained later. The predicted relationship of fracture stress vs fracture mirror radius is shown in Figure 5. Also shown are the ranges of fracture stress observed and typical fiber diameters. It appears that the fracture mirror radius is larger than the fiber diameter itself.

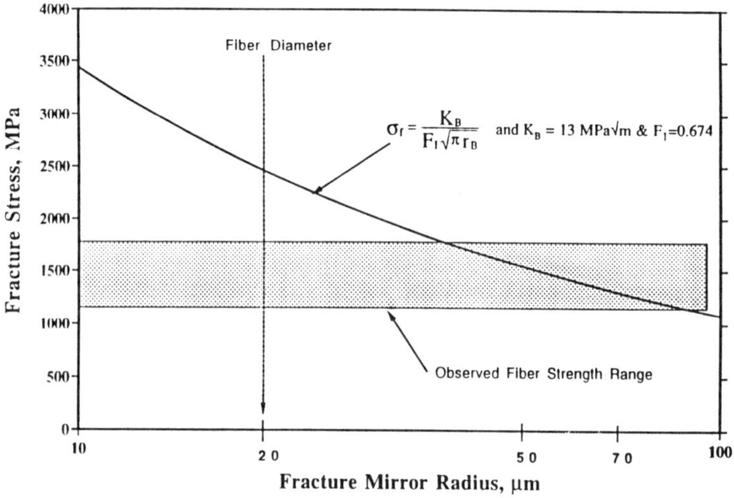

Figure 5. Relationship between fracture stress and fracture mirror radius.

The above discussion assumes a constant F_I. In bulk solids, such as in a bend bar or tensile sample, the ratio of crack size (and also the fracture mirror radius) to the width or diameter (the far free surface or boundary) is very small. Hence the same F_I for both the defect and fracture mirror boundary can be assumed. However, in small-diameter fibers, the crack-size-to-diameter ratio is not small enough to be ignored. Significantly, when the crack is propagating and approaching the free boundary, the crack geometry factor, that is the value of F_I changes continuously. This modification of F_I has to be taken into consideration. The variation of F_I with increasing b/r is plotted in Figure 6 from the data given in reference 5. The stress intensity factor can then be calculated as a function of b/r from equation 1 and using F_I values which are a function of b/r (from Figure 6). Stress intensity factor as a function of the ratio of crack length to fiber radius is shown in Figure 7. When the crack tip approaches the free surface, the stress intensity factor increases exponentially and approaches infinity. So at some distance from the initial crack tip position to the free boundary, the stress intensity should have exceeded 13 MPa √m and hence crack branching would have occurred before the crack reached the far end free boundary. The stress intensity factor exceeds K_B at b/r≅1.175, as observed from Figure 7. The location of the fracture mirror radius is schematically shown in Figure 8. Figure 8 depicts a fiber fracture surface with a semi-elliptical crack. A boundary is also sketched where the stress intensity factor would have reached the stress intensity factor for crack branching. The fracture mirror radius (at θ =0, or the distance OD) for an applied stress of 1517 MPa is predicted to be 9.5 μm for a fiber radius of 8.1 μm from equation 2 and Figure 7. The stress intensity factor is slightly higher near the surface and hence, distance OD is larger than distance OE [8]. The location of the fracture mirror in turn will help to identify the fracture origin and the defect itself.

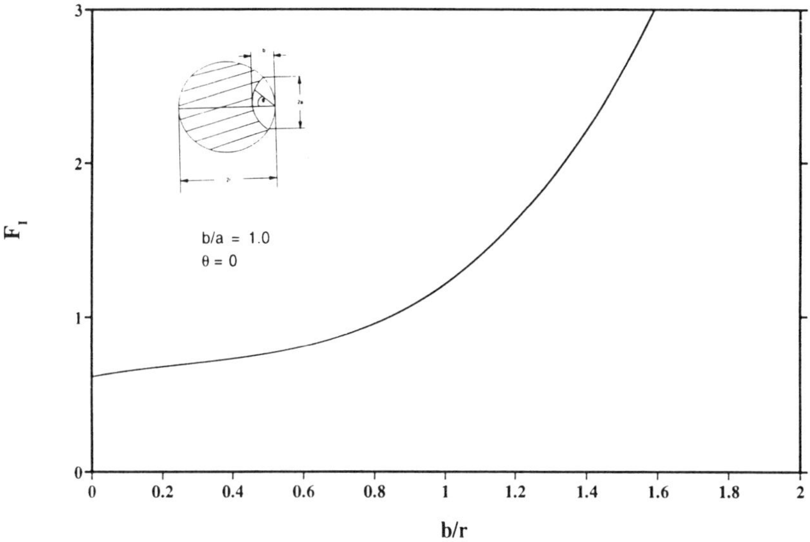

Figure 6. Crack shape factor, F_I, as a function of the ratio of crack length to radius (b/r).

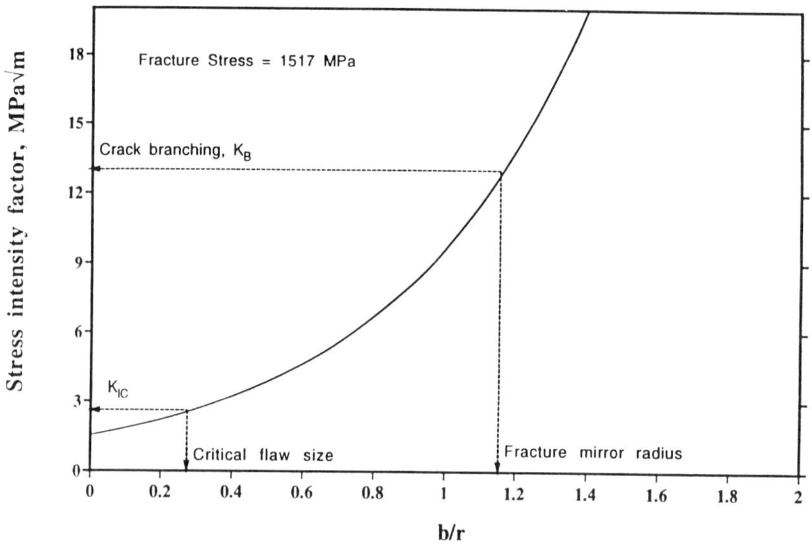

Figure 7. Predicted stress intensity factor (K_I) as a function of the ratio of crack length to radius (b/r).

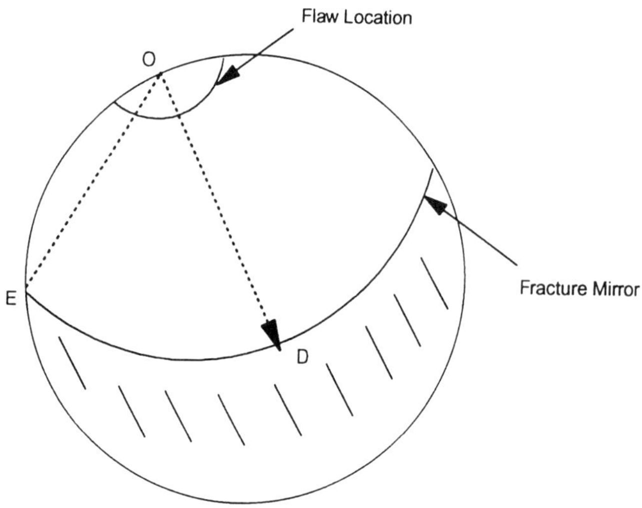

Figure 8. A schematic of a fracture surface illustrating the predicted fracture mirror for a semi-elliptical crack in a finite-diameter fiber.

With the above details of the predicted fracture pattern, the fracture surface shown in Figure 4 is reexamined. The same figure is reproduced again in Figure 9, but with fracture features highlighted. The rougher region where tearing in grains is evident denotes the probable crack branching area. The boundary of crack branching, or the fracture mirror, is marked. The fracture mirror boundary is so close to the outer edge of the sample that it is more difficult to observe the fracture mirror in smaller-diameter samples than in large monolithic samples. The fiber radius is 8.1 μm. The fracture mirror radius (at $\theta = 0$) is measured to be 10 μm. This compares well with the predicted fracture mirror radius of 9.5 μm from Figure 8 for the same failure stress of 1517 MPa.

From the fracture mirror, the site of the fracture origin can be located and is indicated by an arrow in Figure 9. Note that this region does not have any tear marks and appears smooth, characteristic of typical fracture-originating flaws. Closer examination of the flaw shows that it appears to be a larger grain or grain boundary bordering a larger grain.

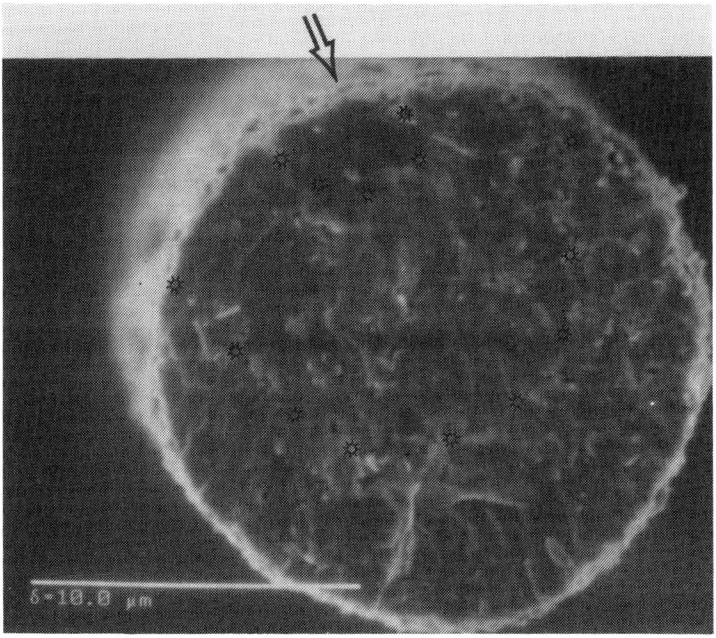

Figure 9. Fracture surface of a fiber highlighting fracture features. (Same as Figure 4).

Considering this defect as a semi-elliptical crack, the crack size and shape parameters are measured to be:

b= 2.7 μm, r=8.1 μm, $\frac{b}{a} = 1.125$ and $\frac{b}{r} = 0.17$

The above parameters yield a value for F_I to be 0.65 from reference 5, which is the same as the value assumed for the plot in Figure 4. The fracture stress calculated using equation 1 is 1426 MPa. This value is in reasonable agreement with the actual failure stress of 1517 MPa. The above quantitative evaluation validates the proposed defect with regards to its location and size.

Now the proposed defect as a grain or grain boundary has to be validated. If the defect is a grain or grain boundary, then the defect size must have a relation to the grain size distribution in the microstructure. This point is illustrated using another experimental result. Fibers were made using the unmilled (as-received) powder and attrition-milled powder. Milling reduces the larger-sized particles and enhances the surface area of the powder, which results in a finer microstructure upon sintering. The strength of the fibers for which the microstructure is the strength-controlling defect should also be higher for the fiber made using the milled powder, if the defect size is related to the grain size. The powder-microstructure-strength relationship is shown in Table 4. The average grain size did not change appreciably, but there is a significant difference in the maximum grain size or distribution near the higher grain sizes, as shown in Figure 10. The average tensile strength (for which the microstructure is the controlling defect) measured at a gauge length of 254 mm (10 inches) did correspond to variations in grain size distribution, as seen from Table 5. This result confirms the proposed strength-limiting defect as a larger grain in the microstructure.

Table 4. Powder, Microstructure, and Strength Characteristics

	Powder	**Microstructure**		**Strength**
	Surface Area*	Mean Grain Size (μm)	Max. Grain Size (μm)	at 254 mm (MPa)
As-Received	1.0	1.4	8.0	1075
Milled	1.5	1.2	4.0	1427

*Surface area normalized with respect to as-received powder.

When finer powder was used, though strength enhancement due to refinement in microstructure was realized, the incidence of process defects also increased. Table 5 compares defect population in fiber made using unmilled powder (as-received) and attrition-milled powder. The process defects were only 15% of the population in the fiber made using the as-received powder, while the process defects were 42% of the population in the milled powder. This result is in agreement with the earlier proposed mechanism of formation of process defects. The higher the surface area of powder, the higher the particle-particle attraction, and hence the difficulty in accomplishing good mixing with the thermoplastic polymer. The incidence of increased process defects reduces the overall strength of the fiber.

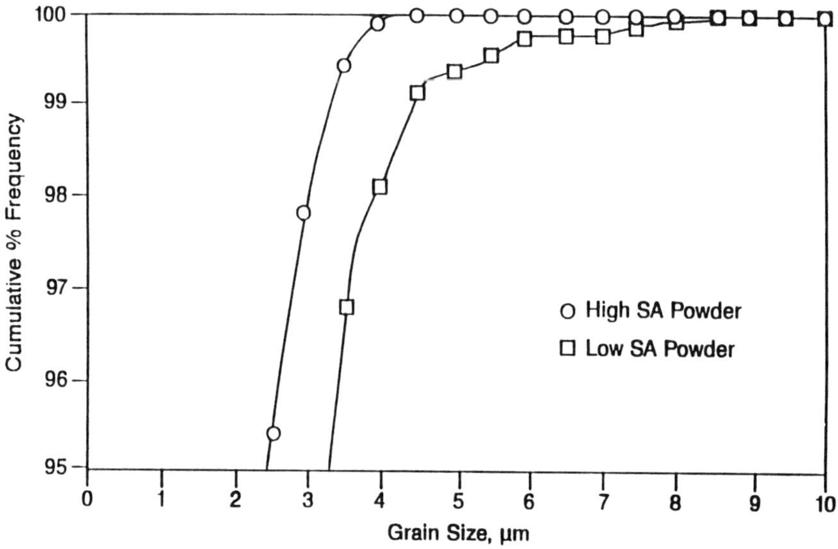

Figure 10. Grain size distribution in fibers made with powders having different surface areas

Table 5. Defects in Fibers Made Using As-Received and Milled Powders

	As-Received	Milled
% Process Defects	15	42
Mean Strength for Process Defects at 254 mm	917 MPa	1090 MPa
Weibull Modulus	3.8	5.9
Mean Strength for Microstructure at 254 mm	1075 MPa	1427 MPa
Weibull Modulus	9.0	12.0
Mean Strength for Overall Population at 254 mm	1054 MPa	1255 MPa

Analysis of Bi-Modal Flaw Distribution

The flaw distribution information obtained at a given gauge length can be extended to predict the strength of fibers at any other gauge length. For this analysis, the flaw distribution information from the fibers made using the milled powder is considered here. Weibull plots for the whole population, for process defects, and for the microstructural (larger grains) defects are shown in Figure 11. The data were taken from strength measured at a gauge length of 254 mm (10 inches).

The strength at any other gauge length can be predicted using respective Weibull parameters for the two types of defects. The generalized Weibull distribution function is given as

$$P_f = 1 - \exp - \frac{L}{L_0}\left(\frac{\sigma}{\sigma_0}\right)^m \quad \text{------------(3)}$$

where P_f is the failure probability up to a stress σ, L is the gauge length of the fiber, L_0 is the reference gauge length at which the Weibull modulus m and scaling parameter σ_0 are obtained. The above equation will adequately describe the entire population if the population is made of one type of defect. If two distinct types of defects are formed by different mechanisms, then the total failure probability at a stress σ at any length L is then described as product of the failure probability of each of the two populations. This is expressed as

$$\ln(1-P_f) = -\frac{L}{L_0}\left[F_1\left(\frac{\sigma}{\sigma_{0,1}}\right)^{m_1} + (1-F_1)\left(\frac{\sigma}{\sigma_{0,2}}\right)^{m_2}\right] \text{---------(4)}$$

where F_1 is the fraction of one of the defects, and $\sigma_{0,1}$, $\sigma_{0,2}$, m_1, m_2 are the corresponding Weibull parameters of the defects. The mean strength σ will correspond to approximately where the probability of failure is 0.5. So the above equation can be used to predict the strength of fiber at any gauge length using the Weibull parameters shown in Figure 11.

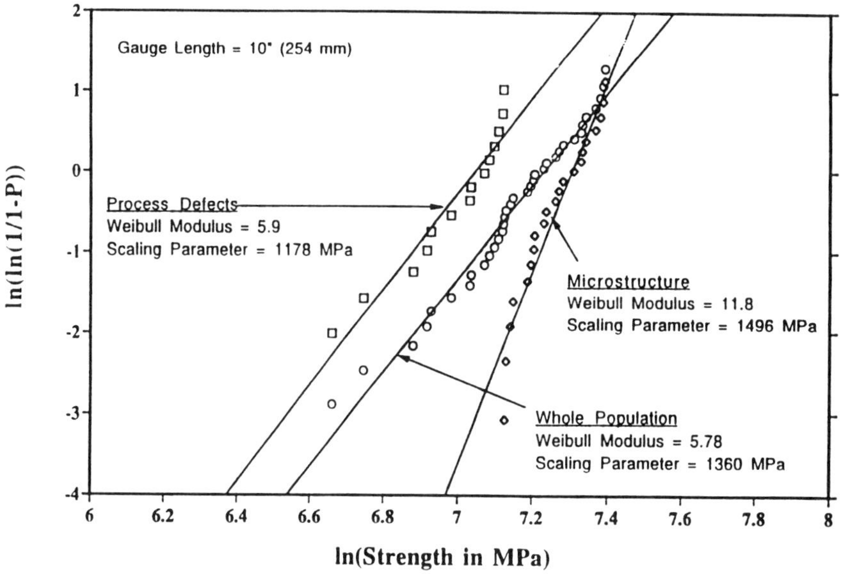

Figure 11. Weibull plots of the whole population, for fibers with only process defects and for fibers with only microstructural defects

The predicted strength as a function of gauge length is shown in Figure 12. This is particularly useful, since the strength of fibers of relevance in a composite is at a gauge length corresponding to the critical distance bridging the matrix crack where the fibers are loaded [9]. This is dependent on the interfacial shear stress, and for typical interfacial shear stresses, this distance varies from 0.5 to 4 mm for Carborundum fibers [10]. The strength of fibers measured from single-fiber composite testing, wherein the strength of fibers were measured at gauge lengths varying from 25 mm to 0.2 mm, correlated well with the extrapolated strength from longer-gauge-length strength data [11,12]. However, it is not practical and economical to directly measure the strength of fibers at these gauge lengths, and the flaw statistics information determined at a convenient gauge length can be reliably used for predicting the fiber strength at these critical gauge lengths.

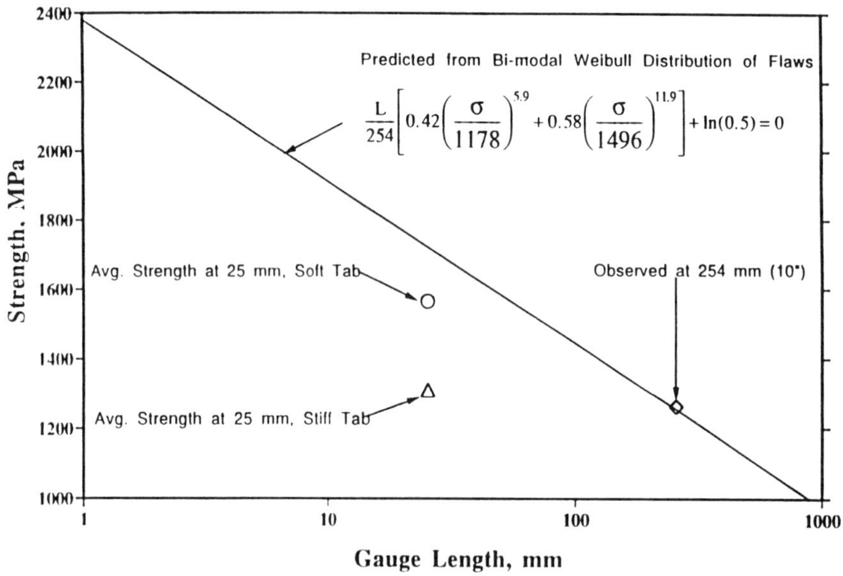

Figure 12. Predicted strength as a function of gauge length from a bi-modal Weibull flaw distribution.

Application of Fractography in Fiber Tensile Testing Methodology

The predicted strength from the bi-modal Weibull distribution of flaws agrees well with the observed average tensile strength at a gauge length of 254 mm (10 inch), as shown in Figure 12. The observed average tensile strengths at a gauge length of 25 mm (~ 1 inch) with different types of paper tabs used are also shown in Figure 12. They are lower than the predicted strength from the flaw distribution. The stiffer the tab, the lower the strength. The fact that strength depends on the stiffness of the tabs itself indicates that the strength measured is a function of test methodology. Fractography was used to further investigate these results.

No petroleum jelly was used on the stiffer tabs; hence, no fractography was done on those fiber samples. However, fractured fragments were collected from the 25.4 mm gauge length test using the softer tab with petroleum jelly. Nearly 70% of the fracture surfaces showed a "step"-like feature, as shown in Figure 13, when tested over a gauge length of 25.4 mm (1 inch). No other obvious flaw could be detected. The sample shown in Figure 13 failed at an apparent tensile stress of 1503 MPa. The "step"-like feature (cantilever curl) in Figure 13 indicates the presence of bending. This is compared with the fracture surface of the fiber shown in Figure 9. No "step"-like feature was observed, and a "flat" fracture surface typical of uniaxial tensile failure was observed. The fiber in Figure 9 failed at a stress of 1517 MPa, but at a gauge length of 254 mm (10 inches). The comparative strength value along with the "step"-like feature suggests that the strength of fiber shown in Figure 13 could be underestimated.

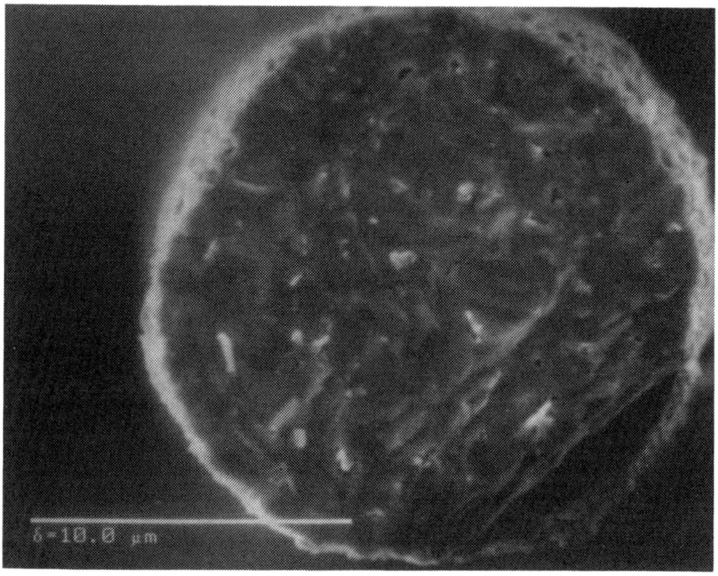

Figure 13. Fracture surface of a fiber showing step-like fracture. Gauge length = 25 mm. Fracture Stress = 1503 MPa.

The maximum stress experienced by the fiber is the sum of applied tensile stress and bending stress resulting from possible misalignment. Misalignments are possible since the fiber is loaded via the paper tab. The weight of the fiber is far lower than the weight of the paper tab used. When the paper is cut in the center, the cut surface is free and it tends to curl up or sway to one side, thus pushing or pulling the fiber to the side. This may induce a bending moment on the fiber. This is schematically illustrated in Figures 14a and 14b. Figure 14a shows the overall setup, while Figure 14b illustrates the bending effect the fiber may experience.

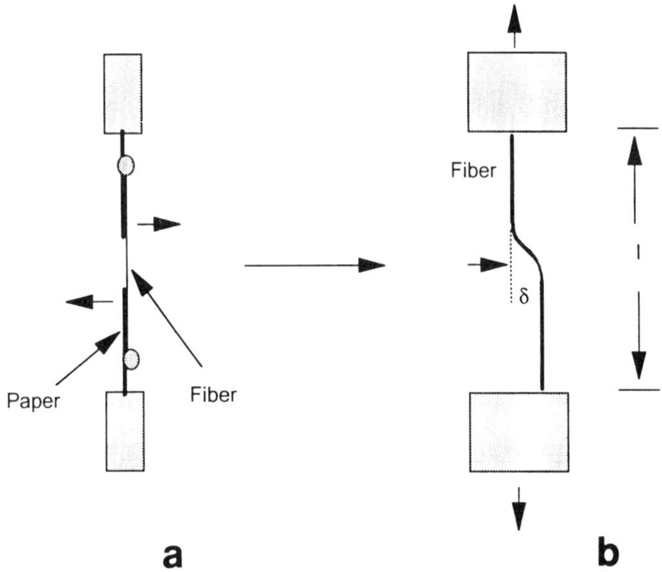

Figure 14. (a) Schematic showing forces from the free end of the paper tab and (b) bending moment as seen by the fiber.

The bending stress in such a situation is given by [13]

$$\Delta\sigma_b = \frac{6ER\delta}{l^2} \quad \text{---------}(5)$$

where $\Delta\sigma_b$ is the bending stress, E is the Young's modulus of the fiber, R is the radius of the fiber, δ is the offset distance or the misalignment distance, and l is the gauge length. The observed tensile stress at a gauge length of 25 mm using softer tabs is lower by 160 MPa compared to the predicted fiber strength at 25 mm from the bi-modal flaw distribution. Using a modulus value of 400 GPa for the fiber and a fiber radius of 10 μm, the offset distance is estimated to be about 4 mm. The fiber could easily be displaced by 2 mm either side due to the inertia of the paper. Stiffer paper is expected to induce more force laterally, and higher misalignment is possible. This is evidenced by the lower strength observed for the same fiber when tested using stiffer paper tabs. Additionally, the use of petroleum jelly softens the tab further by wetting or soaking the paper.

Another important consideration is the effect of gauge length. For the same misalignment or offset distance of 4 mm, by just increasing the gauge length to 254 mm (10 inches), the bending stress is reduced to just 1.5 MPa. Hence, the use of longer gauge lengths yields consistent strength values reflective of defects in the fiber rather than test artifacts. The shorter gauge length tests may not be an issue for small-diameter, low-modulus, Nicalon-type fibers. The modulus for Nicalon SiC fibers is about 200 GPa, and the fiber radius is about 5 μm [14], for which the bending stress is estimated to be only 35 MPa at a gauge length of 25.4 mm. This bending stress is negligible compared to the strength of Nicalon fibers (~2.75 GPa) [14]. However, for Carborundum SiC fiber, the fiber strength could be underestimated by 10 to 20% at shorter gauge lengths.

SUMMARY

Fractography has been used extensively to identify the strength-limiting defects in fibers and through that the source of the origination of these defects. This approach was very useful in identifying the process step which required modification, thereby providing directions for developmental efforts. Using Weibull theory for bi-modal flaws, fractography was also used to extrapolate the strength of the fiber at small gauge lengths using information obtained at a longer gauge length. This is very useful in calculating composite properties. Moreover, fractographic investigation revealed the artifacts (bending moment) associated with the fiber testing and provided directions for minimizing those artifacts.

ACKNOWLEDGMENT

The authors would like to acknowledge the help received from S. Lacki for the SEM work, F. Frechette, M. Andrejcak and L. King in making the fibers, and M. Sylvester, and K. Ziemendorf in testing the fibers.

REFERENCES

1. F. J. Frechette et al., "Non Oxide Sintered Ceramic Fibers," U.S. Patent No. 4 908 340, March 13, 1990.

2. F. Frechette, B. Dover, V. Venkateswaran, and J. J. Kim, "High Temperature Sintered SiC Fibers for Composite Applications," Ceram. Eng. Sci. Proc. 12[7-8] 992-1006 (1991).

3. V. Venkateswaran, F. Frechette, G. V. Srinivasan, and S. K. Lau, "Development of a High Temperature SiC Fiber for Ceramic Matrix Composite," presented at the 19th Annual Cocoa Beach Conference and Exposition, Cocoa Beach, Florida, January 9, 1995.

4. R. W. Ohnsorg and G. V. Srinivasan, "Development and Characterization of High Strength SiC Rotors," Proc. from the 36th ASME/International Gas Turbine Institute Conference, Orlando, June 1991. ASME Publication # 91-GT-191.

5. Stress Intensity Factors Handbook, vol. 2, ed. by Y. Murakami, pp 659-663, Pergamon Press, NY 1987.

6. M. G. Jenkins, A. Ghosh, K. W. White, A. S. Kobayashi and R. C. Bradt, "The Fracture Resistance of a Sintered SiC Using the Chevron-Notch Bend Specimens," Proc. of Symposium on Silicon Carbide, Ceram. Trans., vol. 2, pp 241-252, ed. by J. D. Cawley and C. E. Semler, Amer. Ceram. Soc., 1989 .

7. G. V. Srinivasan, The Carborundum Co., unpublished work on K_{IC} evaluation of α-SiC by the Controlled Flaw Techniques, 1990.

8. H. P. Kirchner and J. C. Conway, "Fracture Mechanics of Crack Branching in Ceramics," in Fractography of Glasses and Ceramics 1986, Advances in Ceramics, vol. 22, p 187-214, Amer. Ceram. Soc., Westerville, OH 1988.

9. W. A. Curtin, "Theory of Mechanical Properties of CMCs," J. Amer. Ceram. Soc., 74[11] 2837-45 (1991).

10. G. V. Srinivasan, V. Venkateswaran, and S. K. Lau, "An Engineering Analysis for Material Design Considerations Using Carborundum's Sintered SiC Fiber

and Slurry Cast/Melt Infiltrated Process," presented at the 19th Annual Conference on Composites and Advanced Ceramic Materials, Cocoa Beach, 1995. To be published in Ceram. Eng. & Sci. Proceedings, 1995.

11. W. A. Curtin and A. Netravali, "Evaluation of Strength of SiC Fibers from Single Fiber Composite Tests," to be published in J. Mat. Sci.

12. G. V. Srinivasan and V. Venkateswaran, "Tensile Strength Evaluation of Polycrystalline SiC Fibers," Ceram. Eng. & Sci. Proc., 14[7-8] 563-572 (1993).

13. R. J. Roark and W. C. Young, "Formulas for Stress and Strain," McGraw-Hill Inc., NY 1975.

14. F. K. Ko, "Preform Fiber Architecture for Ceramic Matrix Composites," Amer. Ceram. Soc. Bulletin, 68[2] 401-414 (1989).

MICROSTRUCTURAL CONSIDERATIONS FOR FRACTOGRAPHIC ANALYSIS OF POLYPHASE CERAMICS VIA TRANSMISSION ELECTRON MICROSCOPY

W. Braue
German Aerospace Research Establishment (DLR)
Materials Research Institute
D-51147 Cologne
Germany

ABSTRACT

The stability of thin amorphous boundaries from polycrystalline structural ceramics in the high temperature and/or high stress regimes is addressed in the light of recent TEM investigations in order to provide a complementary perspective to the engineering approach in fractography. Case studies presented focus on the role of internal interfaces during time-dependent failure of vitreous-bonded Si_3N_4 and SiC ceramics as well as the generation of thin glass lamellae during shock loading of oxides.

INTRODUCTION

The failure mechanisms of polycrystalline structural ceramics in the high temperature/high stress regime involve a variety of nanoscale microstructural features which are beyond the resolution limits of optical microscopy and routine SEM inspections. Other tools are required because of their complex crystallographic structure or chemical composition. These features include thin amorphous layers at grain- and phase boundaries and crystalline secondary phases at triple grain junctions, as well as the nature and distribution of structural defects in the grains, all of which greatly affect the bulk mechanical response under given service or testing conditions.

Today's transmission electron microscopes operating at intermediate acceleration voltages offer a unique combination of high-resolution imaging routinely, down to a point resolution of 0.18 nm, and small-probe capabilities for nanospectroscopy (EDS, EELS) and microdiffraction (CBED) which has since been appreciated as a valuable tool for a nanoscale fractographic analysis of monolithic ceramics as well as ceramic matrix composites [1].

The electron microscopist's natural world is a small one which occasionally fails to interact constructively with the macroscopic approach often adopted in the fractographic analysis. If so, the crucial issue is often found in a mismatch of dimensional scale. While the microscopist may be convinced that the nanoscale feature he has identified after tedious microscope sessions in the dark is certainly related to failure, his fellow fractographer may consider that the critical defect was

larger by several orders of magnitude. Thus, the adjustement to the appropriate grain size range of the prevalent failure mechanism [2] and the identification of the macroscopic defects are mandatory before the transmission electron microscope can add important complementary perspectives to the engineering approach in fractography.
This paper summarizes microstructural aspects from TEM investigations which have been conducted in close interaction with mechanical testing and fractographic analysis and are of general significance in terms of stability of microstructure at high temperatures and/or stresses. The main emphasis is on time-dependent failure in vitreous-bonded Si_3N_4 and SiC materials at intermediate and high temperatures, but as an exiting sidepath in ceramic microstructures, new insights on the formation of thin intragranular amorphous layers during shock-loading of oxide ceramics are provided. Besides scrutinizing recent development in TEM instrumentation,this study addresses (i) the instability of crystalline secondary phases during high-temperature testing of Si_3N_4 thus altering the composition of the continuous amorphous interlayer, (ii) the effect of partial devitrification on static fatigue of Si_3N_4, (iii) equilibrium film thicknesses, structural- and chemical widths of internal interfaces in Si_3N_4-based ceramics, (iv) oxygen segregation at interfaces in SiC as compared to Si_3N_4 materials and (v) the effect of high strain rates in generating thin glass lamellae in mullite polycrystals.

EXPERIMENTAL PROCEDURE

The following materials are involved in this study: (i) a sintered reaction-bonded Si_3N_4 (SRBSN) material doped with 5 wt%. unstabilized ZrO_2, ii) a sintered Si_3N_4 (SSN) from GTE WESCO (grade SNW-1000), (iii) SiC(w)/Si_3N_4 composites employing different types of SiC whisker, (iv) a HIP-SiC material doped with 0.2 wt.% Al (Elektroschmelzwerk Kempten GmbH, Munich, Germany) and a reaction-bonded SiC (RB-SiC) containing approx. 8 wt%. free silicon (Sigri Gmbh, Meitingen, Germany), and (v) a "fused-mullite" refractory brick, supplied by Didier GmbH, Wiesbaden, Germany.
Thin TEM foil specimens were prepared via dimpling and argon ion-beam thinning. Several microscopes were utilized for different microstructural purposes: a Philips EM 430 TEM/STEM (300 kV) for conventional TEM work and microdiffraction, a JEOL 4000 EX (400 kV) for HREM imaging, a Philips 400 ST fitted with a field-emission gun operating at 100 kV and coupled to a Gatan 666 PEELS detector for high-resolution nanospectroscopy and a Zeiss EM 912 (120 kV) with Ω filter for energy-selected imaging (ESTEMI).

RESULTS AND DISCUSSION

Secondary Phase Instability in ZrO_2-doped SBRSN

The common rationale in high-temperature testing of polycrystalline structural ceramics is based on a continuum approach assuming that microstructural integrity with neglectible compositional gradients has been maintained during the testing period. There is, however, TEM evidence from liquid-phase sintered Si_3N_4 materials, that exceeding a threshold level in stress-temperature space, the high temperatures and local stresses initiate partial decomposition of even refractory crystalline secondary phases, as recently demonstrated for m-ZrO_2 in an SRBSN material [3]. It is mandatory to correlate this effect with the actual deformation characteristics of this vitreous-bonded Si_3N_4 material [4].

This evaluation typically gives rise to three regimes: (i) linear-elastic deformation between RT and approx. 900°C (region I), (ii) viscoelastic deformation ("viscoelastic recovery", see [5]) between 900° C and 1150° C (region II) which is reversible as indicated by the closure of the load-displacement hysteresis curve, and (iii) irreversible deformation due to grain-boundary sliding and accumulation of creep damage above 1150°C (region III) thus causing a non-closure of the load-displacement curve.

In qualitative terms, this deformation behavior is supposed to be valid for any kind of liquid-phase sintered Si_3N_4 material. It can be rationalized in terms of a simple spring-dashpot model [6], defined by three interacting elements (Fig.1a). Element 1 defined by the spring S_1, is effective in all three regions listed above. It represents the linear-elastic deformation introduced by both, the ß-Si_3N_4 matrix grains and the crystalline secondary phase (m-ZrO_2). The Kelvin -Voigt element 2 consists of the spring S_2 and the dashpot D_1 accounting for the reversible viscoelastic deformation. Element 3 defined by dashpot D_2 represents the irreversible deformation. The dashpots stand for the response of the residual glassy phase present at triple grain junctions and grain boundaries. Note that D_1 and D_2 exhibit different activation energies with D_1 being effective in region II and III and D_2 in region III only. At the lower testing temperatures (region I) D_1 and D_2 have spring-like behavior due to the high viscosity of the non-crystalline phase.

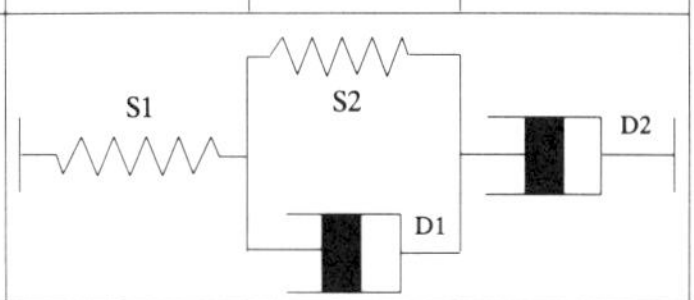

regions I-III	element 1	element 2	element 3
I: RT - 900 °C	elastic	elastic	elastic
II: 900 - 1150 °C	elastic	viscoelastic	elastic
III: > 1150 °C	elastic	viscoelastic	viscous

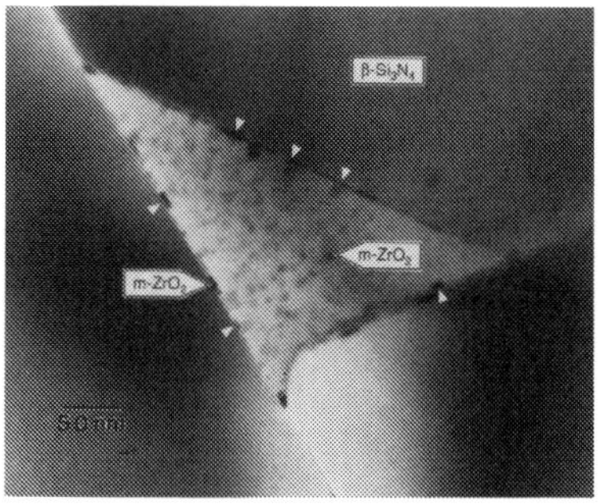

Fig.1 (a) Spring-dashpot model illustrating the high-temperature deformation of vitreous-bonded SRBSN in regions I-III, (b) nucleation and growth of m-ZrO_2 precipitates from triple-grain junctions and ß-Si_3N_4 facets in SRBSN after exposure to testing conditions in region II, from [3].

While the rather moderate testing conditions of region I do not exceed the stability range of m-ZrO_2 in SRBSN, the higher temperatures in conjunction with the local stress distribution during viscoelastic recovery in region II actually caused a noticeable facetting of the previously irregularly-shaped m-ZrO_2 grains followed by their partial dissolution in the pre-existing homogeneous glass. Small probe microanalysis obtained from triple-grain junctions proved an increased Zr-content of the glassy phase for tested specimens in comparison with as-received SRBSN samples [7]. Moreover, these triple-grain junctions exhibited a surprisingly high density of lath-shaped microcrystals which were identified as m-ZrO_2 after cool-down, by means of CBED and from lattice spacings in the HREM images. As emphasized in Fig. 1b, reprecipitation of m-ZrO_2 during cool-down from testing temperatures is controlled by (i) homogeneous nucleation within the bulk glass and (ii) heterogeneous nucleation supported by the low-energy facets of the ß-Si_3N_4 matrix grains [8]. This Zr-rich, high temperature modified amorphous phase is believed to be directly related to the viscoelastic recovery process effective in region II. In region III, the higher testing temperatures and the further reduced viscosity of the amorphous phase enhance the effects of grain-boundary sliding and nucleation of creep cavitation as observed at most m-ZrO_2/ß-Si_3N_4 phase boundaries, similar to other Si_3N_4-materials [9].

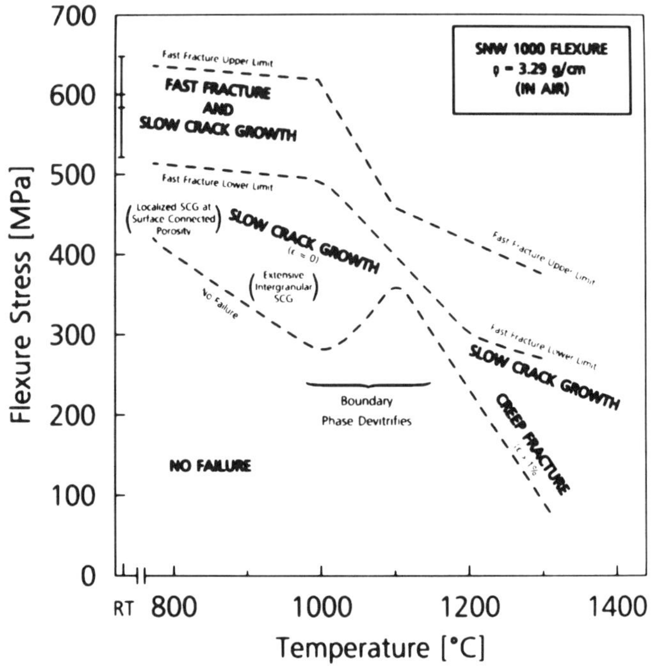

Fig. 2 Fracture mechanism map for SNW-1000 Si_3N_4 tested in air, from [12]

Time-Dependent Failure of Y_2O_3/Al_2O_3-Doped SSN

Due to thermodynamic arguments related to the constrained crystallization of secondary phase microcrystals from very thin amorphous layers [10,11], devitrification of liquid-phase sintered Si_3N_4 is restricted to triple grain junctions leaving a thin non-crystalline film at most grain -and phase boundaries. It has been shown [12], that during static fatigue testing of Y_2O_3/Al_2O_3-fluxed Si_3N_4 (SNW-1000) in air, partial devitrification can occur at a sufficient rate at 1000°C and 1100°C to retard or completely offset time-dependent failure which otherwise is controlled by slow crack growth or creep fracture. A fracture mechanism map (Fig. 2) depicts the stress-temperature regimes where each mechanism (fast fracture, slow crack growth, creep) is dominant and it defines a challenge to the microscopist to provide a decent microstructural basis for the prevalent failure mechanisms [13].

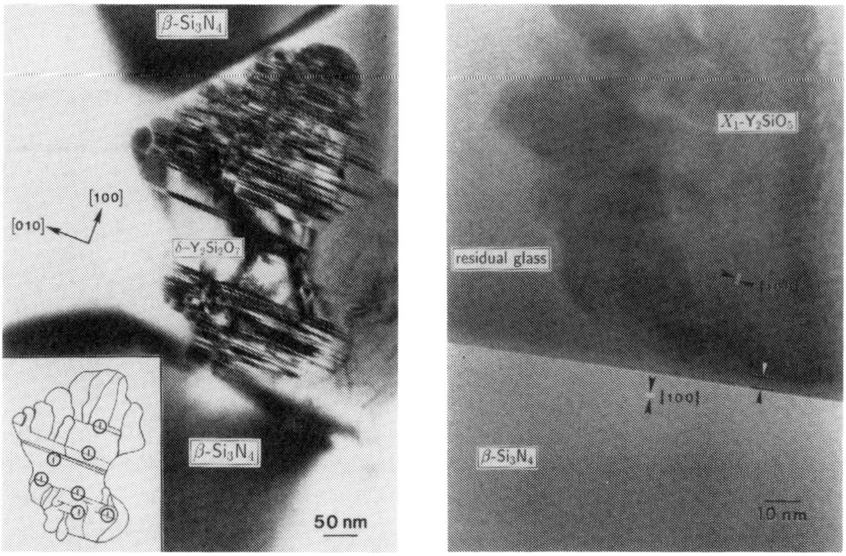

Fig. 3 (a) Intergrowth of δ-$Y_2Si_2O_7$ dendrites at triple grain junction, (b) thin residual amorphous film at X_1-Y_2SiO_5/ß-Si_3N_4 phase boundary

Major devitrification products in the short-term stress rupture experiments of SNW-1000 were yttrium di- and/or orthosilicates. In the initial stage of crystallization, these exhibited dendritic grain morphologies (Fig. 3a) with very small grain sizes, indicative of multiple nucleation sites in the triple-grain junctions [14]. The stress state does not affect the nature of the crystalline secondary phase assemblage formed during testing. However, the amount of crystallization is higher within the tensile region of the flexural specimens than in areas which have experienced compressive stresses. The temporary benefit of partial devitrification during static fatigue testing is balanced by nucleation and growth of creep cavitation (see [15] for review) from residual thin amorphous films (Fig. 3b), causing time-dependent failure to resume above 1200°C.

Structural and Chemical Widths of Intergranular Films in Dense SSN Materials

The early structural and microanalytical information collected in the TEM from thin intergranular amophous layers in Si_3N_4 materials [16-20] have initialized a strong worldwide research effort which continues since the significance of this issue in terms of a fundament assessment of the liquid-phase sintering process in the Si_3N_4 system has been identified almost two decades ago. Following the concept of equilibrium film thickness [21], HREM and high-spatial resolution AEM techniques have been presented indicating that high-angle grain boundaries exhibit a constant intergranular film thickness which strongly depends on the sintering additive system and thus on the interfacial chemistry [22,23].

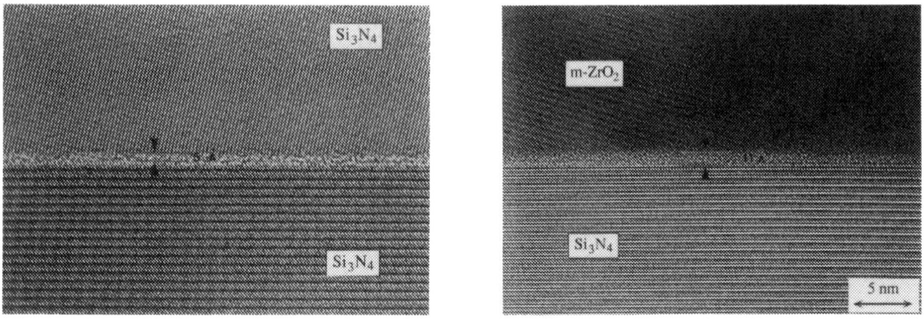

Fig. 4 (a) HREM image of grain boundary in ZrO_2-doped SRBSN revealing an intergranular film thickness of 0.5 nm, (b) HREM image of m-ZrO_2/ß-Si_3N_4 phase boundary in the same material showing an increased film thickness of approx. 1.1 nm, from [7]

Segregation of impurities to the boundary layer alters the force balance between the van der Waals attraction between the matrix grains and the repulsive term due to the combined effects of a steric force (distortions of the $SiO_4{}^{4-}$ tetrahedra in the glassy phase) and an electrical double layer force by introducing additional charges. This results in a wider film thickness for phase boundaries compared to grain boundaries, as shown in Figs. 4 a,b for the SRBSN material discussed before.
From the same argument, increasing the SiO_2 content of the interlayer results in a reduction in film thickness. Formation of crystalline secondary phases that do not incorporate impurity cations, such as Ca, gives rise to increasing solute concentration in the remaining amorphous phase. This substantial increase in the net impurity content results in a wider boundary gap and severe degradation of high-temperature properties [24]

It has recently been proved, for SiC(w)-reinforced Y_2O_3/Al_2O_3-doped Si_3N_4 materials by position-resolved electron energy loss spectroscopy (PREELS) in a field-emission gun TEM, that solid state diffusion from the amorphous interlayer into the bounding crystals can introduce nonequilibrium oxygen gradients, defined as chemical widths, normal to the interface plane of grain- and phase boundaries [25].While PREELS is particularly suited for analysis of low-Z elements, Z-contrast scanning TEM techniques in a dedicated STEM work well to monitor compositional gradients involving heavier elements, e.g. yttrium [26]. The chemical widths, particularly from matrix grain boundaries, can substantially exceed the structural width of the interface, determined by HREM imaging [27].

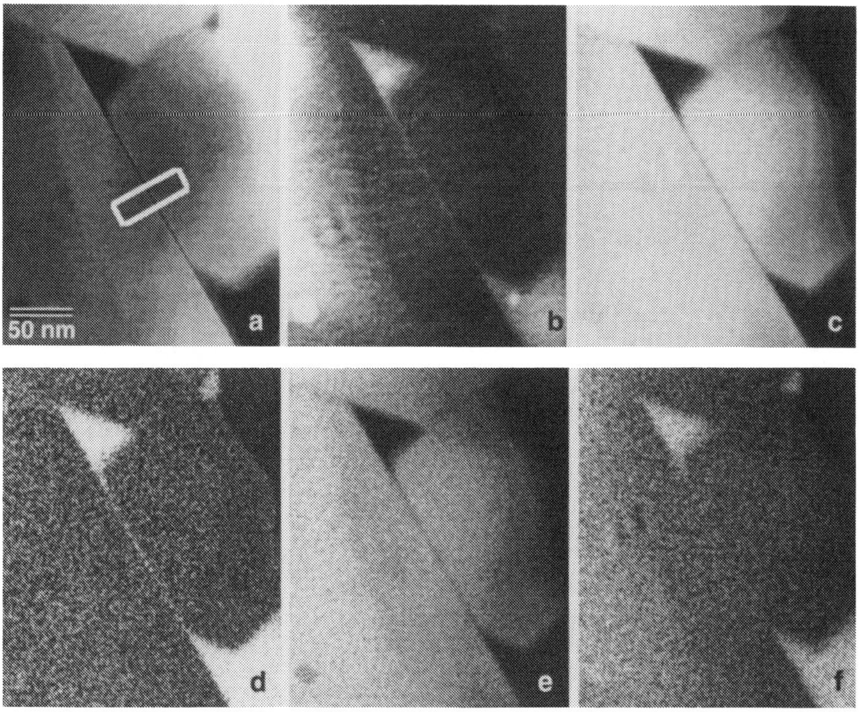

Fig. 5 Bright field and energy-selected images from grain-boundary in vitreous-bonded Si_3N_4, (a) bright field, (b) Al-L, (c) Si-L. (d) O-K, (e) N-K (f) C-K, from [28]

This effect is process-and materials dependent as interdiffusion in the aforementioned composites relies on a SiAlON-type exchange reaction. Quasiperiodic discontinuities involving direct bonding as well as thin amorphous films along SiC(w)/ß-Si_3N_4 phase boundaries resulted from preferential Gibbs-Thompson solution effects of the whisker surfaces [27]. It is anticipated that

chemical widths may actually affect the crack propagation along internal interfaces in ceramic microstructures.
Compared to position-resolved EELS and Z-contrast imaging [28], forming an energy-selected image with inelastically scattered electrons in the Ω filter of the Zeiss EM 912 provides a fast and elegant technique for monitoring chemical distributions over large specimen areas. As shown in Figs. 5 a-f, for a regular bright field image and 5 energy-selected images using the Al-K, Si-L, O-K, N-K and C-K core shell ionization edges, the amorphous phase connecting the grain boundary with the triple grain junctions is Al- and O-rich and Si-and N-depleted with respect to the ß-Si_3N_4 grains. The carbon intensiy probably results from a certain amount of carbon solution in the amorphous phase [29], as no carbon coating was necessary for these specimens to prevent surface charging.

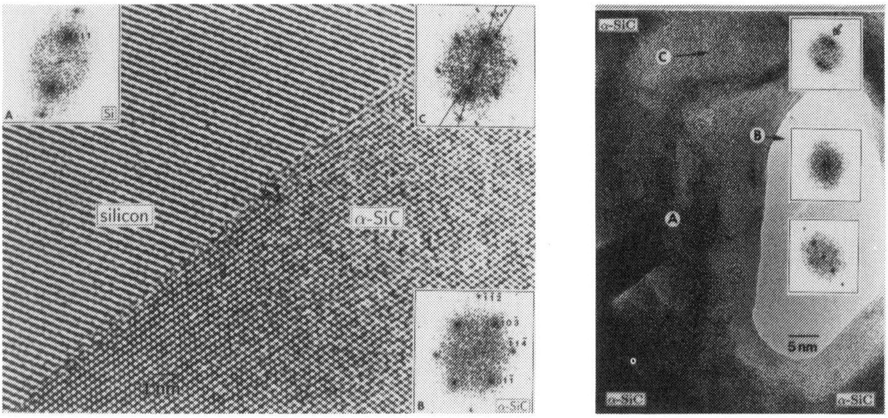

Fig. 6 (a) HREM image of Si/α-SiC boundary from RB-SiC lacking evidence for segregation of an amorphous phase, (b) intragranular "excess- carbon" secondary phase particle in RB-SiC. Optical diffractograms from regions A, B and C refer to SiC microcrystal,amorphous phase and turbostratic graphite, from [32]

Oxygen Segregation in SiC Materials

Our current understanding of oxygen segregation to Si_3N_4 interfaces as briefly summarized above raises the question whether the same mechanisms are operative in the SiC system. Continuous, oxygen-rich amorphous films at grain- and phase boundaries have been identified via TEM techniques for HIP-SiC [30] and other liquid-phase sintered SiC material employing various oxide additons [31]. With (B+C)-doped sintered SiC, attempts to image or otherwise detect amorphous interlayers have failed sofar. The issue of oxygen segregation has been recently addressed to the Si/α–SiC interface in RB-SiC [32]. In terms of interfacial strength, this interface is considered the weakest link in the microstructure, presumably due to oxygen segregation [33]. According to the fractographic analysis however, this material did not fail at Si/SiC interfaces but from large SiC grains instead. In the HREM image (Fig. 6a), the Si/α-SiC interface is rough on atomic scale exhibiting direct bonding between constituents and edge dislocations to accommodate the lattice mismatch along the interface.No oxygen signal could be detected in the position-resolved EELS spectra collected from this boundary (or any type of SiC/SiC boundaries). Instead, significant amounts of oxygen were detected from "excess-carbon" secondary phase inclusions intragranular in the SiC grains (Fig. 6b) which consist of turbostratic graphite and an amorphous phase containing silicon, oxygen and carbon. As discussed elsewhere [32], these particles result from codeposition of SiO_x and graphite on SiC surfaces, where they will be entrapped by mobile Si/SiC interfaces. These polyphase inclusions were also observed in (B+C)-doped SiC [34]. In terms of stability range and structural configuration, this Si-O-C phase bears some striking similarities with oxycarbide glasses [35] and PCS-based nonstoichiometric SiC materials [36]. From stability constraints of Si-O-C phases in other systems, it is anticipated that the Si-O-C phase in SiC materials is a metastabile phase controlled by processing parameters and cool-down conditions [37].

Crystalline-To-Amorphous Transition in Shock-Loaded Mullite Polycrystals

While the previous case studies are all related to the sintering processes and time-dependent failure of vitreous-bonded Si_3N_4 and SiC, the last paragraph of this paper is devoted to another issue of stability of microstructure, which is the generation of defects during shock-compression of structural oxide ceramics.

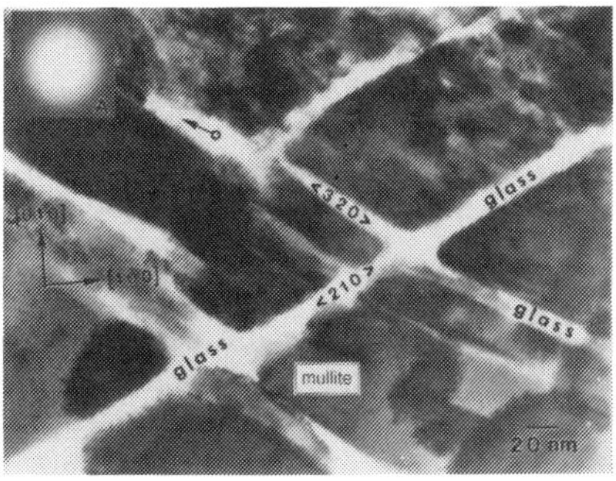

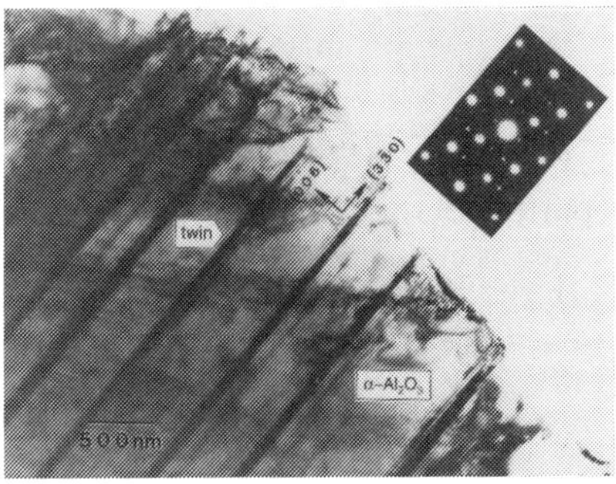

Fig. 7 (a) HREM image of 2:1 mullite grain oriented // to [001], displaying the network of intragranular glass lamellae parallel to low-index crystallographic directions, (b) shock-loaded corundum grain oriented // to $[11\bar{2}0]$, emphasizing the formation of thin twin bands lying on basal planes, from [43].

During shock compression either in the laboratory or in nature, for example during a meteorite impact, strain rates of the order of some $10^6 sec^{-1}$ can give rise to a variety of plastic deformatition features different from those introduced during static experiments [38].

Considerable research effort has been directed to the general issue of defect generation during shock-loading and to the fractographic analysis of shock-loaded alumina in particular [39-41]. Upon loading a crystalline constituent beyond the Hugoniot elastic limit, the shock-wave can cause a thin intragranular glass lamellae to form (see [42] for review). This solid state crystalline/amorphous transition occurs without loss of matter in a very short time, nevertheless allowing the lamellae to form a continuous network extending parallel to well-defined crystallographic directions within the otherwise undeformed host crystal. The formation of shock-wave induced glass lamellae has particularly been reported for SiO_2-polymorphs and feldspars where the amorphization results from straining of Si-O bonds and deformation of the tetrahedral framework on the decompression route [43].
TEM investigations we recently performed on mullite $2Al_2O_3{:}1SiO_2$ crystals (Fig. 7a) recovered from a polyphase refractory brick which had been exposed to dynamic shock experiments up to 40 GPa provided evidence for the first time that the shock-wave induced formation of glass lamellae can occur in the chain silicate mullite as well which is probably related to aluminium coordination changes [44, 45]. The shock-loaded alumina grains coexisting with mullite in the same specimens (Fig. 7b) exhibited the typical twin bands and dislocation activity on basal planes, as observed by other research groups [39, 41].

CONCLUSIONS

From the numerous TEM investigations focused on stability of microstructure of thin boundary layers under the constraints of high temperatures and/or stresses, some conclusions can be derived which are of significance for quantitative fractographic analysis of structural ceramics.
(i) The stability of crystalline secondary phases can change under the combined effects of high temperatures and local stresses during deformation, as shown for the partial dissolution of the m-ZrO_2 in the pre-existing glass thus affecting the macroscopic response of the material.
ii) The static fatigue resistance of vitreous-bonded Si_3N_4 is temporarly improved during partial devitrification of the intergranular phase. The materials performance after devitrification is still controlled by the intrinsic deformation behavior of the residual thin glassy layers via viscous flow or dissolution-reprecipitation processes. At higher temperatures, time-dependent failure resumes due to nucleation and growth of creep cavitation.
(iii) Although ceramic microstructures rarely approach chemical equilibrium conditions due to kinetic constraints, intergranular amorphous layers are characterized by an equilibrium film thickness which depends on the interfacial composition and will affect the high-temperature performance of the material. For sintering additive systems allowing interdiffusion into the bounding crystals, nonequilibrium solute gradients across the interface plane, defined as chemical widths, have been established for grain and phase boundaries which significantly exceed the structural widths as imaged via HREM.

(iv) Two different oxygen sinks have been identified in SiC ceramics, which do not occur in the same material: (a) segregation of oxygen to internal interfaces very similar to Si_3N_4 materials, and (b) "excess-carbon" secondary phase inclusions consisting of turbostratic graphite and an intermediate amorphous Si-O-C phase.
(v) The high strain-rates during shock-loading of ceramic constituents can introduce a crystalline/amorphous transition in the loaded crystal giving rise to an intragranular network of thin glass lamellae. This effect is well established for framework silicates, but has now been confirmed for mullite too.

ACKNOWLEDGEMENT

I gratefully appreciate the support and collaboration contributed over the years by H. J. Kleebe, University of Bayreuth, Germany, J. Goering and H. Schneider, DLR, Cologne, Germany, G. D. Quinn, NIST, Gaithersburg/MD, R. W. Carpenter ASU, Tempe/AZ and K. Das Chowdhury, INTEL, Albuquerque/ NM.

REFERENCES

[1] Principles of Analytical Electron Microscopy, (edited by D. C. Joy, A. D. Romig and J. I. Goldstein), Plenum Press, 1986
[2] T. G. Langdon, Trans. Mat. Res. Jpn. **16B,** 1441 (1994)
[3] W. Braue, H. J. Kleebe and J. Goering , Proc. 4th Conf. Europ. Ceram. Soc. Riccione, Italy, October 1995, in press
[4] J. Goering, W. Braue and H. J. Kleebe, Key Engineering Materials **89-91,** Trans Tech Publications , Switzerland (1994), 641
[5] R. M. Arons and J. K. Tien, J. Mater. Sci. **15**, 2046 (1980)
[6] D. R. Mosher, R. Raj and R. Kossowsky, J. Mater. Sci. **11**, 49 (1976)
[7]H. J. Kleebe, W. Braue and W. Luxem, J. Mater. Sci. **29** , 1265 (1994)
[8] H. J. Kleebe and W. Braue, in preparation
[9] S. M. Wiederhorn, B. J. Hockey, D. C. Cramer and R. Yeckley, J. Mater. Sci. **28**, 445 (1993)
[10] R. Raj and F. F. Lange, Acta Metall. **29**, 1993 (1981)
[11] R. Raj, in Tayloring of Mechanical Properties of Si_3N_4 Ceramics, edited by M. J. Hoffmann and G. Petzow , Kluwer Academic Publishers (1994), 217
[12] G. D. Quinn, W. Braue, J. Mater. Sci. **25,** 4377 (1990)
[13] C. Gandhi and M. F. Ashby, Acta Metall. **27**, 1565 (1979)
[14] W. Braue, G. D. Quinn, Mat. Res. Soc. Symp. Vol. **287**, Materials Research Society, Pittsburgh, PA, (1993), 329
[15] S. M. Wiederhorn, W. E. Luecke, B. J. Hockey and G. G. Long, in in Tayloring of Mechanical Properties of Si_3N_4 Ceramics, edited by M. J. Hoffmann and G. Petzow , Kluwer Academic Publishers (1994), 305
[16] D. R. Clarke and G. Thomas, J. Am. Ceram. Soc. **61**, 114 (1978)

[17] L. K. V. Lou, T. E. Mitchell and A. Heuer, J. Am. Ceram. Soc. **61**, 462 (1978).
[18] D. R. Clarke, N. J. Zaluzec and R. W. Carpenter, J. Am. Ceram. Soc. **64,** 601, (1981)
[19] D. R. Clarke, N. J. Zaluzec and R. W. Carpenter, ibid. **64,** 608 (1981)
[20] D. R. Clarke and N. J. Zaluzec, J. Am. Ceram. Soc. **65**, C132-C133, (1982)
[21] D. R. Clarke, J. Am. Ceram. Soc. **70,** 15 (1987)
[22] H. J. Kleebe, M. Hoffmann and M. Ruehle, Z. Metallkd. **83**, 610 (1992)
[23] M. K. Cinnibulk, H. J. Kleebe and M. Ruehle, J. Am. Ceram. Soc. **76**, 426 (1993)
[24] H. J. Kleebe, M. K. Cinnibulk, I. Tanaka, J. Bruley, R. M. Cannon, D. R. Clarke, M. J. Hoffmann and M. Ruehle, Mat. Res. Soc. Symp. Proc. Vol. **287**, The Materials Research Society, Pittsburgh PA (1993), 65
[25] K. Das Chowdhury, R. W. Carpenter and W. Braue, Ultramicros. **40**, 229 (1992)
[26] J. Liu and J. M. Cowley, Ultramicroscopy **37,** 50 (1991)
[27] K. Das Chowdhury, R. W. Carpenter, W. Braue, J. Liu and H. Ma, J. Am. Ceram. Soc., (1995), paper #193389, in press
[28] R. W. Carpenter, J. S. Bow, M.J. Kim, K. Das Chowdhury and W. Braue, Mat. Res. Soc. Symp. Proc. **357,** The Materials Research Society, Pittsburgh, PA (1995), 271
[29] J. Homeny, G. G. Nelson and S. H. Risbud, J. Am. Ceram. Soc. **71**, 386 (1988)
[30] W. Braue, J. Goering and G. Ziegler, in Ceramic Materials and Components for Engines, edited by V. J. Tennery, The American Ceramic Society, 818 (1988)
[31] L. S. Sigl and H. J. Kleebe, J. Am. Ceram. Soc. **76,** 773 (1993)
[32] K. Das Chowdhury, R. W. Carpenter and W. Braue, Mater. Res. Soc. Sym. Proc. **295**,The Materials Research Society, Pittsburgh, PA, 1993), 183
[33] J. N. Ness and T. E. Page, J. Mater. Sci. **21**, 1377 (1986)
[34] W. Braue and R. W. Carpenter, J. Mater. Sci. **25,** 2943 (1990)
[35] G. M. Renlund, S. Prochazka and R.H. Doremus, J. Mater. Res. **6**, 2716 (1993)
[36] P. Le Coustumer, M. Monthieux and A. Oberlin, J. Europ. Ceram. Soc. **11**, 95 (1993)
[37] W. Braue , K. Das Chowdhury and R. W. Carpenter, Mat. Res. Soc. Symp. Proc. Vol. **327**, The Materials Research Society, Pittsburgh, PA, 1994, 274
[38] M. A. Meyers and L. E. Murr, in Shock Waves and High-Strain Rate Phenomena in Metals, edited by M. A. Meyers and L. A. Murr, Plenum Press 1981, 487-530
[39] C. S. Yust and L. A. Harris, in in Shock Waves and High-Strain Rate Phenomena in Metals, edited by M. A. Meyers and L. A. Murr, Plenum Press 1981, 881-895
[40] J. A. Brusso, D. E. Mikkola, J. E. Flinn and P. V. Kelsey, Scr. Metall. **22**, 47 (1988)
[41] Y. Wang and D. E. Mikkola, J. Am. Ceram. Soc. **75**, 3252 (1992)
[42] D. Stoeffler, J. Non-cryst. Solids **67** , 465 (1984)
[43] G. H. Wolf, S. Wang, C. A. Herbst, D. J. Durben, W. F. Oliver, Z. C. Kang and K.Halvorson, in High-Pressure Research: Application to Earth and Planetary

Sciences, edited by Y. Syono and M. H. Manghnanni, Terra Scientific Publishing Company, American Geophysical Union, 1992, 503-517
[44] W. Braue, H. Schneider and U. Hornemann, Proc. APS Conference on Shock Compression of Condensed Matter, Seattle/WA, 13-18 August, 1995, in press
[45], W. Braue, H. Schneider and U. Hornemann, in preparation

UTILIZATION OF FRACTOGRAPHY IN THE EVALUATION OF HIGH TEMPERATURE DYNAMIC FATIGUE EXPERIMENTS*

Kristin Breder, Metals and Ceramics Division, Oak Ridge National Laboratory, Oak Ridge, TN 37831-6069, Thomas J. Mroz, Advanced Refractory Technologies, Inc., Buffalo, NY 14207, Andrew A. Wereszczak, and Victor J. Tennery, Metals and Ceramics Division, Oak Ridge National Laboratory, Oak Ridge, TN 37831-6069.

ABSTRACT

The slow crack growth properties of six structural ceramics were measured by dynamic fatigue in air and inert atmospheres over a range of elevated temperatures. The material response varied from no strength degradation as a function of stress and environment to significant strength degradation by slow crack growth (SCG) and by a combination of SCG and creep. The fractographic investigation showed that SCG was evidenced by growth of isolated cracks and often by an intergranular fracture mode, while creep was evidenced by accumulated damage such as void formation and opening of the microstructure at grain boundaries and triple junctions. For the materials in which the strength was unaffected by the stress and environment, the fracture surfaces were essentially indistinguishable from the inert fracture surfaces.

INTRODUCTION

The dynamic fatigue method, in which strength is measured as a function of stressing or loading rate, is well suited to the measurement of the slow crack growth (SCG) properties of ceramics.[1-4] Knowledge of the fast fracture, creep and SCG parameters are necessary for design and life-time prediction of ceramic components. The regimes of stress, temperature, and environment in which SCG exists vary from material to material, and SCG may exist simultaneously with other strength-degradation mechanisms such as creep and cyclic fatigue. In

* Research sponsored by the U.S. Department of Energy, Office of Fossil Energy, Pittsburgh Energy Technology Center, Advanced Combustion Technology Program, DOE/FE AA 20 10 000, under contract DE-AC05-84OR21400 with Lockheed Martin Energy Systems.

terms of lifetime predictions, it is necessary to understand the relative contributions from the different mechanisms, as each of them may develop differently with time. The fracture mechanics framework utilized in lifetime predictions during SCG assumes that linear elastic fracture mechanics is valid, but if creep occurs simultaneously, then corrections must be made. The best way to decouple the mechanisms is to perform experiments such that only one process is active at any time, e.g., creep and SCG in Si_3N_4 was studied by performing experiments in air, argon (Ar) and nitrogen (N_2) atmospheres at elevated temperatures.[5-7] These experiments, combined with microstructural analyses, indicated significant differences in the damage mechanisms due to creep and SCG. Zeng et al.[8,9] performed stressing-rate experiments on Al_2O_3 and SiC whisker-reinforced Al_2O_3 in water at room temperature and found a clear difference in fracture mode between SCG and fast fracture. Differences in fracture propagation paths were also observed by Pletka and Wiederhorn[1] who compared slow crack growth in several ceramics and glass ceramics by the dynamic fatigue method and by crack velocity measurements using double torsion experiments. The ceramics with the coarser microstructures (presumably the ones exhibiting R-curve behavior) had different crack velocities measured by the two methods, showing that the crack propagation mode, e.g. transgranular vs. intergranular, is important and must be determined in these types of experiments. These examples show the importance of the fractographic analysis for determination of SCG. In the present work, several ceramic materials have been compared and different SCG paths will be shown.

MATERIALS AND EXPERIMENTAL PROCEDURE

Six ceramic materials were compared in this work. They were Siliconized SiC (Si-SiC) NT230 from Saint-Gobain Norton,[10-12] β-SiC from Coors Ceramics Company,[12] Lanxide DIMOX, a SiC particulate reinforced Al_2O_3 from Lanxide Composites Inc.($SiCp/Al_2O_3$),[12] hot isostatically pressed Si_3N_4 with yttria as a sintering aid (PY6) from GTE Laboratories Inc.,[5,6] and AlN and SiC whisker reinforced AlN from Advanced Refractory Technologies Inc.[13] Further details about these materials can be found in the cited references. These materials are candidate structural ceramics for applications in which lifetime predictions are important.

The dynamic fatigue experiment is commonly referred to as a stressing rate or a strength technique, to distinguish it from direct crack velocity measurement methods such as the double torsion experiment.[1] The measurements can be performed by testing in three- or four-point bending, biaxial bending or in tension. The basic principle is the assumption that there is a relationship between the crack velocity and the applied stress intensity, such as is expressed in the power law form:

$$v = A \left(\frac{K_a}{K_{IC}}\right)^n , \qquad (1)$$

where v is the crack velocity, A and n are environmentally determined material constants, K_{IC} is the material fracture toughness and K_a is the applied stress intensity. The crack velocity exponent n is determined in the dynamic fatigue experiment by measuring the fracture stress as a function of stressing rate:

$$\sigma = B \dot{\sigma}^{\frac{1}{n+1}} , \qquad (2)$$

where σ is the strength in the given environment, $\dot{\sigma}$ is the stressing rate and B is a constant that is a function of the material toughness, crack velocity, inert strength, and crack geometry. With the knowledge of n and A (from B) failure prediction diagrams can be determined from well established fracture mechanics procedures.[3,4]

The present experiments were conducted in ambient air at temperatures between 1100 and 1400°C. SiC four-point-bend fixtures with 20 mm inner and 40 mm outer span were used in a hydraulic flexure test system with load control. The loading rates were set so as to correspond to stressing rates ranging between 40 MPa/s and 0.0001 MPa/s. Resulting times to failure ranged from a few seconds up to 1200 h. In order to decouple the effects of SCG, oxidation, and creep, the Si_3N_4 was also tested in argon and nitrogen atmospheres at 1370°C.[5,6] The flexure strengths were calculated according to classical beam theory, and in the cases where creep was observed by virtue of observed specimen curvature after the test, the stress was as a first approximation adjusted according to the analysis by Hollenberg et al.[14] The fractography was performed using a stereo microscope with a video set-up, a digital image analysis system and a SEM (Hitatchi S-800).

RESULTS AND DISCUSSION

The dynamic fatigue graphs, i.e., strength as a function of stressing rate, are shown in Figs. 1 through 3. Figure 1 shows Si-SiC, β-SiC and $SiCp/Al_2O_3$ measured at 1400°C. The β-SiC exhibited no strength degradation as a function of time at this temperature. This result is consistent with results from the literature for this material measured in a dry, oxidizing environment.[15,16] Further, the strength level of this SiC was unchanged from room temperature, which is also consistent with previous results.[15,16] The Si-SiC ceramic exhibited a reduction of fracture strength with decreasing stressing rates. Other work[11] showed increased variability in the strength as the temperature and time-to-failure were increased. The n-value for this material, calculated according to Eq. (2), was 15.5. The strength of the SiCp/Al_2O_3 was only measured at stressing rates of 0.01 MPa/s and above. At lower stressing rates creep became the dominant mechanism, and the specimens crept to the bottom of the four-point flexure fixture before fracture (a deflection of > 2 mm). The specimens which failed at 0.01 MPa/s at 220 MPa also exhibited significant creep, and the reported strength overestimated the stress at this condition. Adjusting the stress according to Hollenberg[14] using a creep exponent of 3 results in a strength reduction of 17% as compared to the fast fracture strength at this temperature.[12,17]

Fig. 2 shows the strength as a function of stressing rate for HIP'ed Si_3N_4 at 1370°C. This material was tested in air, Ar, and N_2, and it is seen that strength alteration was apparent in all cases. When the strengths were adjusted for creep, the specimens measured in Ar and N_2 had no significant strength degradation while the specimens measured in air showed significant loss of strength, presumably due to SCG.

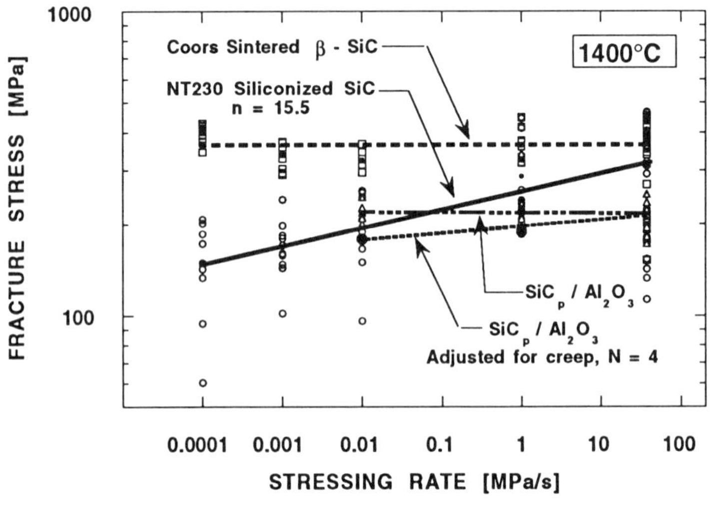

Figure 1. Dynamic fatigue results for β-SiC, Si-SiC and SiC particulate reinforced Al_2O_3 measured in four-point flexure in air at 1400°C.

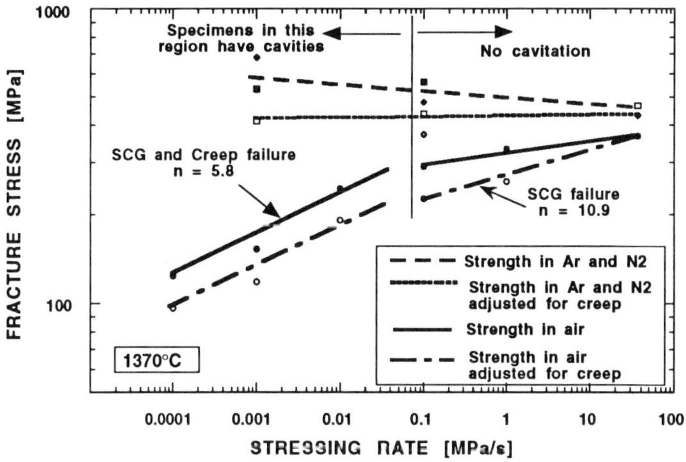

Figure 2. Dynamic fatigue results for HIP'ed Si_3N_4 measured in four-point flexure in air and inert environment at 1370°C.

Figure 3 shows strength as a function of stressing rate for AlN and AlN reinforced with 20% SiC whiskers measured in air at 1200°C. The addition of SiC whiskers did not strengthen the AlN significantly at this temperature, but the creep deformation was considerably lower. The creep exponents for these experimental materials were not known, so a strength adjustment to account for the creep was not performed. The presence of SCG must be sought in the fractography results as the dynamic fatigue results showed no evidence for such.

Figure 4 shows the fracture surface of the β-SiC fractured at 420 MPa at a stressing rate of 0.0001 MPa/s at 1400°C. The fracture initiation point was a pore and there was no evidence of SCG around this. This is consistent with the strength results, which showed no strength degradation with time and no difference in strength between room temperature and 1400°C.

A typical fracture surface of the Si-SiC fractured at 40 MPa/s at 1400°C is shown in Fig. 5 a). The fracture initiation point was a pore. Figure 5 b) shows the fracture surface of the Si-SiC fractured at 1400°C and 0.0001 MPa/s. It can be seen that the fracture origin was a Si-rich area which enlarged during the test. The enlargement was probably due to subcritical (slow) crack growth. This type of failure origin was typical for the Si-SiC tested at 1400°C, and the increased variability in the strength results was believed to be due to the variability in the size and composition of these Si-rich areas.[11]

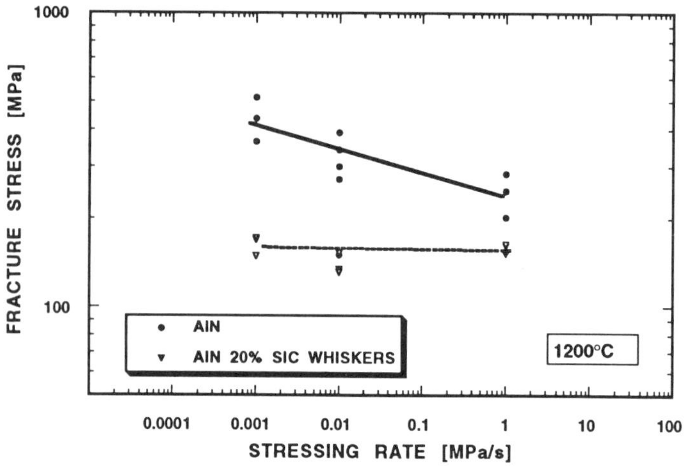

Figure 3. Dynamic fatigue results for AlN and SiC whisker reinforced AlN measured in four-point flexure in air at 1200°C.

Figure 4. Fracture surface of β-SiC, measured at 0.0001 MPa/s at 1400°C. The initiation point was a pore with no signs of SCG.

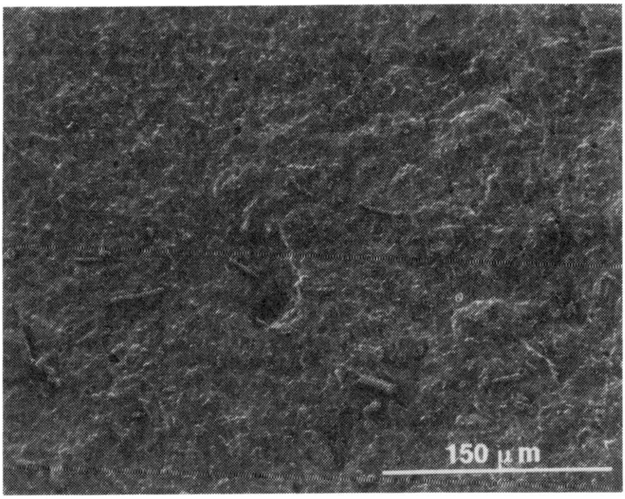

Figure 5. a) A pore acting as fracture initiation point in Si-SiC tested at 40 MPa/s at 1400°C.

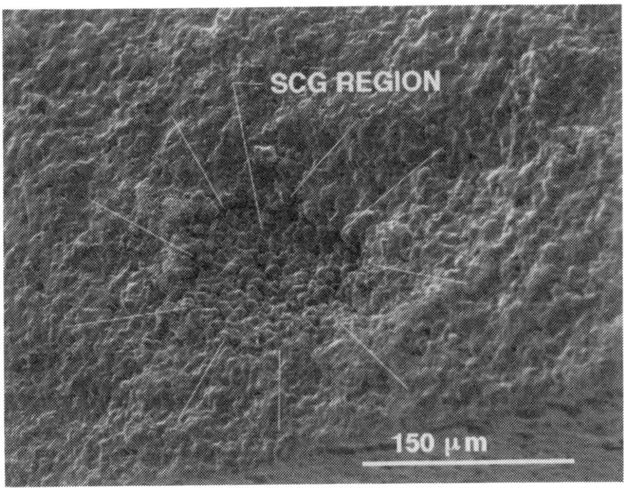

Figure 5.b) SCG zone in a Si-SiC specimen tested at 0.0001 MPa/s at 1400°C.

Figure 6 shows the fracture surface of a SiCp/Al_2O_3 specimen which failed at 215 MPa (uncorrected for creep) at a stressing rate of 0.01 MPa at 1400°C. The fracture surface was rough (Fig. 6 a) near the tensile surface at the corner and contained a series of microcracks which were interpreted as an opening up of the microstructure due to creep (Fig. 6 b). None of the specimens tested at this condition exhibited failures from a well defined origin as was the case for the Si-SiC ceramic, although the fast fracture failures for these two materials were similar, with failure from metal-rich areas in the microstructure.

Typical fracture surfaces for the Si_3N_4 are shown in Fig. 7 a) and b). Fig. 7 a) shows a micrograph of the fracture surface of a specimen tested in air at 1370°C at a stressing rate of 0.0001 MPa/s. The SCG zone is clearly visible as the much rougher area. For these specimens it was found that the details of the SCG zone were not very different from the fast fracture zone: no accumulation of microcracks was observed, and, in both cases, the fracture was intergranular. Several cracks were found to grow simultaneously in the gage section and local microstructural variations could account for which one would grow to a critical size first. The interpretation that the damage shown in Fig. 7 a) was a SCG zone was verified by measuring the strength as a function of stressing rate in two inert atmospheres, Ar and N_2. As shown in Fig. 2, a strength alteration was initially observed for these two conditions as well, but if the strengths were adjusted for creep, no severe degradation was seen. A typical fracture surface after testing in N_2 is shown in Fig. 7 b); the specimens tested in Ar showed similar features. Numerous cavities can be seen, indicating that creep was an active mechanism at that temperature, and that in air the material undergoes both creep and strength degradation due to other mechanisms. In the present case, it was determined that the SCG was environmentally assisted, i.e., the crack growth was a result of oxidation, temperature, and stress, and the creep was due to temperature and stress. The ultimate failure in air was due to the SCG, indicating that this mechanism was the dominating one at those conditions. It is however conceivable that other environment-temperature-stress conditions could result in a different controlling mechanism.

Figure 8 a) shows the fracture origin in the AlN tested at 1 MPa at 1200°C in air. The fracture origins in this material were mainly pores, however, a closer look at the pore in this figure indicates the presence of a small zone of intergranular fracture before the transgranular fast fracture became predominant. A similar transition from intergranular to transgranular fracture has been observed in Al_2O_3 and SiC whisker reinforced Al_2O_3 as these materials underwent transition from sub-critical to critical crack velocities while being tested in water.[9] This result shows that this material will undergo SCG under certain circumstances, but in this particular experimental material, this was not observed in the strength data because the specimens failed in fast fracture mode due to the existence of large (50-100 μm) pores, or underwent substantial creep before failing from a pore. The SiC-whisker-reinforced AlN showed improved creep properties, but not higher strength values. A typical fracture surface for this material is shown in Fig. 8 b), where the fracture origin is seen to be a cluster of SiC whiskers. Due to the existence of these whisker clusters, this material did not exhibit improved strength over the AlN ceramic. Neither of the AlN-based materials exhibited significant SCG, but as shown there were indications in the AlN that SCG may have existed, and the present results show that in improved versions of these materials it will be necessary to characterize the various time dependent failurc mcchanisms thoroughly.

As can be seen from the above results, what may be interpreted as SCG may manifest itself in different ways depending on the material and environment combination. However, a general trend seems to be that SCG is growth of a isolated crack (or several independent cracks), while creep damage occurs at several sites that eventually coalesce. The SCG process is the true growth of cracks with breaking of bonds due to a combination of the applied stress and the environment, while the creep damage is a plastic flow mechanism with the opening of voids, triple junctions and grain boundaries by material diffusion mechanisms.[18]

Figure 6. a) Fracture surface with creep damage in a SiC particulate reinforced Al_2O_3 specimen tested at 0.01 MPa/s at 1400°C.

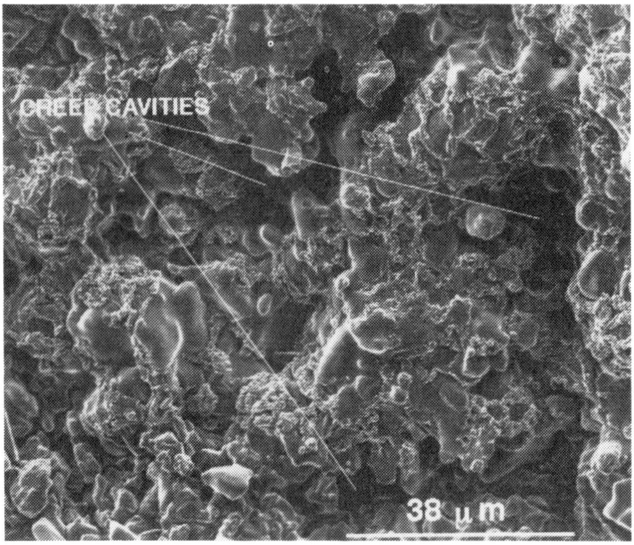

Figure 6. b) The creep damage region in a crept SiC particulate reinforced Al_2O_3 specimen.

Figure 7. a) Transition between SCG and fast fracture zones in Si_3N_4 tested in air at 1370°C.

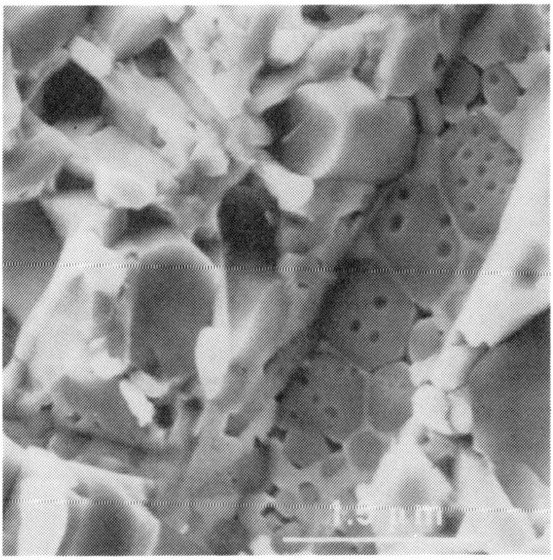

Figure 7. b) Creep cavities in Si_3N_4 tested in N_2 atmosphere at 1370°C.

Figure 8. a) Typical fracture initiating pore in AlN tested in air at 1200°C. Note the transition from intergranular to transgranular fracture indicating the presence of SCG.

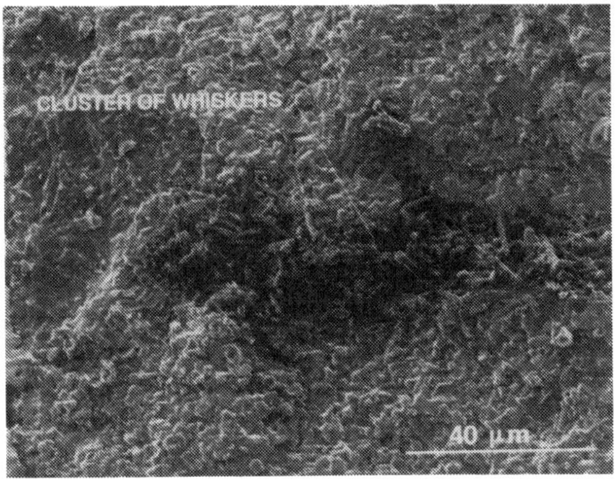

Figure 8. b) Whisker clusters acted as fracture initiation points in the SiC whisker reinforced AlN.

CONCLUSIONS

The present results show that it is necessary to perform thorough fractography in order to correctly interpret the dynamic fatigue results to determine the slow crack growth properties of a ceramic material. It is also necessary to distinguish between the different strength degradation mechanisms that may be active because the fracture mechanics framework which is used for life-time predictions is different depending on which one is operative. Hence, observing strength degradation in a dynamic fatigue experiments and using these results for lifetime predictions may lead to erroneous results if the strength degradation is due to other mechanisms such as creep.

The ceramic materials in the present work exhibited a range of behaviors. β-SiC showed no strength degradation as a function of temperature or stressing rate in air. Hence, design with this ceramic material at the temperatures under considerations can be performed utilizing fast fracture data and the appropriate statistical variability. The Si-SiC in this study showed a loss of strength as a function of stressing rate, and the fractographic evidence showed SCG-zones originating from metal-rich areas. The creep of this material was limited, evidenced by little permanent deflection of the specimens after the test. The SiC particulate reinforced Al_2O_3 exhibited creep at the test temperature, evidenced by large permanent deflection of the flexure bars as well as fractographic evidence of creep damage on the fracture surface. The most complicated behavior was seen in the HIP'ed Si_3N_4 ceramic. This material showed significant strength degradation as a function of stressing rate, and SCG zones were observed on the fracture surfaces. However, this material underwent creep as well as SCG, and by testing in inert atmospheres as well as in oxidizing atmospheres the mechanisms were shown to be additive. The two AlN-based materials exhibited a fast fracture to creep transition, however evidence of limited SCG was observed in AlN at the lower stressing rate, indicating that a stronger version of this material might be prone to SCG.

REFERENCES

1. B. J. Pletka and S. M. Wiederhorn, "A Comparison of Failure Prediction by Strength and Fracture Mechanics Techniques," *J. Mater. Sci.*, **17** [5] 1247-1268 (1982).

2. J. E. Ritter Jr., S. M. Wiederhorn, N. J. Tighe, and E. R. Fuller Jr., "Application of Fracture Mechanics in Assuring Against Fatigue Failure of Ceramic Components," NBSIR 80-2047 (1980).

3. J. E. Ritter Jr., "Engineering Design and Failure of Brittle Materials," pp. 6676-686 in Fracture Mechanics of Ceramics **Vol. 4** Edited by R. C. Bradt, D. P. H. Hasselman and F. F. Lange, Plenum Press, New York, NY, 1978.

4. S. M. Wiederhorn and J. E. Ritter Jr., "Application of Fracture Mechanics Concepts to Structural Ceramics," pp. 202-214 in Fracture Mechanics Applied to Brittle Materials, ASTM STP 678. Edited by S. W. Freiman, ASTM, Philadelphia, PA, 1979.

5. A. A. Wereszczak, K. Breder, and M. K. Ferber, "Role of Oxidation in the Time-Dependent Failure Behavior of Hot Isostatically Pressed Silicon Nitride at 1370°C," *J. Am Cer. Soc.* **76** [11] 2919-22 (1993).

6. A. A. Wereszczak, T. P. Kirkland, K. Breder, M. K. Ferber, and P. Kandelwal, "High Temperature Dynamic Fatigue Performance of a Hot Isostatically Pressed Silicon Nitride," *Mat. Sci. and Eng. A* **191** 257-266 (1995).

7. A. A. Wereszczak, T. P. Kirkland, and M. K. Ferber, "Differences in Creep Performance of a HIP'ed Silicon Nitride in Ambient Air and Inert Atmospheres," to be published in Cer. Eng. Sci. Proc. **16** (1995).

8. K. Zeng, K. Breder, and D. J. Rowcliffe, "Dynamic Fatigue of an Al_2O_3/SiC Whisker Composite in Water," *Cer. Eng. Sci. Proc.* **12** [9-10] 2233-50 (1991).

9. K. Zeng, K. Breder, and D. J. Rowcliffe, "Comparison of Slow Crack Growth Behavior in Alumina and SiC-Whisker-Reinforced Alumina," *J. Am. Cer. Soc.* **76** [7] 1673-80 (1993).

10. B. J. McEntire et al., "Ceramic Component Processing Development for Advanced Gas-Turbine Engines," *J. Eng. Gas T.,* **115** [1] 1-8 (1993).

11. K. Breder, "Time-Dependent Strength Degradation of a Siliconized Silicon Carbide Determined by Dynamic Fatigue," *J. Am. Cer. Soc.*, in press 1995.

12. K. Breder and V. J. Tennery, "Dynamic Fatigue Behavior of two SiC and a SiC particulate Reinforced Al_2O_3 at Elevated Temperatures," Submitted for publication, 1995.

13. T. Mroz, C. Mroz, J. Fernando, M. K. Ferber, and K. Breder, "Evaluation of Ambient and High Temperature Properties of AlN-SiC Composite," Presented at the 19. Annual Conference on Ceramic and Ceramic Composites, Cocoa Beach, FL. 1995.

14. G. W. Hollenberg, G. R. Terwilliger, and R. S. Gordon, "Calculation of Stresses and Strains in Four-Point Bending Creep Tests," *J. Am. Cer. Soc.*, **54** [4] 196-99 (1971).

15. K. Y. Chia and K. S. Lau, "High Toughness Silicon Carbide", *Ceram. Eng. Sci. Proc.* **12** [9-10] 1845-1861 (1991).

16. D. E. McCullum, N .L. Hecht, L. Chuck and S. M. Goodrich, "Summary of Results of the Effects of Environment on Mechanical Behavior of High-Performance Ceramics", *Ceram. Eng. Sci. Proc.*, 1886-1913 **12** [9-10] (1991).

17. H. T. Lin and K. Breder, "Creep Deformation in an Alumina SiC Composite Produced via the Directed Metal Oxidation Process," Submitted to *J. Am. Cer. Soc.*, 1995.

18. M. K. Ferber and M. G. Jenkins, "Evaluation of the Strength and Creep-Fatigue Behavior of Hot Isostatically Pressed Silicon Nitride," *J. Am. Ceram. Soc.*, **75** [9] 2453-62 (1992)

THE USE OF LASER RAMAN MICROPROBE SPECTROSCOPY FOR INVESTIGATING RESIDUAL SURFACE STRAIN IN SELECTED CERAMICS

Norman L. Hecht, University of Dayton Research Institute, 300 College Park, Dayton, OH 45469-0172
Perry P. Yaney, Department of Physics, University of Dayton, 300 College Park, Dayton, OH 45469-2314
George A. Graves, University of Dayton Research Institute, 300 College Park, Dayton, OH 45469-0172

ABSTRACT

Machining, various processing procedures, and property measurements can generate residual strains in ceramic components. Depending on the composition, crystal structure, and application, the residual strain generated can enhance or degrade the ceramic. Laser Raman microprobe spectroscopy has been found to be an effective tool for investigating residual strain in ceramic components. Strain induced by stresses in a ceramic material can alter the Raman spectra by (a) the generation of additional spectral features, (b) relative changes in intensity or widths of the Raman lines, and (c) shifting or splitting of lines. Specifically, Raman spectroscopy is being used to investigate the residual surface strain generated during selected machining processes and flexural testing. These studies are centered on silicon carbide (SiC) and silicon nitride (Si_3N_4) ceramics and the residual surface strains introduced by different machining methods. The results show a strong correlation between the Raman linewidths and the residual surface stress.

INTRODUCTION

The University of Dayton (UD) has participated in an Oak Ridge National Laboratory (ORNL) sponsored project investigating the thermal-mechanical behavior of candidate ceramics for heat engine applications since December 1984. As part of this project, Raman microprobe spectroscopy is being used to monitor residual strains in candidate ceramics subject to different machining treatments, mechanical loading, and microindents. In the initial work, the Raman microprobe was used to study transformation toughened ZrO_2-based ceramics. The Raman microprobe identified the extent and amount of phase transformation (tetragonal → monoclinic) around cracks extending from Vickers indents [1]. In more recent work, the Raman microprobe was used to study the strain around Vickers indents and along the tensile surface of Si_3N_4 flexural test bars in a four-point bending fixture [2]. In the current work, the Raman microprobe is being used to study residual strain in SiC and Si_3N_4 flexural test bars machined by different methods. The correlations between machining methods, residual strain, and fracture strength were evaluated and the results obtained are presented.

BACKGROUND

The phenomenon of frequency-shifted scattered light was first predicted theoretically by A. Smekal in 1923 and was first observed experimentally in liquids by C. V. Raman in 1928 [3]. This effect is now known as Raman scattering. Raman spectroscopy provides information over the visible spectrum and requires no specimen preparation in contrast to the typical requirements of infrared absorption spectroscopy. However, before the discovery of the laser, the limitations of light sources available to induce scattering and factors resulting in degradation of signal levels, such as fluorescence, phosphorescence, and sample imperfections, restricted the use of Raman spectroscopy. Developments in dispersing and detection systems helped to improve signal levels and maintain interest in Raman spectroscopy as a laboratory curiosity. The advent of the laser in 1960 and its application to spectroscopy resulted in a resurgence of Raman spectroscopy. Laser sources provide a number of highly monochromatic excitation lines capable of producing Raman spectra with good signal-to-noise ratio (SNR) from small sample areas. Laser Raman spectroscopy/microspectroscopy has since developed into an important analytical and complementary spectroscopic technique. A schematic of the UD laser Raman microprobe system is shown in Figure 1.

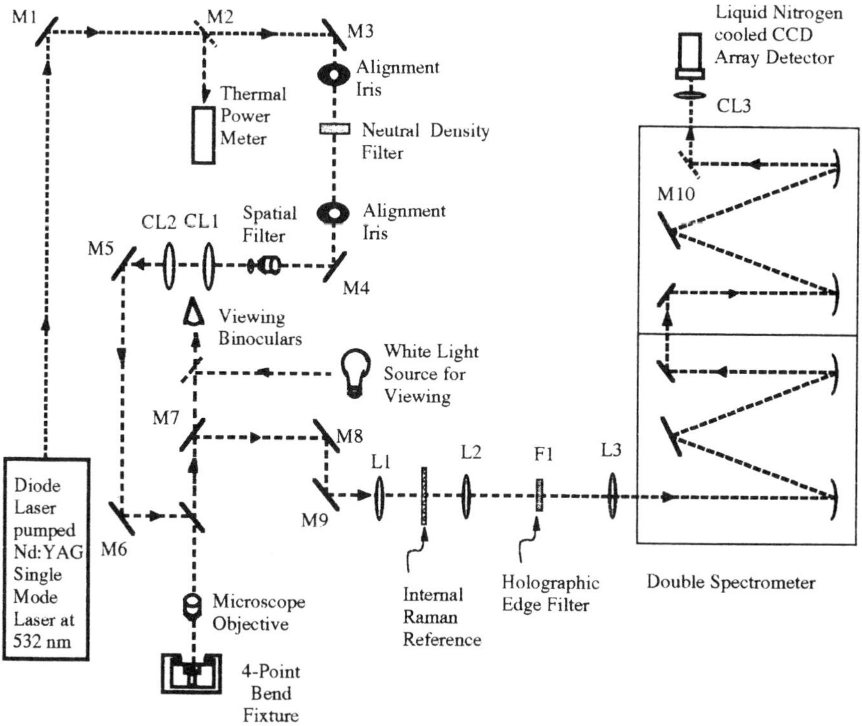

Figure 1. Laser Raman microprobe system in the Physics Department of the University of Dayton.

Monochromatic light interacts with the molecules or lattice structure of a probed material producing observable inelastically scattered light at shifted frequencies (Raman lines) and elastically scattered light at the same frequency as the incident beam (Rayleigh line). The frequencies of the Raman lines (about 10^{-12} times less intense from opaque solids than the incidence light) in the scattered light are shifted from the incident frequency by precise values characteristic of the specific material (see Figure 2). These Raman spectra are due to the vibrations and/or rotation of the molecules or lattice vibrations in the solid. Both internal and external vibrations occur in crystalline solids. The internal vibrations involve molecular-like stretching and bending of the chemical bonds and the external vibrations involve partial rotations and translations of molecular units in the crystal lattice. Raman lines or "modes" can be assigned to these molecular or lattice vibrational mechanisms and are used for material identification.

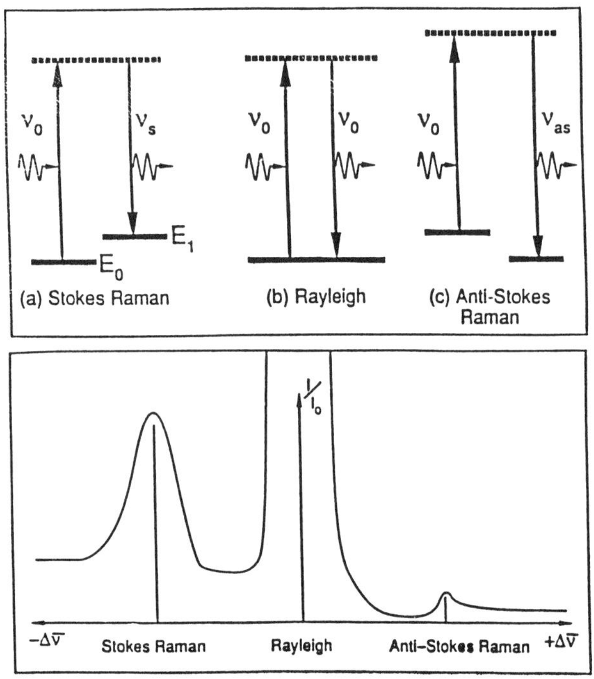

Figure 2. Light scattering process showing (a) Stokes Raman scattering, (b) Rayleigh scattering, and (c) anti-Stokes Raman scattering, and the corresponding spectrum. (The dotted energy "levels" are virtual states.)

Strain in a crystal can alter the Raman spectrum in the following ways [4]:

(1) the generation of additional lines due to lowering of crystal symmetry,

(2) intensity variations of the lines,

(3) broadening of the line widths, and

(4) shifting or splitting of the lines.

Strain in a crystal occurs in a number of situations for which changes in characteristic Raman spectra have been used as an investigative tool. In semiconductors, strain tends to occur by mismatches of the crystal parameters (lattice spacings, thermal expansion conditions) at epitaxial boundary layers [5]. For sapphire, silicon, and other materials strain is induced in the surface crystallites during polishing [6]. These strains are built into the lattice (referred to as intrinsic or residual strain) and produce Raman line shifts or broadenings. Externally applied stress induces strain, and flaws in a material (foreign inclusions, pores, and cracks) can be sites for stress concentrations and can cause Raman line shifts. A model to relate strain to Raman line shifts has been developed at UD [7] as well as at other institutions. The model can be used to determine shifts of internal lattice modes due to a given strain in a specific crystalline lattice.

Grinding and polishing of ceramic materials can introduce residual surface strain. These processes usually distort, to some extent, the crystalline lattice in the surface grains, resulting in a more rapid damping of the lattice vibrations, which in turn results in a broadening of the Raman lines. Different machining processes can introduce different levels of residual strain in the specimen surfaces resulting in different Raman linewidths. Inclusions of foreign atoms in the crystal lattice and other defects in the crystal structure can also introduce residual strains which may be reflected in the Raman spectra. As an example, Figure 3 shows a typical Raman line shift due to flexural strain in a Si_3N_4 test bar. Surfaces which have been only cut with a diamond saw and not ground typically give the narrowest lines. This is shown in Figure 4(a) for cubic CVD SiC. The change in this line after grinding treatments is shown in Figure 4(b). A more dramatic change due to built-in strain in the bulk material is the line splitting shown in Figure 4(c).

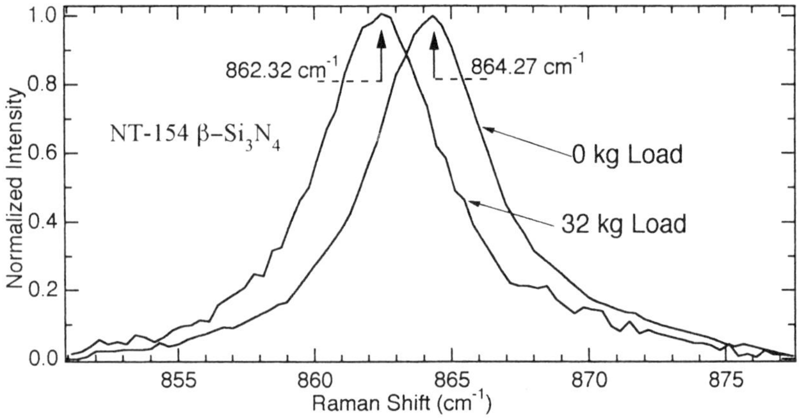

Figure 3. Shift of the 864 cm^{-1} line on the tension surface of a Si_3N_4 ceramic test specimen in a four-point bend jig due to a 32-kg load.

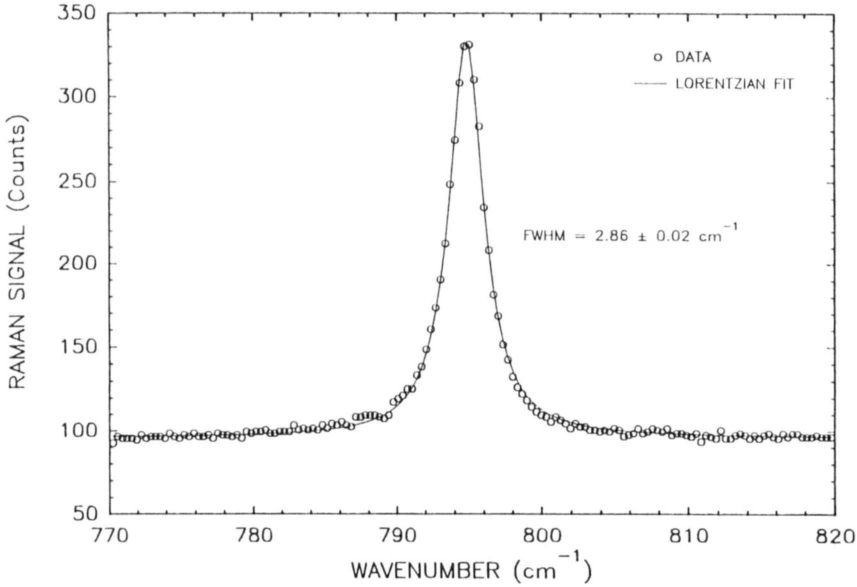

4(a)

Figure 4. Raman TO line of cubic CVD SiC: (a) from a diamond cut, unpolished end of a Group 26 specimen.

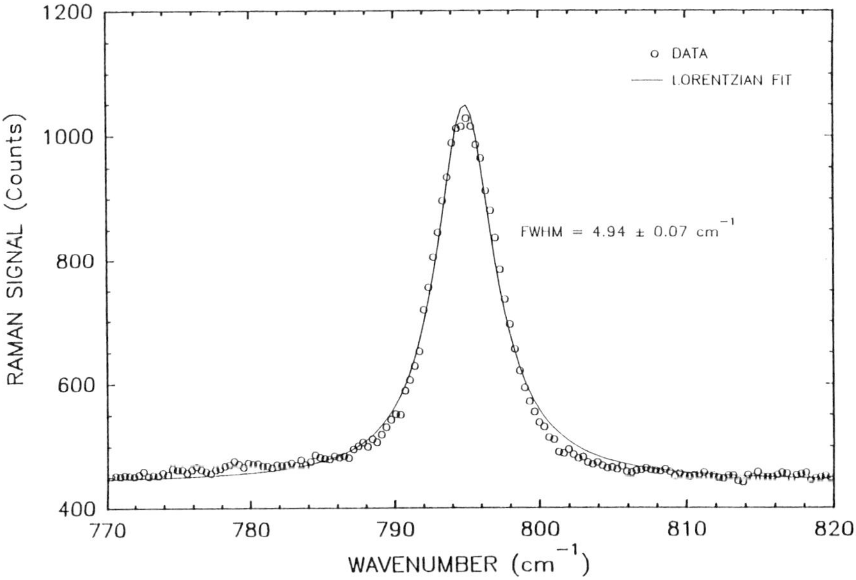

4(b)

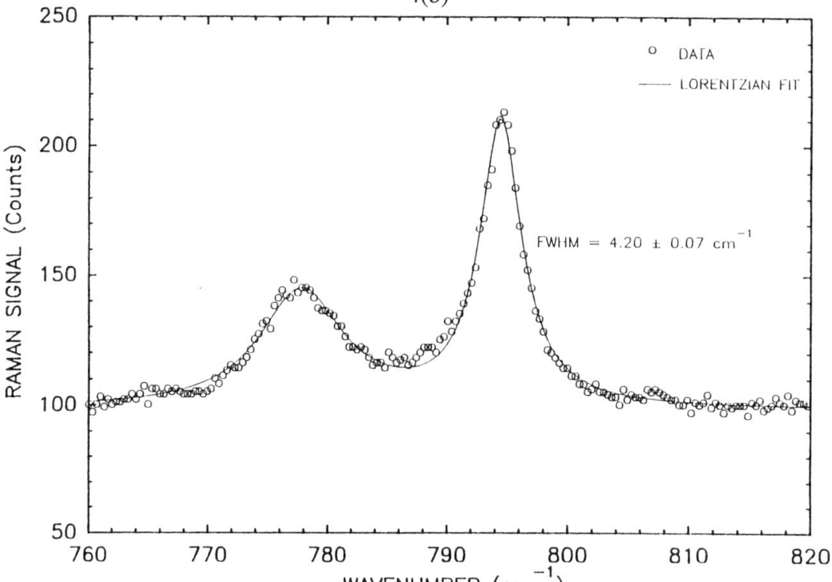

4(c)

Figure 4. Raman TO line of cubic CVD SiC: (b) from the machined surface of a Group 23 specimen (see Table II) and (c) from one edge of a Group 27 specimen showing the presence of bulk strain.

EXPERIMENTAL PLAN

In our study to investigate selected machining effects, three different sets of four-point bend flexural test specimens (3 x 4 x 50 mm - ASTM C1161B) were used for evaluation: (a) four Allied Signal GS-44 Si_3N_4, (b) 12 Morton International CVD SiC, and (c) 11 ORNL provided Si_3N_4.

The different machining methods employed for the three test sets are outlined in Tables I through III. The residual strain in these flexural test specimens machined by different methods was investigated by Raman microprobe spectroscopy. The data obtained were acquired using a diode-laser pumped, doubled Nd:YAG laser. This laser provides 60 mW of highly stable, single-mode CW power at 532 nm. It has a bandwidth of less than 2 MHz. This allows us to acquire high quality data in a relatively short period of time. The spectrometer resolution and CCD array detector pixel size are the factors limiting the spectral resolution of the detection system. This amounts to an effective system bandwidth of about 0.6 cm^{-1}. High SNR and computer fitting techniques allow us to detect small changes in the widths of Raman lines, often less than 0.1 cm^{-1}. The data obtained for each specimen typically represents the average of 20 recorded positions spaced 5 to 15 μm apart.

Table I. Machining Treatments for GS-44

Number	Machining	Organization	Machine
1*	Baseline	Chand Kare Inc.	-
2**	Longitudinal	ORNL	Harrig NC
3	Transverse	ORNL	Harrig NC
4	Longitudinal Creep Feed	ORNL	Nicco

*The baseline specimen was final ground following ASTM C1161 but with a wheel that used a grit size finer 320.

**Final grind followed ASTM C1161.

Table II. Machining Treatments for CVD SiC Specimens*

Group	Wheel Type	Wheel Grit Size (μm)	Removal Rate mm/Pass	No. Passes	Total Amount Removed/Side (mm)
22	Norton SD150 R75B99E-1/4	100	0.005	5	0.025
23	Norton SD320 R75B99E-1/4	45	0.005	40	0.20
24	General Dia. CGD600 R75B-1/4	25	0.0025	10	0.025
25	Norton SD800 R50-1/32	18.5	0.0025	10	0.025
26	Norton ASD800 R75B56-1/4	18.5	0.0025	10	0.025
27	Norton D 3/6 mic R75B-1/4	3 to 6	0.0025	10	0.025

*All specimens longitudinally ground at Chand Kare Inc.

Table III. ORNL Si_3N_4 Specimens*

Billet Designation	Wheel Grit Size		Average Grinding Force (Newton)
	No.	(μm)	
13D21	320	44	38
07B21	80	177	90
06B21	320	44	75
06D21	150	105	93
13A21	500	33	135
07D21	150	105	101
12C22	320	44	28
09B21	600	30	122
03D21	320	44	33
03A21	320	44	65
14C21	500	33	237

*Specimens longitudinally ground at ORNL.

The spectral data were analyzed using a least squares fit of the Lorentzian line shape function to the data. For each data set, the peak height above background, the center frequency position, the full width at half maximum (FWHM), and the background level were fit as free parameters. The background level is designed to work with tilted backgrounds (i.e., it has a slope and y-intercept as free parameters). The fitting uncertainties in the widths were typically 1 to 5 percent. The poor optical quality of some specimens and relatively weak signal strengths account for the higher uncertainties.

Flexure strength was measured for all of the specimens used in these studies. The specimens were measured at 20°C in an Instron Universal Testing Machine 55RL23 following ASTM C1161. The strength values reported represent the average obtained for seven to ten specimens machined by the different methods cited.

Selection of Raman Lines for Analysis

For the GS-44 specimens, the width and the position of the 864 cm^{-1} line was investigated. Due to difficulties with fitting the width and the position of the 864 cm^{-1} line, the study was divided into two parts. In part one, the linewidths of 864 cm^{-1} line were measured. In part two, the positions of the lines with respect to a reference line of nitromethane were determined. In previous studies on Si_3N_4 [2,7], it was found that the 864 cm^{-1} line shifted up with applied compressive stress. The position measurements on the GS-44 specimens showed that the positions were directly correlated with the linewidth broadening. This confirmed the expectation that the broadening was due to residual compressive strains introduced by specimen machining.

The spectra from GS-44 Si_3N_4 specimens showed at least five additional lines that were absent from the spectrum of a typical β-Si_3N_4 specimen, such as Norton/TRW NT-154 Si_3N_4 (see Figure 5). These extra lines are present in the spectrum of α-Si_3N_4 and correlate with the data of Wada et al. [8]. The spectra suggests that these specimens are a mixture of α- and β-phases.

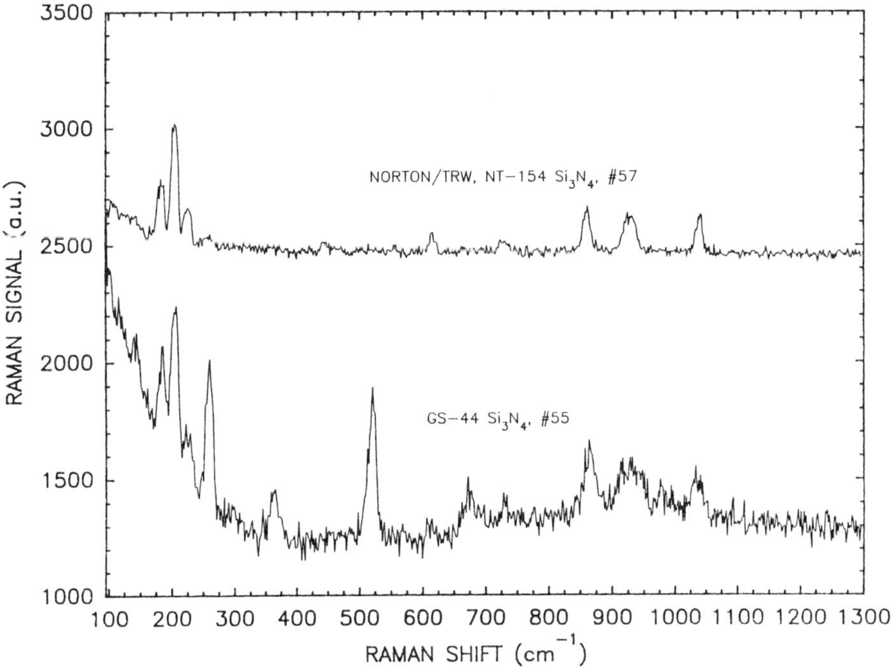

Figure 5. Survey Raman spectra of Norton/TRW NT-154 Si_3N_4 (upper trace) and GS-44 Si_3N_4 (lower trace).

For the CVD SiC specimens, the transverse optical (TO) mode at 795 cm^{-1} was studied. Two specimens from each of the groups were evaluated. Each specimen was probed on the two surfaces near one edge of the long dimension. All of the specimens probed showed a splitting of the TO line along the edge. The largest splittings were observed in specimens from Group 27 (see Figure 4c). Since both splitting and line broadening depend on the character and intensity of a bulk strain, correlations between the various parameters of the two split lines were studied. These correlations supported the contention that the splitting was due to residual bulk strain introduced prior to machining.

For the ORNL Si_3N_4 specimens, the Raman line at 449 cm^{-1} was used for evaluating line broadening. For these specimens, the Raman line for free silicon at 523 cm^{-1} was found and also used in this study. To better compare the results in this phase of the study, a baseline spectrum was obtained using an unstrained silicon wafer. The silicon nitride Raman line at 449 cm^{-1} and the silicon Raman line at 523 cm^{-1} for each specimen were measured. In the past, the Raman line at 864 cm^{-1} was used for the measurements on Si_3N_4. For the current specimens, the Raman line at 864 cm^{-1} was very broad and very weak. The broadening and weakening of this line is believed to be due, in part, to the presence of both α- and β-Si_3N_4 in the specimens. Therefore, a series of measurements were made to determine if there were any other Raman lines with the same or better sensitivity as that of the 864 cm^{-1} Raman line. The measurements were carried out on two specimens which were chosen because they were representative of the coarsest (80) and finest (600) grinding grits. The findings indicated that the sensitivity of the 449 cm^{-1} Raman line to residual strain was approximately the same as the 864 cm^{-1} Raman line. The signal levels from these two lines were quite similar, and the 449 cm^{-1} line was narrower and free from any interferences.

DISCUSSION OF RESULTS

The Raman line broadening measured for each category of flexural test specimens, the average flexural strength, and the corresponding machining variables used are compiled in Tables IV through VI.

Table IV. Summary Results for GS44*

Specimen No. and Machining Procedure	Raman Line Width (cm^{-1})	Average Flexural Strength (MPa)
#1 longitudinally ground at Chand Kare Inc.	13.68 ± 0.61	1160 ± 162
#2 longitudinally ground at ORNL	15.46 ± 0.81	1135 ± 55
#3 creep feed ground at ORNL	13.63 ± 1.29	1183 ± 71
#4 transversely ground at ORNL	15.61 ± 0.75	871 ± 71

*All specimens probed at ~0.7 mm from edge.

Table V. Summary Results for CVD SiC

Specimen Group No.	Grit Size (μm)	Raman Linewidth (cm^{-1})	Average Flexural Strength (MPa)
22	100	3.20 ± 0.19	352 ± 29
23	45	3.45 ± 0.15	383 ± 23
24	25	3.11 ± 0.10	454 ± 17
25	18.5	3.05 ± 0.15	389 ± 32
26	18.5	2.87 ± 0.14	478 ± 52
27	3-6	2.83 ± 0.12	297 ± 65

Table VI. Summary Results for ORNL Si_3N_4

Billet Number	Wheel Grit No.	Average Grinding Force (Newton)	Average Flexural Strength MPa (m)*	Average Raman Linewidth	
				Si_3N_4**	Si***
07B	80	90	571 (26)	4.35 ± 0.08	4.52 ± 0.04
07D	150	101	610 (24.4)	4.10 ± 0.04	3.82 ± 0.06
06D	150	93	700 (18.3)	4.24 ± 0.05	3.73 ± 0.08
03D	320	33	672 (30.2)	3.95 ± 0.03	3.66 ± 0.06
12C	320	28	758 (24.7)	4.10 ± 0.04	3.50 ± 0.08
06B	320	75	802 (47.9)	4.27 ± 0.04	3.76 ± 0.08
03A	320	65	816 (12.2)	3.14 ± 0.04	3.28 ± 0.07
13D	320	38	893 (26.4)	4.42 ± 0.05	3.68 ± 0.08
13A	500	135	892 (16.7)	4.15 ± 0.07	3.60 ± 0.07
14C	500	237	989 (30.9)	4.15 ± 0.04	3.79 ± 0.10
09B	600	122	1016 (20)	4.00 ± 0.09	3.77 ± 0.09

*(m) = Weibull modulus

**449 cm^{-1} line

***523 cm^{-1} line

GS-44

The GS-44 specimens were probed at two locations, one close to the chamfered edge and one at some distance from the edge (typically about 0.7 mm). These measurements were taken to determine if there was any difference between the tensile surface and the chamfer. The linewidths tended to be smaller at the chamfered site due to less machining.

The Raman linewidths given in Table IV for the GS-44 specimens were unusually large. This effect is believed to be due, in part, to the presence of the α-Si_3N_4 phase. Additionally, the 848 cm^{-1} line of α-Si_3N_4 interferes with the 864 cm^{-1} line of the β-phase to give a doublet. Moreover, energy dispersive spectroscopy (EDS) of these specimens showed the presence of free silicon, magnesium, iron, aluminum, and chlorine. The presence of a sufficient concentration of these impurities can cause stress in the test specimens, which can contribute to the observed linewidths. Nevertheless, the data in Table IV show that, as expected, the linewidth is independent of the machining direction and that the finer machining procedures produce the narrower lines and the highest strengths. Since narrower lines imply lower residual strain, this correlation indicates that machining can degrade the ceramic specimens and reduce their strength. Thus, it is observed that the flexural strength tends to be inversely related to the residual surface strain caused by machining effects.

CVD SiC

The high symmetry of the cubic lattice of CVD SiC gives a Raman spectrum with a two-fold degenerate TO line at 795 cm^{-1} and singly-degenerate longitudinal optical (LO) line 975 cm^{-1}. If the lattice symmetry of the CVD SiC is strongly deformed during processing or by the presence of residual strain, lattice symmetry can be lowered. The lower symmetry can cause additional lines to appear with the possible removal of the degeneracies of existing lines. Splitting of the TO line was observed, as described in the Experimental Plan section. At the same time, the LO line showed no splitting, as expected. This splitting was found on both ground and cut surfaces indicating that this behavior was independent of the machining indicating that it was due to a residual bulk stress and the corresponding strain introduced during the growth of the material.

A bar graph showing the mean and standard deviations (of the mean) of the linewidth data measured for the SiC specimens from the different test groups is presented in Figure 6. The bars are plotted in the order of grinding fineness. The two bars on the extreme right side are for etched specimens from Group 23 and Group 27, respectively. Except for Group 23, the linewidths in Figure 6 show decreasing strain for the finer grit grinding wheels. Group 23 has the widest linewidth [see Figure 4(b)] and the highest residual surface strain which is attributed to the 40 passes used in the finishing process (four to eight times those of the other groups). The high number of passes introduced significantly more strain in the surface of the test specimens resulting in a reduced flexural strength, as shown in Table V.

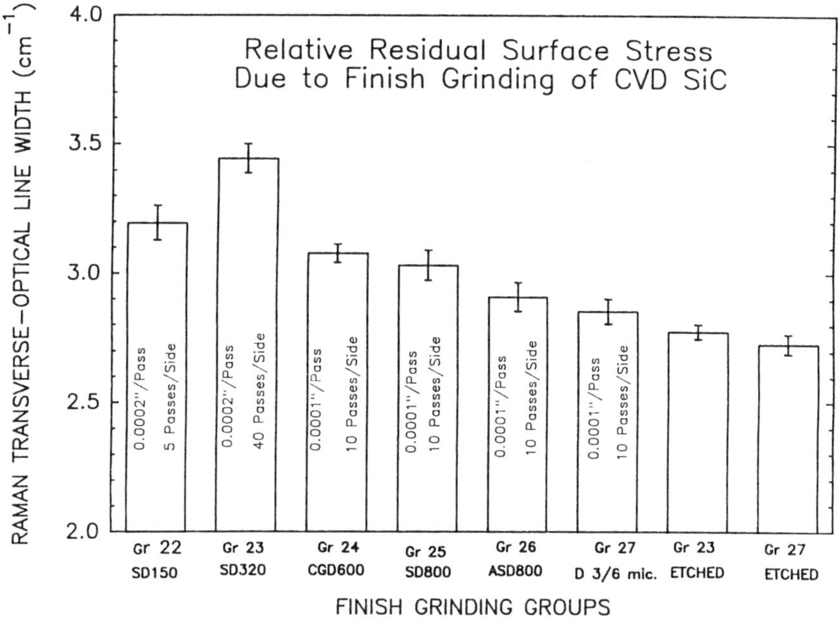

Figure 6. Average linewidths of the 795 cm^{-1} line of CVD SiC showing relative residual surface stress due to the machining treatments given in Table III.

In order to better determine the source of line broadening, specimens from each group were etched to remove the surface layer. The widths for the etched specimens were lower than the widths measured for the machined specimens. These results strongly support the conclusion that the variation in linewidth is due to residual surface strain introduced by the grinding process. Currently, no theoretical model or data directly relating linewidth to actual residual strain values is available. Empirically, however, it is observed that line broadening due to residual strain is accompanied by line shifts. Since line shifts can be theoretically related to strain [7] and experimentally to stress [2], then it may be possible to calibrate line broadening in terms of residual stress for specific lines.

From measurements of the splitting of the 795 cm^{-1} line, it was found that the specimens from Group 27 showed the largest TO splitting. These findings suggest that the specimens from Group 27 contain considerably more bulk strain than the remaining specimens. This large bulk strain is seen as the reason for the low flexural strength reported in Table V for Group 27. Moreover this bulk strain may also be responsible for the scatter in the flexural strengths across the remaining groups.

ORNL Si_3N_4

The results from the ORNL Si_3N_4 were more difficult to evaluate. Although the flexural strength increased with the finer grinding treatment, the line broadening for both the Si and Si_3N_4 Raman lines did not show similar trends. The presence of free-Si with its strong Raman line at 523 cm^{-1} was not expected for this material. From the Raman data it was determined that the free-Si represents less than three percent of the bulk. The location of the Si in the Si_3N_4 could not be determined from scanning electron microscope (SEM) micrographs.

The implication is that the presence of the free and 'softer' silicon greatly complicates the degree to which strain is introduced into the Si_3N_4 grains. Within a given grit size, the Si linewidths follow (with one exception) the grinding force. Moreover, the largest two Si linewidths occur with the coarsest grits.

CONCLUSIONS

Laser Raman microprobe spectroscopy provides a means for detecting the presence of strains in solid specimens. This strain is revealed typically by broadening, shifting, or splitting of Raman lines. The presence of residual surface strain introduced by machining treatments is revealed in the broadening of the Raman lines. Studies show that this broadening is accompanied by slight line shifts. Systematic measurements of shifts of specific Raman lines due to applied stress show that these line positions increase with applied compressive stress. The line broadening due to machining is directly correlated with line positions which confirms that the residual surface strain is compressive.

The measurements required to reveal residual surface strains are accomplished by time-consuming, high-resolution Raman spectroscopy. There is considerable need for enhanced data acquisition techniques in order to begin the development of a larger database that includes a variety of materials. This database should include correlations with mechanical behavior and fractography.

ACKNOWLEDGMENTS

The authors would like to acknowledge the support of Dr. D. R. Johnson and Dr. M. E. Ferber of Martin Marietta at ORNL and the work of our graduate students N. D. Wolfe and M. Arif. We would also like to acknowledge the technical support of D. E. McCullum, D. W. Grant, and J. D. Wolf.

REFERENCES

1. M. Lander, S. Pruchnic, and P. P. Yaney, "Application of Laser Raman Macro-Microprobes Using Cylindrical Optics to the Study of Zirconia Ceramic Materials," *Bull. Am. Phys. Soc.* 32:2163, 1987.
2. P. P. Yaney and S. L. Ernst, "Investigation of Strain Mapping in Silicon Nitride Ceramic Using a Laser Raman Microprobe," *Bull. Am. Phys. Soc.* 37(9):1892, 1992.

3. R. S. Krishnan, "Historical Introduction," Chapter I in The Raman Effect, Vol. 1: Principles, edited by A. Anderson (New York: Marcel Dekkar, Inc.), p. 2, 1971.
4. P. M. A. Sherwood, Vibrational Spectroscopy of Solids (London: Cambridge University Press), p. 3, 1972.
5. S. Nakashima and M. Hangyo, "Characterization of Semiconductor Materials by Raman Microprobe," *IEEE J. of Quantum Electronics* 25(5):965, 1989.
6. J. Rydzak and W. R. Cannon, "Fourier Transform Infrared Microscopy Method to Determine Stress in Sapphire," *J. Amer. Ceram. Soc.* 72(8):1559, 1989.
7. T. D. Fang and P. P. Yaney, "A Numerical Simulation of the Stress-Induced Shifts in the Raman Spectrum of β-Si_3N_4 Ceramic," *Bull. Am. Phys. Soc.* 38:(7):1663, 1993.
8. N. Wada, S. Solin, J. Wong, and S. Prochazka, "Raman and IR Absorption Spectroscopic Studies on α, β, and Amorphous Si_3N_4," *J. Noncrystalline Solids* 43:7, 1983.

FRACTOGRAPHY, FRACTURE MECHANICS AND FRACTAL GEOMETRY: AN INTEGRATION

J. J. Mecholsky, Jr.
Department of Materials Science and Engineering
University of Florida
Gainesville, FL 32611

ABSTRACT

It is apparent from observations of fracture surfaces that there are features which appear self-similar and scale invariant These features can be related to other properties such as fracture toughness, elastic modulus and atomic bond lengths. Fractography identifies the useful features created during the fracture process. With the use of fracture mechanics, quantitative fractography relates the crack sizes and surrounding topography to the failure stress and energy released during fracture. Fractal geometry is a non-Euclidean geometry which is characterized by self-similar and scale invariant structures and which can be used to quantitatively relate the tortuosity of the surface to the fracture toughness. The combination of fractography, fracture mechanics and fractal geometry, then, can be used to relate the creation of the surface during bond rupture and crack propagation to the fracture energy. The purpose of this paper is to show the relationships between fractography, fracture mechanics and fractal analysis.

INTRODUCTION

Observations of the fracture surface can relate to us something of the process of brittle fracture. Ravi-Chandar and Knauss[1] noted the self-similarity of fracture surfaces on PMMA although they did not use the term "self-similar." Poncelet[2] showed that the "mirror" region on glass surfaces was not smooth, but rather showed features which appeared to be similar in geometry to mist features. The characteristic features on brittle fracture surfaces contain information about the stress at failure, the size and type of defect causing fracture and the nature of the fracture process[3]. These characteristic features appear on glasses, glass ceramics, polycrystalline ceramics and single crystals[4]. Dickinson and co-workers[5] have shown that particles are emitted during crack propagation. For example, light is emitted on most of the materials examined. The light bundles are emitted in a set (fractal) pattern. Single crystal silicon has features reminiscent of the hackle regions in glasses[6]. The hackle features are self-similar to other features on the surface and are scale invariant, as well, i.e., magnified regions appear the same as smaller magnified regions. Of course, macroscopic crack branching can occur as well. Sakai et al.[7] showed that the branching traces can be represented by a fractal pattern.

It is apparent from observations of the fracture surface over the years that there are features which appear self-similar and scale invariant[8]. The significance of this is that using fractography and fracture mechanics principles, these features can be related to other properties such as fracture toughness, elastic modulus and atomic bond lengths. This knowledge then will help described the creation of the surface during bond rupture and crack propagation. The purpose of this paper is to show the relationships between fractography, fracture mechanics and fractal analysis. Using these principles, I will show that there are useful relationships which can aid in analysis of fractures and help describe the fracture process from a fundamental viewpoint.

FRACTURE MECHANICS

The principles of fracture mechanics have been used to analyze the fracture surfaces of brittle materials for many years[9]. An excellent review of the applications is presented in the ASM Handbook of Ceramics and Glasses.[10.] The analysis is based on the fundamental equation of fracture mechanics which relates the stress at failure, σ_f, and the size of the fracture initiating (sharp) crack, c, to the toughness of the material in the form of the critical stress intensity factor, K_{IC}:

$$K_{IC} = Y \sigma (c)^{1/2} \qquad (1)$$

where Y is a geometric and load factor which accounts for the shape of the fracture-initiating crack and the loading condition. For small cracks relative to the thickness, which fail in either tension or flexure, the Y factor is ~1.24 if the crack is the equivalent semi-circular crack of radius c [$c = (a\,b)^{1/2}$ where a is the semi-minor axis and b is the semi-major axis of an elliptical crack.]. Randall[11] describes the details of being able to approximate non-ideal crack shapes to those that can be analyzed. Notice that ASTM defines toughness[11] in either of three ways, as K_{IC} noted above, G_{IC}, the strain energy release rate or γ_f, the fracture energy. Thus, even though G_{IC} and γ_f have different units (those of energy, usually J/m^2) than those of critical stress intensity, they are all expressions of toughness.

The application of equation (1) can be quite useful if properly applied. Generally it can be misleading to use the equation to calculate the crack size. In the instances where this equation has been used in this fashion, the fracture toughness or stress was not as well known as suspected. It is much more useful to measure the crack size and either calculate the stress at fracture or toughness knowing the other parameter, respectively[13].

FRACTOGRAPHY

Fractography has been used for years to identify the cause of failure.[13,14] It has been used occasionally to aid the understanding of the fracture process. However, the potential for determining the sequence during the fracture process is great and should be utilized in the future. There are some very useful and comprehensive summaries of fractography in the literature,[4,15-17] so I will not reiterate these here. However, what I want to describe are the equations which are often not used to their full potential during analysis of field and laboratory failures.

The stress at failure can be related to the radius of the mirror-mist, r_1, mist-hackle, r_2, or crack branching, r_3, boundaries measured from the origin of fracture through a fracture mechanics-type equation :[12]

$$K_{bj} = Y \sigma r_j \qquad (2)$$

where K_{bj} are the crack branching stress intensities associated with each of the boundaries, rj, respectively, i.e., j=1.2 or 3, Y is a geometric and load factor similar to that in EQ (1), but can be a different value.[18-20] Notice that the form of Eq. (2) is similar to Eq.(1). Thus, the relationships are similar in the condition necessary to start a crack to propagate and that of the condition to cause it to deviate out of the plane perpendicular to the principal tensile stress. Also note that Eq. (2), which is really a set of three equations, is valid for isotropic materials. In the case of anisotropic materials, such as single crystals, the criterion for mirror boundary formation is a critical value of a fracture energy that appears to be a multiple value of the critical fracture energy to cause the crack to propagate.[21] Further notice that the existence of Eqs (1) and (2) implies that the crack-size-to-mirror-size ratios are constant, i.e., $c/r_j = A_j$. This relationship is quite useful, because if an origin is difficult to identify, the equivalent semi-circular crack size can be determined from the size of the mirror boundary.[12,22] It is really the equivalent size, c, that is important in any analysis . The effective size is that which affects strength and fracture behavior.

FRACTAL GEOMETRY

Fractal geometry is a non-Euclidean geometry which describes irregular shapes and is often characterized by self-similar or self-affine scaling and by scale invariance.[23] A self-similar surface is one in which the scaling is isotropic, i.e., one feature on the surface is similar to another feature on the surface in any direction. If this characteristic holds for different magnifications, then this trait is called scale invariant. Although there is some controversy in the literature about the nature of fracture surfaces,[24-28] i.e., whether or not they are self-similar or self-affine, I will argue in this paper that there are some fracture surfaces which are self similar and some which are self affine. Fracture surfaces which are formed in structures which are considered isotropic should be self-similar. Self-similar fracture surfaces would be expected with glasses and fine-grained polycrystalline ceramics.[29] Natural and man-made composites can be either self-similar or self-affine. Cements , rocks, such as westerly granite, and geological minerals would be expected to be self affine.[30] Single crystals and glass ceramics could be either.[6] The surfaces that I will describe here are considered self-similar. Self-similar surfaces are described by their fractal dimension, D. For surfaces, the dimension will lie between 2 and 3. A value close to two would be relatively smooth, and a value close to three would be near volume filling and appear very tortuous.

INTEGRATION

It has been observed in the past that fracture surfaces of tougher materials appeared rougher.[4,31] Fractal dimensions are a convenient numerical measure of this roughness. It is unreasonable to expect one universal value to describe all fracture surfaces, as proposed by some investigators[32]. Several other investigators have shown a direct correlation between the toughness and the fractal dimension of self-similar fracture surfaces during brittle fracture.[29,33-36] What will be described here is the relationships that can be developed and which are useful in failure analysis.

Mecholsky and Freiman[37] showed that the flaw-to-mirror size ratio is equal to the fractal dimensional increment, D*:

$$c/r_1 = D^* \qquad (3).$$

Figure 1 is a graph of equation (3) for some of the materials listed in Table 1. [For sake of clarity, only the mantissa of the fractal dimension will be considered in evaluating mechanical properties; it is the fractal dimensional increment which contains the structural geometric information.] This relationship implies that the scaling which exists in approximately the plane of fracture (perpendicular to the principal stress direction) is related to the scaling in the perpendicular direction. Thus, the amount of tortuosity is directly related to the energy involved in the fracture, since the mirror radius is related to the stress (and strain energy) during fracture. The stress level is directly related to the stored energy (through the elastic modulus.)

Table 1 - Material and Fractal Properties

MATERIAL	K_{IC}(MPam$^{1/2}$)	E(GPa)	D*	a_0(nm)
Silica Glass	0.73	70	0.11	1.0
Borosilicate Glass	0.71	64	0.09	1.4
$3BaO\text{-}5SiO_2$ Glass	0.72	64	0.1	1.3
Pyroceram Glass Ceramic	2.5	120	0.13	3.4
Zinc Silicate Glass Ceramic	1.6	90	0.05	6.3
Lithia Disilicate G.C.	3.0	137	0.24	2
Lithia Borosilicate G.C.	2.7	90	0.18	5
$3BaO\text{-}5SiO_2$ G.C.	1.8	90	0.22	1.8
Li2O-2SiO2 G.C. (A)	1.3	72	0.18	1.8
Li2O-2SiO2 G.C. (B)	3.0	137	0.24	2.0
Alumina (AD 99)	2.9	390	0.21	0.3
Alumina (AD999)	3.9	406	0.31	0.3
Alumina (GEND)	3.9	283	0.23	0.8
Alumina (UCC)	2.5	393	0.15	0.3
Zinc Selenide	0.9	69	0.3	0.4
S. C. Silicon (100){110}	1.23	169	0.09	0.6
S. C. Calcium fluoride	0.3	114	0.07	0.1
S. C. Spinel	1.2	240	0.09	0.3

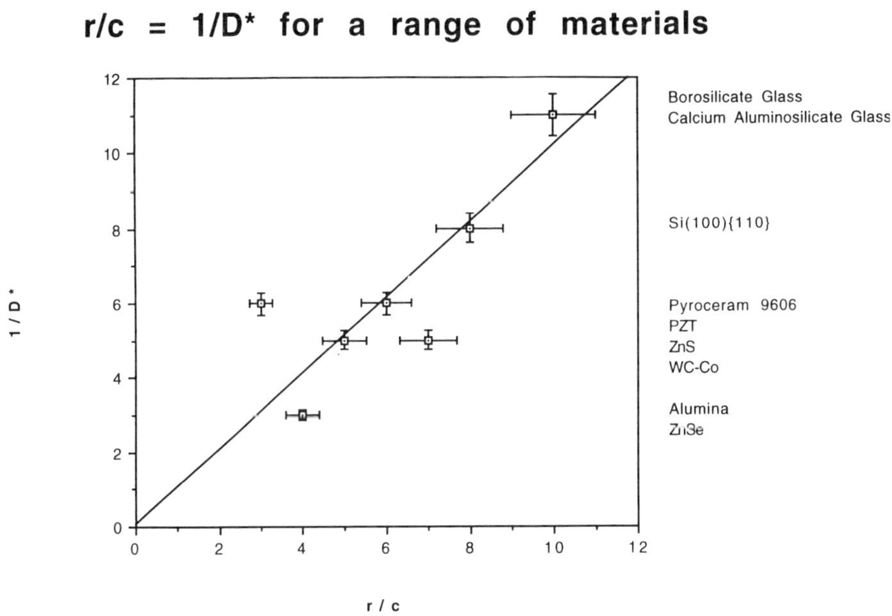

Figure 1 - 1/D* As A Function of r/c. Notice that for a range of materials D* = c/r.

The relationship between the toughness in the form of the critical stress intensity factor, K_{IC}, and the fractal dimensional increment, D*, was experimentally found to be:

$$K_{IC} = K_0 + [(1/2)\, a_0 \;\; E]\, D^{*1/2} \qquad (4)$$

where E is the elastic modulus, a_0 is a characteristic atomic dimension, and K_0 allows for the possibility of Euclidean (flat) fracture. The implications of this relationship will be discussed in the following paragraphs, but first another representation provides more insight into the meaning of the relationship between fractography, fracture mechanics and fractal analysis. If we assume that a plane-stress condition exists during fracture, then the fracture energy, γ_f, can be related to D*:

$$\gamma_f = (a_0/2)\;\; ED^* \qquad (5)$$

If we graph the fracture energy as a function of the elastic modulus times the fractal dimensional increment, the slope is $(a_o/2)$. Figure 2 is an example of the graph for many glass ceramics. Table 2 shows a set of values for a_o for several classes of materials. Notice that single crystals have values between 1-10 A, and glasses have values between 10 & 20 A, glass ceramics have values between 20 and 80 A, and fine-grain polycrystalline materials have values between either 1 and 5 or 10 and 20 A. Although the exact meaning of a_o is not known at this time, there can be a reasonable hypothesis of the meaning of these values. Single crystals fracture due to the breaking of bonds between atoms, so that a_o would be reasonably assumed to be the average stretched bond distance just before fracture. In some cases, this should enable us to determine which bonds are fracturing first. In the case of glasses, the a_o value could represent the average free volume in the vicinity of the crack tip at the time of fracture. This interpretation is consistent with molecular dynamics modeling of fracture by Simmons et al.[38] and Tsai et al.[39] For glass ceramics, the results shown in Table 2 and Figure 2 can be interpreted to mean that these materials fail by the fracture of glass-crystal clusters. The polycrystalline materials fall into two categories: bond breaking of crystals and bond breaking of the glassy grain boundaries. The values of a_0 between about 1 and 5 A represents the breaking of crystalline bonds as in the case of single crystals. The values of a_o between about 10 and 20 A represents those cases in which fracture occurs from the glassy grain boundaries and as such fracture just as the glass phases, i.e., from the free-volume regions.

Table 2 - D* and a_0 Values for Different Classes of Materials

Material Class	D*	a_0	Interpretation+
Glasses	0.07-0.1	10-20	free volume
Glass Ceramics	0.06-0.3	20-80	glass-crystal cluster
polycryst. ceramics	0.06-0.35	3-20	stretched bonds or glass phase
single crystals	0.07-0.12	1-10	stretched bonds

+ $\gamma_f = (a_0/2)\, ED^*$

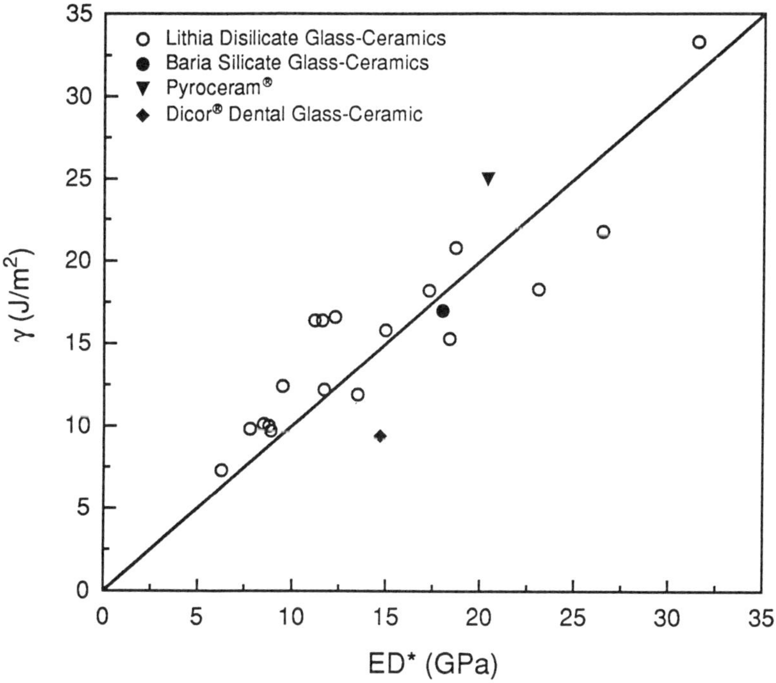

Figure 2 - Fracture Energy,γ, As a Function Of The Elastic Modulus,E, Times The Fractal Dimensional Increment, D*, For Glass Ceramics.

CONCLUSIONS

Fractography, fracture mechanics and fractal geometry are related to each other. This fact is evident from observing that the flaw-size-to-mirror-size ratio is the fractal dimensional increment and the fractal dimensional increment is directly related to the fracture energy which can be shown to be related to the critical stress intensity factor. Fractography relates the fracture surface features which have been shown to be self-similar for many ceramics and intermetallics, to the stress at failure and the amount of energy during failure.

REFERENCES

1. K. Ravi-Chandar and W. G. Knauss, "An Experimental Investigation into Dynamic Fracture: II. Microstructural Aspects," Inter. J. Fracture, 28 65-80 (1984).
2. E. F. Poncelet, "The Markings on Fracture Surfaces," Trans. Soc. Glass Tech., 52 279-288 (1958).
3. J. J. Mecholsky, Jr."Quantitative Fractography: An Assessment" in Ceramic Trans. 17, Fractography of Glasses and Ceramics II. Frechette and Varner eds., Am. Cer. Soc., Westerville, OH, 413-451 (1991).
4. R. W. Rice, "Ceramic Fracture Features, Observations, Mechanisms and Uses," in Fractography of Ceramic and Metal Failures, ASTM STP 827, J. J. Mecholsky, Jr. And S. R. Powell, Jr., eds. American Soc. For Testing and Materials, Phila. PA, 1 - 100 (1984)
5. S. C. Langford and J. T. Dickinson, "The Emission of Particles and Photons from the Fracture of Minerals and Inorganic Materials," in ACS Symposium Series: Spectroscopic Characterization of Minerals and Their Surfaces, L. M. Coyne, et al., eds., 415 224-244 (1990).
6. Y. L. Tsai and J. J. Mecholsky, Jr., "Fractal Fracture of Single Crystal Silicon," J. Mater. Res. 6, 1248 - 1263 (1991).
7. T. Sakai, M. Ramulu A. Ghoshand R. C. Bradt, "A Fractal Approach to Crack Branching (Bifurcation) in Glass, in Ceramic Transaction 17 op. Cit. (1991)
8. J. J. Mecholsky, Jr., T. J. Mackin and D. E. Passoja, "Self-similar Crack propagation in Brittle Materials," Adv. In Ceramics 22: Fractography of Ceramics and Glasses Varner and Frechette, eds.,The Am. Ceramic Soc (1988).
9. S. W. Freiman, J. J. Mecholsky, Jr., and P. F. Becher, "Fractography: A Quantitative Measure of the Fracture Process," in Ceramic Transactions 17, op. Cit., 55-78 (1991).
10. I. Bar-on, Applied Fracture Mechanics, pp. 645-651 in Engineered Materials Handbook (Ceramics and Glasses) ASM International The Materials Information Society (1991).
11. P. N. Randall, "Plain Strain Crack Toughness Testing of High Strength Metallic Materials", in ASTM STP 410, pp. 88-126 (1967).
12. ASTM Annual Book of Standards, E616, p.674, The American Society of Testing and Materials, 1916 Race St., Phila. PA 19103 (1984).
13. J. J. Mecholsky, Jr., Quantitative Analysis of Fracture Surfaces, Cer Trans, Am. Cer. Soc. (1992).
14. e.g., F. W. Preston, "A Study of the Rupture of Glass," J. Soc.of Glass Tech. 10 234-269 (1926).
15. V. D. Frechette, Fractography of Glass and Ceramics, The American Ceramic Soc. (1990).
16. G. D. Quinn, J. J. Swab and M. J. Slavin, "A Proposed Standard Practice for Fractographic Analysis of Monolithic Advanced Ceramics," MTL TR 90-57, Army Materials Laboratory, NTIS Access No. ADA-231989 (1990).
17. R. W. Rice, "Perspective on Fractography," Adv. In Ceramics 22, in Fractography of Glasses and Ceramics Varner and Frechette, eds. The Am. Ceramic Society, Inc. 3-56 (1988).
18. H. P. Kirchner and J. C. Conway, Jr.,"Fracture Mechanics of Crack Branching in Ceramics," in Fractography of Glass and Ceramics, Advances in Ceramics, Vol. 22, The American Ceramics Society, Westerville, OH, pp. 187-213 (1988).
19. J. C. Newman and I. S. Raju," Analysis of Surface Cracks in Finite Plates under Tension and Bending Loads," NASA TP 1578 (1979).

20. D. Broek, Elementary Engineering Fracture Mechanics, Sijthoff and Nordhoff, The Netherlands (1978).
23. B. B. Mandelbrot, "The Fractal Geometry of Nature," W. H. Freeman Co., New York (1982).
24. E. Bouchaud, G. Lapasset, J. Planes and S. Naveos, "Statistics of branched fracture surfaces" Phys. Rev. B (1995)
25. A. V. Neimark, "Characterization of Self affinity in the global region" Phys. Rev B 50 [20]435-438 (1994)
26. C. Y. Poon, R. S. Sayles and T. A. Jones," Surface Measurement and Fractal Characterization of Naturally Fractured Rocks, J. Phys. D:Appl. Phys. 25 1269-1275(1992).
27. B. Dubuc, J. F. Quiniou, C. Roques-Carmes, C. Tricot and S. W. Zucker, "Evaluating the Fractal Dimensions of Profiles", Phys. Rev. A39(3) 1500-1512(1989).
28. A. V. Neimark and K. K. Unger, "Method of Discrimination of Surface Fractality," J. Colloid. & Interface Sci. 158, 412-419 (1993).
29 J. J. Mecholsky, Jr. , D. E. Passoja and K. C. Fineberg-Ringel, "Quantitative Analysis of Brittle Fracture Surfaces Using Fractal Geometry, J. Am. Ceramic Society (1990)
30. P. A. Scott, T. Engelder and J. J. Mecholsky, Jr., "The Correlation Between Fracture-toughness Anisotropy and Crack Surface Morphology in the Ithica Formation, Appalachian Basin," in Fault Mechanics and Transport Property of Rocks, ed. B. Evans and T. Wong, Academic Press NY 341-370 (1992)
31. R. W. Rice, "Topography of Ceramics," in Surfaces and Interfaces of Glass and Ceramics, eds. V. Frechette, W. LaCourse and V. Burdick, pp. 439-472, Plenum Press (1974).
32. E. Bouchaud, G. Lapasset, J. Planes, Europhys. Ltrs. 13, 73 (1990).
33. Y. Fahmy, J. C. Russ and C. C. Koch, "Application of Fractal Geometry Measurements to the Evaluation of Fracture Toughness of Brittle Intermetallics. J. Mater. Sci. 6, 1856-1861 (1991).
34. J. J. Mecholsky and T. J. Mackin Fractal Analysis of Fracture in Ocala Chert J. Mater. Sci Ltrs. 7, 1145-47 (1988).
35. J. Y. Thompson, J. J. Mecholsky and K. J. Anusavice," The Effect of Microcracking on the Toughness and Fractal Dimension of Lithia Disilicate Glass Ceramics," J. Am. Ceramic Soc. Accepted for Pub. (1995).
36. B. B. Mandelbrot, D. E. Passoja and J. Paullay, Fractal Character of Metals Nature 308:721-722.
37. J. J. Mecholsky, Jr. And S. W. Freiman, " Relationship Between Fractography and Fractal Geometry," J. Am. Cer. Soc.74[12]3136-38 (1991).
38. J. H. Simmons, T. P. Swiler and R. Ochoa, J. Non-Crystalline Solids 134, 179-182 (1991).
39. Y. L. Tsai, T. P. Swiler, J. H. Simmons and J. J. Mecholsky, Jr., Fractal Analaysis of Molecular Dynamics Simulations of Brittle Fracture, Peer Reviewed Proceedings of the Theoretical Modeling Symposium at the Annual Meeting of the American Ceramic Society in Cincinnati during May 1995.

Fractal Analysis of Fracture Surfaces Using the Patchwork Method

Isa Bar-On and Christopher A. Brown
Associate Professors
Mechanical Engineering Department
Worcester Polytechnic Institute
Worcester, MA 01609-2280

William A. Johnsen
Manufacturing Engineering Program
Mechanical Engineering Department
Worcester Polytechnic Institute
Worcester, MA 01609-2280

Anthony M. Calomino and David N. Brewer
NASA Lewis Research Center
21000 Brookpark Road
Cleveland, OH 44135

presented at:
the Conference on Fractography of Glasses and Ceramics III
9-12 July 1995
New York State College of Ceramics
Alfred University, Alfred, New York

ABSTRACT

Topographic data sets have been acquired by confocal microscopy from surfaces of silicon nitride fractured in stable and unstable manners. The patchwork method of fractal analysis calculates the actual surface area as a function of the scale of measurement by repeated, virtual tiling of the surface, represented by the topographic data set, with triangular patches. The triangle represents the scale of measurement, or observation, and with each repetition of the virtual tiling exercise a new triangle size is used. The tendency is for the smaller triangles to measure greater areas. The intermediate result is a scale-area plot: relative areas (measured divided by projected areas) versus corresponding triangular patch areas, on log-log axes. The plots are characterized using smooth-rough crossover scales, the scale above which the relative areas are close to one and the surfaces appear smooth, and below which the log of the relative areas increase in a regular fashion with respect to the log of the triangular areas. At scales below the smooth-rough crossover scale the plots are characterized by the slope, from which a patchwork fractal dimension can be determined. There appears to be some potential to use the slopes as a means of discerning, on the fracture surface, regions of stable from regions of unstable crack growth.

1. INTRODUCTION

The objective of this work is to advance our understanding of fracture phenomena in ceramics by determining scale-area relations and crossover scales on several portions of fractured surfaces of silicon nitride. This is a preliminary investigation to test the patchwork method (Brown et al. 1993) for this type of fractal analysis of topographic data sets acquired from fracture surfaces using confocal microscopy. In this preliminary investigation crossover scales will be determined, as well as the complexity of the geometries limited by these scales. The crossovers and complexities of several fracture regions on silicon nitride for stable and unstable crack growth are determined.

In metallic materials different types of fracture - cleavage, ductile or fatigue fracture - can usually be distinguished by careful fractography. In ceramic materials the difference in the fracture appearance for different fracture regimes such as unstable fast fracture versus stable slow fracture can sometimes be discerned optically. However observations in the scanning electron microscope do not give indications of differences which are readily discernible. Thus a quantitative description of the topography of two different fracture regimes in a silicon nitride using the patchwork method is investigated here.

In a larger sense this work may lead to a better understanding of the relationship between fractographic analysis and topographic data acquisition from fracture surfaces, and assist the move towards a more reproducible quantitative evaluation of fracture surfaces. This kind of fractal analysis appears to have the potential to produce fractographic characterization parameters with clear physical interpretations for understanding fracture-related phenomena.

Fractal analysis by the patchwork method (Brown et al. 1993, 1994) has been applied to topographic data sets acquired from a number of engineering and natural surfaces (Brown 1993) to understand what sort of scale-area relations and crossover scales these surfaces demonstrate. The area of a rough surface, and essentially all surfaces appear rough when viewed at a sufficiently high magnification, is not unique. The surface area varies, generally increasing, as the scale of observation decreases. This scale-area relation is similar to the traditional scale-length problem of "how long is the coast of Britain" (Mandelbrot 1977).

The scale-area and scale-length relations can be used to define the fractal dimension, and much of the research on fractal properties of surfaces has been to determine the fractal dimension, by one of a variety of methods (Johnsen and Brown 1994). The fractal dimension can be estimated from the slope of log-log plots of area or length versus scale (Mandelbrot 1977, Brown and Savary 1991, Brown et al. 1993). The more negative the slope, the greater is the fractal dimension and the greater is the geometric complexity or intricacy of the surface.

Aspects of fractal analysis which have received less attention are crossover phenomena and crossover scales. A crossover occurs when there is a change in slope on the log-log plot of area or length versus scale. These crossover scales indicate the limits for which a particular fractal dimension is an appropriate description of the geometry of the surface, and they are important for clarifying the physical interpretations of the surface geometry for understanding relations between surface geometry and behavior or surface geometry and substrate microstructure.

Most surfaces of engineering interest demonstrate a smooth-rough crossover (Brown et al. 1994). The smooth-rough crossover is the scale of observation above which the surface appears smooth, below which it appears rough. At scales above the smooth-rough crossover, the surface geometry is most easily described using Euclidean geometry, and at scales finer than the smooth-rough crossover the surface geometry may be most aptly described using fractal geometry. An alternative approach to defining a smooth-rough crossover is to consider the surface as a self-affine fractal (Mandelbrot, 1985). While this approach may be superior in its mathematical elegance, the physical interpretation for engineering applications is often less clear with self-affine fractal analyses than it is using a smooth-rough crossover.

At fine scales there may be additional crossovers, and often these contribute to the physical interpretation of topographic data. Fine-scale crossovers may lead to fractal dimensions and geometric complexities that are either greater or less than the larger-scale fractal dimension which began at scales just below the smooth-rough crossover. Occasionally these fine-scale crossovers can be interpreted in terms of the topographic data set and its failure to represent the actual surface geometry. This may be because of noise in the data acquisition (Johnsen 1995), or because limits on the resolution of the data acquisition system (Brown and Savary 1991).

It is not intended to provide a substantive review of the literature on the fractal analysis of fracture surfaces. It will be instructive to discuss a few examples.

Two early groups of researchers used metallographic sectioning and digitizing of micrographs. This method of data acquisition is tedious, has spatial resolution limited by the patience of the digitizers and necessarily misses topographic data in the direction orthogonal to the sectioning direction. Underwood and Banerji (1983) used a scale-length, coastline, or compass, method to evaluate fracture surfaces on several steel specimens which had received different heat treatments before fracture. The topographic data sets were acquired by metallographic sectioning perpendicular to the fracture surface, then digitizing micrographs from the section. The work is important in that they used the projected length of the measured portion of the profile normalized at each scale evaluated to normalize the measured length at that scale. They also showed that the scale-length evaluation was sufficient to distinguish the heat treatment to which the fracture specimens had been subjected. Mandelbrot *et al.* (1984) used the slit-island method to analyze fracture surfaces. The fracture surface was coated and mounted metallographically and then polished metallographically in sections approximately parallel to the fracture surface so that islands of the fracture surface appear in the coating and mounting material. The shapes of these islands were evaluated to determine the fractal properties of the fracture surface geometry and distinguish the treatments of the materials that were fractured. Since these two methods look at sections which are intended to be orthogonal, there is mathematically no reason why they should be related, although both methods were able to distinguish properties of the material and addressed the potential for fractal geometry in the analysis of fracture surfaces.

Fracture in ceramic materials usually occurs rapidly with little or no stable crack extension. Stable crack extension can be observed when rising crack growth resistance is encountered or under conditions of sustained-load crack growth (static fatigue). Recently Bar-On et al. have shown that stable or unstable fracture can be observed in fracture toughness tests depending on the details of the test conditions. The stable and unstable fracture toughness tests gave different types of load/displacement records, as shown in Fig. 1. Correspondingly, readings could be discerned optically which identified the different fracture regimes, as shown in Fig. 2.

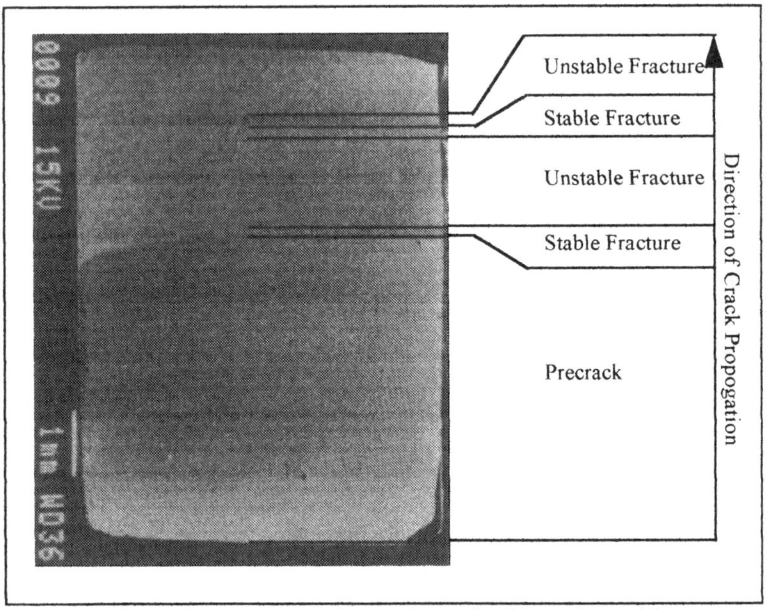

Fig. 1: An SEM micrograph showing regions of stable and unstable crack growth, as indicated.

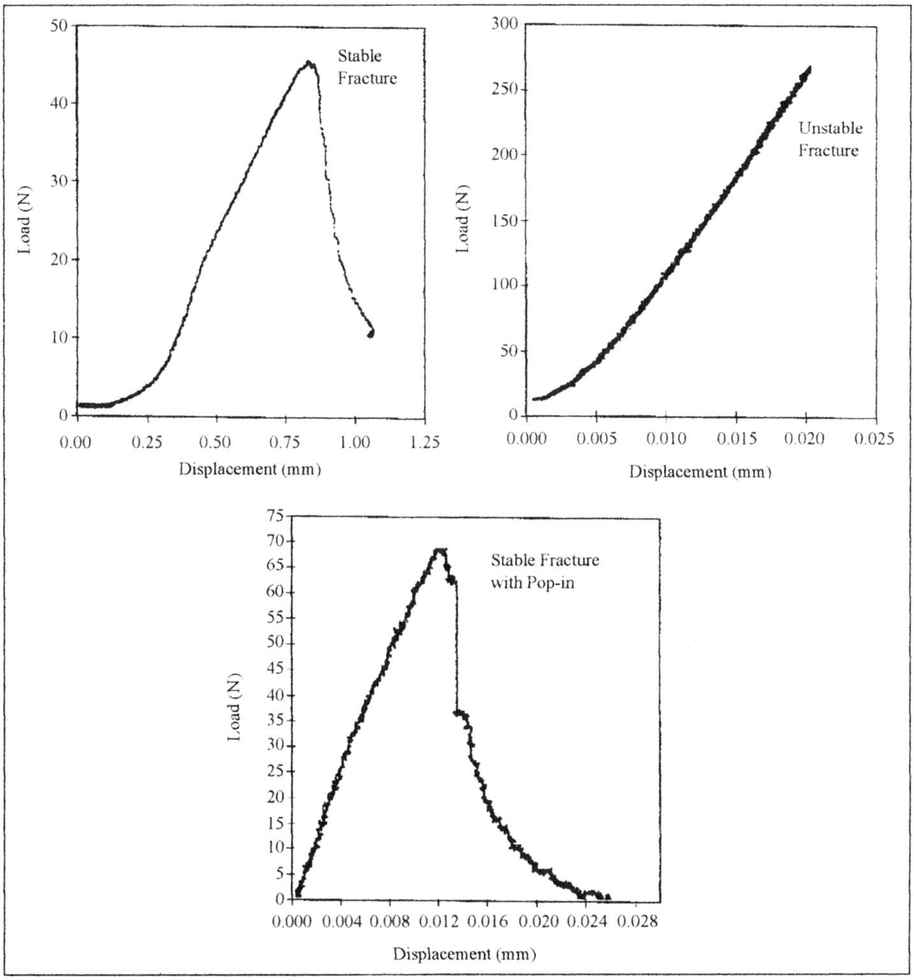

Fig. 2: Plots of stable and unstable fracture regimes.

Our approach is to evaluate 3d topographic data sets acquired from fracture surfaces using the patchwork method as performed by SURFRAX (Burlington Computer Systems, Waterbury Center, Vermont). The patchwork method has been developed to make use of the elevations and their relative positions in the data sets, so that there is not, as with the coastline and slit-island methods, orthogonal views which have the potential to produce unrelated characterization parameters. We apply this method to learn about the potential to discern stable and unstable growth.

2. METHODS

2.1 Fracture Tests

A hot-pressed silicon nitride (Norton NC 132) was used for this study. The material has a maximum grain size of 3 μm. The room-temperature properties as reported by the manufacturer are as follows: the density is 3.25 g/cm^3, the modulus of rupture is 8.25±137 MPa, the Vicker's hardness is 16 GPa, the Young's modulus is 3.2 x 10^3 MPa, and Poisson's ratio is 0.27. The material was cut into 6 x 8 x 45 mm bend-bar specimens with the long axis perpendicular to the hot-pressing direction.

Fracture toughness tests were performed using the single-edge precracked beam method. In this method a straight precrack is introduced using a special fixture. An effort was made to obtain some stable crack extension prior to final fracture. This was made possible through the careful selection of a very stiff test set-up. Details of the testing are described elsewhere (Bar-On et al.). The fracture toughness values measured from the test with stable crack extension were lower (4.0 Mpa√m) than those for tests with unstable crack extension (4.6 Mpa√m). Some of the tests exhibited repeated pop-in behavior, as shown in Fig. 1. The pop-ins are a result of alternate stable and unstable crack extension. This type of crack growth behavior left markings on the surface which could be discerned optically, but not in the scanning electron microscope.

2.2 Topographic Data Acquisition

The topographic data sets, consisting of elevations from the surface on a regular grid, were acquired using an Olympus confocal microscope. The basic principle of confocal microscopy is to move the specimen relative to the objective lens and use the focal distance to determine the elevations at a large number of locations on the image. While the specimen is moving, the images at a large number of positions are processed. Each pixel on the images is examined to determine at which position, or distance from the lens, it is the brightest. The position of the specimen when a point is the brightest is used to determine the elevation of that point.

The elevations were acquired on several 249μm-by-160μm sections of the fracture surface. A total of 187,300 elevations were acquired over each section, on a grid of 780 by 240 points. The spatial resolution is therefore different in the two directions and is 320 nm/pt and 670 nm/pt respectively. Because there were artifacts from the data acquisition on the edges of the topographic data sets, the data were cropped, using a pre-processing feature on SURFRAX. The final size of the data sets was 660 by 160 elevation points over a surface 211μm by 107μm. The approximate locations on the fracture surfaces where the data sets were acquired are shown in Fig. 1. The regions of stable and unstable crack growth are also shown.

2.3 Topographic Data Analysis

The patchwork method (Brown et al. 1993), as performed by SURFRAX, evaluates large topographic data sets, generally 100x100 elevation points or larger, by measuring the surface area as a function of the scale of measurement. The surface is measured at each scale by virtually tiling the surface in a patchwork fashion, with triangular patches, whose area represents the scale of observation. At each scale, and therefore for each tiling exercise, the triangular patches all have the same area in three dimensions, although the shapes can vary within limits.

The virtual tiling is done by the computer which locates the corners of the triangles at or along linear interpolations between the topographic data points. At the largest scales examined, there can be as few as two triangles on the data set, and they are generally insensitive to the variations or roughness in the topographic data. At the finest scales examined, there can be many tens of thousands of triangles tiled over the data set, each with an area which is half the square of the spacing between the elevations in the data set. This is the finest scale from which meaningful information can be extracted from the topographic data.

The measured area at each scale is the number of triangles tiled onto the surface at that scale times the area of the triangles. The measured area at each scale is normalized to calculate a relative area by dividing the measured area by the projected of the surface covered by the triangles at that scale. This is similar to the procedure used by Underwood and Banerji (1983). The normalization is necessary, because in general the projection of the tiled area will be smaller than the projection of the entire data set. Only full triangles are used in the tiling, so any triangle that would hang off the edge of the data set is ignored. The normalization is convenient, because it allows for direct comparison of different data sets as a function of scale, including the comparison of crossover scales. This was not the case in the comparison of coastline lengths (Mandelbrot 1977). The normalized values are called the relative areas. At large scales, above the smooth-rough crossover, they are equal to one. The relative areas at each scale can also be interpreted as a weighted average of one over the cosine of the slopes that the triangular patches make with the reference plane, nominally parallel to the fracture surface (Fig. 3):

$$\text{relative area} = \sum \left(\frac{p_i}{p_t} \right) \left(\frac{1}{\cos \Theta_i} \right)$$

where p_i is the projected area of one triangle, i is the summing index extending over the total number of triangles in that virtual tiling exercise, p_t is the total projected area for that virtual tiling exercise, and θ_i is the angle between that triangle and the nominal horizontal plane.

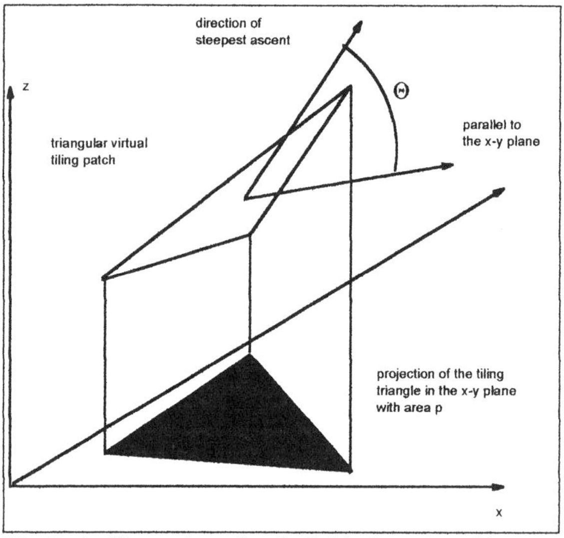

Fig. 3: An illustration of an individual patch and its projected area for showing the relation between the relative area and patch orientation. The X-Y plane is nominally horizontal.

The slopes and crossovers are calculated from plots of the logs of the relative areas versus the corresponding logs of the scales of observation, i.e., the areas of the triangular patches. This plot shows the scale-area relations over a range of scales of observation. The smooth-rough crossover is, when proceeding from the large scales towards the fine, the scale, or triangular patch size, which corresponds to the first relative area to exceed a certain threshold in relative area, which is set as a percentage (10%) of the smallest relative area to the largest relative area calculated on that data set. The slope, which can be related to the fractal dimension by adding two to its absolute value, is determined by searching the plot to find the two-decade-wide region in relative area (other widths can be used) that has the smallest regression coefficient (alternatively the region with the steepest slope can be used). These two parameters are used to characterize the plots and thereby the geometry of the surfaces.

Using SURFRAX to perform the patchwork analysis, the largest and smallest triangles in the virtual tiling are determined from the size of the data sets and the spacing of the digitized points, the minimum spatial resolution. In this study, the largest triangles fit to the data sets were approximately 3,000μm^2, and the smallest were 0.223μm^2.

3. RESULTS

A series of three virtual tiling exercises on a topographic data set, acquired from a section of the fracture surface representing a region of stable crack growth, is shown in

Fig. 4. These three have been selected from 200 virtual tiling exercises that were performed on each topographic data set.

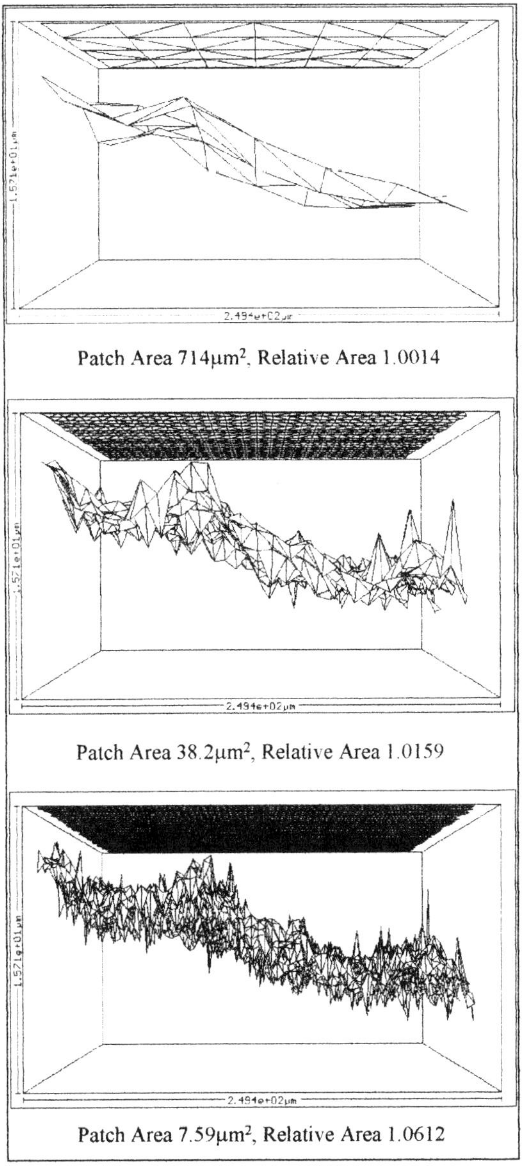

Patch Area 714µm², Relative Area 1.0014

Patch Area 38.2µm², Relative Area 1.0159

Patch Area 7.59µm², Relative Area 1.0612

Fig. 4: A series of three virtual tiling exercises on a topographic data set acquired from a region of unstable crack growth.

Typical log-log scale-area plots, one from a region of stable crack growth and one from a region of unstable crack growth, are shown in Fig. 5. Note that the relative areas for each triangular patch size are consistently larger for the stable-growth region as compared with the unstable growth region, and, consequently, the smooth-rough crossover and the slope are also larger for the stable-growth region.

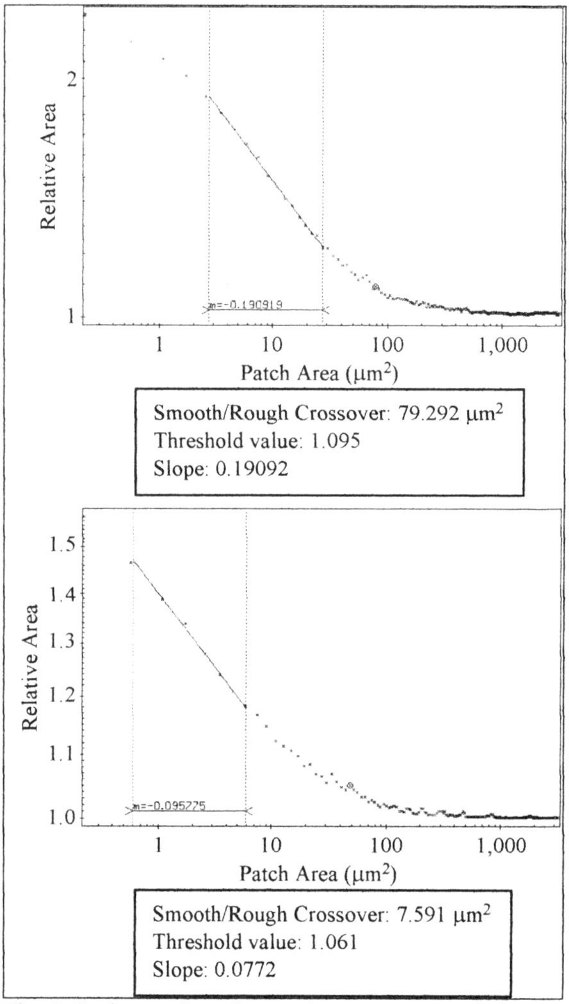

Fig. 5: Scale, or triangular patch areas, versus relative areas plotted on logarithmic axes for regions of (a) stable F09-LAS1 (b) unstable F09-LAR2 crack growth.

The slopes of the log-log scale-area plots and smooth-rough crossover scales from all the regions analyzed, divided into stable and unstable crack growth regions for the specimens, are shown in Tables 1A and 1B. The smooth-rough crossover scales have units of length squared, as they correspond to triangle areas. The slopes are unitless and have been multiplied by -1000. (Therefore, to determine a fractal dimension, one needs to divide by 1000 and add 2 to the values for the slope shown in the table.)

One specimen, F12, had only stable crack growth beyond the pre-crack region. The means and standard deviations of the slopes and crossover scales for each region and specimen are also shown in the table. There appears to be a tendency for the geometric intricacy, as indicated by the slope, to be larger near the edges of the specimen and smaller in the middle.

Table 1A: Slopes and crosssover scales from the unstable and stable regions analyzed.

F-09					
Unstable			Stable		
	X	Slope		X	Slope
LAR1	21.98	132	LAS1	79.29	191
LAR 2	19.42	103	LAS2	44.3	160
LAR3	27.57	144	LAS3	33.79	152
LAR4	10.78	80	TRI1	33.79	157
SMA1	84.38	181	TRI2	30.6	143
SMA2	19.42	100	mean	44	160
SMA3	100.58	209	std.dev.	20	18
SMA4	30.6	146			
SMA5	30.6	145			
mean	38	138	X= crossover (all in mm^2)		
std.dev.	32	40	Slope = value from plots times -1000		

Table 1B: Slopes and crossover scales from a section across the specimen parallel to the crack front.

F-12		
Group 3	X	Slope
Near - edge	40.64	160
	44.3	155
	30.6	136
Center	27.57	133
	24.69	134
	44.3	136
Near - edge	79.29	202

4. DISCUSSION

The patchwork method appears to be applicable to topographic data sets acquired from fracture surfaces. The tendencies observed in the scale-area plots show the same basic character as seen in other surfaces of engineering interest (Brown et al. 1993, Brown 1993). Similar sorts of interpretation, i.e., smooth-rough crossover scale and geometric intricacy, can be applied to understanding the log-log scale-area plots.

The edge regions and regions of stable growth appear, in general, to have a greater geometric intricacy and larger smooth-rough crossover scale than do the center regions and regions of unstable crack growth. The presence of more frequent crack branching in the edge regions which tend towards plane stress, and during the slower, stable crack growth, is one appealing interpretation that could be placed on this result. It is also intriguing to consider that more energy may be required to drive the crack through these regions with higher smooth-rough crossovers and greater geometric intricacy. In other words, considering a Griffith criterion for energy imbalance driving crack extension, as a consequence of the branching and geometric tortuousness of the crack path, more energy is expended in creating new surface as the fracture surface tends towards higher smooth-rough crossovers and higher slopes.

The consideration of fracture energy calculated as work per unit area, combined with the observation that the actual fracture surface area is a function of scale of observation and can double in three decades of areal scale, raises interesting issues. The nominal cross-sectional area of the test specimen is the area traditionally used for the calculation of the fracture toughness or more specifically the critical crack driving force. The fracture surface area could potentially be correlated to the macroscopic fracture resistance of a ceramic. It could also explain increases in fracture resistance. The question arises if there exists a scale of observation at which an inherent fracture resistance for a given material can be defined which uses the actual surface area instead of the nominal cross sectional area and how this scale would be determined.

One concern in the interpretation of the differences in the slopes and crossovers between the two regions is in the high variability in the results of the analyses. The coefficient of variation (standard deviation divided by the mean) is more than two and a half times greater for the regions of unstable growth than for the regions of stable growth. It may be that the higher level of variability is a characteristic of the unstable crack growth regions. In any event this higher variability makes it more difficult to discern, fractographically, regions according to the crack growth rates.

The size of the sections over which the topographic data were acquired is a concern. Some variability is expected from section to section, provided they are small enough. Indeed, the variability that is observed should be closely tied with the size of the section from which the topographic data sets are acquired and with the spatial resolution of the topographic data. This cannot be demonstrated here, because in this study only one size and one spatial resolution are used. Obviously, larger data sets would include a wider variety of smaller features. The relative areas at each scale on a larger data set would represent a mean over some number of smaller features, with potentially different character. On the other hand, smaller data sets would be sensitive to the individual character of the smaller features. This suggests that the size of the section at which the variability between the sections becomes obvious could be used as an indication of the scale of the heterogeneity of the surface topography.

Clearly more experience with the patchwork method will be necessary to understand how it can be applied to best advantage to characterize and discern different regions on a fracture surface, as well as in advancing our understanding of the fracture process. With more systematic work, it might become a useful tool for the failure analysis of brittle or other materials helping to identify the type of fracture or crack history causing failure.

5. CONCLUDING REMARKS

The patchwork method of fractal analysis, as performed by SURFRAX, gives results that appear to have reasonably clear physical interpretations. It also appears that the regions of stable and unstable crack growth in a ceramic material can be distinguished by the differences in their geometric complexities, as indicated by the slopes of the log-log plots of relative area versus scale of measurement. It is intriguing to suppose that the greater complexity of the stable growth regions are due to more prolific crack branching, expending more energy. More work is required to test these preliminary observations and suppositions.

6. REFERENCES

C. A. Brown, P. D. Charles, W. A. Johnsen, S. Chesters. *Fractal analysis of topographic data by the patchwork method.* WEAR, Vol. 161, pp 61-67, (1993).

C. A. Brown, P. D. Charles, W. A. Johnsen. *Method of Quantifying the Topographic Structure of a Surface.* US Patent #5,307,292, (1994).

C. A. Brown. *Scale-Area Relations by Fractal Analysis of Diverse Types of Topographic Data Using the Patchwork Method.* Intrl. Conf. on Quant. Microscopy and Image Analysis, Charleston, South Carolina, USA. July 19-21, (1993).

B. B. Mandelbrot. Fractals Form Chance and Dimension. W. H. Freeman, San Fransisco (1977).

W. A. Johnsen, C. A. Brown. *Comparison of Several Methods for Calculating Fractal-Based Topographic Characterization Parameters.* FRACTALS, Vol. 2, No. 3, pp 437-440, (1994).

C. A. Brown, G. Savary. *Describing ground surface texture using contact profilometry and fractal analysis.* WEAR, Vol. 141, pp 211-226, (1991).

B. B. Mandelbrot. *Self-Afine Fractals and Fractal Dimension.* Physica Scripts, Vol. 32, pp 257-260, (1985).

W. A. Johnsen. *Design and Analysis of a Device to Acquire Detailed Topographic Information From Pavement Surfaces*. Worcester Polytechnic Institute, Worcester, MA, USA, Master's Degree Thesis, (1995).

K. Banerji, E. E. Underwood. *Quantitative analysis of fractographic features in a 4340 steel*. ACTA STEREOL 2/SUPPL I, pp 65-70, (1983).

B. B. Mandelbrot, D. E. Passoja, A. J. Paullay. *Fractal character of fracture surfaces of metals*. NATURE, Vol. 308, pp 721-722, (1984).

I. Bar-On, F. I. Baratta, K. Cho. *Crack Stability and its Effect on Toughness of Hot Pressed Silicon Nitride Beam Specimens*. To be published in J. Am. Ceram. Soc.

ACKNOWLEDGEMENTS:

The authors would like to gratefully acknowledge the generous support of NASA/Langley Research Center Grant #NAG-1-1606 (1994). Some of this work was performed while Isa Bar-On was a NASA/OAI Faculty Fellow at NASA Lewis Research Center. Their support is gratefully acknowledged.

MECHANISMS FOR HACKLE FORMATION AND CRACK BRANCHING

Edwin K. Beauchamp
Beauchamp Associates; Albuquerque, NM

ABSTRACT

Optical, SEM, and TEM microscopy and interference microscopy were used to conduct a detailed examination of fracture surfaces and macroscopic branching events produced in high velocity fracture of soda-lime glass. Results of that examination were compared with various models for the generation of hacklc and branching. Most observations in this study and in the literature are consistent with a model in which the edge of the crack surface remains completely connected, i.e., no microcracking in advance of the crack tip. Concepts are presented that explain the existence of the mist/hackle boundary, the crack branching process, and the low terminal velocity commonly observed for fracture in silicate glasses.

INTRODUCTION

During dynamic fracture of brittle materials, as the crack velocity increases and approaches the terminal velocity, the fracture surface morphology changes from a smooth, flat "mirror" at lower velocities to a surface that is progressively rougher. The fracture surface features that are characteristic of these rough zones are termed "velocity hackle". Rice (1) has provided detailed descriptions of velocity hackle features in a wide variety of ceramics and glasses and has documented the differences between fracture surfaces in very fine-grained and coarse-grained ceramics as well as single crystals. The presence of velocity hackle on a fracture surface is a very valuable aid to the fractographer in analyzing the circumstances of the failure(2,3). The sequence of the zones of "mist", "hackle" and "branching" provides information on the direction of fracture and leads to the location of the fracture origin. The shape of the boundaries between the zones provides information on the local stress distribution and the dimensions of the zones (the distance of the zone boundary from the origin) provides a measure of the failure stress. Kirchner *et al.* (4,5) have shown that, in general, the equation

$$\sigma_f = \Phi K_i / r_i^{1/2} \qquad (1)$$

where Φ depends on the loading geometry and K_i is the stress intensity for the material at a boundary, is a valid relationship, in isotropic materials, between the failure stress, σ_f, and r_i, the distance from the origin to the boundary. Commonly, the product ΦK_i is replaced with the parameter A_i, whose value is determined from measurements of r_i at known failure stresses

There is an obvious similarity of the equations relating the mirror/mist, mist/hackle and branching radii to the Griffith equation, as modified by Irwin (6)

$$\sigma_f = \frac{1}{Y} K_{1c} / c^{1/2} \tag{2}$$

where σ_f is the stress at failure, Y is the geometric correction factor, K_{1c} is the fracture toughness of the material, and c is the crack length. That similarity suggests that, like K_{1c}, the parameters A_i are constants of the material. Attaining A_i, or K_i is, then, a condition for the onset of mist, hackle or branching. The alternative equivalent is that there is a particular value of the strain energy release rate, G, at which mist is generated, with higher values for hackle and branching. Tsai and Mecholsky (7) showed that, for single crystals and other anisotropic materials, the energy criterion is the only one that is valid for estimating the fracture stress from branching. There is an implication that attaining a particular value of K_1 or G is a necessary and sufficient condition for generation of hackle. However, Rice (8) noted that, under certain conditions such as fracture on easy cleavage planes in single crystals, velocity hackle is not generated before terminal velocity is attained. He concluded that G may be a necessary condition for hackle generation in those instances, but not a sufficient condition.

In spite of the application of velocity hackle and branching in fractography, there has been no general understanding of how the hackle features are formed or why apparent boundaries exist. Various models describe the processes for generation of these features and a number of assumptions have been made about the dependence of those processes on K_1 or G. The intent in this paper is to describe some of the previous models and to present information that, in some instances, substantiates those models and, in others, suggests an alternative interpretation.

CHARACTERISTICS OF MIST AND HACKLE

To provide a basis for understanding the generation of mist and hackle, consider the set of micrographs in Figures 1 through 5, which show the progressive change

Fig. 1 Fracture surface in mirror region near origin. Black dots are artifacts of the Pt deposition.

Fig. 2. Fracture surface in mirror region (at ~1/3 r_m).

Fig. 3. Fracture surface in mirror region near mist boundary. Small curved arrows indicate the path of lateral crack growth.

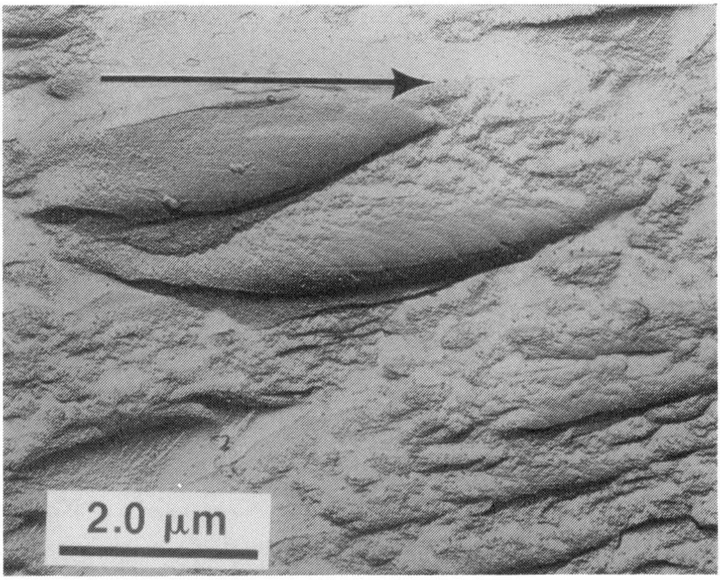

Fig. 4. Fracture surface in mist region. Note the twist hackle features in lateral crack growth.

in the fracture surface as the crack velocity, and strain energy release rate, increases. This fracture surface was on a “dicing” fragment from a lightly tempered soda lime glass sheet of 6.4 mm thickness. For these micrographs, the fracture surface was replicated with cellulose acetate and then the replica was shadowed with evaporated platinum. The replicas were examined in a transmission electron microscope.

Figure 1 shows a portion of the fracture surface near the point at which the crack went critical, i.e., the mirror region near the origin for this particular branch of fracture. The small dark spots are artifacts of the Pt deposition used in the shadowing process. The surface appears relatively featureless at this magnification; however, the coarseness of the Pt deposits prevents observation of features smaller than 20 nm. The only discernible characteristics are roughly circular features of diameter ~50 nm scattered over the surface. These may have been produced by local inhomogeneities in the glass composition. However, the shadowing indicates that all of the features protrude from the surface. Because this is a replica of the glass fracture surface, it implies that depressions were present in the glass surface. These depressions may be a result of the reaction of the glass with atmospheric water. An attempt was made to prevent formation of such products by replicating immediately after the fracture; however, a freshly formed fracture surface is highly reactive.

Figure 2 shows a portion of the surface about 1/3 of the distance from the origin to the mirror/mist boundary. At this point the surface is covered with long, slender, rounded ridges or folds oriented in the direction of crack growth. Optically, these features cannot be observed so the surface is still mirror-like. Figure 3 shows the surface near the mirror/mist boundary but still within the mirror region. The ridges are now longer and wider than those in Figure 2 and appear to be asymmetric. They display zones that appear relatively smoother than the surrounding area. Some of the relatively smooth zones show substructure in the form of lines curving away from the direction of the overall crack growth.

Figure 4 shows the surface just beyond the mirror/mist boundary. The features have the same general character as those in Figure 3, but they are now large enough that they can be observed optically. Within the hackle zone, as shown in Figure 5, these features have grown so that they now exceed 10 μm in length, but they still have the same character as those in the mist and mirror zones.

In selecting the mist/hackle boundary, there is some subjectivity as well as a dependence on the magnification used in observing the boundary. In the present paper, the boundary is defined, somewhat arbitrarily, as the point at which features

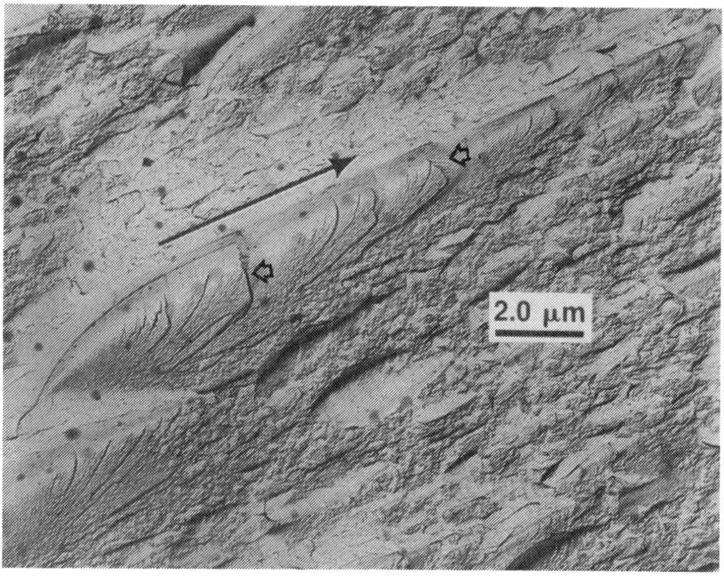

Fig. 5. Fracture surface in the hackle region.

greater than 10 μm in length appear. Features greater than 10 μm will be referred to as "major hackle".

THE GENERATION OF MIST AND HACKLE FEATURES

Criteria for Hackle Generation

A valid model for the generation of mist and hackle features must explain how the individual features are produced, i.e., what happens at the crack tip that results in a portion of the crack moving out of the primary crack plane. It must also explain the increase in size of the features as G increases and the changes in characteristics of features that are associated with the mirror/mist and mist/hackle boundaries. It should also be consistent with observations that, in fracture on easy cleavage planes in certain crystals, mist and hackle do not appear before branching, and branching does not occur until very high velocities are attained (8).

There are several factors that are generally accepted as contributing to the generation of velocity hackle. Certainly, the increase in strain energy release rate is of paramount importance. Most models of hackle formation and branching also cite the analyses of the stress field around a rapidly moving crack as the basis for expecting the crack to leave the primary crack plane. Yoffe (9) and Erdogan (10) showed that the maximum in $\sigma_{\theta\theta}$ shifts from $\theta = 0$ degrees to $\theta = 60$ degrees as the crack velocity increases to 0.5 V_l (the longitudinal sound velocity). At velocities less than 0.4 V_l, the peak stress is at $\theta = 0$, but the peak is much broader than in the static case. Thus, in the range of K values at which mist and hackle appear, the crack becomes unstable to the extent that it becomes relatively easier to leave the primary crack plane.

The similarity in the features in the mirror, mist, and hackle zones, which was observed in Figures 2 through 5, has been cited by others. Johnson and Holloway (11) noted that, although the features in the mist zones were much smaller than those in hackle zones, they had very similar character. They remarked that the processes of generation were probably the same for the two types of features. Ravi-Chandar and Knauss (12) arrived at the same conclusion for features produced in the fracture of a glassy polymer, Homalite-100. Using the scanning tunneling microscope to examine glass fracture surfaces, Kulawansa *et al.* (13) concluded that the features in the mirror and mist regions were very similar in character and were probably generated by the same process.

Earlier Models for Hackle Formation

Most of the models for formation of mist and hackle, and also for macroscopic branching, have been based on the supposition that microcracks exist in the materials and that microcracks ahead of the moving crack front grow under the influence of the stress field of the crack tip (14, 15, 16, 17). In those models, the mist and hackle features are produced by the intersection of the primary crack with the growing microcracks. As the magnitude of the stress field increases, more extensive growth of the microcracks occurs before the primary crack intersects them. As a consequence, the size of the features increases with increasing strain energy release rate.

Johnson and Holloway (11) described the hackle formation process in terms of the bifurcation of the crack into two or more elements growing from the same line segment as shown in Figure 6. They described the generation of hackle features in terms of intersection of those elements of fracture during forward and lateral growth. The assumption of a literal bifurcation has recently reappeared in theoretical models (18, 19, 20) that examined high velocity instability. Those models, in two-dimensional space, predict instabilities in the lattice at high velocity that permit the simultaneous failure of more than one bond at the crack tip. Thus, the crack splits and leads to microcracking out of the primary plane.

Fig. 6. Schematic of velocity hackle formation showing bifurcation along an element of the crack front (after Johnson and Holloway (11)).

An Alternative Model (Twist-Type Features)

The general character of the features shown in Figures 2 through 5 suggests an alternative process for their generation. This model does not depend on either the activation of microcracks or the bifurcation of the crack at a line segment of the crack front. Instead, it is based on the similarity of the features to twist hackle. That similarity is evident particularly in Figure 4, but even for the features in the mirror region.

Frechette (21) has described the generation of twist hackle in terms of splitting of the crack at a point on the crack front and propagation of the crack from that point on two noncoplanar surfaces. A schematic of that process is shown in Figure 7. In the generation of twist hackle, the driving force for splitting the crack into two noncoplanar portions is a change in the direction of the stress that drives the crack. As indicated in the schematic, the change in direction is a rotation about the vector along the direction of the crack growth. Sommer (22) has shown that a rotation of at least 3 degrees is necessary to produce a twist hackle feature during slow crack growth.

The twist hackle feature is a result of the attempt of the crack to run perpendicular to the new stress direction. Apparently, for small twist angles and low crack propagation velocities, the growing crack can accommodate by twisting like a ribbon. In that event, the crack surface remains smooth and the crack front remains a nearly straight line. The accommodation may also involve locally large deformations in the form of ripples in the surface. For twist angles greater than 3

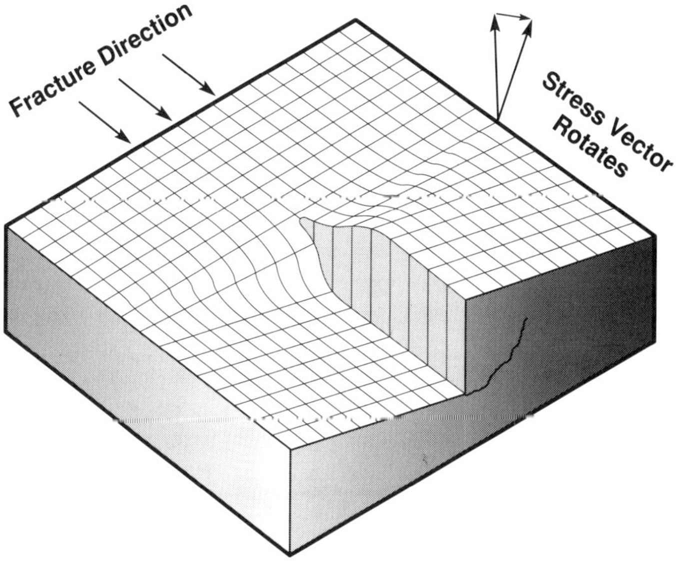

Fig. 7. Schematic of twist hackle formation. The grid pattern is added to illustrate local deformation (after Frechette (21)).

degrees, the accommodation is obtained by ribbon-like rotation of segments of the fracture surface that are separated by twist hackle. As shown in the schematic, this is accomplished by confining the twist to a short segment of the crack front. The interference fringes in Sommer's micrographs (22) show how rapidly that accommodation takes place. In Figure 1 of Sommer's paper, the micrograph shows that, while the crack travels less than 5 μm, it changes its elevation relative to the adjacent surface by one fringe (540 nm). The "splitting" of the crack front is apparently a result of slowing the growth of that region of high twist relative to the growth of the rest of the front. The crack front appears to be discontinuous. However, in fact, the boundary of the fracture surface is still fully connected although portions of it are now essentially perpendicular to what would ordinarily be construed as the the crack front.

The slowing of the region of high twist is a result of the dependence of the crack velocity on the local Mode I stress intensity. The crack surface in the region of high twist is at a large angle to the applied stress so its velocity is lower than in those regions where the crack surface is normal to the applied stress.

It is the lateral growth of the crack front from one of the advancing portions that completes the separation of the material on either side of the crack and produces the rapidly curving surface that characterizes the twist hackle feature. Lateral growth is common to all of the fracture processes in which intact ligaments are left in the wake of the advancing crack front. The relative smoothness of the lateral growth indicates that the local stress intensity during that growth is substantially lower than at the primary crack front. Occasionally, lateral crack growth takes place from both "edges" so that a sliver of material is separated from the fracture surface (23).

The marked similarity of the mist and hackle features in Figs 2 through 5 to twist hackle suggests that mist and hackle are generated in the same way as twist hackle, i.e., by "splitting" the crack into two noncoplanar portions and then completing the fracture by lateral growth from one of the segments to the other. It is apparent in Figure 5 that the lineaments on the hackle features, in the form of curved lines that first appeared in Figure 3, are essentially fine twist hackles created when the crack grew laterally and down to intersect the adjacent fracture surface. The rapid change in direction of the stress that drives that lateral growth is responsible for the generation of those lineaments.

There is an implication, in the comparison of the velocity hackle features with twist hackle, that local twist exists where the velocity hackle features initiate, i.e., the local stress vector has rotated. A discussion of how that can occur will be given in a later section. Note that, in certain V *vs.* K regimes, the slowing of the crack at the twist in the crack front will be more pronounced than in other regimes. As will be shown, at high velocity, where the change in velocity is more pronounced, the twist angle necessary to produce a hackle feature may be much less than 3 degrees.

Kulawansa et al (13) presented a model for velocity hackle that is similar to that presented above. They described the process in terms of crack fingering, in which the crack front breaks up into fingers that advance more rapidly than the contiguous front that separates the fingers. In their model, unlike that presented here, the slower portions of the crack front remain perpendicular to the applied stress. The spacing of the fingers is determined by the "energy gained by finger growth and the energy lost due to the creation of the additional fracture surface." To create the features, these fingers of fracture then rotate under the influence of transient changes in the direction of the maximum tensile stress. They likened the instability of the crack front to meniscus instability in the flow of viscous liquids (Taylor instability).

"Mesa" Features

Although most of the features in Figures 2 through 5 show similarity to twist features, there is another type of feature found in the velocity hackle region with slightly different character. That type of feature can be the dominant type in certain circumstances. The feature, shown in Figure 8, has the appearance of a mesa with a raised zone having a relatively flat top separated from the rest of the fracture surface by lateral crack growth. The surface produced in that lateral growth has the same appearance as that associated with the twist-like features in Fig. 2 to 5, i.e., secondary twist hackle lines curving in the direction of the lateral

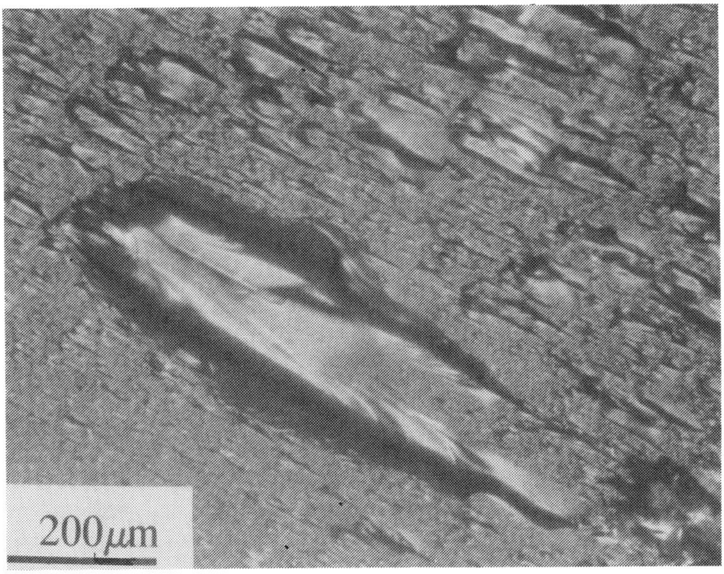

Fig. 8. Mesa features in the hackle region. The larger feature extended from two, or perhaps three, tongues. Crack ran from left to right.

growth. This is the type of feature that Johnson and Holloway described in Ref. 11. In what follows, these features will be referred to as "mesa" features, while those that have the character of twist hackle will be referred to as twist features. In spite of the apparent differences between mesa features and twist features, their formation is very similar.

The formation of mesa features is shown schematically in Figure 9. Essentially, the process begins when a short segment of the crack front tilts out of the primary crack plane, then tilts back. As a consequence, that portion of the crack grows

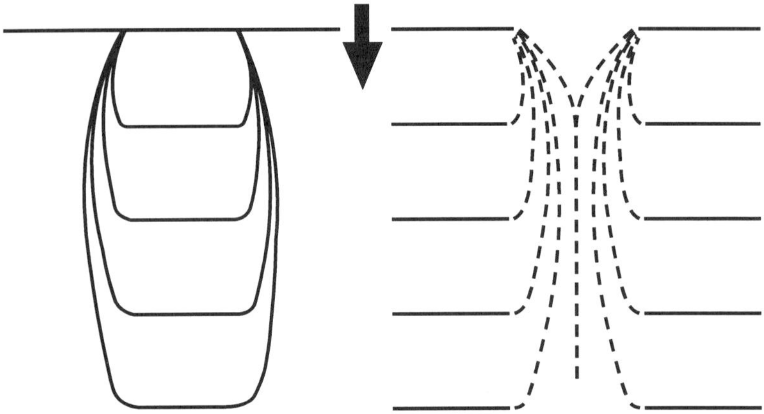

Fig. 9. Schematic of formation of mesa feature: a) development of fracture from the tongue, b) development of fracture on primary surface.

nearly parallel to the primary but on a different plane. As is the case with twist features, the sections of the crack front that connect this segment of the crack with the primary crack are at an angle to the applied stress. As a result, the stress intensity driving their growth is lower. Consequently, the crack velocity is lower in those connecting regions and the central portion grows like a tongue. That tongue tilts down toward the primary fracture surface in the example in Figure 8 and eventually reconnects with the primary crack. Most mesa features are completed by lateral growth from the tongue to the adjacent fracture surface as shown in Figure 9b. Kulawansa *et al.* show a schematic diagram similar to that in Figure 9 in describing the generation of these features, which they described as "localized crack branching events".

In most examples of mesa features, the growth that completes the separation of the two fracture surfaces is from the tongue to the primary crack. Fracture surfaces like those in Figure 8 are then exposed to view. However, occasionally, lateral growth occurs from the primary crack to the tongue as shown in Figure 9b and a feature like that in Figure 10 is created. The preponderance of features that look like 9a suggests that separation by lateral propagation from the tongue is easier than by lateral crack growth on the primary plane. Johnson and Holloway described these two features in terms of two different mechanisms of formation but, in fact, they are created in essentially the same way. They designated the features in Figure 8 as type A while they designated that in Figure 10 as type B.

Fig. 10. Mesa feature after alternative separation (lateral fracture from primary plane). Fracture direction, left to right. Arrows mark initiating tongues.

The feature in Figure 8 does not correspond exactly to the schematic in Figure 9. In Figure 8, at least two, and probably three, small tongues of fracture coalesced to form one large tongue. This is a common occurrence in mesa features whose length exceeds about 20 μm. As exemplified in the smaller features in Figure 8, mesa features less than 5 μm in length usually issue from a single tongue of fracture.

The elongated "ripples" that appear very early on in the mirror region (Figure 2) seem to have the same character as the mesa features. They are also apparently produced by the tilt of a small element of the crack front out of the crack plane. In the case of the features in Figure 2, however, there is no obvious retardation of the portions of the ripple that are not oriented normal to the applied stress. The crack front moves ahead with a nearly straight front (other than the ripple).

ENERGETICS OF MIST AND HACKLE FORMATION

As noted in the Introduction, mirror/mist and mist/hackle boundaries occur at specific values of A_i (in Equation 1) or equivalent specific values of K_I or G. There is an implication in those observations that a certain level of G is a necessary condition for generation of hackle. Various explanations (1,8,24) of how velocity hackle is formed are based on the assumption that it is a necessary condition. For example, in the model involving the growth of microcracks in advance of the crack tip, generation of mist and hackle requires a certain level of strain energy release rate to grow the microcracks to a given size before the primary intersects them (1).

There are indications that major hackle formation may occur at substantially lower values of K_I than that corresponding to the mist/hackle or even the mirror/mist boundary. Mist and hackle zones are often not uniform arrays of features. Rather, as in the example in Figure 11, larger features are lined up in the direction of fracture while the surface between them contains only smaller features.

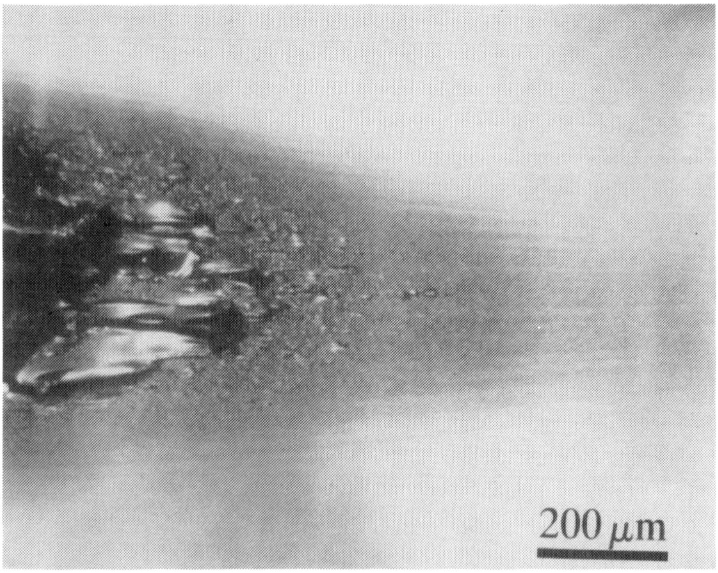

Fig. 11. Lines of mist and hackle features on fracture surface of diced tempered glass. (Crack propagated from right to left.)

Occasionally, also, streaks of much larger features are found in a zone which otherwise consists only of mist or might even appear to be mirror. Rice (1) remarked that these hackle streaks were sometimes associated with an inclusion or pore, or with the interaction of the crack front with a surface flaw. He suggested that the generation of those features might result from the stress riser associated with the inclusion. In that view, the local stress intensity along the line where the streaks formed would have been higher than in the surrounding area. However,

Rice also noted that the streaks ran far beyond the point where the stress field could have any direct effect.

The location of hackle streaks in the wake of inclusions and pores suggests that streaks may be generated in lieu of wake hackle features when the G value approaches that for the onset of mist. As noted in the previous section, many of the features in the mist and hackle region have a character that is very similar to that of twist hackle. A reasonable premise is that distortion at the crack front, such as the twist that might ordinarily have produced a wake hackle feature, can be an agent in the generation of velocity hackle.

Characteristics of Hackle Streaks

The premise that distortion at the crack front can be a factor in hackle generation was tested by a detailed examination (25) of hackle streaks extending from large twist hackle features on fracture surfaces in soda-lime glass. The fracture surfaces had been produced in dicing of a lightly tempered glass sheet. Light tempering results in a residual stress profile with a maximum tensile stress in the midplane of the sheet but a relatively small gradient in stress near that maximum. Consequently, K_1 is expected to be relatively constant in a band near the center of the plate. Figure 12 shows one of the surfaces that were examined. The hackle streaks are to the right of the points marked "A" and the twist hackle features are to the left. The chevron markings along the streak are Wallner lines generated at the streak.

Figure 13 shows a portion of the hackle streak at higher magnification. This view shows that the streak is actually a band of hackle features consisting of very large features along the center line with smaller features on either side. The orientation of the smaller features indicates the local direction of the crack propagation (an arrow shows that direction for the features above the central line).

Characterization of the fracture surfaces described in Ref. 25 included determination of local crack velocity along the streak and on either side of the streak, measurement of the distortion of the fracture surface in the vicinity of the streak and a tabulation of the number and size of hackle features along the streak. The local velocity was calculated using two types of Wallner lines, those emanating from the hackle streak with a chevron shape and those produced by far-field stress pulses. The velocity determinations employed an analysis similar to that described by Tsirk (26), in which the local velocity along the streak was calculated from the angle between Wallner lines emanating from the hackle streak.

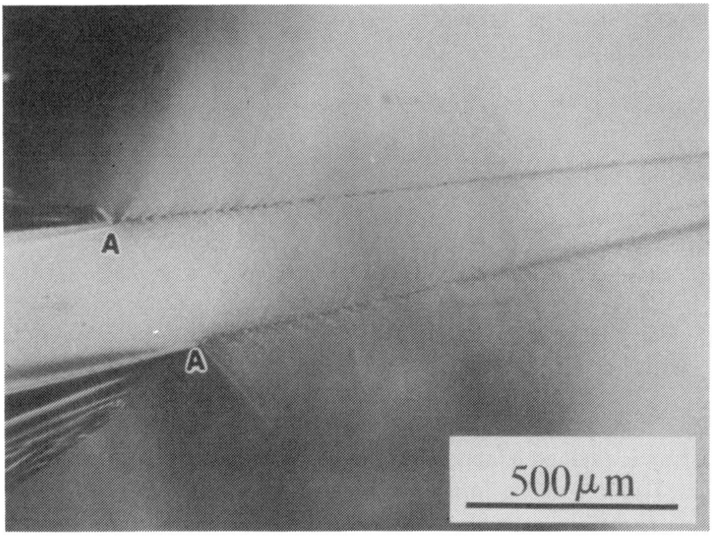

Fig. 12. Hackle streaks extending from twist hackle features, which terminated at points A. Chevron Wallner lines extend from streaks at 60° angle.

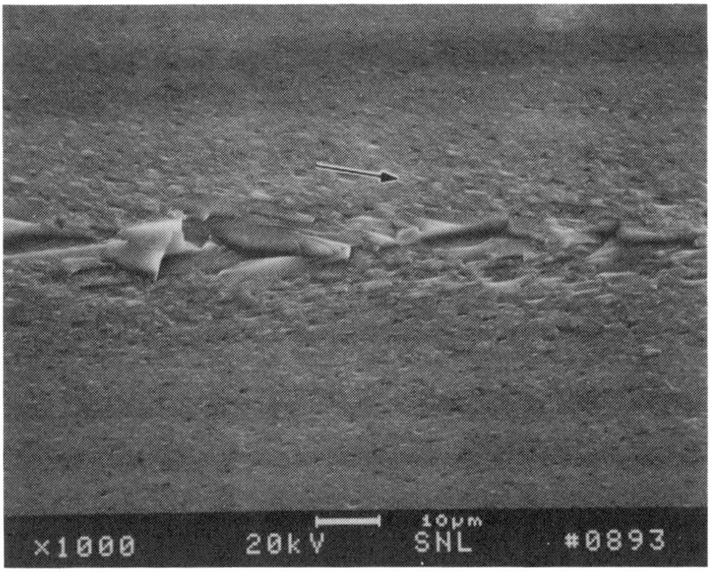

Fig. 13. A portion of the streak in Fig. 12 at higher magnification. Arrow indicates local direction of fracture (from orientation of small hackle features).

In addition, the shape of the Wallner lines was used to infer changes in velocity in the vicinity of the streak.

The crack velocity was nearly constant along the streak and in a band extending approximately 300 μm on either side of the streak, indicating that the strain energy release rate, G, was essentially the same at the streak as it was in the mirror regions bordering the streak. In fact, a cusp in the far-field Wallner lines revealed that the velocity was slightly lower at the streak than away from it. The implication from the velocity data is that mist and major hackle (>10 μm in length) were generated along the streak at G values below that for either the mirror/mist or mist hackle boundaries. The conclusion is that G_h is not a necessary condition for the generation of hackle.

Effect of Crack Front Distortion on Hackle Formation

Two sets of observations provided substantiation that localized twist or some other form of distortion at the crack front was the agent that contributed to the generation of mist and hackle. The first of these was a correlation between the surface roughness (size and number of mist and hackle features) along the streak and the local distortion of the fracture surface. Crack surface distortion was measured using interference microscopy. Figure 14 is a composite plot of the measured distortions in terms of twist angle and vertical displacement across the hackle streak as a function of position along the streak. The histogram showing the surface roughness was generated by counting features in areal segments of 200 μm length and 100 μm width along the streak.

The correlation between size and numbers of hackle features and local distortion in Figure 14 argues that the distortion is a contributing factor for the generation of hackle. Somewhat more compelling evidence is provided by examination of hackle features originating at Wallner lines. Figure 15 is an example of hackle features originating at a Wallner line produced at the hackle streak. The Wallner line was generated by a stress pulse a short distance behind the crack front near the termination of a major hackle feature that extends from the left of the micrograph. A major hackle feature initiates at the Wallner line. In addition, several smaller twist features emanate from the line below the streak and three mesa type features emanate from the line above the streak. The smaller features will be referred to as "satellite" hackle features because of their clear association with the major feature.

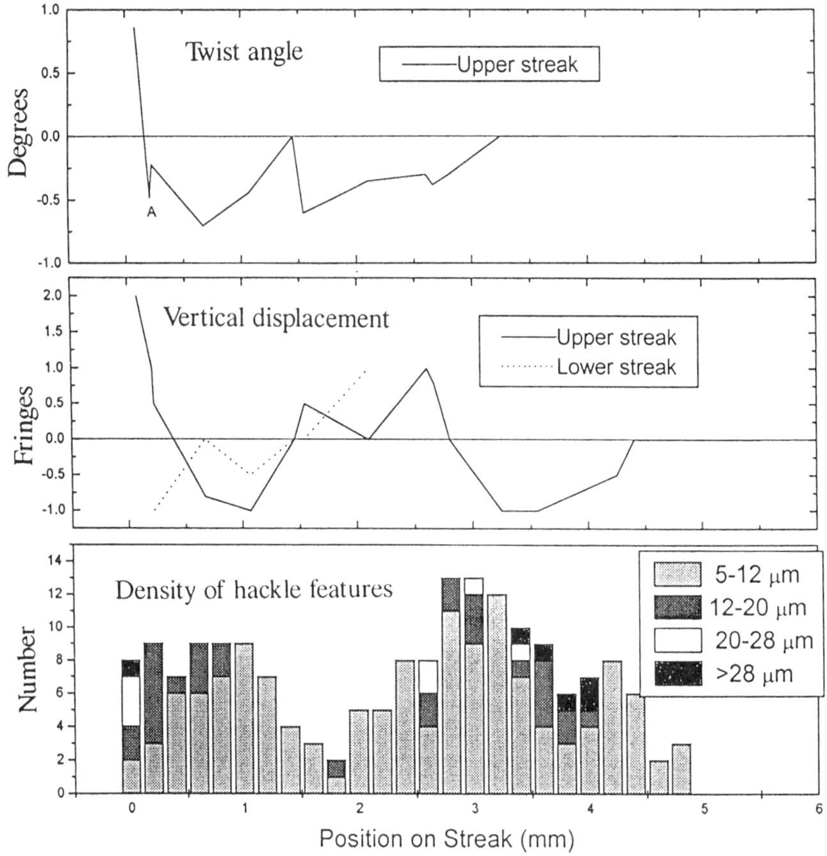

Fig. 14. Roughness and distortion vs position along streaks in Fig. 12.

Measurement of the distortion of the fracture surface along the more prominent chevron lines issuing from the hackle streaks in Figure 12 showed displacement of the order of 80-100 nm. In effect, the stress pulse that produced the Wallner line created a shear strain. Apparently, shear strains of that magnitude or less are sufficient to produce twist type features when G approaches G_m.

Figure 16 shows another example of hackle features issuing from a Wallner line. In this case, the Wallner line was produced by a stress pulse from a distant source so that the wave front of the pulse was nearly flat. Passage of the pulse resulted in a permanent tilt of the fracture surface of about one degree; hence, this is really an "impulse line". Most of the hackle features issuing from the Wallner line are of the mesa type. In addition to the features originating at the impulse line, smaller ones

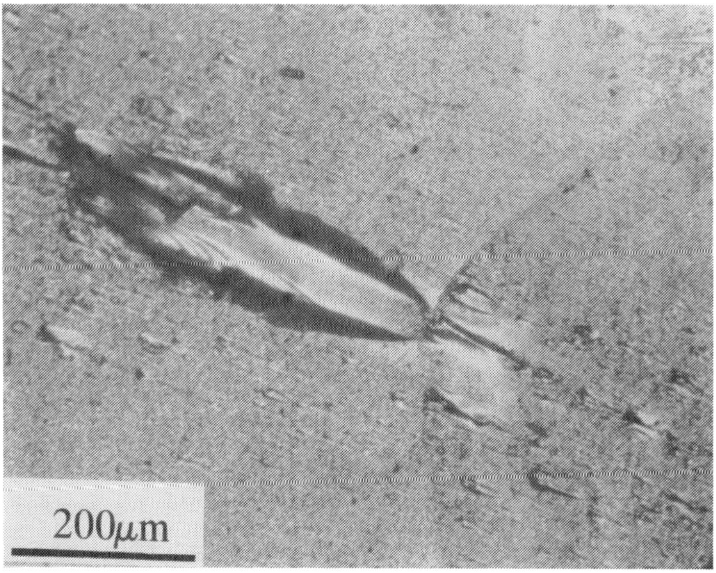

Fig. 15. Satellite hackle features on a Wallner line that extends from mesa feature.

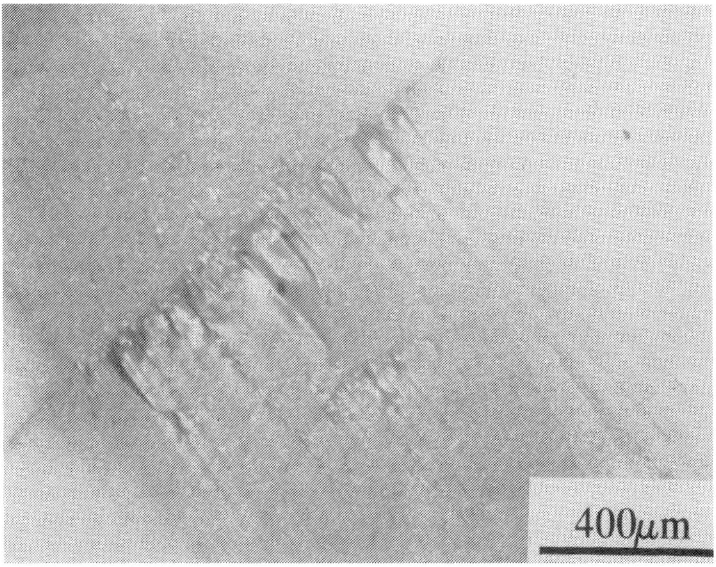

Fig. 16. Hackle at a far-field Wallner line. Most are mesa features. A chevron Wallner line extending from the largest feature also has satellite features.

can be seen extending from chevron Wallner lines. These Wallner lines were generated at the ends of some of the hackle features issuing from the impulse line.

The association of twist-type features with the chevron Wallner lines and of mesa features with the impulse line suggests how the nature of the distortion at the crack front influences the nature of the features that are produced. Along the chevron Wallner line, the distortion is essentially a shear, i.e., a twist, across the crack front. Thus the most likely form of hackle feature would be one that represented a response to the shear, i.e., a twist-type feature. On the other hand, the distortion at the impulse line is a tilt. That tilt would be expected to enhance the tendency for a portion of the crack front to leave the primary crack plane, i.e., it creates a condition that is conducive to formation of a mesa feature. A careful examination of both types of Wallner lines reveals that there is not a perfect correlation between the type of deformation at a Wallner line and the type of hackle feature produced. Of course, in the examples shown, the character of the distortion present at the crack front was not entirely that of the Wallner line. The net distortion is a combination of that from the Wallner line and that associated with the hackle streak.

The major conclusion from this examination of hackle streaks was that major hackle features can be generated at K, or G, values substantially below the value corresponding to the mist/hackle boundary provided that some form of distortion exists on the crack front. The larger the distortion, for a given G value, the larger the features that are generated. Similarly, the higher the G value, the larger the features, for a given level of distortion. Figure 17 shows a schematic of hackle formation that this dependence suggests. The curved line in the graph separates the zone where combinations of distortion and G values can create features of a given size, from the zone where features of that size are not generated. For larger features the line would move up and to the right. The line marked G_H corresponds to the hackle boundary, while that marked G_T is the threshold value of G below which no hackle of the indicated size will be formed.

In Figure 14, the largest measured twist angle is one degree. Clearly, most of the twist-type hackle features were produced with considerably smaller twist angles than the 3 degrees that Sommer (22) found to be necessary to produce twist hackle at lower velocities. It is not clear why that should be. The governing factor in formation of a twist hackle at low velocity or T-type velocity hackle appears to be the retardation of one portion of the crack front relative to the adjacent portions. The crack then moves forward as two separated segments. That retardation is apparently caused by the reduction of the K value along the twisted

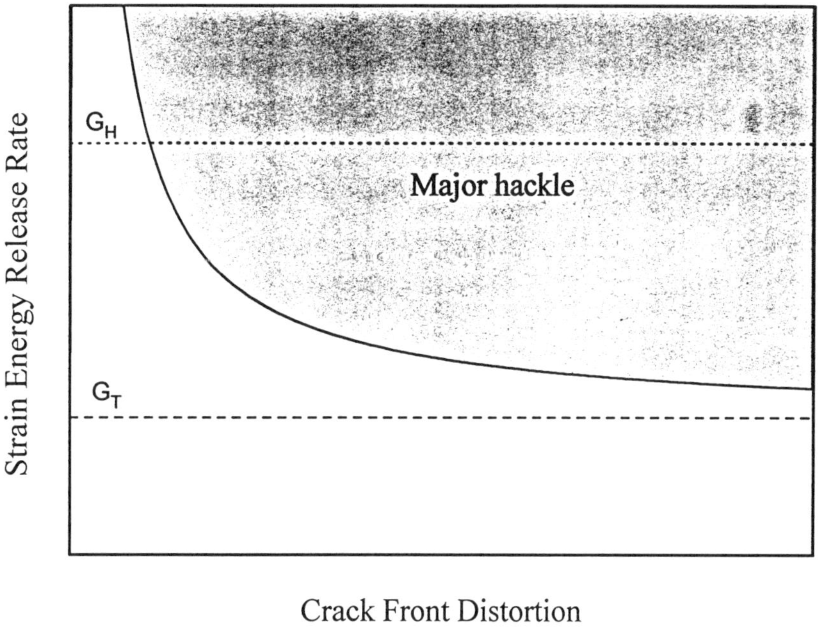

Fig. 17. Schematic of dependence of hackle size on energy and distortion. The line represents onset of formation of hackle of a particular size, say 10 μm length.

portion of the crack front and the resultant reduction in the local crack velocity. It seems reasonable to expect that the level of distortion ought to determine the amount of reduction in K so that the ΔK is smaller for velocity hackle than for low velocity twist hackle. If that is true, it would seem that velocity hackle could form only if the lower ΔK is more than compensated by a much stronger dependence of velocity on K. However at the K values where velocity hackle forms, the slope of the V-K plot is relatively shallow compared with that for K that is only slightly greater than K_{1c}.

It is not clear why streaks persist as they do. Twist hackle terminates relatively quickly because the two noncoplanar surfaces that form the feature tilt toward one another and eventually close to form a continuous front. Similarly, individual twist-type velocity hackle features in a streak terminate quickly, indicating that the crack front distortion that engendered them has disappeared or diminished. In appparent contradiction, new features continue to be generated, suggesting that the distortion along the streak line is still present.

The Mist/Hackle Boundary

The generation of satellite hackle along the Wallner line in Figure 15 and the schematic in Figure 17 suggest an explanation for the hackle boundary. At the relatively low value of G along the streak where the feature in Figure 15 appeared, the satellite features are small compared with the feature that produced the Wallner line. However, for $G > G_h$, the stress pulses produced at the termination of one hackle feature may be able to produce satellite features of comparable size. Thus, the hackle boundary may correspond to the G value at which a single, random perturbation at the crack front can produce a major hackle feature that becomes the nucleus for generating a band of major hackle features. The band is created by lateral transfer of crack front distortion as stress pulses issue from the first feature and satellite features.

That explanation for the hackle boundary also suggests why Snowden (27) found no mist or hackle in his experiments with shock pulses. He used the pulses to accelerate a crack in borosilicate glass very rapidly to velocities approaching the Rayleigh velocity, i.e., much higher than the terminal velocity in most glass fracture. Perhaps, in his experiments, the time to attain that high velocity was so short that there was no opportunity for a stress wave to interact with the crack front. Even if one hackle feature had been formed, there would have been no opportunity for lateral transfer of the distortion necessary for generation of satellite features.

THE EFFECT OF HACKLE ON CRACK VELOCITY

As noted above, the presence of a cusp in the crack front at the hackle streak indicated that the velocity at the streak was slightly lower than away from the streak. At the end of the streak, where the features were less than 2 μm long, the cusp disappeared. That lowering of velocity in the presence of the larger hackle features suggests that the hackle features were responsible. The likely explanation for the effect on velocity is that the hackle features behave as crack interface bridges, reducing the local K value and, hence, reducing the velocity. Mai and Lawn (27) have described the effect of bridging on the K value. The bridging effect may also be one of the reasons that the strain energy release rate at the hackle boundary is so much greater than the energy required to create the additional fracture surface associated with the hackle formation.

It should not be surprising that hackle features act like bridges. In the models for their formation in Figures 7 and 9, the separation of the crack into noncoplanar segments leaves a region of intact material behind the main crack front. The lateral crack growth that runs through that intact material and completes the fracture process at the feature is slow enough that the nearby crack surfaces cannot separate as they would without the feature present. In effect, the unbroken ligament acts as a spring, i.e., a crack interface bridge.

The generation of Wallner lines at the ends of major hackle features indicates that the stress pulse responsible for the Wallner line is produced in the last stage of formation of the hackle feature. It seems reasonable that the pulse is generated at the completion of the lateral crack growth that opens the bridge. Until that moment, the crack is held together by the ligament. When the ligament breaks, there is a sudden separation of the crack faces. That separation produces the acoustic emission that has been measured in high velocity fracture. That stress pulse also creates the Wallner line.

The bridging effect of hackle features may explain why the terminal velocity in a typical fracture in glass is so low compared with the Rayleigh velocity. Most experiments have given values of about 1500 m/s for the terminal velocity in glass. In those experiments, the velocity rose asymptotically to the terminal value as the crack progressed through the mist zones and hackle zones. Rose (16) argued that the low velocity was a result of the opening and closing of microcracks in front of the crack tip. That process dissipated energy from the moving crack in the form of heat.

However, in certain circumstances, velocities much greater than 1500 m/s can be attained in glass. As noted above, in experiments by Snowden (27), in which an existing crack was shock loaded so that the terminal velocity was attained after propagating only a short distance, velocities of over 3000 m/s were attained. Snowden found no mist or hackle on those surfaces even when branching occurred. Similarly, Field (14) found that, when propagation occurred on easy cleavage planes of single crystals of MgO and other materials, no mist and hackle were formed and velocities approaching the Rayleigh velocity were attained. Hull and Beardmore (28) obtained similar results in tungsten. On the other hand, when fracture ran on other planes in which branching into alternate planes was relatively easy, mist and hackle formed at relatively low K values, and much lower terminal velocities were observed. Field also measured small changes in the velocity of glass when "serious roughening" appeared on the fracture surface of glass samples. He attributed that decrease to the use of released strain energy in the forming of extra fracture surface in the hackle features. In view of the bridging effect of major hackle observed here, it seems reasonable to attribute the lower terminal velocities to hackle bridging.

MACROSCOPIC CRACK BRANCHING

The Nature of Branching

In its simplest terms, branching is a process in which a rapidly moving crack splits to form two or more fracture surfaces. The angle between the branches is limited by the nature of the applied stress. For a uniform biaxial tension, it can be as large as 180 degrees, while for lath bending, it is limited to about 45 degrees. Any model of branching must explain that dependence of maximum branching angle. It must also describe how the secondary crack originates at the primary crack plane and what determines its growth to the point at which it becomes critical in the far field stress.

Lawn (30) describes three possible mechanisms for branching. The first of these is based on the analyses of Yoffe and others (9,10) who showed that, for fast fracture, $\sigma_{\theta\theta}$ has a maximum at other than $\theta = 0$. That leads to a natural tendency for the crack to move out at an angle from the crack plane. Lawn points out, however, that branching occurs at smaller angles than expected from the Yoffe hypothesis. Also, branching occurs at lower velocities than those predicted on the Yoffe hypothesis. In the second mechanism, microfractures occur in front of the crack tip and eventually are linked to the primary. Lawn notes that this mechanism seems to explain the generation of mist. As K increases, these microcracks get larger before the primary crack arrives so that the hackle features increase in size. However, as Lawn observes, microfractures are not observed in glasses and homogeneous crystals where there are no intrinsic flaws. The third mechanism that Lawn describes involves stress-wave branching. Stress pulses generated at the crack front, and reflected back to it, interact to change the local direction.

To understand how branching occurs in glass, consider the micrograph in Figure 18. This micrograph was taken with transmitted light passing through a plate of lightly tempered glass. The camera axis was normal to the surface of the plate. The lower section of the figure is a continuation of the upper section. The plate had been covered with transparent tape prior to fracturing to keep the dicing fragments together. The heavy dark line running horizontally was a primary crack while the heavy, dark, relatively straight lines running off at 60 to 90 degree angles were secondary (branching) fractures.

In addition to the heavier lines there are a number of fine curved lines issuing from the primary crack. Some of these are connected to the secondaries while others

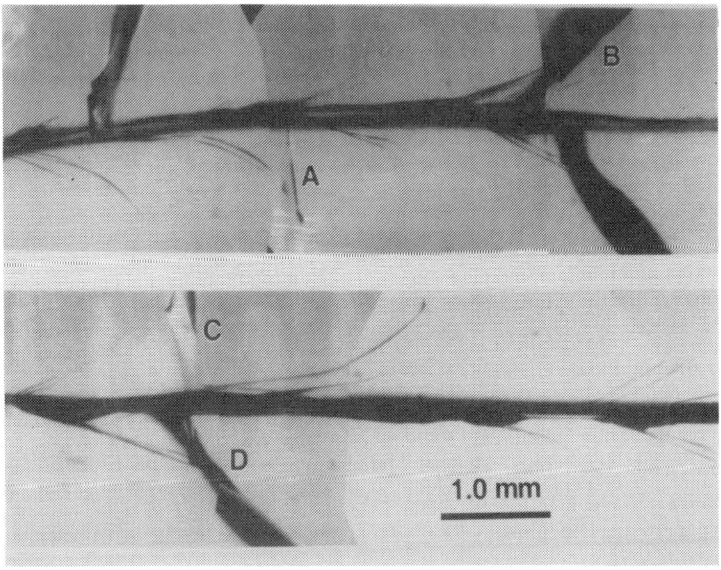

Fig. 18. Macroscopic branching (micrograph taken with transmitted light).

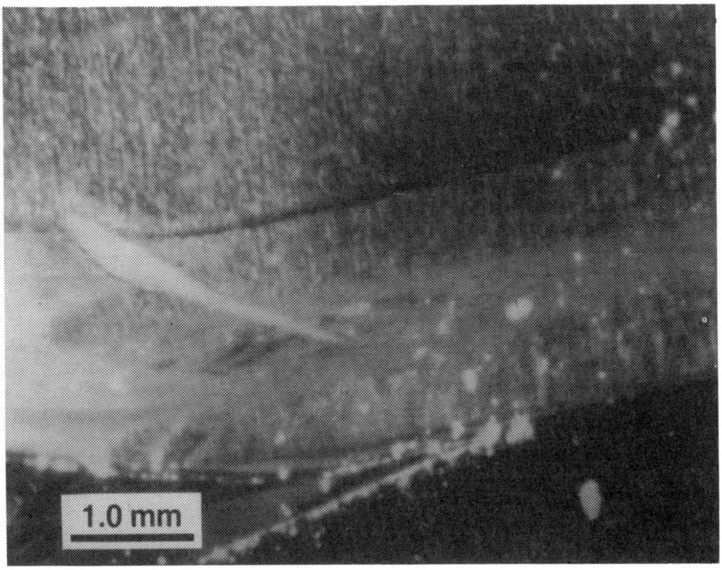

Fig. 19. Ribbon crack curving away from primary crack. Viewed through the glass plate surface at an angle of 45 degrees to the primary plane.

terminate within the glass. These fine lines are extended ribbons of fracture that initiated at the primary crack. One such ribbon is shown in Figure 19. It has the appearance of a feather whose width increases with distance away from its initiating point at the primary crack.

A Model for Branching

These ribbons of fracture are the precursors to macroscopic branching. Figure 20 is a schematic showing the branching process. Immediately after its creation, the precursor is a narrow tongue exiting the primary. As it grows forward and away from the primary, it becomes more ribbon-like and it also grows laterally. That growth is driven initially by the stress field surrounding the primary crack tip. However, as the tip of the ribbon moves away from the primary crack plane, the effect of the applied stress field becomes more dominant. The branching fracture is produced when the lateral growth of the ribbon is sufficient that the tip of the ribbon attains criticality in the far field stress. As indicated, the "secondary" crack that issues from the tip of the ribbon grows back toward the primary as well as away from the primary.

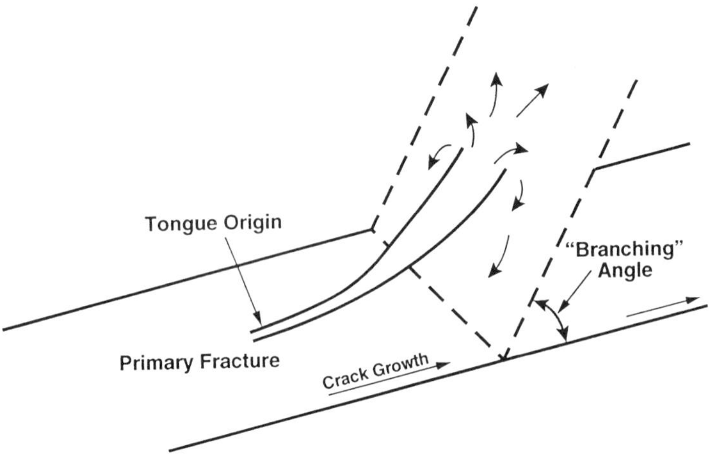

Fig. 20 Schematic of the macroscopic branching process.

It is clear, in this model, why the "branching angle" does not correspond to the angle at which $\sigma_{\theta\theta}$ is a maximum in the Yoffe analysis. The branching angle is determined by the orientation of the tensile stress field after the tip of the ribbon reaches criticality. As noted above, the stress field contains a component from the Irwin stress field around the primary crack. However, at the distance from the primary where criticality is attained, the contribution from the far field stress is comparable to, or even greater than, the primary crack stress field. Thus, the branching angle is primarily a function of the direction of the far-field stress.

The variation in branching angles for a given loading situation exists because the far-field stress varies with time and with details of the local crack pattern. To a large extent, this variation results from the introduction of a free surface by the primary crack; however, the creation of other secondaries also contributes. Moreover, because the rate of growth of a ribbon depends on details of the combined stress field, the time required for a given ribbon to reach criticality will vary.

With this model in mind, consider the details of the crack branching in Figure 18. Note the variation in the trajectories of the fracture ribbons. Generally speaking, the shorter the distance the ribbon travels before it reaches criticality, the smaller the branching angle. Branches C and D, for example, with $\theta \approx 60$ degrees, were generated when the ribbon had moved about 0.6 mm from the primary, while branch D, at 90 degrees, was produced when the tongue had extended about 0.95 mm from the primary. The implication is that, at shorter times where branching occurs at the smaller distance from the primary, the primary crack has advanced only far enough to produce a 60 degree orientation of the far field stress. However, when the ribbon moves 0.95 mm from the primary before reaching criticality, the elapsed time is enough that the primary crack has extended to the point that the remanent stress is oriented at 90 degrees.

To understand how a branching angle of 90 degrees occurs in Figure 18, recall that the tempering process produces a uniform biaxial tensile stress field. However, after the primary crack runs through the sheet, the remaining tensile stress field is parallel to the primary crack surface. Thus, the branches at 90 degrees are created when the primary crack tip has advanced well beyond the ribbon at the moment the tip of the ribbon becomes critical.

Occasionally, a ribbon that is subcritical in the remanent field can still produce a secondary. For the tempered plate, that can happen when the primary crack extends laterally to the surface of the plate, admitting water vapor to the tip of the ribbon and promoting subcritical crack growth. Because the residual stress in the vicinity of the tip of the ribbon is parallel to the primary crack surface during that growth, the branch will be oriented at 90 degrees to the primary. In fractured tempered glasses, the formation of secondaries by subcritical crack growth is evidenced by audible clicks well after the catastrophic fracture of the sheet is complete.

Examination of the surface produced by the delayed branching fracture will often show an arrest line where the subcritical crack growth began. Figure 22 shows an example of one of these arrested ribbons (marked with an arrow). The branching

Fig. 21. Branch surface, oriented at angle of 60° to plane of micrograph, showing arrest line (marked with small arrow.

fracture surface generated from that ribbon is the light-shaded band through the horizontal center of the micrograph. It is oriented at about 60 degrees to the plane of the micrograph. The point on the primary surface where the ribbon originated is above the top edge of the micrograph. The stress driving the growth of the branching crack, assuming a circular crack of radius 0.5 mm and a K value appropriate to subcritical crack growth, was 28 MPa. That matches the value obtained from the branching distance during dicing of the lightly tempered glass sheets used in this study.

Obviously, from Figure 19, not every ribbon of fracture produces a macroscopic branch. It is only when the ribbon grows large enough to become critical that a macroscopic branch forms. Moreover, ribbons of fracture are generated well before the branching boundary is reached. Every mesa feature that is completed by extension of the primary crack, as in Figures 9b and 10 leaves a tongue or a partially developed ribbon penetrating beneath the surface. At low G values, the growth of those ribbons is too slow for them to develop into macroscopic branches. In terms of the model presented here, the criterion for macroscopic branching is that the velocity of the crack growth at the tip of the ribbon be high enough that criticality is achieved before the primary crack front moves far enough from the ribbon so that it stops growing. In the vicinity of K_b, the velocity of the primary does not change much with increasing K. Thus the criterion for branching appears to be that the G value at the tip of the ribbon exceeds a critical value at which growth of the ribbon can result in attaining criticality.

In an investigation of dynamic fracture in a polymer glass (Homalite-100), Ravi-Chandar and Knauss(12) observed a similar branching behavior. They remarked that branches were formed by a precursor turning smoothly away from the primary crack plane. They also saw that many of these precursor cracks failed to develop into complete branching cracks. They concluded that the branching process does not involve the splitting of the crack into two cracks along a common front. Unlike the model to be presented below, however, they argued that the branching process began with the growth and interaction of microcracks that were oriented parallel to the primary crack plane.

Doyle (31) observed ribbon precursors to branching in PMMA and attributed the initiation of those ribbons to the growth of craze-cracks in advance of the primary crack. He postulated a model for branching in nonpolymeric glasses that involved "embryonic branches" growing into ribbons and noted that not all ribbons result in branches. As one of the possible mechanisms for the formation of the embryonic branches, he suggested that the crack front deflected out of the primary crack plane because of interaction with a microstructural defect. He noted that the tendency for a sector to deviate out of the primary plane increases with velocity because of the Yoffe effect.

The velocity of the primary crack does not change rapidly with K near K_b. Thus, whether the ribbon tip grows to criticality depends on the magnitude of the K value at the ribbon tip rather than on the time that is available for growth. There is a clear inference that, as the K value increases at the primary crack front, it also increases at the ribbon tip. There is also an implication that the ribbon is undergoing stable growth up to the point of criticality, at least in the sense that the length of the ribbon does not determine the K value at its tip.

Origin of Macroscopic Branching

In this model of macroscopic branching, one of the key issues is how the ribbon of fracture originates (32). Figure 22 is a micrograph of a fracture surface at the base

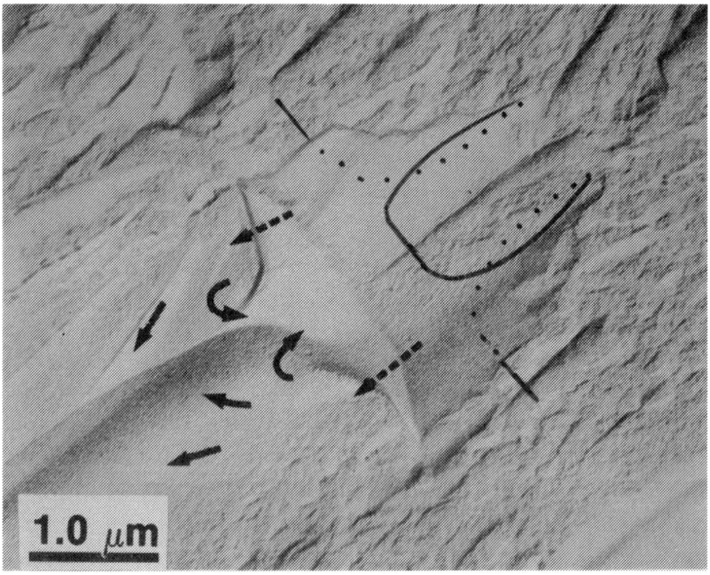

Fig. 22. Origin of a macroscopic branch. Crack propagation from upper right to lower left. Arrows show fracture direction on primary plane and up to intersect tongue surface.

of one of the ribbons. In this instance, the ribbon did not develop into a macroscopic branching fracture. Examination of the details of the feature reveals that it has the same general character as that of a mesa feature. At its initiation point, a short segment of the crack left the primary crack plane by tilting away from the primary plane. Note that there is a smooth transition as the segment leaves the primary, i.e., there is no suggestion in this micrograph that the stress field in front of the primary crack tip had activated a microfracture that was then intersected by the crack front. A tilt in the opposite direction resulted in a short section in which the crack growth grew essentially parallel to the primary crack plane. The tongue, in its early growth, is sketched on the figure. Dotted lines show the position of the nearby primary crack at that same moment and arrows show the direction of growth of primary beneath the tongue and then up to intersect the tongue. Unlike the mesa feature in Figure 8, the feature in Figure 22 was not terminated by having the tongue tilt back to the primary. Instead the tongue grew away from the primary. The fracture surface to the left shows what is essentially a wake hackle feature that was produced as the primary crack grew laterally and up towards the tongue to complete the separation across this fracture surface.

Not all of the origins of branching cracks look exactly like that in Figure 21. As noted earlier, often the tongues of fracture from two or more adjacent mesa features coalesce to form one large tongue. All other factors being equal, it should be much easier to attain criticality when starting the ribbon with such a broad feature than to start with a single narrow tongue.

The branching shown above, including the originating feature in Figure 21, all occurred in tempered glass. In branching of tempered glass, the tensile zone is located within the glass sheet. For typical glass sheet processing, intrinsic flaws in that zone would not be expected to be activated at stresses less than 1 GPa. Consequently, it may not be surprising that branching in tempered glass does not involve activation of the available internal flaws. On the other hand, in uniaxial tension or flexural loading of a bar, the sample surface is in tension. It seems reasonable to expect, in that case, that flaws on the surface of the sample could be activated by the stress field at the crack tip and that secondary cracks would issue from those flaws. In glass whose surface has not been severely damaged, that apparently does not occur. Figure 23 shows branching fractures in a glass microscope slide. The micrograph is of the plane of the slide using transmitted light. As in the case of branching in tempered glass (biaxial tension), the branching process begins with a fine ribbon of fracture curving out of the primary plane and

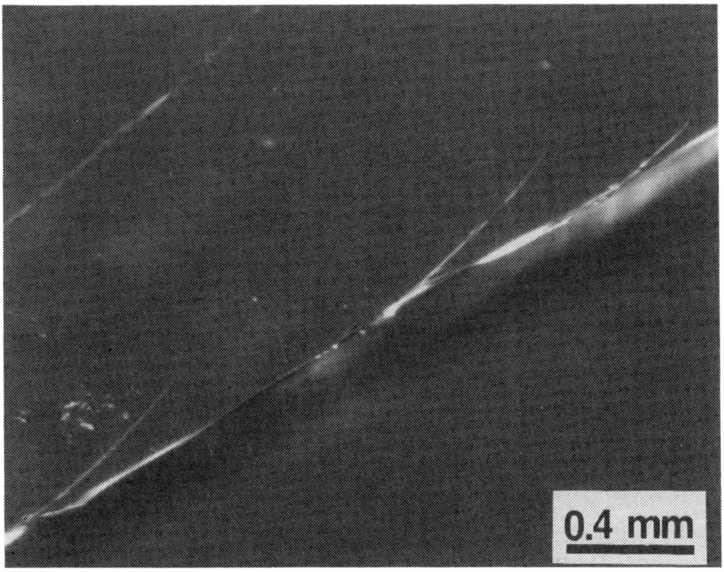

Fig. 23 Fracture ribbons produced in flexural failure. Crack propagation from lower left to upper right.

eventually becoming critical in the far field stress. As in the tempered glass, ribbons of fracture are generated well before macroscopic branching occurs.

Figure 24 is an optical micrograph of a fracture surface at the base of one of the ribbons of fracture. The origin of that ribbon is indicated by the arrow. Just as in the tempered glass branching process, the ribbon begins at a mesa feature. That

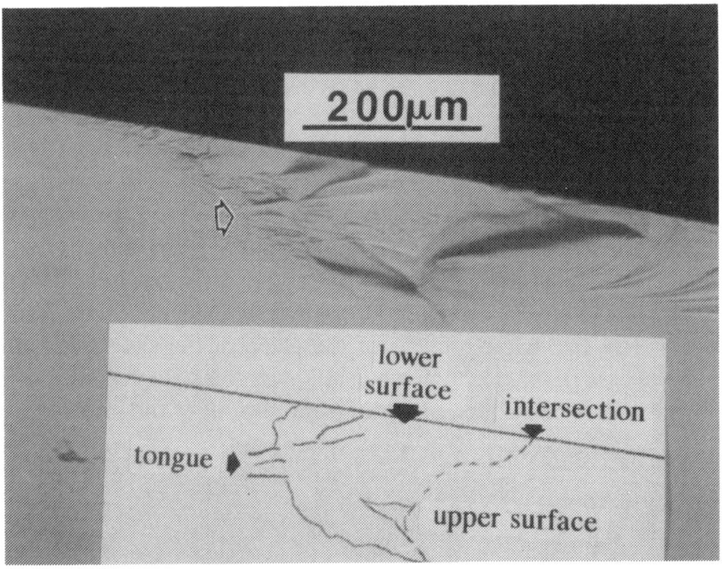

Fig. 24. Origin of one of the fracture ribbons in Fig. 23. Crack propagation from left to right.

feature is not produced at the surface of the slide but is 60 μm beneath the surface. Thus, as in the tempered glass case, branching in this flexural failure began with a tongue of fracture moving out of the primary crack plane and growing in the stress field of that crack tip.

These observations do not necessarily mean that surface flaws are never activated by the stress field at the crack tip. However, even when a slide is abraded with silicon carbide paper, lowering the strength to less than 70 MPa, branching initiates by the tongue growth process rather than activation of one of the numerous indentation flaws produced by the abrasion.

To determine whether the initial tongue of fracture is oriented at an angle corresponding to the Yoffe prediction, several tongues of fracture were examined with the SEM at a viewing angle of 45 degrees to the fracture surface. Typical measured values of the angle were 14 to 20 degrees. Erdogan's calculations (10) showed no maximum in $\sigma_{\theta\theta}$ at $\theta \neq 0$ for crack velocities less than 0.4 of the longitudinal sound wave velocity, V_l. Nevertheless, the Yoffe effect is probably responsible for the deviation out of the primary crack plane that results in generation of mesa features and the precursors to microscopic branching. That contribution comes, not from the appearance of a maximum in $\sigma_{\theta\theta}$ at $\theta \neq 0$ at lower velocities, but from the broadening of the peak about $\theta = 0$. Essentially, even at 0.25 of V_l, that peak is sufficiently broad that deviation out of the primary plane at, say, a 20 degree angle is relatively easy compared with the quasistatic case.

CONCLUSIONS

The overall message from the examples shown here is that, in soda-lime silicate glass at least, all of the processes for forming mist and hackle and for macroscopic branching involve continuous growth at the crack front. There is no step in any of those processes that depends on activation of microfractures in advance of the crack front or the conjunction of such microfractures with the crack front. Neither is there any process that requires the generation of two or more crack surfaces from a single line segment at the crack front.

Other conclusions are that:

1. Two general types of hackle features are found in mist and hackle zones. One type is very similar to twist hackle produced at lower velocities and is associated with a twist or shear displacement at the crack front. The other type begins with a short segment of the crack front tilting out of the primary plane and forming a mesa-like feature.

2. G_h, the strain energy release rate ascribed to the mist/hackle boundary, is neither a necessary nor sufficient condition for the generation of hackle.

3. Major hackle can be generated at $G < G_h$, provided there is a suitable level of distortion at the crack front.

4. Stress pulses that produce Wallner lines can also aid in generation of major hackle.

5. The rapid separation of the crack faces when lateral crack growth completes a hackle feature produces a stress pulse that can result in a Wallner line extending from the end of the hackle feature.

6. G_h may be the value at which Wallner lines produced at one major hackle feature can generate other major hackle features; consequently, the zone is filled rapidly with hackle by lateral stress-pulse transmission.

7. The unbroken ligament present during formation of a hackle feature can act as a crack-interface bridge. The suppression of the K value at the crack tip in the vicinity of the hackle results in a slowing of the crack.

8. The bridging effect of the hackle in the mist and hackle zones may be responsible for the low terminal velocity common to fracture in silicate glasses.

9. Macroscopic bridging begins with a tongue of fracture extending from a mesa-type hackle feature. That tongue grows initially in the stress field of the primary crack tip, developing into a ribbon that curves away from the primary plane. When the tip of the ribbon attains a critical size, as a result of lateral growth, in the far-field stress, branching occurs.

10. The macroscopic branching angle is determined by the orientation of the far-field stress at the ribbon tip at the moment the tip attains a critical G value.

ACKNOWLEDGMENTS

Much of this work was accomplished while the author was at Sandia National Laboratories, supported by US DOE Contract DE-AC04-94AL8500.

REFERENCES

1. R. W. Rice, "Ceramic Fracture Features, Observations, Mechanisms and Uses", pp. 1-103 in ASTM Special Technical Publication, Vol. 827, *Fractography of Ceramic and Metal Failures*. Edited by J. J. Mecholsky and S. R. Powell, American Society for Testing and Materials, Philadelphia, PA, 1984.

2. J. J. Mecholsky, S. W. Freiman, and R. W. Rice, “Fractographic Analysis of Ceramics,” Fractography in Failure Analysis, ASTM STP 645, B. M. Strauss and W. H. Cullen, Jr. Eds., American Society for Testing and Materials, 1978, pp. 363-379.

3. J. J. Mecholsky, S. W. Freiman, R. W. Rice, "Fracture Surface Analysis of Ceramics," J. Matls. Sci. **11** (1976) 1310-1319

4. H. P. Kirchner and J. W. Kirchner, "Fracture Mechanics of Fracture Mirrors," *J. Am. Ceram. Soc.* **62** [3-4] 198-202 (1979).

5. H. P. Kirchner and J. C. Conway, "Comparison of the Stress-Intensity and Johnson-and-Holloway Criteria for Crack Branching in Rectangular Bars," *J. Am. Ceram. Soc.* **70** [8] 565-69 (1987).

6. G. R. Irwin, "Fracture", pp. 551-90 in *Encyclopedia of Physics*, Vol. 6. Edited by S. Flugge. Springer-Verlag, Berlin, 1958.

7. Y. L. Tsai and J. J. Mecholsky, Jr., "Fracture Mechanics Description of Fracture Mirror Formation in Single Crystals," *Int. J. Fract.* **57** 167-182 (1992).

8. Roy W. Rice, "Perspective on Fractography", Advances in Ceramics, Vol.22; Fractography of Glasses and Ceramics, The American Ceramic Society, Inc (1988) pp. 3-56.

9. E. H. Yoffe, "The Moving Griffith Crack," *Phil. Mag.* **42**, (1951) 739-50.

10. F. Erdogan, "Crack Propagation Theories," Chap. 5 in *Fracture*, Vol. 2. Edited by H Liebowitz, Academic Press, New York, (1968).

11. J. W. Johnson and D. G. Holloway, "Microstructure of the Mist Zone on Glass Fracture Surfaces," *Phil. Mag.* **17** 899 (1968).

12. K. Ravi-Chandar and W. G. Knauss, " An Experimental Investigation into Dynamic Fracture: II. Microstructural Aspects," *Int. J. Fract.* **26** (1984) 65-80.

13. D. M. Kulawansa, L. C. Jensen, S. C. Langford, and J. T. Dickinson, "Scanning Tunneling Microscope Observations of the Mirror Region of Silicate Glass Fracture Surfaces," *J. Mater. Res.*, **9** [2] 476-485 (1993).

14. J. E. Field, H. S. Dobbs, and M. J. Twigg, "The Velocity of Crack Propagation in Brittle Materials," in Fracture 1969, H.L. Pratt, ed., Chapman and Hall, Ltd., London.(1970).

15. J.E. Field, "Brittle Fracture: Its Study and Application," *Contemp. Phys.*, **12** [1], 1-21 (1971).

16. L. R. F. Rose, "Recent Theoretical and Experimental Results on Fast Brittle Fracture," *Int. J. Fract.* **12** (1976) 799-813.

17. A. S. Kobayashi and M. Ramulu, "A Dynamic Fracture Analysis of Crack Curving and Branching," *J. de Physique*, Colloque C-5 197-206 (1985).

18. M. Marder and S. Gross, "Origin of Crack Tip Instabilities" *J. Mech. and Phys. Solids* **43** [1] 1-48 (1995).

19. Y.-H. Taguchi, "Dynamical Modelling of Fracture Propagation," *Matls. Sci. and Eng.* **A176** 295-98 (1994).

20. K. S. Kim, "Elastic-Plastic Analysis of Crack-Propagation with Branching," *Eng. Fract. Mech.* **47** [4] 599-612 (1994).

21. V. D. Frechette, Advances in Ceramics, Vol. 28, *Failure Analysis of Brittle Materials*, American Ceramic Society, Westerville, OH, 1990. V. D. Frechette, "Characteristics of Fracture- Exposed Surfaces", Proc. Of Brit: Ceram. Soc. 5 97-106 (1965)

22. E. Sommer, "Formation of Hackle 'Lances' in Glass", *Eng. Frac. Mech.*, **1** 539-546 (1969).

23. T.-Y. Pan, R. E. Robertson, and F. E. Filisko "Similarities in the Fracture Surface Features of Borosilicate and Polymer Glasses," *J. Mat.. Sci.* **24** (1989) 3635-42.

24. A. I. A. Abdel-Latif, R. C. Bradt, and R. E. Tressler, "Dynamics of Fracture Mirror Boundary Formation in Glass," *Intl. J. Fracture*, **13** [3] (1977) 349-359.

25. E. K. Beauchamp, "Crack Front Stability and Hackle Formation in High Velocity Glass Fracture," *J. Am. Ceram. Soc.* **78** [3] 689-97 (1995).

26. A. Tsirk, "Formation and Utility of a Class of Anomalous Wallner Lines on Obsidian"; pp. 57-69 in Advances in Ceramics, Vol. 22, *Fractography of Glasses and Ceramics*. Edited by V. D. Frechette and J. R. Varner, American Ceramic Society, Westerville, OH, 1988.

27. W. E. Snowden, "Crack Growth in Glass Subjected to Controlled Impacts," PhD thesis, U. California, Berkeley, CA; Lawrence Berkeley Laboratory Report for US EKDA Contract W-7405-ENG-48, November, 1976.

28. Y. W. Mai and B. R. Lawn "Crack-Interface Grain Bridging as a Fracture Resistance Mechanism in Ceramics: II. Theoretical Fracture Mechanics Model," *J. Am. Ceram. Soc.*, **70** [4] 289-94 (1987).

29. D. Hull and P. Beardmore, "Velocity of Propagation of Cleavage Cracks in Tungsten," pp. 629-637 in *Proceedings of the First International Conference of Fracture* (Sendai, Japan, 1966), Vol. 2. Japanese Society for Strength and Fracture of Materials, Japan, 1966.

30. Brian Lawn, Fracture of Brittle Solids, Second Edition, Cambridge University Press, 1993, pp.95-99.

31. M. J. Doyle, "A Mechanism of Crack Branching in Polymethyl Methacrylate and the Origin of the Bands on the Surfaces of Fracture," *J. Matls Sci.* **18** (1983) 687-702.

32. E. K. Beauchamp, "Fracture and Strain Energy Release in Prestressed Glass and Glass-Ceramics," Fracture Mechanics of Ceramics, Vol. 3, edited by R. C. Bradt, D. P. H. Hasselman and F. F. Lange (Plenum Press, New York 1978) p. 823.

HACKLES REVISITED

Are Tsirk
MTA Bridges and Tunnels
Randall's Island
New York, NY 10035

ABSTRACT

Hackle-related markings, as well as effects on or by hackles in the presence of water, are explored in obsidian.

"Hackles" is used here in a broad sense to include twist hackles, tails or wake hackles, and the fine and coarse hackle marks of mist and hackle regions. The hackle-related markings considered were all encountered in experimental knapping or flintworking of obsidian, a volcanic glass, from Glass Buttes, Oregon, and unknown sources in Mexico. References on fracture markings in this context are cited in (1,2).

Tails are commonplace. But multiple tails at the same inclusion occur less frequently. These are explored, with an emphasis on double tails.

Two fundamentally different types of localized mist or hackle manifestation are considered. Mist lines in the direction of fracture propagation were already reported in 1944 by Appelt for quartz glass (3). They were also illustrated in (2), and studied in depth recently by Beauchamp (4,5). Configurations of localized mist or hackle manifestation not in the direction of fracture propagation, due to "Wallner effects," were investigated by Andrews (6,7) for carbon-filled, frozen rubber, and discussed in (4) for glass. Configurations characterized by localized inhibition of mist are considered as well.

Observed effects related to hackles and water are also explored.

As a background, some aspects of twist hackles and (single) tails are reviewed first.

The study is based essentially on optical microscopy, with magnifications usually not exceeding 200X because of the curved surfaces considered. References to mist and its visibility are in this context. Existence of finer surface roughness, as in a mirror region, is well known (3,4) but not considered here except by reference.

TWIST HACKLES

It is well known that twist hackles (8) are formed due to the direction of the principal tensile stress rotating about a line in the direction of fracture propagation (9). This twist in the tensile stress direction (Mode III contribution) may occur very locally or over a large region. At times this change can be considered in a static sense. Yet at other times it must be viewed in terms of stress waves. The changes can be due to external loads, as well as material and geometrical effects.

Recurrent twist hackle patterns have been produced in glass by stress waves using ultrasonic modulation, as in Figs. 5.19 and 5.20 in (10). Counterparts of such recurrent families of twist hackles also occur in knapping of obsidian.

Geometrical effects can lead to localized lagging of the fracture front, to Mode III participation and to twist hackles (11). This may also be expected for some material effects.

"Premature" manifestation of a twist hackle in obsidian has been observed due to an inclusion near but away from the fracture plane (12). When a fracture is deflected around an inclusion, but being still completely in the matrix, a short hackle sometimes occurs on one or both sides of the inclusion. Groups of hackles are sometimes observed to the side of but not contiguous with an inclusion. Short twist hackles are occasionally also seen just upstream from an inclusion. All of these twist hackles can be explained by localized stress changes (Mode III contributions) caused directly or indirectly by an inclusion.

A single twist hackle in a group of twist hackles sometimes starts earlier than others, at an inclusion or other irregularity. In some obsidians with many inclusions, every twist hackle in a family of long twist hackles is sometimes observed to start at an inclusion. All of these can be viewed as "premature" manifestations of twist hackles or as tails or wake hackles persisting due to encountering a Mode III stress environment. It is expected that, in at least some such cases, a Mode III environment contributed to the initial manifestation of the tails.

TAIL OR WAKE HACKLE

A tail or a wake hackle at an inclusion is formed due to a surface mismatch of two partial fracture fronts being produced at the inclusion. A surface mismatch to produce a tail can occur when two partial fracture fronts pass around the sides of an inclusion (8). There are other possibilities, however, when fracturing of or at an inclusion occurs before passage by the partial fracture fronts.

In obsidian, a tail has also been observed with a fracture passing through or partly through an inclusion, as in Fig. 1A. That production of surface mismatches at an inclusion is possible without the fracture running around the inclusion is also evident from the observations in the next sections on multiple tails. It is not the intent here to discuss the specific nature and detailed mechanisms for production of such surface mismatches in obsidian. The possibilities can be seen to be numerous when one keeps in mind the

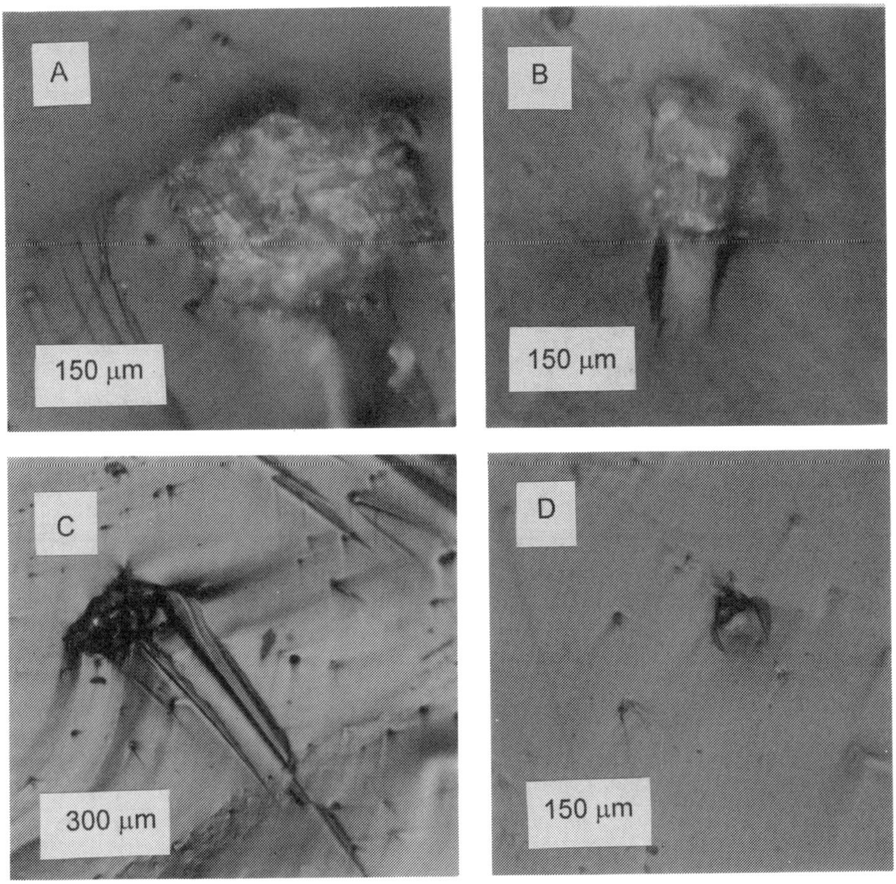

Figure 1. Tails: (A) Tail (bottom, out of focus) formed by fracturing partly through inclusion, seen by hackles extending from it. Double (B) and triple (C) tails. (D) Double tails with and without trailing mist lines, and mist line (at left, out of focus).

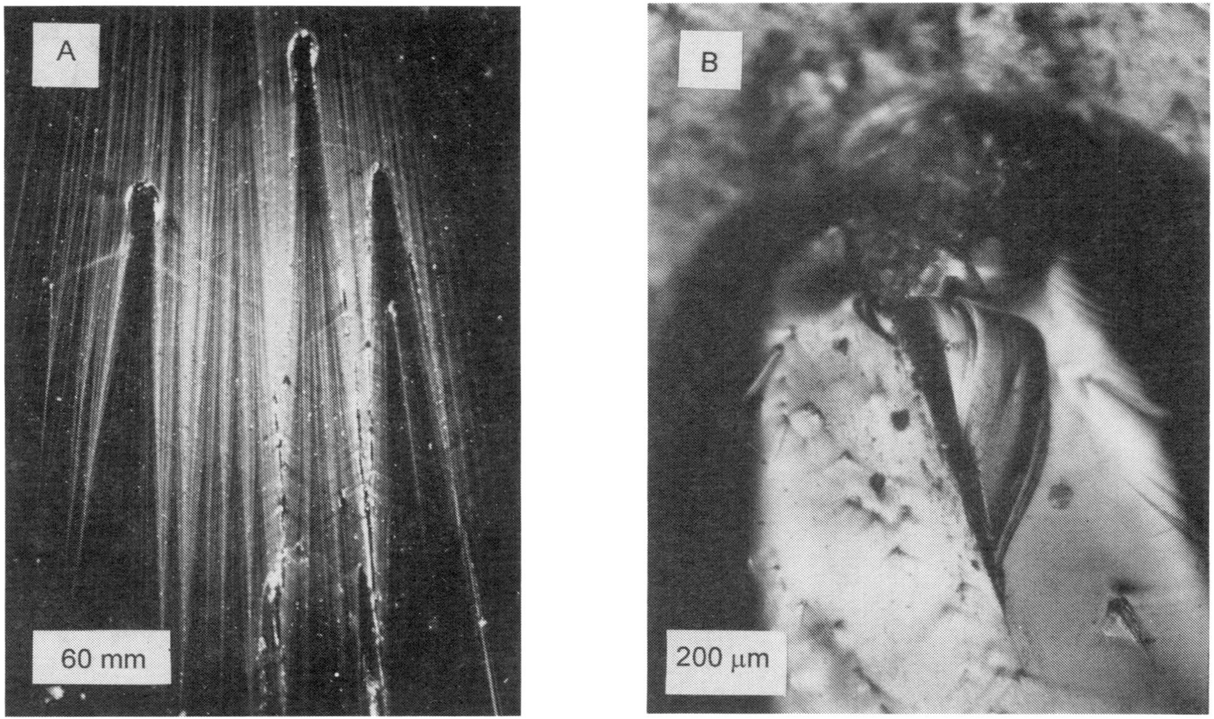

Figure 2. (A) Double tails, mist lines and Wallner mist lines (Photo by Prof. V.D. Fréchette, 1982). (B) Detail at head of double tail at top center of (A).

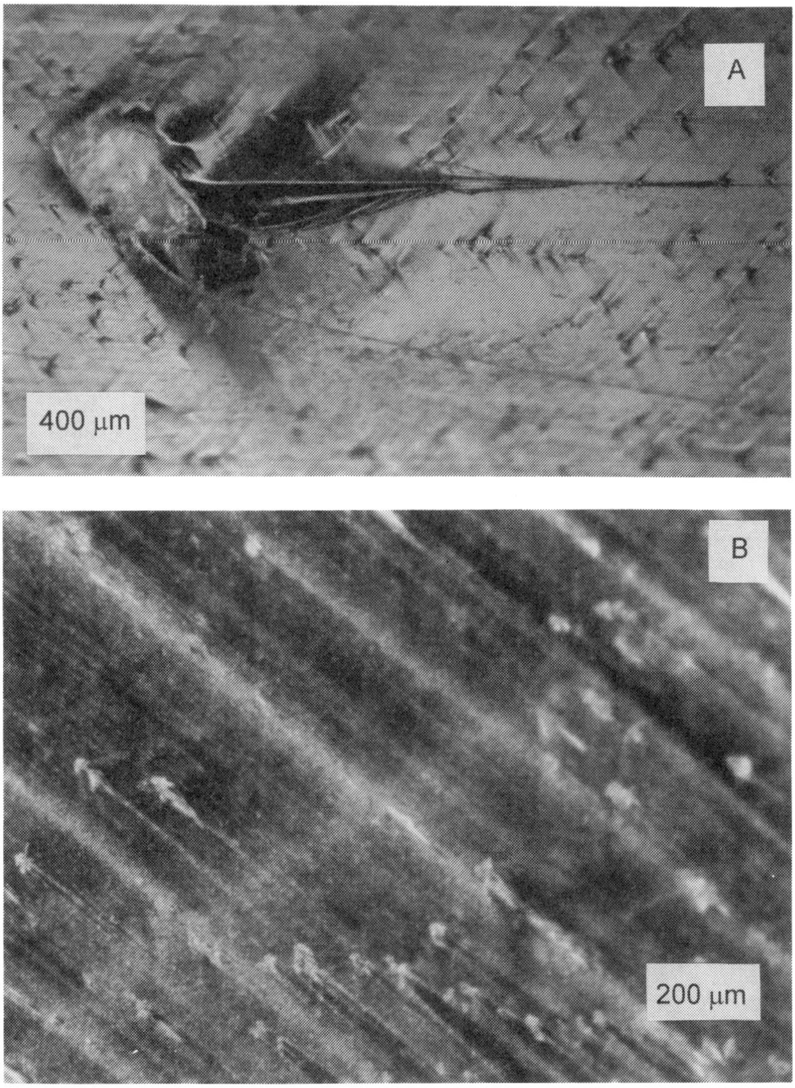

Figure 3. (A) Double tail formed by fracture passing through inclusion onto different plane. (B) Mist lines in fracture direction (down towards right), a Wallner mist line (detail of white arc at upper right in Fig. 2A), and numerous double tails with trailing mist lines (near bottom).

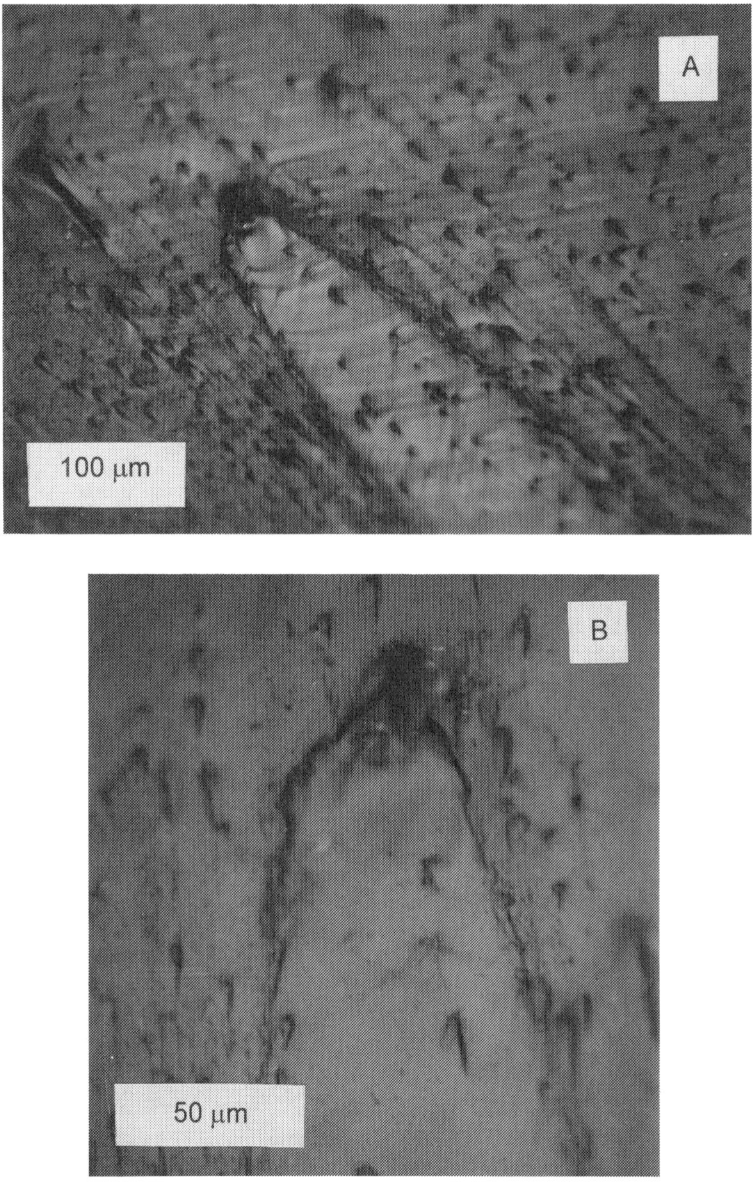

Figure 4. Double tail (A) with detail (B) (Photos by Prof. V.D. Fréchette, 1982).

variability of the structure and other properties of the inclusions, their geometry, orientation and locations with respect to the fracture plane (See Figs. 1 to 4). Many polycrystalline inclusions in obsidian are complex, with discontinuities and irregularities, and some even appear to have flaws, cracks or "cleavage" planes. The fracture paths through some of them are rather tortuous. Discontinuities and steps, tilting and twisting of the fracture plane, as well as fracturing in various directions are sometimes evident in an inclusion in obsidian.

It seems practical to ascribe an all encompassing, robust mechanism for the formation of a tail, as stated in the beginning of this section, but with the understanding that the associated surface mismatch can occur with a fracture passing around, through or partly through an inclusion, depending on the timing of the fracturing of or at the inclusion.

MULTIPLE TAILS

Multiple tails or wake hackles, associated with the same inclusion, have been observed fairly frequently in obsidians (Figs. 1 to 4). A multiple tail at an inclusion is formed due to multiple partial fracture fronts producing multiple surface mismatches at the inclusion. For example, a triple tail as in Fig. 1C is associated with three surface mismatches and four partial fracture fronts.

The multiple fracture fronts and surface mismatches can be viewed to be due to irregularities within the inclusion, a localized twisting (Mode III) stress environment due to the inclusion or a combination of both. Of course, fracturing at the inclusion must occur before partial fracture fronts can pass completely around it.

Single and multiple tails as well as twist hackles near a mist region are usually observed with trailing mist lines (Fig. 1D). Multiple tails are also seen as some of the constituent markings of mist and hackle regions.

VARIABILITY OF DOUBLE TAILS

A double tail, a variant of multiple tails, is produced by the formation of three partial fracture fronts as a fracture encounters an inclusion. It is manifested due to the surface mismatches where the adjacent partial fracture fronts meet.

Some variations of double tails, with some of their details, are shown in Figs. 1 to 4. In obsidian, they are usually observed with pronounced "shoulders" (Figs. 1D, 2, 3A). Those near general mist regions commonly have mist lines as "arms" extending from the shoulders (Figs. 1D, 2, 3A), while those further away do not (Figs. 1B, 1D). Some of the markings, as the unusual ones in Fig. 2A photographed at 2.9X magnification, are rather large. Smaller ones, as in Figs. 1B, 1D and 4, are more common, especially with trailing mist lines near mirror boundaries. Some of those with trailing mist lines resemble markings known as conic lines (7) or fracture hyperbolas and related markings (10,13). Their formation, however, is very different from fracture hyperbolas in materials such as PMMA (polymethyl methacrylate), as seen from comparison of the figures here with Fig.

6.6 in (7), for example. Unlike the markings in PMMA, each of the double tails has an inclusion at its "head" (the apex of the "hyperbola").

One or both of the shoulders of a double tail may be curved (Figs. 2A) or straight (Fig. 1B). This geometry depends on the time the partial fracture fronts have to propagate independently. Some double tails appear to have no such shoulders (Fig. 4A), but at higher magnification (Fig. 4B) are still seen to have tiny "tails" or hackle marks extending from an inclusion (or irregularity). Depending on the viewer's scale of interest, it may be said that some double tails have, instead of the pronounced shoulders, mist lines extending from the inclusion.

Double tails may be symmetrical (Fig. 2A) or quite distorted (Figs. 3A). This depends on factors including the orientation of the inclusion and the fracture plane within it. Occasionally, the mist lines trailing from the shoulders of a double tail are convergent.

Double tails seem to be most common adjacent to general mist regions, either upstream of or to the side of such regions. In regions associated with slower fracture velocities, they are seen less frequently though by no means rarely (Figs. 1B). As already noted, double tails can also be found as the occasional constituent markings of mist and hackle regions (Fig. 5).

Instead of the trailing mist lines, the arms of some double tails are rarely seen to consist of recurrent large hackle marks resembling a linear series of twist hackles.

Some of the larger double tails in Fig. 2A have maximum level differences ranging from about 0.05 to 0.2 mm. Sometimes the central plane is tilting so that the maximum level difference occurs well downstream from the inclusion at the head of the marking. Occasionally, the central plane is twisted, sometimes being higher than the outer plane on one side and lower on the other.

Most commonly, the double tails discussed are formed with a fracture passing through an inclusion and onto a different fracture plane, forming the central partial fracture front (Fig. 3A). This partial fracture front is also expected to form occasionally by an inclusion separating from the matrix, away from the primary fracture plane. The timing of all these secondary fractures is, of course, highly relevant to the formation and independent propagation of the central partial fracture front.

The formation of a double tail is sometimes apparently associated with an independent fracture starting at an inclusion away from the primary fracture plane, as probably in Fig. 2B. It is of interest to ask whether such secondary fracturing occurred ahead of the primary fracture front. Despite determined efforts, no convincing evidence has been found of any propagation backwards from an inclusion through the matrix, as with fracture hyperbolas in PMMA. However, a fracture starting in an inclusion and propagating backwards within that inclusion is occasionally expected. This may be the case for the double tail in Fig. 2B, as suggested by some hackle marks within the inclusion (not shown). After this internal fracturing, incidentally, the fracture at the inclusion in this case

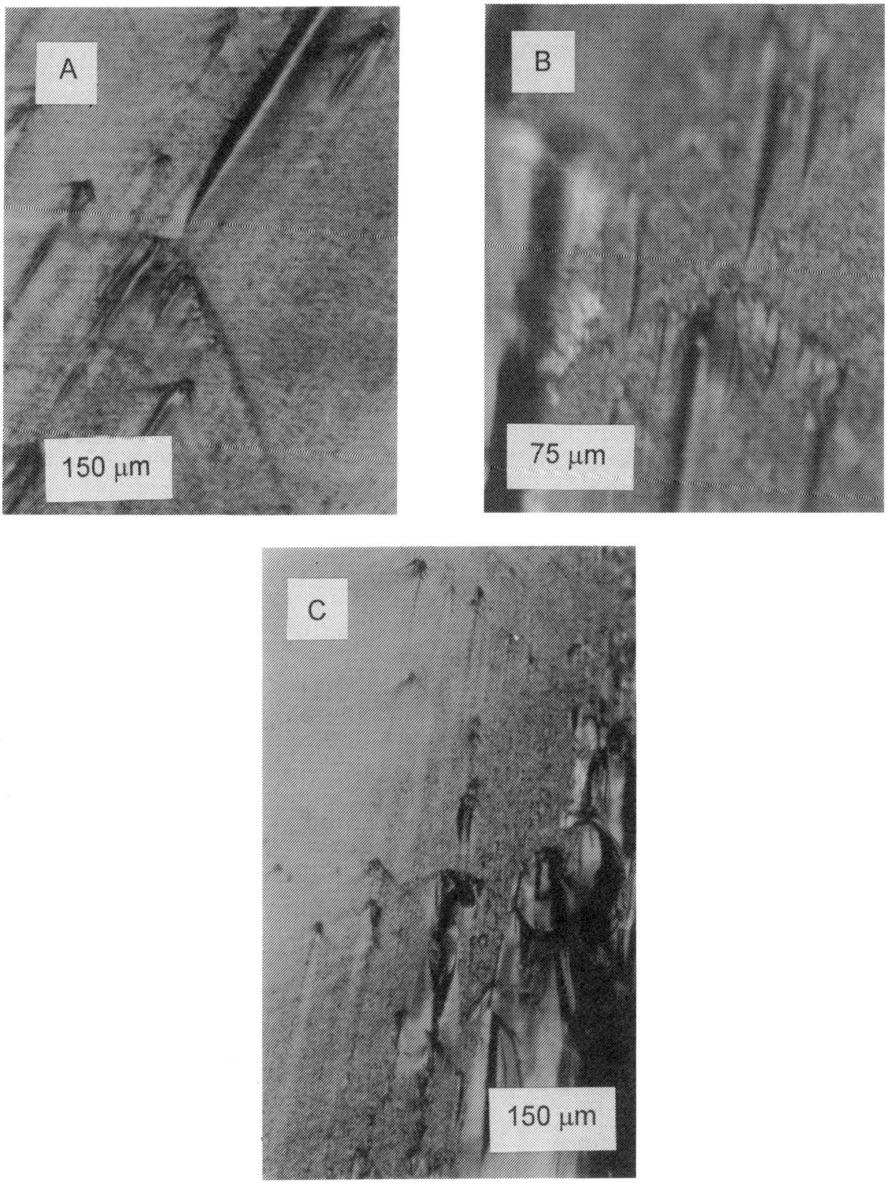

Figure 5. Wallner mist and hackle configurations.

still keeps "diving" away from the primary fracture plane, to finally spread and form the surface between the two tails.

Double tails with their trailing mist lines are rather conspicuous near mirror boundaries because of the absence of mist within the configurations, at least near their heads. This absence of mist relates to the lower stress intensities expected along the central partial fracture front, because its leading the other parts of the fracture locally reduces the mechanical loading on it.

WALLNER MIST AND HACKLE CONFIGURATIONS

Markings referred to here as Wallner mist, mist-hackle or hackle configurations or lines (collectively, as Wallner mist configurations or lines) are shown in Fig. 5. A number of the white linear features extending diagonally across Fig. 2A, and probably the broad white arc in the upper right portion of that figure, are Wallner mist lines or configurations as well, representing bands or "lines" of mist (or hackle) enhancement due to stress pulses. A portion of the white arc referred to is enlarged in Fig. 3B. On the Wallner mist-hackle or hackle configurations in Fig. 5, larger hackle marks are initiated due to a stress pulse. In Fig. 2A, there are many more Wallner mist configurations that are much smaller than those visible in the figure.

In obsidian, the Wallner configurations noted have usually been observed in association with the ripple of a Wallner line. However, it is usually the narrow band of mist or the larger hackle that makes the markings more conspicuous. For some of the configurations studied, the author was not able to detect a ripple with an optical microscope. Similar "Wallner markings" in carbon-filled, frozen rubber were reported by Andrews (6,7) to have no ripple associated with them. In very recent research with lightly tempered soda-lime glass, Beauchamp (4) has discussed the greater size and density of hackle features occurring on Wallner lines. For one of such Wallner lines studied in depth, he reports that "it does not represent a ripple on the fracture surface where the crack returns to the original plane after passage of the stress pulse. Rather, it is an 'impulse line' formed by a more permanent change in the surface orientation." A plane shift is also seen to occur with the ripples associated with at least some Wallner mist or mist-hackle configurations in obsidian. "Ripples" is used loosely here, to include the "impulse lines" in (4).

Just as the usual Wallner lines, Wallner mist and hackle configurations can take on various forms and be used for calculation of relative fracture velocity (2,10). (From their association with Wallner lines, it is apparent that the shear wave velocity is relevant here.) This is of particular interest in mist regions where the usual Wallner lines may be difficult to see. From a tangent to the Wallner mist-hackle configuration in Fig. 5A, for example, a lower bound for the average fracture to shear wave velocity is calculated to be about 0.595 for that mist region. This compares reasonably with the highest values reported previously for obsidians (2).

If it were not for their association with Wallner lines, it would not at all be evident whether shear or longitudinal stress waves are associated with the formation of the Wallner mist configurations. However, it is expected that increased strain energy is made available along these configurations to enable the enhanced formation of hackle marks along them. The sources of the stress pulses involved may be at the fracture front, as for Fig. 5B, or away from it, as for Fig. 5C and the referenced white arc in Fig. 2A.

In Figs. 5A and 5B, the constituent hackle marks of the Wallner mist-hackle configuration are relatively smooth and vanish gradually downstream. Fig. 5C, in contrast, shows the hackle marks initiated along the Wallner mist-hackle configuration becoming more complex, apparently due to significant stress pulses arriving from sources to the right of the markings. Incidentally, this complex configuration is seen as a premature hackle manifestation, being intrusive to the mist region from the mist-hackle boundary

Fig. 5B shows a Wallner hackle configuration with a number of its constituent markings tending to merge gradually.

Caution is needed in observation of Wallner mist configurations, as in Fig. 2A, especially when the coarser hackle marks are absent. In optical microscopy, a ripple can increase or decrease the apparent prominence of the typical mist in a region, potentially leading to misinterpretation of a usual Wallner line as a Wallner mist line. Not all of the white diagonal lines in Fig. 2A could be verified, by optical microscopy, to be Wallner mist configurations. This cautionary remark is even more important for the configurations considered in the next section.

ANTI-WALLNER LINES

On the right half and especially the lower part of Fig. 2A, a number of black or dark bands or "lines", some narrow and others broad, extend diagonally and parallel to the Wallner mist lines. Near the head of the double tail at the top of the figure, an ill-defined, wide and dark band extends diagonally to the left. These features are visible because of the apparent inhibition or suppression of mist associated with them. Optical microscopy with higher magnification indicates that a number of them are at least partly associated with minute ripples and also suggests that, at least with some, the inhibition of mist may indeed be real. However, confirmation of the latter point was still considered hazardous in light of the cautionary remark at the end of the last section and Fig. 6 in (4).

Therefore, a more general investigation was made of regions of mist and especially at mirror-mist boundaries, for localities where mist formation is inhibited, seen by its being less abundant or less pronounced. A number of features, most frequently and conspicuously near mirror boundaries, were identified where mist inhibition was due to dynamic (rather than quasi-static) effects. Examples are shown in Figs. 6 and 7. Such bands of mist inhibition are termed "anti-Wallner lines." It is suggested that they are formed due to the effects of localized obstruction, or effective obstruction, to the regular progress of fracture being manifested at appropriate "points" along successive fracture fronts.

During usual fracturing in a region of mist or light mist without an exceptional irregularity at the fracture front, strain energy is released and "commmunicated" to appropriate "points" at subsequent fracture fronts downstream, tending to contribute to mist formation. A

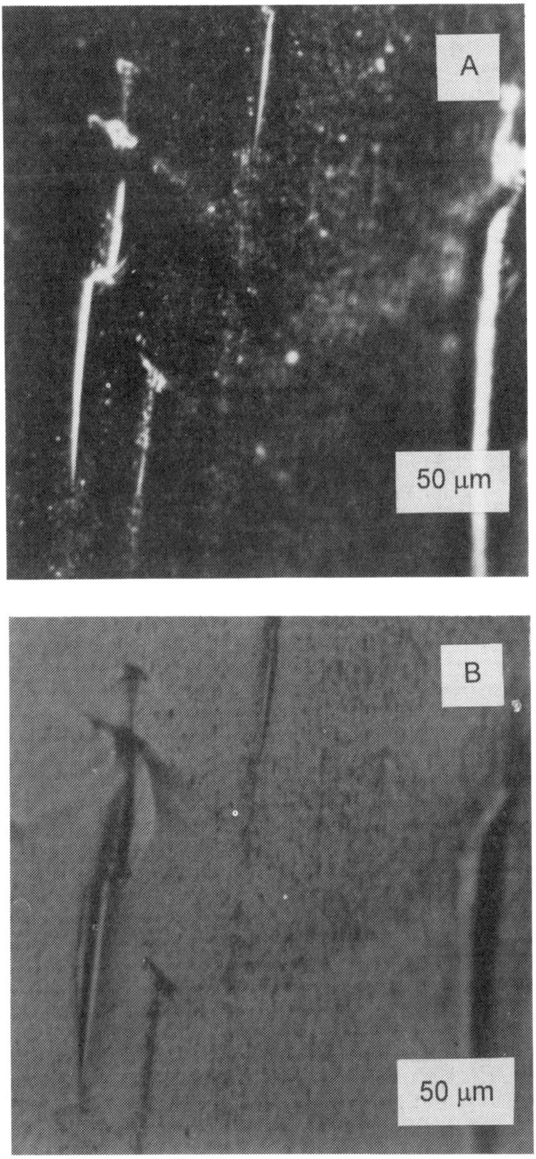

Figure 6. Anti-Wallner configuration (upper left) near mirror boundary: Darkfield (A) and brightfield (B) lighting. Mist inhibition seen at intersection with mist line.

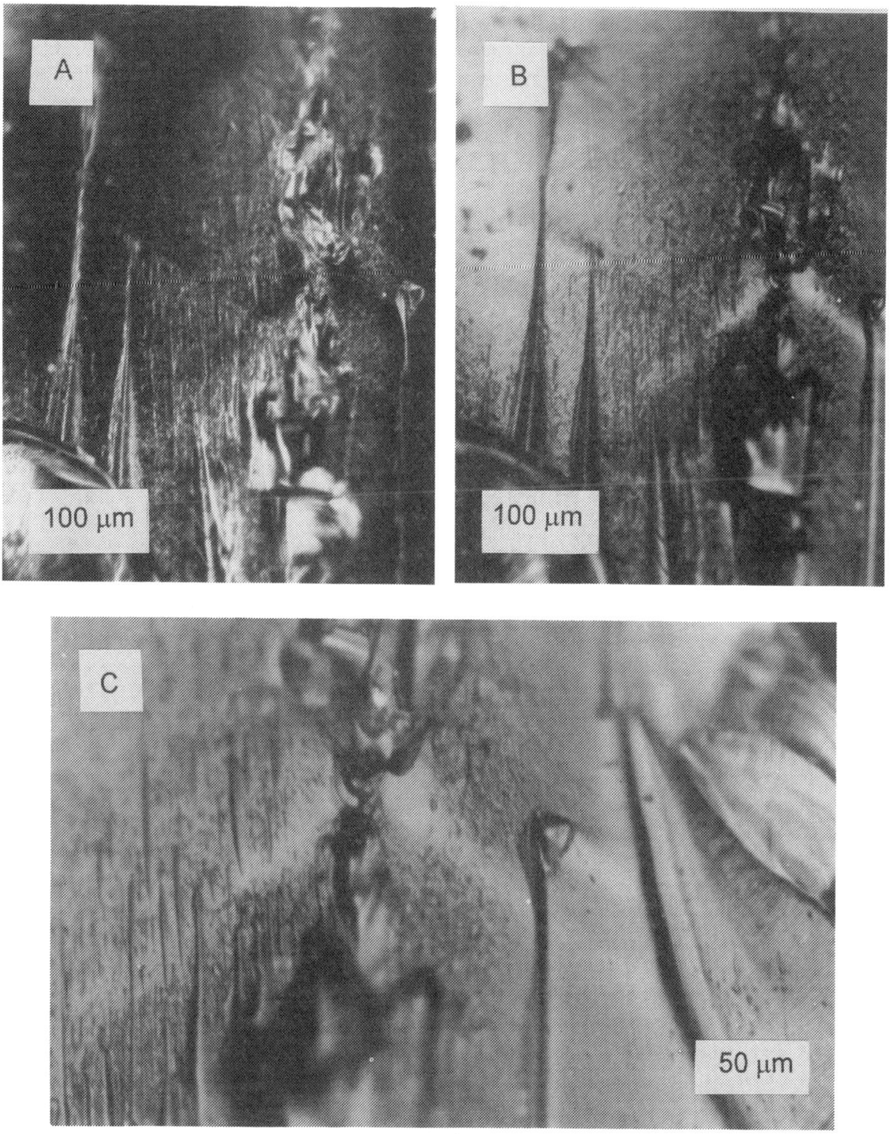

Figure 7. Anti-Wallner line in recurrent mist-hackle region: Darkfield (A) and brightfield (B,C) lighting.

sudden increase in strain energy released locally is associated with a stress pulse that is expected to enhance mist manifestation further and to lead to a Wallner mist line. A local obstacle to fracture, on the other hand, is associated with a sudden decrease in the strain energy released locally, or at least communicated to appropriate "points" along successive fracture fronts. This, it is suggested, tends to inhibit mist manifestation at the appropriate points on the subsequent fracture fronts.

Anti-Wallner lines, Wallner lines and Wallner mist lines are analogous for the purpose of defining their geometry in fracture plan. However, with Wallner mist configurations viewed as being due to "Wallner effects", anti-Wallner lines may be considered as being due to "anti-Wallner effects". Anti-Wallner lines are expected to have various forms, and can be used for calculation of relative fracture velocity. For example, the anti-Wallner lines in Fig. 7 give a reasonable (2) average fracture velocity of about 0.616 with respect to the shear wave velocity that is likely to be relevant here.

Some anti-Wallner configurations are seen as well-defined bands or "lines". Yet some other such configurations are more obscure. It is not clear whether suppression of longitudinal stress wave effects may sometimes contribute to the manifestation of the latter configurations.

The potential manifestation of "anti-Wallner lines" or "anti-Wallner effects" by inhibition of very fine surface roughness in mirror regions has not been explored. However, an implication of a local obstruction or effective obstruction on a ripple profile, though not the subject here, is of interest. Both mist enhancement and inhibition have often been observed in association, or partly in association, with a ripple in fracture profile. This suggests that some of these ripples correspond not to a sudden strain energy release, but rather to a sudden decrease or a combination of a sudden decrease and increase in strain energy release. A sudden decrease in strain energy release can be due to an obstruction or an effective obstruction to the regular progress of fracture.

MIST LINES

The term "mist lines" is used here for the narrow bands or "lines" of mist (with fine hackle) or mist and hackle (with coarse hackle) in the local fracture direction. These bands are manifested by the constituent hackle marks forming or becoming more abundant or pronounced within them. In obsidian, mist lines occur very often near mirror boundaries, most commonly extending from single or double tails, but occasionally directly from an inclusion or irregularity. Reference was already made to double tails with trailing mist lines (Figs. 1 to 4). Numerous mist lines are seen in Figs. 4 and 5 in (2), as well as Figs. 1D, 2A, 4A, 5, 6 and 8 here. Mist lines in quartz glass were reported by Appelt already in 1944 in unpublished research, shown as Figs. 5 and 6 in (3). Reference was already made to what the authors have called "fracture hyperbolas" (3,10) with mist lines extending as their arms. Most recently, a rigorous study of mist lines was made by Beauchamp (4,5), as "streaks" or "hackle streaks". The focus of that study is on the formation of mist and hackle regions more generally.

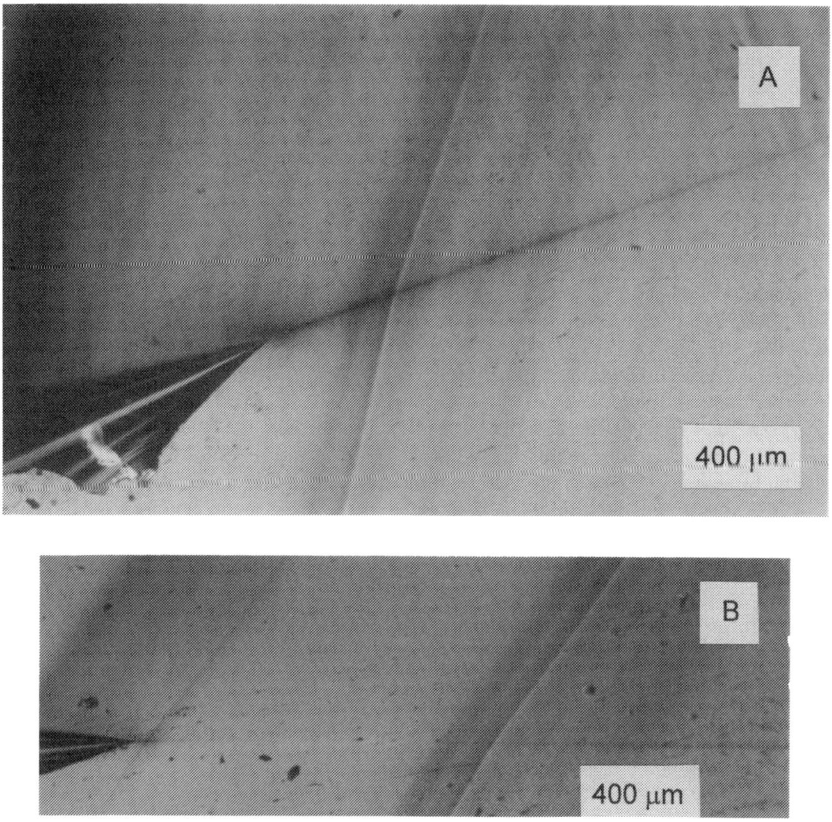

Figure 8. Mist line trailing a twist hackle vanishing gradually in (A) and becoming more pronounced in (B).

Potential mechanisms for mist line formation are suggested here. The most challenging aspect of their manifestation, emphasized here, is their persistence over great distances. While the onset of mist lines could be explained by some localized stress conditions, their persistence can not. A similar statement has been made by Rice regarding early onset of mist more generally (14). Another aspect of mist line manifestation that calls for an explanation is the persistent confinement of mist (at some scale) to a band, rather than its spreading out laterally.

Mist lines in obsidian have been observed in mirror regions near their boundaries, as in (3,10) for other glasses, and in mist regions. Further from the side of a mist region, mist lines tend to become shorter and more faint, some just barely visible with optical microscopy. Mist lines often tend to become weaker (Fig. 8A) and vanish gradually. Some, however, start out as barely visible and become more pronounced towards a mist region (Fig. 8B). In one observed instance, a mist line became gradually visible with optical microscopy only some distance downstream of a wake hackle that was apparently responsible for its initiation. The presence of a mist line with hackles too fine to be detectable with optical microscopy (3,4) can be inferred for the apparently missing section in the latter case. Similarly, it is likely that mist lines not detectable with an optical microscope occur further from mist regions than the faint ones observed. The above observations suggest mist lines can occur only when the conditions are "ripe" or "almost ripe", approaching those for the occurrence of mist in general. Any additional factors can be viewed as pertubations of the existing conditions, already near instability, along the "line" where a mist line is manifested. Such pertubations may become increasingly subtle towards a general mist region, and would have to be more severe in mirror regions further from a mirror boundary.

Observations suggest that local lagging or "pinning" of the fracture along the "lines" of mist plays an important role in the manifestation of mist lines. This localized lagging or "pinning" tends to make more strain energy available for fracture and, at least in some cases, is expected to contribute to a mixed mode stress environment along the mist "lines". Such localized lagging or "pinning" is particularly evident at some mist lines trailing a wake hackle (a single or a double tail), where the mist "line" forms along a line not parallel to the fracture direction at either side of it. The smooth region within and near the head of a double tail is apparently mist-free because of the locally leading fracture front reducing the stress intensity. Local lagging is also evident in Fig. 2A from convergent mist lines, with more pronounced mist lines at those locations. The locally leading fracture fronts between the convergent mist lines lead to inhibition of mist in the black or dark bands seen in Fig. 2A. (The author prefers to avoid "increased stress intensity" with the local lagging, because of the complications with this concept in the presence of mist and hackle.) The directionality of the hackle marks towards the center of a mist band, observed in (4) and sometimes evident also here, and the irregularities or "kinks" in Wallner lines crossing a mist line also indicate localized lagging or "pinning" of the fracture front at mist lines. Such "kinks" or irregularities, however, may be quite obvious, very subtle or not apparent at all, as seen from Figs. 2A and 8, and from Figs. 4 and 5 in (2). In earlier research, Smekal (3) reported all Wallner lines to be deformed in the vicinity of a mist line, and even kinked at the intersections. For a mist line trailing a capillary in Fig. (6) in (3), the Wallner line "kinks" are seen to flatten out gradually, with the mist line widening as they do so.

For a mist line trailing the wake hackle of a double tail, sometimes the coarse hackle is observed to start at the wake hackle termination, with the fine hackle of the mist line already starting upstream of that location (also inside the double tail feature). In general, some mist lines are very narrow (Fig. 8), while others are wider (Fig. 3B). This may relate to the nature of the local lagging or "pinning", as in Fig. 6 in (3) already discussed, and to other effects.

Some mist lines are seen to consist of essentially fine hackle, as in Fig. 3B, while others also have coarse hackle and even "linear" macrobranching features near their central portions, as in Fig. 2A. Such macrobranching features have only been observed when mist lines converge (Fig. 2A). Increased hackle size near the central portion of a mist band is not only expected from the mist line formation mechanism suggested. It also relates to another consideration. Stress wave effects, which play an important role in any mist or hackle region, are expected to be the strongest there from sources at and behind the fracture front, due to secondary fracturing. The suggested "pinning" effect, together with those from stress waves is also expected to serve as a mechanism in confining the hackle marks (visible with optical microscopy) to the band of a mist line.

For multiple tails in a mist region, mist lines are sometimes observed to trail the outermost but not all the innermost tails, as expected.

The above observations lend support to the suggested significance of the local lagging or "pinning" of the fracture fronts in mist manifestation.

The following additional, seemingly self-generative mechanism is suggested. The local lagging or "pinning" of a fracture front could contribute to local manifestation of fine and coarse hackle which, in turn, could promote the local lagging or "pinning" because of the greater energy demands associated with production of the hackle. Formation of the hackle, of course, leads to blurring of the notion of a fracture front and to a disarray in the local state of stress, with a mixed mode stress environment inevitable. The self-generative mechanism of mist lines suggested involves increased strain energy release rates as well as a mixed mode stress environment. The relative significance of these, with respect to the suggested self-generative mechanism, is not clear. As with any mist and hackle, stress wave effects are also relevant. Researchers have reported level differences (3,4) and a slope change (4) across a mist line. These are accommodated piece-wise by the hackle marks of a mist line. It is of interest to note that occasionally at a location where a mist line would be expected, a line of recurrent long hackle marks resembling a series of long tails or twist hackles is observed. The role of Mode III stress environment is expected to be significant in such a case.

Instead of the "self-generative" mechanism suggested above, there could in principle be some "external" factor, conceivably a “global” Mode III stress environment, promoting the persistence of mist lines. After all, mist lines can be observed as extensions of tails or twist hackles, and two of the mist lines in Fig. 2A revert to twist hackles near the bottom of the figure. Also, persistence of tails as twist hackles appears to be conspicuously absent in mirror regions where mist lines occur. As already noted, a mist line has been associated with angular changes across it (4). However, the above hypothesis seems unlikely in general, since mist lines often persist over great distances in regions expected to have no Mode III stress environment.

In summary, the following three necessary ingredients for explaining the manifestation of mist lines over great distances are suggested: conditions approaching those for formation of general mist regions, localized lagging or "pinning" of the fracture making more strain energy available for fracture as well as sometimes contributing to a mixed mode stress environment locally, and the "self-generative" mechanism noted above with the localized mist (and hackle) manifestation promoting localized lagging or "pinning" of the fracture.

Incidentally, some aspects of the formation of the very common bands of mist and hackle near the centerline of thermally tempered glass plates, as in Fig. 7.16 in Kerkhof (10), appear to support the explanation offered for the persistence of mist lines. In part (B) of the cited figure, the fracture appears to lag locally at the mist band, despite the residual tensile stresses being highest in that region.

Mechanisms have been suggested above for the persistence of mist lines over surprisingly great distances. Those for mist line initiation and termination call for additional explanation. More than one specific mechanism can be responsible for the initiation of mist lines. Level and angular changes, as well as local Mode III stress environment, are expected to be important here (4,5). This is evident from a mist line occasionally starting at an inclusion or irregularity, rather than at the termination of a tail or a twist hackle. Similarities to the formation of a tail are relevant here.

A mist line may run into a general mist region. As it does, it has been observed to widen, and coarser hackle near its center tends to appear or become more abundant, persisting some distance into the general mist region. The widening may relate to the nature of the "pinning", as already noted in reference to (3). The increase in the coarser hackle apparently relates to the conditions approaching those for the formation of the mist region, combined with any "pinning" effect and the "self-generative" mechanism suggested. Mist lines visible in optical microscopy often become gradually less pronounced and disappear (4). Localized stress wave effects associated with the already existing mist along a mist line may be relevant to such terminations.

In general, further research is clearly needed for a better understanding of mist lines.

HACKLES AND WATER

Observations related to hackles and water are considered in this section. Reference is not made here to the subcavitation hackles upstream of a Michalske cavitation scarp or downstream of a deceleration scarp (8,15), nor to the twist hackles sometimes initiating at a cavitation scarp (8) where the subcavitation hackles terminate.

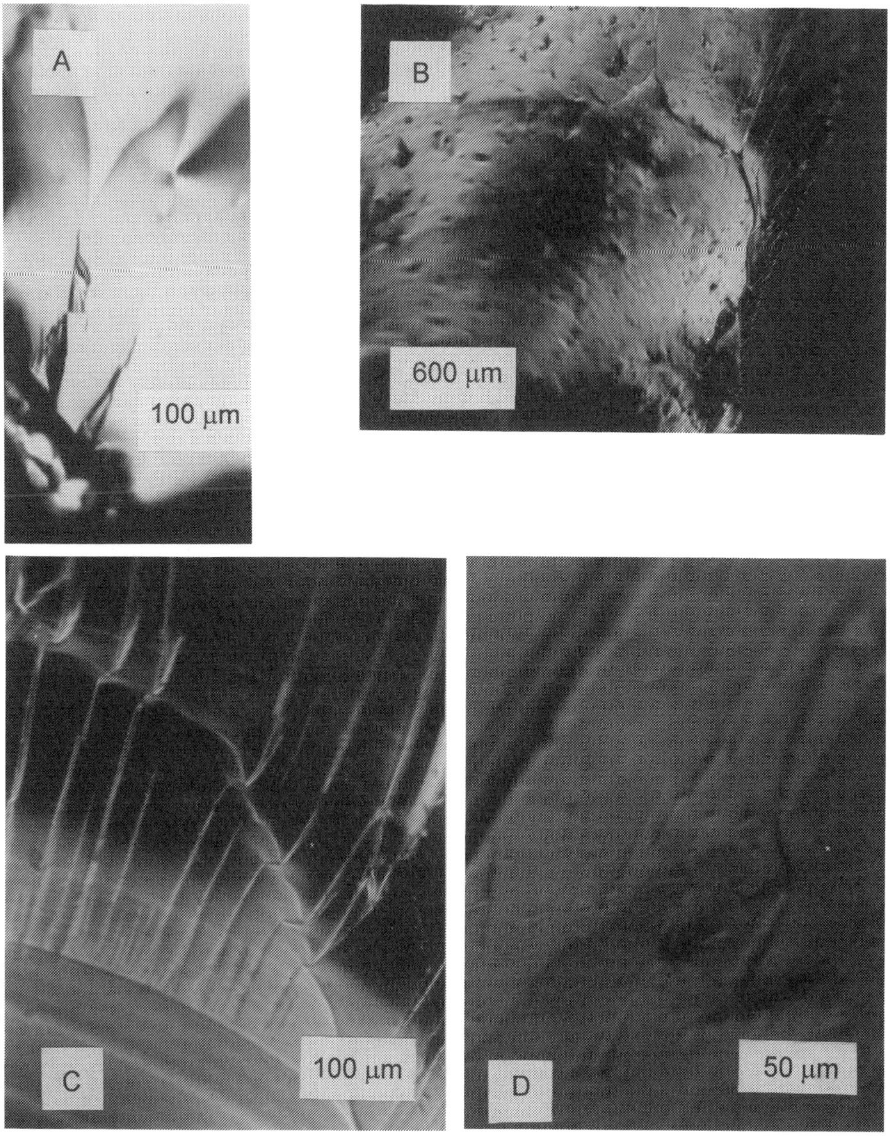

Figure 9. Hackles with wetted fractures: (A) "Wake scarps" at a hackle and an inclusion. (B) Hackle scarp extending upward from right edge. (C,D) Twist hackle zig-zags at depletion scarps. (D) Offset at hackle zig-zag formed due to differential water supply at its sides.

Interaction of twist hackles with a cavitation scarp has been discussed in (16,17), with a kink noted in the hackle at the scarp location indicating a sudden change in the fracture direction. This has been observed when cavitation occurs on one portion of a fracture front, which then quickly sweeps around other slower portions of the fracture that are "still wet and subject to the effects of viscous drag"(17). Such kinks, as in Fig. 5 in (16), have also been occasionally observed in pressure flaking obsidian with a wetted surface.

Twist and wake hackles are sometimes observed to have pronounced effects on the manifestation of scarps in obsidian. Locally, a scarp is sometimes manifested prematurely, as a "wake scarp" downstream of a hackle (Fig. 9A), apparently due to the latter features limiting access of water to a wetted fracture front. Premature manifestation of scarps is also observed at inclusions with or without a wake hackle. In one case, incidentally, a dry fracture front was seen to form in the wake of an inclusion, only to be closed in by encounter scarps converging from the sides to form a triangular "wake scarp". The inclusions and wake or twist hackles in an obsidian, with their associated "wake scarps", can have a pronounced effect on the overall configuration of a scarp in obsidian. Similarly, the overall configuration of scarps is observed to be affected drastically sometimes by the presence of long twist hackles. These hackles are sometimes observed to limit water available to a wetted fracture front locally, suggesting that the wetting and advance at the overlapping partial fracture fronts at a twist hackle require more water locally. At a family of twist hackles, (cavitation or depletion?) scarps sometimes tend to form earlier nearest the hackles, and separated "ovoidal" Sierra scarps (8) tend to be centered between adjacent twist hackles.

A twist hackle (or a persistent tail) is associated with partial fracture fronts in overlapping fracture planes that are initially not interconnected. Therefore, the spread of water laterally along a fracture front is not possible from one plane to another before a lateral breakthrough occurs. Especially with irregular fractures in an inhomogeneous glass like obsidian, and with quite variable paths of transport of water to the fracture front, such lateral redistribution of water is expected but inhibited by hackles. This partitioning of the paths of water transport helps to explain why a scarp is sometimes observed to form on one side of a hackle before the other.

Some effects of water on hackle formation are seen in Figs. 9B to 9D. Fig. 9B shows a linear feature extending diagonally upward from the right edge of the specimen. Such a marking, termed a "hackle scarp," is a scarp manifested with a "hackle" instead of the usual escarpment between the two fracture planes, in this case associated with a wet (right side) and dry (left side) fracture front. Near the top center of the cited figure, there is a partial Sierra scarp (8). Some of its upstream portions, having a "hackle" instead of the usual escarpment, are examples of hackle scarps as well. Most of the offset portions at the zig-zag twist hackles in Fig. 9C and 9D are also hackle scarps. All of the referenced hackle scarps are characteristically in a direction very different from that of the more usual hackles and the general fracture direction in their vicinity, just away from the marking.

A similar though not an identical fundamental mechanism is suggested for the formation of all of the aforementioned hackle scarps. As already noted, water cannot spread laterally along a fracture front across a hackle when the overlapping partial fracture planes are not connected at a hackle initially. Suppose that the partial fracture front on one of these planes is dry and the other one, on the other fracture plane, is wet. When propagation of the wetted fracture front at a given velocity requires a lower stress intensity than the dry one, and water is readily transported to the wetted front, as in the case of an encounter scarp (8), the partial fracture front that is wet is expected to propagate more readily than the partial fracture front that is dry (15). This can lead to an extension of the wet and a retraction of the dry fracture front by the overlapping fracture planes at a hackle, and also promote faster propagation of the wet fracture in the downstream direction. Consequently, the "hackle" tends to manifest itself in the direction of a hackle scarp, as in the cases referenced above.

For some of the twist hackle zig-zags in Fig. 9C, the upstream (lower) and downstream (upper) sides of the offset at the zig-zag were associated with a dry and a wet fracture front, respectively. The formation mechanism of the hackle scarp at this zig-zag offset is similar to that described above for those in Fig. 9B.

For the twist hackle zig-zag in Fig. 9D, on the other hand, the offset was formed with a wet fracture front on each side of it. Yet the hackle does make a sharp turn. It is suggested that the mechanism for the initial turning at this zig-zag relates to differential water transport to the fracture front on the two sides of a hackle, corresponding to what has been referred to as the Region II (15). The upper side with more water requires a lower stress intensity for fracture propagation (at a given velocity), enabling greater extension of the fracture on that side of the hackle. In a sense, the mechanism for the formation of such "partition scarps" is similar to those for the other hackle scarps considered above except for the presence and absence (previous cases) as opposed to the differential amounts of water (this case) as well as different fracture velocities involved. An escarpment type scarp is seen in Fig. 9D as the dividing line between the lighter and the darker regions associated with dry and wet fracture fronts, respectively. The orientation of this (probably a depletion) scarp suggests that water is becoming more abundant rightward. This is consistent with more water being available on the right than the left side of the twist hackle, and the mechanism suggested above.

In a region associated with a wet fracture front, one of several parallel twist hackles is occasionally observed to take a surprising turn, to run obliquely with respect to the others. Such a twist hackle can be understood as a partition scarp discussed above.

Several mechanisms could contribute to or be responsible for the initiation of hackle scarps. The partial fracture fronts in different planes may already exist, as in the case of the twist hackles in Figs. 9C and 9D, leading to the turning at the hackle zig-zag as already indicated.

For other types of hackle scarps, another initiation mechanism akin to that most often ascribed to a tail can be suggested. Suppose an inclusion (or other irregularity) is encountered along a scarp with the usual escarpment connecting the planes to each side of it. As the partial fracture fronts pass around the inclusion, in different planes to start with, and as the wet partial fracture front extends further along the front, the dry and wet partial fracture fronts will tend to meet with a surface mismatch that extends in the direction of the hackle scarp. In a number of cases, a hackle scarp has indeed been observed to start at an inclusion. Reflection on tails discussed in previous sections suggests additional possibilities in principle.

Another conceivable contributory initiation mechanism for hackle scarps may be of interest to note. Suppose that the usual normality relation for principal stress and fracture plane is considered to hold also for the formation of the usual escarpment of a scarp. This implies rotation of the stress in such a way as to contribute to formation of a hackle along a scarp provided the scarp is not parallel to a fracture front. This effect, combined with a singularity along a scarp, may perhaps contribute to initiation of a hackle scarp by changing a usual scarp to a hackle scarp. A combination of these two effects, as in the formation of tails, may sometimes be relevant.

A hackle scarp is sometimes observed to revert to an escarpment scarp (scarp with the usual escarpment), with the hackle extending separately, usually tending to line up with the general fracture direction. The general mechanism already described for the extension of a hackle scarp is evidently lacking or not dominant in such cases. The continuing escarpment scarp has sometimes been observed to revert back to a hackle scarp, with the reversion occasionally confirmed to occur at an inclusion.

The presence of water, by both hackle scarps and escarpment scarps, can serve to initiate twist hackles running in their usual direction. The partial Sierra scarp near the top center of Fig. 9B is formed when water reaches the fracture front from behind it (8). The associated cavitation or depletion scarp downstream from it is not shown in the figure. This figure shows a characteristic case where two hackle scarps are seen to merge and continue as a twist hackle further downstream beyond all scarps. Some hackle scarps, as in Fig. 9B, are seen to form and then turn as a "twist hackle" in the more usual direction. On a series of Sierra scarps, such occurrences were observed to be more frequent with increased Mode III stress environment. In some cases with a strong Mode III environment, hackles are seen to form at a Sierra scarp and apparently extend directly in the more usual direction rather than follow a scarp at first.

The formation mechanisms for twist hackles are well known, and those for hackle scarps were suggested here. The direction for the extension of a hackle, either as a hackle scarp along the scarp or as a twist hackle in its more usual direction, apparently depends on which of the competing mechanisms associated with these is predominant.

Some effects of twist or wake hackles on the water transport to a fracture front were already noted. In addition, cases have been observed with the fracture outrunning the effects of water and later an encounter scarp starts from a twist hackle. In such cases, the twist hackles evidently served as "channels" of water transport to the fracture front. But more generally, the presence of various hackles can still pose questions and provide clues on the paths of water transport to a fracture front.

"Hackle zig-zags" serve as convenient clues to location of cavitation scarps. Hackle scarps in general, when better understood, can be useful in distinguishing different types of scarps, such as a depletion from a cavitation scarp, and a deceleration from an encounter scarp.

DISCUSSION

Investigation of fracture markings in experimental knapping of obsidian, despite obvious limitations, has some distinct advantages. The irregularities in "loading", geometry and material characteristics permit observation of a great variety of markings in their diverse forms. The markings discussed are but a sampling of this.

In the context of hackle-related markings, having numerous and variable inclusions in obsidian is of particular interest in the above sense. However, generalizations from limited observations in obsidians with inclusions warrant caution. For example, the observation of each twist hackle in a family sometimes starting at an inclusion would obviously be misleading for glasses in general.

A similar but more intricate question is posed by some of the observations for double tails in obsidians with numerous inclusions, and their relation to what have been called fracture hyperbolas and related markings in industrial glasses (3,10).

Photographs of markings in glass from unpublished research by Klemm already in 1945 appear in (3,10), termed fracture hyperbolas by the authors. These very small markings, shown at magnifications of 450X and 890X, each resemble a double tail with mist lines appearing to extend from an inclusion. According to the authors, they occur at the edge of a fracture mirror and are associated with a fracture starting ahead of the primary fracture front, as in the case of a material such as PMMA. Kerkhof (10) specifically states that they do not originate from the joining of a partial fracture passing through an obstruction. The latter, however, is observed to be the common though not the only mechanism in obsidians. The observed differences may relate to the nature of the variable inclusions in obsidian, including their elastic, strength and thermal characteristics.

However, double tails in obsidian sometimes appear to be formed with a secondary fracture starting at an inclusion away from the primary fracture plane, with some backward propagation within the inclusion expected at least occasionally. This agrees with or is at least closer to the referenced observations reported by Kerkhof (10) for industrial glass.

With double tails in obsidian, no convincing evidence has been found for a fracture starting at an inclusion ahead of the primary fracture and propagating backwards within the matrix as in PMMA. This possibility may exist theoretically, however, if allowance is made for differential elastic, thermal and strength characteristics of the matrix and an inclusion. In a silica glass reinforced with continuous fibers, fracture initiation ahead of the matrix fracture has been reported (18). If such a mechanism were common or likely in glass more generally, it would be expected but has not been observed for double tails in obsidians with numerous and variable inclusions. This suggests such a mechanism to be at least uncommon or unlikely in glass. The observation that Wallner wakes in obsidians with inclusions are associated with secondary fracturing behind rather than ahead of the primary fracture front (2) lends further support to this.

Although double tails in obsidian also occur with relatively slow fracture velocities, they appear to be most common at higher velocities near mirror boundaries. The observed manifestations at slower fracture velocities, apparently not reported for industrial glasses, are expected to relate to the variability of the inclusions encountered in obsidian. Double tails in regions of slow fracture velocity are much more common in some obsidians than others. The apparent increased frequency of double tail occurrence at greater fracture velocities needs further clarification. It may perhaps be related to dynamic stress fields at higher velocities. Incidentally, some constituent markings in an obsidian mist region are double tails (Figs. 5C).

As to self-consistency of some of the observations, it is of interest to note that from occurrence of multiple tails at an inclusion it could be inferred that single tails could occur also with a fracture passing through or partly through an inclusion, and vice versa.

That stress waves can locally enhance the formation of mist and hackle is shown by Wallner mist and hackle configurations. That their lack or relative weakness can inhibit mist manifestations locally is seen from anti-Wallner lines. These observations are consistent with and support the significance of stress wave effects in a general explanation for mist and hackle regions (4,8,19). A Wallner mist or hackle configuration, especially in a general mist region, can be seen as a well ordered manifestation of such effects. Such configurations within a general mist region can sometimes be seen as less orderly and almost obscure.

Some Wallner hackle configurations provide examples of orderly merging of hackle marks that may tend to lead to macrobranching (Fig. 5B). This is consistent with a possible macrobranching mechanism suggested in Rice (14).

Localized enhancement and inhibition of mist, such as the cases considered, can serve to test more general theories or criteria on mist and hackle manifestation (4).

The considerations on hackles and water provide examples of how observations on the interplay of different types of markings can be relevant to each of them, and improve the understanding of fracture processes more generally. The study in (4) is such an example with mist lines (streaks) and Wallner mist-hackle configurations.

The formation of a hackle along a scarp can be related to the detailed mechanisms for the curving of hackles in a dry environment. Both cases relate to the required and available stress intensities at the partial fracture fronts at the overlapping fracture planes of a hackle. Also, both of these cases suggest a clarification is needed on what precisely is meant by the usual statement that a hackle is in the local fracture direction.

A hackle scarp is not always as subtle as a scarp with an escarpment. This suggests the possibility of their manifestation in materials other than glass.

Hackle scarps, just as scarps in general, are of interest to archeologists for manufacture as well as use of obsidian tools. In use-wear studies, such markings can provide clues to the materials worked. In studies on manufacture of artifacts, they can provide clues to the working environment or to intentional wetting of a working platform. Such wetting can reduce the forces required to start the fractures and thereby can enhance the control exercised.

It is of interest to note that Figs. 1 to 5 are for Glass Buttes and others for Mexican obsidian. Figs. 1C, 1D, 5B to 7 and 9A are for blade or biface breakages, with all others for flake surfaces. All the flakes were produced by direct percussion with a hammerstone, except those in Figs. 9B to 9D that were pressure flaked with the working platform region wetted with saliva. Fig. 9A is for an intentional blade breakage under water.

ACKNOWLEDGEMENTS

For his interests leading to this study, the author gratefully acknowledges his indebtedness to Dr. Don Crabtree, the pioneer in the art and science of knapping. The author is grateful for the kind efforts of Dr. Stephen Freiman of the National Institute of Standardization and Technology and especially Prof. V.D. Fréchette of Alfred University for stimulating and guiding inquiries on fracture mechanics of knapping over a number of years. Prof. Fréchette's discussions of many ideas on fractography, his assistance in some of the photography (Figs. 2A and 4), and suggestion for the term "anti-Wallner lines" is acknowledged with gratitude. The author is grateful to Dr. Terry A. Michalske of Sandia National Laboratories for discussions of scarps and knapping already in 1978. The author would also like to thank him and Dr. Roy W. Rice, Consultant, for their efforts and for pointing out relevant research.

REFERENCES

1. B. Cotterell and J. Kamminga, Mechanics of Pre-Industrial Technology, Cambridge University Press, Cambridge, England, 1990, 147-151.

2. A. Tsirk, "Formation and Utility of a Class of Anomalous Wallner Lines on Glass," in Fractography of Glasses and Ceramics, ed. by V.D. Fréchette and J.R. Varner, The Am. Ceram. Soc., 57-69, Westerville, Ohio, 1988.

3. A. Smekal, "Verfahren zur Messung von Bruchfortpflanzungs-Geschwindigkeit an Bruchflächen (Methods for measuring Crack Propagation Velocity on Fracture Surfaces)," Glastechnische Berichte 23, (3) 57-67 (1950).

4. E.K. Beauchamp, "Crack Front Stability and Hackle Formation in High Velocity Glass Fracture," J. Am. Ceram. Soc., 78, (3) 689-697 (1995).

5. E.K. Beauchamp, "Mechanisms for Hackle Formation and Crack Branching," this volume.

6. E.H. Andrews, "Stress Waves and Fracture Surfaces," J. of Appl. Phys. 30, (5) 740-743 (1959).

7. E.H. Andrews, Fracture of Polymers, American Elsevier, New York, 1968, 186-189.

8. V.D. Fréchette, Failure Analysis of Brittle Materials, The Am. Ceram. Soc., 1990, 7-42.

9. E. Sommer, "Formation of Fracture 'Lances' in Glass," Engineering Fracture Mechanics 1, 539-546 (1969).

10. F. Kerkhof, Bruchvorgänge in Gläsern (Fracture Processes in Glasses), Deutsche Glastechnische Gesellschaft, Frankfurt, 1970.

11. A. Tsirk, "On a Geometrical Effect on Crack Front Configuration," Int. J. of Fract. 17, R185-R188 (1981).

12. T. Del Bene, Personal communication, 1982.

13. J.A. Kies, A.M. Sullivan, and G.R. Irwin, "Interpretation of Fracture Markings", J. of Appl. Phys. 21, 716-720 (1950).

14. R.W. Rice, "Perspective on Fractography," in Fractography of Glasses and Ceramics, ed. by V.D. Fréchette and J.R. Varner, The Am. Ceram. Soc., 3-56, Westerville, OH, 1988.

15. T.A. Michalske, "Dynamic Effects of Liquids on Crack Growth Leading to Catastrophic Failure in Glass," Ph. D. Thesis, Alfred University, Alfred, 1979.

16. T.A. Michalske, V.D. Fréchette, and R. Hudson, "Dynamic Effects of Liquids on Surface Crack Extension in Glass," Advances in Fracture Research, Vol. 2, ed. by D.Francis, Pergamon, 1091-1097, New York, NY, 1981.

17. T.A. Michalske, "Fractography of Stress Corrosion Cracking in Glass," in Fractography of Glass, ed. by R.C. Bradt and R.E. Tressler, Plenum Press, 111-142, New York, 1994.

18. T.A. Michalske and J.R. Hellman,"Strength and Toughening of Continuous-Alumina-Fiber-Reinforced Glass-Matrix Composites," J. Am. Ceram. Soc. 71, (9) 725-731 (1988).

19. V.D. Fréchette, "Characteristics of Fracture-Exposed Surfaces," Proc. of the Brit. Ceram. Soc. 5, 97-106 (1965).

INDENTATION CRACK SHAPES AND THEIR EVOLUTION DURING SUBCRITICAL CRACK GROWTH IN ION-EXCHANGED GLASSES

Priya J. Dwivedi* and David. J. Green

Department of Materials Science and Engineering,

The Pennsylvania State University,

University Park, Pennsylvania 16802

ABSTRACT

The influence of surface residual stress on indentation crack shape and its evolution during subcritical crack growth was studied in ion-exchanged soda-lime and sodium alumino-silicate glasses. Crack shapes were determined optically from crack-arrest markings which were generated on the fracture surfaces during crack extension. The shapes, both after indentation and subsequent subcritical growth, were found to be very different from the semielliptical shapes observed in stress-free glasses. It was observed that high residual surface compression pinned the cracks near the failure origin (indentation site) and the growth occurred totally below the surface. The influence of the extent and the magnitude of the surface compression on crack-shape evolution was examined. It was found that the depth of the crack in ion-exchanged glasses was larger as compared to stress-free glasses. Furthermore, the depth of the crack increased with an increase in the compression layer thickness.

INTRODUCTION

In order to study the fracture behavior of surface-stressed materials, indentation flaws have been widely used since they are known to simulate "natural" surface contact flaws (1-6). The shape of the indentation cracks is usually assumed to be semicircular in these previous studies. In a recent study (7), it was observed

---

* Presently at Corning Incorporated, Corning, NY.

Supported by the US Department of Energy under Grant No. DE-FG02-86ER45252 for stress-free (annealed) glasses that the crack shape after indentation is semicircular or semielliptical, and the ellipticity could change significantly during the subcritical crack growth process. The introduction of a surface stress is expected to distort the crack shape particularly when the magnitude of the compressive stress is high and the layer is less than the crack size. In particular, one may expect the crack dimensions to be reduced in the vicinity of the compressive layer. Some indications of this type of crack shape distortion have been suggested in previous studies on glasses containing thin surface compression layers (8,9).

The importance of crack shape in materials can be seen from previous studies on the determination of surface-stress (4,5,8) using indentation cracks. These studies utilized the fracture mechanics analysis of Lawn and Fuller (2) for thin-layer surface-stressed materials with varying degree of success. In one study (4), the estimated and observed surface stress values were in good agreement, but in the other two studies (5,8), the indentation crack length measurements considerably underestimated the surface residual stress. It was speculated that the disagreement was due to the crack shape being different from the assumed semicircular shape (5). Fractographic evidence indicated the radial crack fronts in materials containing thin layers of residual compressive stresses were being deflected near the surface resulting in a shorter crack length on the surface (8,9). A schematic of this type of crack shape is shown in Fig. 1. Clearly, for a proper study of the fracture behavior of brittle materials with surface residual-stresses, it is important to know the crack shape as it plays an important role in the determination of stress intensity factors and the failure behavior. For example, the changes in crack shape during subcritical crack growth may be important in understanding the long-term strength and reliability of surface-stressed materials.

The aim of this work was to investigate the indentation crack shapes and their evolution during subcritical crack growth in ion-exchanged glasses. Soda-lime silicate and sodium alumino-silicate glass bars were ion-exchanged at temperatures below their respective glass transition temperatures. The process involved replacing sodium ions in the glass structure with potassium ions from the molten salt. The bars were then indented and subjected to four point bending for the generation of crack-arrest markings during subcritical crack growth. Crack-shapes were determined optically from the crack-arrest markings by post-mortem observation of the fracture surfaces. This method has been explained in detail elsewhere (7). The primary focus of this study was to relate the changes in crack shape to the magnitude and depth of the surface stresses. This was accomplished by varying the temperature and duration of the ion-exchange process and determining the effect of these changes on crack shape during crack growth.

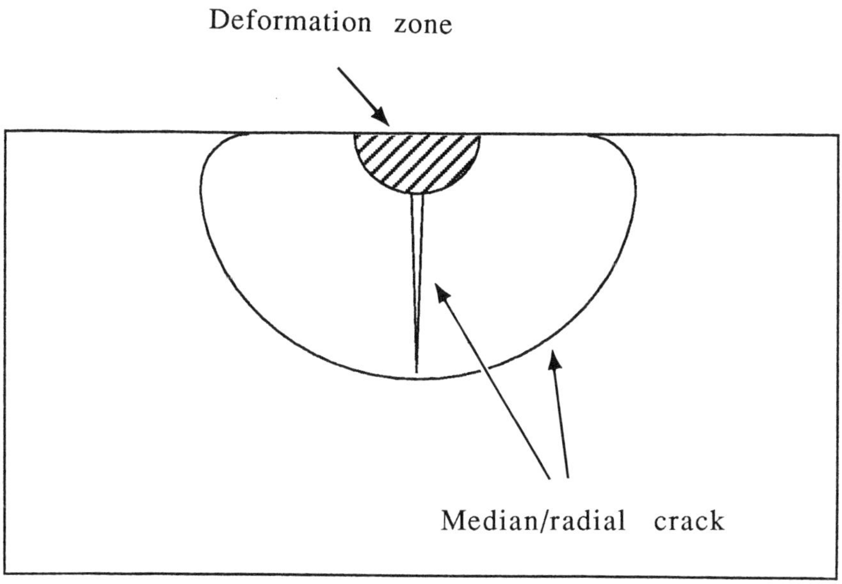

Fig. 1. Schematic of crack shape suggested by Tandon and Green (5) for an indentation crack interacting with a thin surface compressive layer.

EXPERIMENTAL PROCEDURE

Soda-lime glass (SLS)[†] and sodium aluminosilicate glass (SAS)[§] bars of nominal sizes 60 x 7 x 2.2 mm, and 60 x 7 x 2.6 mm, respectively, were used. The bend bar specimens were beveled, cleaned, and annealed to remove extraneous residual stresses. The specimens were subsequently cleaned again and ion-exchanged in molten potassium nitrate bath[#] to replace sodium ions in the glass surface with potassium ions (K^+) from the molten salt. The

---

[†] Centre Glass Co., State College, PA
[§] Code 0319, Corning Inc., Corning, NY
[#] Kirk Optical Co., Inc., Freeport, NY

temperatures of ion-exchange in this study were below the glass transition temperature, T_g, of the SLS and SAS glasses, which are 520°C and 625 °C respectively. The exchange temperatures were maintained within ±2°C and the sodium ion (Na^+) concentration in the salt bath was kept below 0.3 wt%. To obtain uniformity in the ion-exchange process, the specimens were suspended in the bath in such a way as to ensure they did not touch each other. SLS glass specimens were ion-exchanged at temperatures, T, of 415°C, 450°C, and 500°C for different periods of time, t, ranging from 1 to 144 h. The SAS glass specimens were ion-exchanged at 450°C and 500°C for 1 h each. After ion-exchange, the specimens were removed from the bath, cooled in air at room temperature, washed, and dried. This was followed by Vickers indentation¶ in air with a contact duration of 15 s at the peak load. The specimens were aligned during indentation such that one set of radial cracks was perpendicular to the edges. This ensures these cracks will be in mode I loading during the application of the bending stresses. Indentation loads, *P*, ranging from 9.8 N to 68.6 N were used to introduce the cracks.

After indentation, the ion-exchanged specimens were subjected to four-point bending stresses using an *in situ* crack growth observation device which was constructed to fit on top of an inverted microscope£ (10). This arrangement allowed direct observation of crack growth in indented specimens under an applied load. Indentation cracks were subcritically extended and intermittently partially unloaded to generate crack-arrest markings. The specimens were subsequently fractured and the fracture surfaces examined under the reflection microscope. Crack-arrest markings were evident where the stress field was perturbed, and this provided a 'fingerprint' of the crack-shape evolution process. The relevant crack size measurements were made directly on the fracture surfaces. To determine the initial crack-shapes, some specimens were fractured after indentation i.e., without any subcritical crack growth.

For each ion-exchange condition, about 3 specimens were used with the exception of one exchange condition of SAS (500°C, 1 h) where the subcritical crack growth experiment could not be performed easily due to the presence of high residual stresses in the interior of the specimen. These stresses are tensile and can lead to crack growth even in the absence of the applied stresses (self-fatigue)(11).

The compression at the surface for three ion-exchange conditions, (i) SLS, T = 450 °C, t = 1 h, (ii) SLS, T = 450°C, t = 12 h, and (iii) SAS, T = 500°C, t = 1 h, was estimated from the potassium ion concentration profile at the surface (12) and was confirmed by a strain gage-surface removal technique (12).

¶ Model 3212, Zwick Inc., East Windsor, CT

£ Nikon Epiphot, Nikon Inc., Melville, NY

RESULTS AND DISCUSSION

The residual stress state for some of the ion-exchanged glasses is summarized in Table I (12). The data show that exchange depth increases with exchange time and the SAS glass can be exchanged to greater depths. The stress obtained by surface removal technique are found to be lower than those obtained from the potassium ion concentration profile and this was ascribed to stress relaxation (12).

Crack-Shape after Indentation

When a Vickers diamond indenter is pressed into a stress-free glass surface and then removed, the surface deforms and median cracks are formed when the applied indentation load is above a threshold level. Alternatively, radial cracks may form during the unloading process. The two crack types are often difficult to distinguish from surface measurements as median cracks usually extend to intersect the surface during unloading.

It is well known that the indentation process is accompanied by a change in the stress field upon transition from loading to unloading stages. When an indenter is loaded into a stress-free material, the stress field below the indentation deformation zone is tensile whereas near the surface around indentation site it is compressive. Upon unloading, however, the stress field reverses its sign, i.e. it becomes compressive below the deformation zone and tensile near the surface. During loading, the tensile stress field below the indenter may result in the formation of median cracks. During or after unloading, due to the presence of tensile stresses near the surface, the median cracks extend towards the surface to form median/radial cracks. Alternatively, radial cracks may form during or after the unloading process if the net stress intensity factor, K, reaches the critical stress intensity factor, K_c at the surface.[@] The radial cracks, during formation extend below the surface and may join to form a single crack front. This is usually accompanied by a lance mark formed by the intersection of the two radial crack planes.

Now, if high compressive stresses are introduced into a thin layer at the surface, e.g., in an ion-exchange process, the effect of the indentation applied and residual tensile stresses near the surface will be reduced. One would expect this to result in three possibilities.

(1) If the indentation-induced tensile residual stress at the surface is of lower magnitude than the compressive stress induced by ion-exchange, the radial cracks originating from the deformation site at the surface may form such that

[@] For some materials, indentation load, and indenter geometry, the stress field near the surface can be tensile even during loading which may result in the formation of radial cracks (13).

Table I. Residual Surface Compressive Stress, Internal Tension and the Layer Thickness Values for Various Ion-Exchange Conditions Obtained from Different Techniques.

Glass / Ion-Exchange Condition	Surface Stress (Microprobe) (MPa)	Surface Stress (Strain Gage) (MPa)	Internal Tension (MPa)	Compressive Layer Thickness (μm)	
				K^+ Profile	Optical
SLS glass					
450°C / 1 h	- 940	- 790	2	5	5-7
450°C / 12 h	- 1030	- 610	5	17	15-20
SAS glass					
500°C / 1 h	- 1010	- 880	16	50	50-55

the crack front is totally below the surface. This type of crack-shape is termed Type I.

(2) On the other hand, if the indentation-induced tensile residual stress exceeds the residual surface compressive stress, the radial cracks will have surface traces but crack length at the surface will be shorter than at some distance below the surface. This type of crack-shape is termed Type II.

(3) A third possibility is that median cracks could form but without the formation of the radial surface traces. This type of crack-shape is classified as Type III. Schematics of each type of crack shape are shown in Figs. 2(a), (b) and (c).

In Fig. 3(a), an optical micrograph of the shape of the crack after indenting (P = 49 N) a specimen of SAS glass ion-exchanged at 450°C for 1 h is shown. It is seen that the residual surface stress field disturbs the regular geometry of the cracks and the crack shape deviates severely from the semicircular or semielliptical geometry, which is usually obtained in stress-free silicate glasses. It is observed that indentation introduces radial cracks below the surface rather than ones that intersect the surface. Further, during formation these radial cracks do not join to form a radial/median crack. The formation of radial cracks below the surface must be the result of the high compressive stress at the surface. The condition in Fig. 3(a) appears to be a Type I geometry. The indentation-induced tensile stress at the surface appears to be lower than the surface compressive stress since the radial cracks do not have surface traces. Thus, the radial cracks have undergone a pinning effect during their formation at the surface near the deformation zone. The Type I crack-shape, was also observed in SLS glass for some combinations of ion-exchange schedules and indentation loads.

In order to determine whether Type II cracks could occur, experiments were performed to determine crack shape for specimens in which stress relaxation near the surface was substantial. For this purpose, SLS glass specimens were ion-exchanged at a higher temperature of 500°C (only 20°C below the T_g of the glass) for a longer duration of 6 h. An indentation load of 9.8 N was used to produce cracks. It is seen in Fig. 3(b), that the radial cracks do have surface traces under such conditions.

For some ion-exchange conditions in which the duration of exchange was much longer, instead of the formation of Type I cracks, median cracks formed below the deformation zone (Type III). This is demonstrated in Fig. 3(c), where a median crack (P = 9.8 N) is formed in SLS glass exchanged at 415°C for 144 h.

Crack-Shape Evolution during Subcritical Crack Growth

Now, let us consider the evolution of cracks in these ion-exchanged glasses under subcritical growth due to an applied bending stress. In this work, the indentation cracks in the SLS and SAS glasses were grown under applied stresses ranging from 40 to 90 MPa. This is substantially less than the

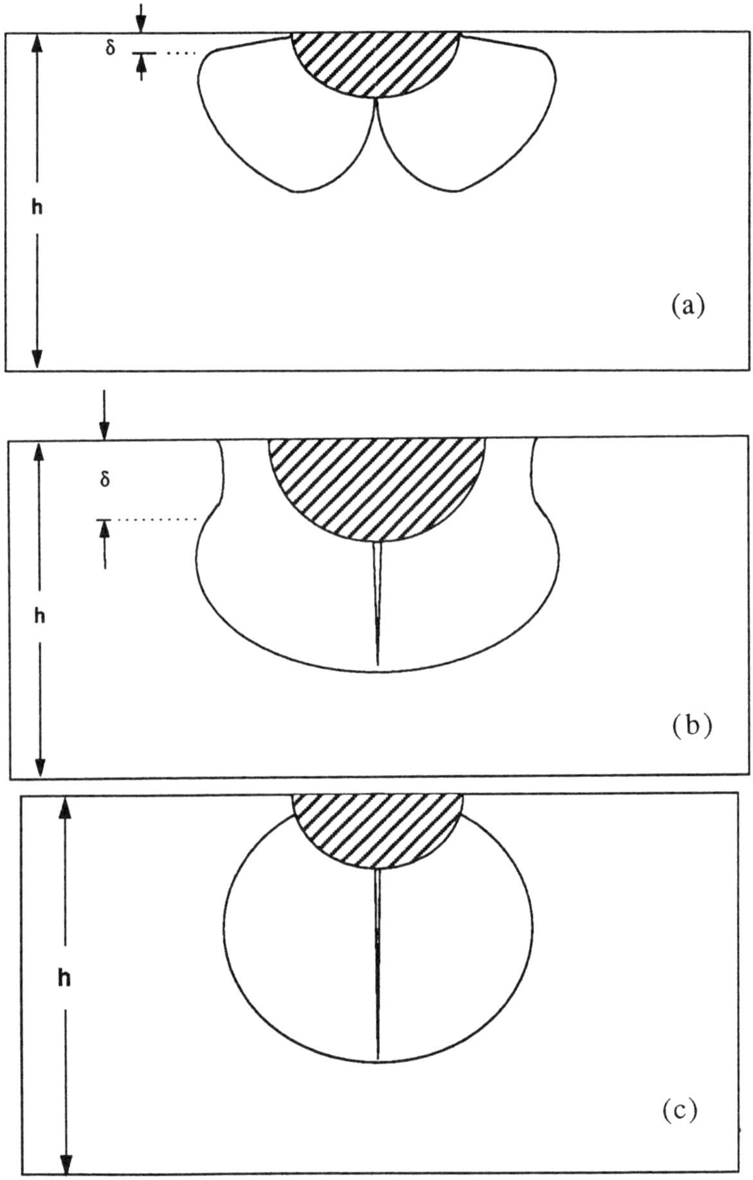

Fig. 2. Schematics of indentation crack shapes which may form in materials with residual surface compressive stresses, (a) Type I, (b) Type II, and (c) Type III.

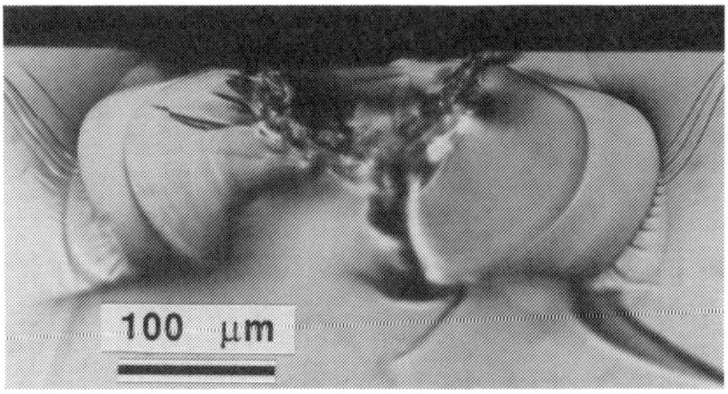

(a)

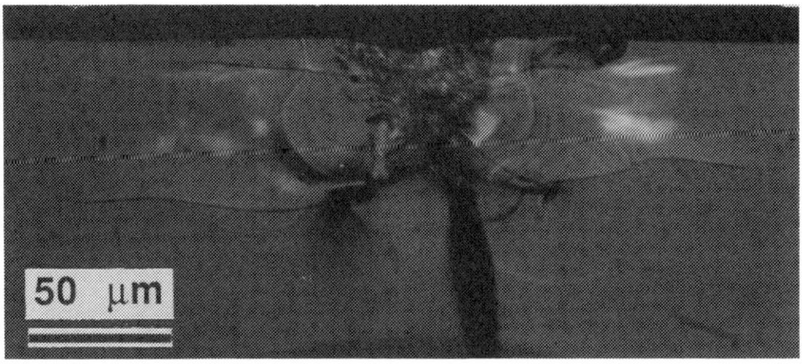

(b)

(c)

Fig. 3. Optical micrographs of the indentation crack shapes formed in ion-exchanged glasses. (a) Type I, (b) Type II, and (c) Type III.

measured compressive stresses (Table I) in these glasses. Once the cracks became 3-4 times larger than the initial crack lengths, the applied stress could be further reduced with magnitudes similar to those used for the stress-free glasses (15-20 MPa). Fig. 4 shows the crack-shape evolution in a specimen of SLS glass ion-exchanged at 450°C for 12 h and indented at P= 9.8 N. The crack-shape after indentation for this exchange condition was found to be of Type I which is expected due to the high surface compression (Table I). It is evident that the crack shape changes significantly as the crack grows subcritically. Furthermore, it appears that the crack fronts are deflected near the surface due to compressive stresses such that the crack is shorter at the surface than at some distance below the surface. From a careful examination of a smaller crack (under identical ion-exchange and indentation conditions) at higher magnification in Fig. 5, however, it can be seen the crack was growing subcritically below the surface. At this point, one may pose the question about the availability of moisture to the crack for its subcritical crack growth. It appears that the deformed zone and the crack opening in this vicinity must be large enough to enable moisture to have access to the crack tip. Furthermore, the high compressive stress at the surface (Table I) appeared to have prevented the crack growth at the surface. Under these condition, the residual tension in the interior of the specimen could also play a role in the subcritical crack growth. For some of the exchange conditions the previous estimates of the residual tension (1-5 MPa) indicate these stresses may be as high as 25% of the applied stress. For conditions in which the residual tension is low compared to the applies stress, one expects the crack to be growing in the region of maximum tension, i.e., slightly below the compressive layer. For conditions in which the internal tension is higher, crack growth into the depth of the specimen is expected to become more predominant.

In order to quantify the change in crack shape for a crack growing below the surface, the crack geometry was idealized as shown in Fig. 6. It is postulated that the crack propagates below the surface at a depth approximately equal to δ, at which the net stress intensity factor is approximately zero. The crack front below the points A in Fig. 6 can be approximated by part of an ellipse, with the major axis being equal to $2c$ and minor axis, $2a$. The deepest point along the crack front is denoted by point D and the total depth of the crack is a^*.

Some very useful information can be obtained from a careful observation of the crack-arrest markings in Fig. 4. For example, the growth of the crack is uniform along its front, and one can compare the ratio of the increases in the major and minor axes of the crack, $\Delta c/\Delta a$, i.e. points B and D in Fig. 6. It can be readily seen in Fig. 4 that the crack grew more along its major axis than into the depth indicating that the velocity at points B, v_B, was higher than that at point D, v_D. Therefore, the stress intensity factor at point B, K_B, was larger than that at point D, K_D, i.e. $K_B/K_D > 1$. Moreover, with crack growth, the ratio $\Delta c/\Delta a$ increased, i.e. v_B/v_D increased. This suggests that the ratio v_B/v_D

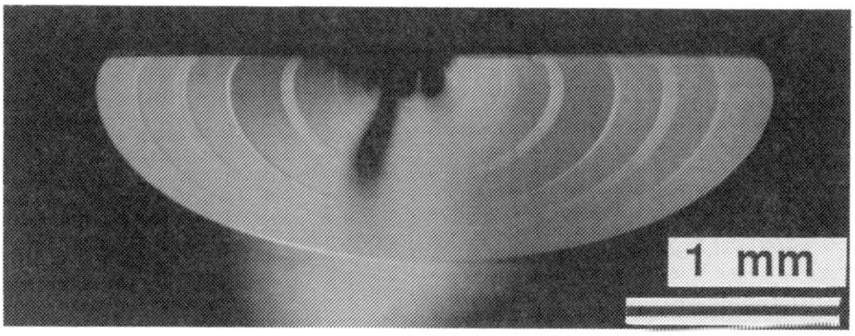

Fig. 4. Optical micrograph of the evolution of indentation crack shape in SLS glass ion-exchanged at 450°C for 12 h (P = 9.8 N).

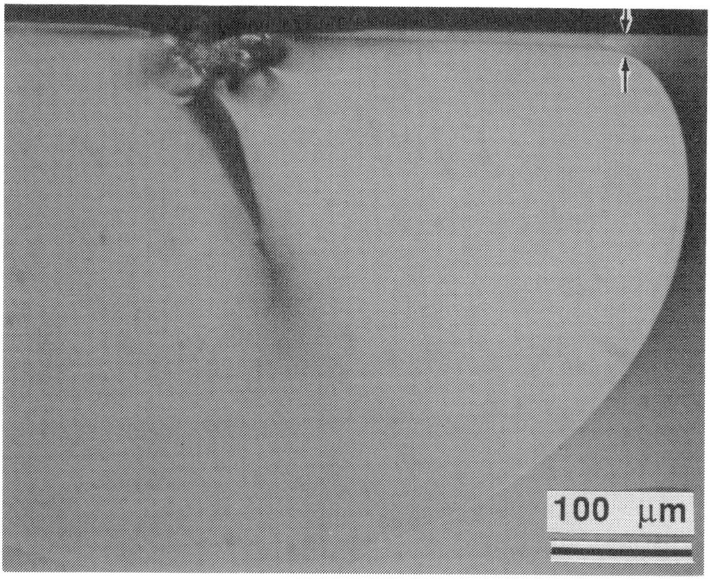

Fig. 5. An optical micrograph showing the shape of a subcritically grown crack in ion-exchanged SLS glass (450°C, 12 h). Note that the crack grows below the surface. The approximate depth of the compressive layer, δ, is indicated by arrows.

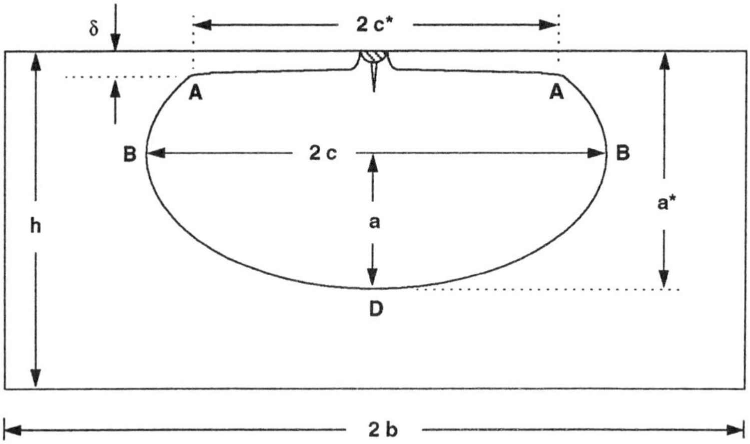

Fig. 6. Schematic of a subcritically grown Type I indentation crack in a glass with a thin layer of residual surface compression.

can be estimated by measuring the distances Δc and Δa between consecutive arrest-markings (at points B and D, respectively, in Fig. 6). As the crack growth at points B and D takes place essentially below the ion-exchange layer i.e. in the base glass, the subcritical crack growth relationship for these points can be considered to be the same as that for the base glass. Therefore, once the relative crack velocity v_B/v_D is known, the ratio K_B/K_D can be obtained readily. Basically, an increase in $\Delta c/\Delta a$ with crack extension suggests an increase in the K_B/K_D ratio.

The thickness of the compressive layer, δ, lies between points A and B in Fig. 6, and for thin layers should be approximately equal to the depth of point A. This thickness was estimated optically for some ion-exchange conditions by viewing the fracture surfaces e.g. in Fig. 5, the arrow indicates the approximate depth of the compressive layer. These thicknesses were compared to the values obtained from the K^+ ion concentration profiles and good agreement was obtained (12). Some examples of these measurements are included in Table I. This suggests that the thickness of the compressive layer in an ion-exchanged material can be estimated from the crack-arrest markings on the fracture surfaces.

During the subcritical growth of Type II cracks, a crack may or may not extend along the surface depending upon the magnitude of compression at the surface. If the compression is fairly low such that $K > K_{th}$ (the net stress intensity factor is greater than the threshold value for subcritical growth), the crack will be expected to grow along the surface. On the other hand, if the compression is high, the applied stress may be only high enough to grow the crack below the compressive layer. Under such conditions, a crack will remain pinned at the surface and the crack length at the surface will not increase during subcritical crack growth. This type of behavior was observed for the crack shown in Fig. 4. Thus for all the cases studied here, the subcritical growth occurred, only below the surface.

Influence of Ion-Exchange Duration and Temperature on Crack-Shape Evolution

In order to study the effect of time and temperature of ion-exchange on crack-shape evolution, several ion-exchange conditions were studied. These were (a) SLS glass — 450°C, 1, 24, 144 h, and (b) SLS glass — 415°C, 1, 24, and 144 h. The optical micrographs for the crack-shape evolution for these exchange conditions are shown in Fig. 7. For each ion-exchange condition, it was found that the crack-arrest marking pattern was reproducible irrespective of the magnitude of the applied bending stress used for subcritical crack growth.

Let us, first discuss the influence of ion-exchange time on crack-shape evolution. For this purpose, conditions shown in Figs. 7(a-c), i.e. SLS glass exchanged at 450°C for 1, 24, and 144 h, respectively, are considered. It is observed that as the duration of ion-exchange at a given temperature increases, the total crack depth a^* increases for a given amount of crack extension parallel to the surface. This can be seen by comparing Figs. 7(a) through (c), and is further illustrated in Fig. 8, where total crack depth a^* is plotted as a function of maximum crack length c for the three ion-exchange conditions. In addition, the shape of the crack, defined by its aspect ratio a/c, changes with extension, as shown in Fig. 9, where a/c ratios for exchange conditions presented in Figs. 7(a-c) are plotted as a function of the relative crack depth, a/h (h is the specimen thickness). In Fig. 9, the data on stress-free SLS glass is also included for comparison. It can be readily seen that the aspect ratio a/c for cracks in glasses with surface compressive stresses are higher than those in stress-free glass. The internal residual tensile stress in ion-exchanged glasses acts as an additional driving force at the crack tip at point D (see Fig. 6) resulting in higher K_D/K_B values. Thus the velocity at point D, v_D (in the depth direction) is increasing relative to that at the ends of the major axis, v_B, giving rise to higher a/c ratios as compared to the stress-free glasses.

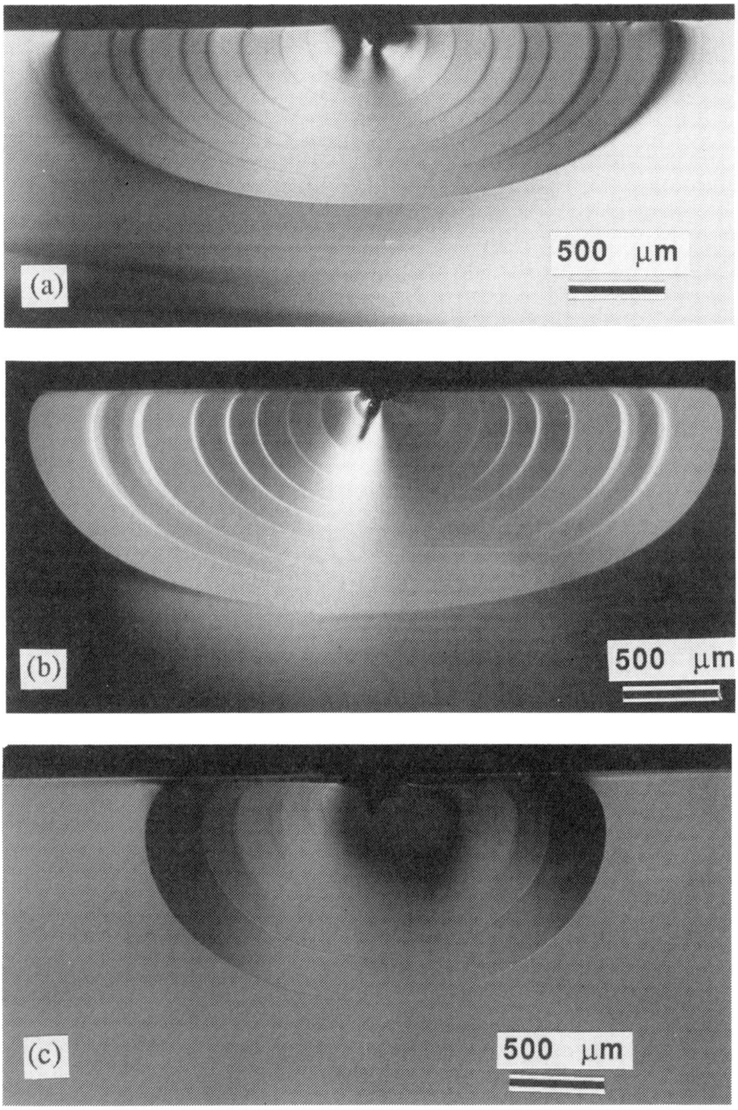

Fig. 7. Optical micrographs of the crack-shape evolution during subcritical crack growth in glasses with different ion-exchange schedules. The ion-exchange conditions for SLS glass are: (a) 450°C, 1 h, (b) 450°C, 24 h, (c) 450°C, 144 h.

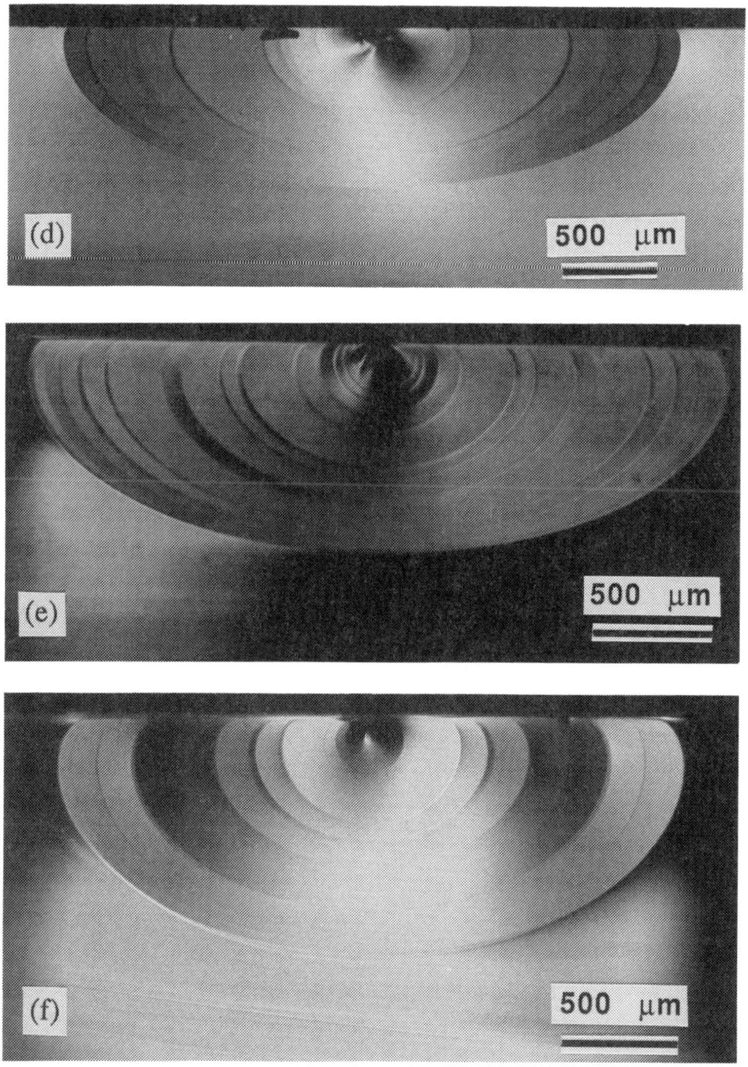

Fig. 7. Optical micrographs of the crack-shape evolution during subcritical crack growth in glasses with different ion-exchange schedules. The ion-exchange conditions for SLS glass are: (d) 415°C, 1 h, (e) 415°C, 24 h, and (f) 415°C, 144 h.

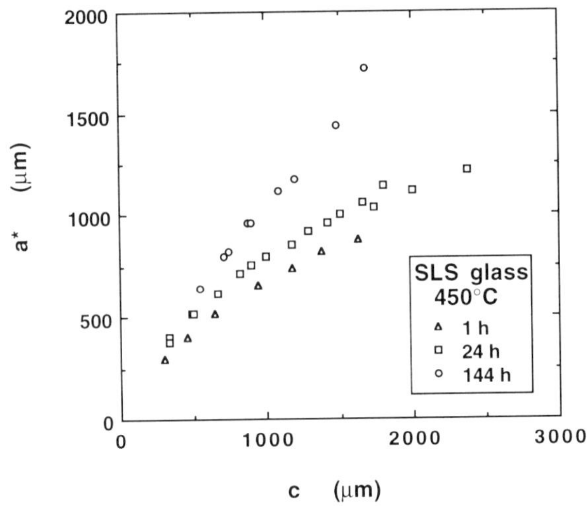

Fig. 8. Total crack depth a^* as a function of maximum crack length c in SLS glass ion-exchanged at 450°C for 1, 24, and 144 h, respectively.

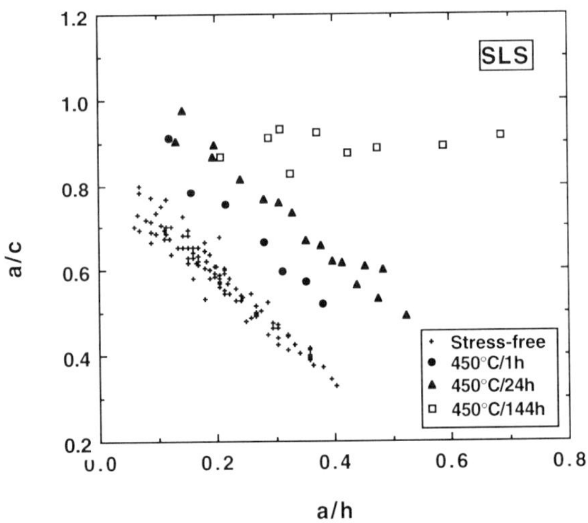

Fig. 9. Aspect ratio a/c as a function of relative crack depth a/h in SLS glass ion-exchanged at 450°C for 1, 24, and 144 h, respectively. For comparison, data for stress-free glass is also included.

In Fig. 9, it can be further seen that for shorter exchange times (e.g. 1 h), the *a/c* ratio decreases with crack extension, i.e. the ratio $v_D/v_B < 1$ and decreases with crack extension. However, as the exchange time increases, the decrease in *a/c* ratio reduces, e.g. for 144 h exchange, and the *a/c* ratio remains almost constant as the crack grows. The decrease in aspect ratio for the stress-free glass has been discussed elsewhere (7) and is a reflection of the stress-intensity factor being highest at the surface. The effect of the ion exchange is to increase the aspect ratio and for the longest exchange time, the aspect ratio remains almost constant. This data implies the stress-intensity factor ratio K_B/K_D must be decreasing and suggests that it is the residual tension in the interior of the specimen that causes the change in crack shape. The residual tension is expected to increase as the ion exchange time is increased. Thus, one concludes the residual tensile stresses promote crack extension into interior of the specimen (away from the surface). Similar behavior is seen also in Fig. 7(d-f).

SUMMARY AND CONCLUSIONS

In summary, the shape of the indentation cracks formed in ion-exchanged glasses are found to be extremely distorted as compared to semicircular or semielliptical geometry in stress-free glasses. The size and the shape of indentation cracks on residually-stressed surfaces are governed by a combination of the macroscopic residual stress profile, and indentation induced residual stress profile.

It was observed that in glasses containing high compressive stress at the surface, cracks were pinned at the surface near the origin and their growth occurred below the compressive layer. The crack shape below the surface could be approximated as a truncated ellipse. The maximum crack length occurs at a depth below the surface where the net stress intensity factor is maximum. It was found that the aspect ratio for the cracks in ion-exchanged glasses are higher as compared to those in stress-free glasses. It is suggested that the residual internal tension in ion-exchanged glasses leads to an increase in velocity in the crack depth direction, which results in higher aspect ratios as compared to stress-free glass. Furthermore, with an increase in ion-exchange layer thickness, the aspect ratio increased, and the maximum crack dimension occurred at greater depths away from the surface for a given crack length. Thus, one concludes the residual tension in the interior of the specimen promotes subcritical crack growth in this region and leads to significant changes in crack shape.

REFERENCES

1. D. B. Marshall and B. R. Lawn, "Strength Degradation of Thermally Tempered Glass Plates," *J. Am. Ceram. Soc.*, **61** [1-2] 21-27 (1978).
2. B. R. Lawn and E. R. Fuller, "Measurement of Thin-Layer Surface Stresses by Indentation Fracture," *J. Mater. Sci.*, **19** [12] 4061-67 (1984).
3. D. J. Green and B. R. Maloney, "Influence of Surface Stress on Indentation Cracking," *J. Am. Ceram. Soc.*, **69** [3] 223-25 (1986).
4. M. F. Gruninger, B. R. Lawn, E. N. Farabaugh, and J. B. Watchman, Jr., "Measurements of Residual Stresses in Coatings on Brittle Substrates by Indentation Fracture," *J. Am. Ceram. Soc.*, **70** [5] 344-48 (1987).
5. R. Tandon and D. J. Green, "Indentation Behavior of Ion-Exchanged Glasses," *J. Am. Ceram. Soc.*, **73** [4] 970-77 (1990).
6. M. M. Chaudhari and M. A. Phillips, "Quasi-static Indentation Cracking of Thermally Tempered Soda-lime Glass with Spherical and Vickers Indenters," *Philosophical Magazine A*, **62** [1] 1-27 (1990).
7. P. Dwivedi and D. J. Green, "Indentation Crack-shape Evolution During Subcritical Crack Growth," *J. Am. Ceram. Soc.*, **78** [5] 1240-46 (1995).
8. P. H. Kobrin and A. B. Harker, "The Effects of Thin Compressive Films on Indentation Fracture Toughness Measurements," *J. Mater. Sci.*, **24** [4] 1363-67 (1989).
9. D. J. Green and R. Tandon, "Surface Stress Effects on Indentation Fracture"; pp. 2757-68 in Advances in Fracture Research, Vol. 4. Edited by K. Salama, K. Ravi-Chandar, D. M. R. Taplin, and P. R. Rao, Pergamon Press, Oxford, U.K., 1989.
10. P. Dwivedi and D. J. Green, "Determination of Subcritical Crack Growth Parameters by In Situ Observation of Indentation Cracks," *J. Am. Ceram. Soc.*, in press.
11. M. F. Bakioglu, F. Erdogan, and D. P. H. Hasselman, "Prediction of the Self-Fatigue of Surface Compression-Strengthened Glass Plates"; pp. 787-97 in *Fracture Mechanics of Ceramics*, Vol. 4, Edited by R. C. Bradt, D. P. H. Hasselman, and F. F. Lange, Plenum Press, New York, 1978.
12. P. J. Dwivedi, Crack-shape Evolution During Subcritical Crack Growth in Glasses; Ph.D. Dissertation. The Pennsylvania State University, 1994.
13. R. F. Cook and G. M. Pharr, "Direct Observation and Analysis of Indentation Cracking in Glasses and Ceramics," *J. Am. Ceram. Soc.*, **73** [4] 1787-817 (1990).

Crack Initiation During Hertzian Indentation of Normal and Anomalous Glasses- A Study Using Fractography and Acoustic Emission

Thomas M. Cleary, Ford Motor Company-Glass Division, Dearborn, MI
and
James R. Varner, NYS College of Ceramics at Alfred University, Alfred, NY

ABSTRACT

The strength of glass is primarily controlled by surface flaws (cracks). The majority of these flaws are introduced into the glass surface by contact with other hard objects during processing or in service. The purpose of this research was to obtain a better understanding of how mechanical damage to glass surfaces influences deformation and crack initiation behavior in "normal" (non-densifying) and "anomalous" (densifying) glasses. Hertzian indentation in conjunction with acoustic emission testing and fractography was used to study the cracking behavior under controlled conditions for several silicate glasses. A "normal" (soda-lime-silica) and two "anomalous" (vitreous silica and borosilicate) glass compositions were investigated. Various levels of mechanical surface damage were generated by grinding and polishing of the samples. For all three of the glass compositions studied, it was found that the application of high levels of mechanical surface damage greatly reduced the load at which cracking was initiated and affected the types of cracks formed. For the "normal" glass, low levels of damage led to the formation of cone cracks during indentation; whereas for the highly damaged specimens, both cone and median cracking was observed. For the "anomalous" glasses, the specimens with low levels of surface damage exhibited cone cracks and precursors for radial cracking; whereas the highly damaged samples only demonstrated cone crack formation. Changes in inelastic deformation during indentation are proposed as the causes of the changes in cracking behavior.

1. INTRODUCTION

Previous investigations into elastic/plastic indentation deformation and cracking behavior of silicate glasses have shown that indentation cracking sequences are composition dependent, falling into two types of behavior, based on whether the glasses are "normal" or "anomalous" [1-5]. Normal glasses do not exhibit significant densification during elastic/plastic indentation, but deform in a manner more closely described as plastic flow [1,3,5]. This shear-dominated, plastic-flow process is reconstructive in nature; i.e., bonds are continually broken and remade with neighbors. In contrast, anomalous glasses do significantly densify during elastic/plastic indentation. This densification is believed to be due to the bending of the Si-O-Si linkages bordering interstices in the network [3,4]. This is a pressure-induced densification which does not entail the breaking of bonds. The process simply involves the collapse into a more close-packed structure by minor bond rotations and changes in bond angles [5].

Hertzian indentation, however, is considered to be a purely elastic contact in silicate glasses at room temperature, at least up until the point of cracking [6]. Therefore one might expect normal and anomalous glasses to react in similar manners, since mechanisms of permanent deformation are not expected to be involved. During loading with a Hertzian indenter, a circular area of contact develops as a result of local deformations in both the sphere and the sample [7]. The radial stress is the only tensile stress at the surface, and it reaches a maximum value just outside the contact region. When this stress encounters a critical flaw, a ring crack develops, which then propagates into the glass to form a cone crack, assuming there is a sufficient level of stored elastic energy. Past investigators have also noted the formation of radial, median, and lateral cracks during Hertzian impact studies of silicate glasses [6,8-9].

Most of the past studies into cracking behavior of glasses, have utilized in-situ optical microscopy in conjunction with indentation. Another technique which shows great promise for these types of investigations involves the use of acoustic emission analysis during indentation. Acoustic emission testing is the detection of elastic waves generated by sudden deformation or cracking in a stressed material. Several investigators have used this type of testing to study cracking behavior during sharp indentation trails in ceramics and glasses [10-12]. To date, however, there has been little mention of this approach for studying Hertzian cracking behavior in glasses.

II. EXPERIMENTAL PROCEDURES

Three glasses were used in the experiments: (1) a soda-lime-silica (float glass from AFG Ind., Inc., 5 mm thick), (2) a vitreous silica (7940 from Corning, Inc., 3.5 mm thick), and (3) a borosilicate (7740 from Corning, Inc., 4 mm thick). Specimens were cut into approximately 40 x 40 mm sizes and annealed before any treatment and testing. The atmosphere (non-tin) side of the float glass was used for these studies. The surfaces of some of the samples were ground or polished using various abrasives (see Tables I & II for details). Other specimens were acid "polished" by placing them in a solution of 5% HF-5% H_2SO_4 for 20 minutes, and others were tested "as-received".

Hertzian indentation trials were done using a tungsten carbide sphere having a diameter of 780 μm on an Instron model 8562 universal testing machine in conjunction with a Physical Acoustics Corporation LOCAN AT acoustic emission detection unit. A piezoelectric transducer was mounted to the surface of the samples with high-vacuum grease for the purpose of detecting acoustic emission (AE) signals during testing. The AE signal was processed by a PA 1220A acoustic emission preamplifier in single input mode, with a gain of 40 dB. The output voltage from the strain gauge of the Instron unit was linked into the LOCAN AT unit by means of a coaxial cable, so that the load at which acoustic activity occurred was recorded as a parameter. The sample was placed onto a thin compliant layer of rubber to eliminate background noise produced by the sliding of the sample on the platform stage during indentation (Note that in this arrangment, the actuator moves the platform and, hence, the sample; the indenter is stationary). Figure 1 shows a diagram of the experimental arrangement.

Loading and unloading was carried out at a rate of 30 N/minute with a hold time of 10 seconds at maximum load. A maximum indentation load of 100 Newtons was used for the anomalous (vitreous silica and borosilicate) glasses, whereas a maximum load of 150 Newtons was used for the normal (soda-lime-silica) glass. This was due to the fact that the soda-lime-silica samples required a greater load to initiate cracking than did the anomalous samples. A minimum of 10 trials per sample were completed, and AE data were analyzed in terms of the number of AE "hits" occurring during loading and unloading, the load at which the first AE "hit" occurred, and the energy/hit ratio. Characterization of the indentation sites was completed using optical and scanning electron microscopy. Both top and cross-sectional views of the indentation sites were examined. Cross-sections were produced by indenting the samples in front of a pre-fabricated scratch , and then fracturing the sample. Earlier investigators have reported that this technique does not alter the final crack morphologies [3,4,13].

III. RESULTS AND DISCUSSION

A. Indentation Cracking:

Since Hertzian indentation is considered to be a purely elastic contact, at least up until cracking, one might expect "normal" and "anomalous" glasses to behave similarly, due to the fact that mechanisms of permanent deformation are not expected to be involved. This assumption proved to be invalid for the testing completed in this study. Details are discussed below.

i. Normal Glass (Soda-Lime-Silica): Table I shows the acoustic emission (AE) data collected for Hertzian indentation trials in the soda-lime-silica (SLS) samples. Statistical t-tests showed some of the differences to be significant. The load required to initiate the first AE event was much higher for the samples with little or only minor mechanical damage (as-received, acid-polished, diamond-polished) than for the highly damaged (ground) specimens. This is expected , since the highly damaged specimens have a wider distribution of larger flaws; therefore, the tensile radial stress at the surface has a greater opportunity to encounter a critically sized flaw for the formation of ring and cone cracks.

In theory, the mildly damaged samples should require a significantly higher strain energy before any of these smaller flaws can become unstable. This would lead one to expect the mildly damaged samples to require a much higher load to initiate cracking, to have an inner-most ring crack of a larger diameter, and to release a greater amount of energy per cracking event than for the highly damaged specimens. Acoustic emission data (Table I) and microscopic characterization supported this theory. The "as-received" sample had an inner-most ring crack diameter of approximately 180-200 μm, whereas the 120-grit "ground" sample had an inner-most ring crack diameter of only approximately 100 μm. Figures 2-3, which show top views of Hertzian indentation sites in "as-received" and 120-grit "ground" SLS samples, demonstrates the differences in diameters of the ring cracks formed. Table I shows that the mildly damaged samples had significantly higher "energy/hit" ratios than did the highly damaged specimens.

Table I. Acoustic Emission Results for Soda-Lime-Silica Glass

Sample Treatment	Load (N) of 1st AE Event	Number of Loading Events	Number of Unloading Events	Energy per Event (volt.sec)
As-Received	101 ± 28	20.1 ± 10.2	5.9 ± 2.2	199
Acid-Polished	113 ± 22	7.6 ± 3.4	8.7 ± 2.9	240
Polished, 3 μm	101 ± 32	6.9 ± 8.1	12.1 ± 2.5	189
Polished, 30 μm	81 ± 27	6.1 ± 5.7	17.1 ± 7.1	232
Ground, 600 grit	41 ± 6	18.6 ± 8.9	4.2 ± 4.0	177
Ground, 320 grit	48 ± 8	38.7 ± 14.1	7.6 ± 2.1	164
Ground, 120 grit	56 ± 9	28.5 ± 11.4	4.9 ± 1.9	112

In addition to the typical ring and cone cracks, the highly damaged samples also exhibited median and lateral cracks. In contrast, the mildly damaged samples had only ring, cone, and lateral cracks (no median cracks). Figure 4 shows a cross-sectional view of a Hertzian indentation site for an "as-received" SLS sample, which shows the cone and lateral cracking behavior. Figures 5a-b are cross-sectional views (at different magnifications) of a Hertzian indentation in a 120-grit "ground " SLS sample. The cone, median, and lateral cracks are easily seen in Figure 5a. In addition, there is evidence of material displacement at the surface of the 120-grit "ground" sample (Figure 5b), which would indicate that some plastic flow was involved in the deformation during the Hertzian indentation. This contradicts the theoretical assumption that permanent deformation mechanisms are not involved during Hertzian indentation. The mildly damaged samples did not demonstrate any noticeable material displacement (i.e., plastic flow).

There was a need to explain the occurrence of median cracking and the material displacement in the highly damaged ("ground") samples. Previous investigations have demonstrated that during elastic/plastic (sharp) indentation in "normal" glasses, shear faults are generated directly below the indentation sites due to the plastic flow behavior [3,5]. Arora, et. al. [3] credited the intersection of these shear bands as the source for median crack formation during elastic/plastic indentation in SLS. Therefore, for normal glasses, it was hypothesized that more severe surface damage somehow promotes the formation of shear bands during loading with a Hertzian indenter, and that these shear bands cause the formation of the median cracks.

ii. Anomalous Glasses (Vitreous Silica, Borosilicate): Data will only be shown for the vitreous silica specimens since the results for the borosilicate specimens were similar. Acoustic emission (AE) data in Table II show that the vitreous silica specimens behaved in a similar manner to the soda-lime-silica samples, in that, the highly damaged "ground" samples required lower loads to initiate the first AE event during loading, and demonstrated lower "energy/hit" ratios than for the mildly damaged specimens. Once again, this is probably due to the fact that there are many large surface flaws available to form the inner ring crack. In addition to these trends, the "anomalous" glasses showed another characteristic; the highly damaged specimens showed a drastic reduction in acoustical activity during unloading in comparison to the mildly damaged specimens.

Table II. Acoustic Emission Data for Vitreous Silica

Sample Treatment	Load (N) of 1st AE Event	Number of Loading Events	Number of Unloading Events	Energy per Event (volt.sec)
As-Received	40 ± 5	33.2 ± 12.9	13.5 ± 5.8	253
Acid-Polished	48 ± 8	19.1 ± 13.2	10.1 ± 4.5	179
Polished, 3 μm	28 ± 8	21.0 ±11.5	7.4 ± 3.1	314
Polished, 15 μm	47± 12	7.6 ± 9.9	9.9 ± 6.2	270
Polished, 30 μm	46 ±16	13.8 ± 6.9	9.4 ± 5.1	326
Ground, 600 grit	18 ± 2	20.0 ± 3.5	0.9 ± 1.0	167
Ground, 120 grit	23 ± 8	11.2 ± 6.7	0.4 ± 0.7	131

Microscopic evaluation revealed that the mildly damaged samples showed precursors to radial cracking which were probably responsible for the unloading AE events, whereas the highly damaged samples did not have the radial precursors. This can be seen in Figures 6-7, which show top views of Hertzian indentations in "as-received" and 120-grit "ground" vitreous silica samples respectively. For anomalous glasses, it was hypothesized that more severe surface damage (e.g., from grinding), produces a disturbed region at the surface which deforms, thereby reducing densification of the glass; ultimately leading to reduced residual stresses during unloading. It is believed that this is the cause for the reduction of unloading AE events (assumed to be related to radial precursor cracking). Figures 8-9 show cross-sectional views of the "as-received" and 120-grit "ground' silica samples respectively. As shown in these figures, there was no apparent material displacement (related to plastic flow) or median cracking like that found in the indentation sites of the highly damaged soda-lime-silica specimens.

IV. CONCLUSIONS

Acoustic emission testing in conjunction with Hertzian indentation and fractography proved to be a very effective method for studying cracking behavior in glasses. These studies demonstrated that deformation and crack-initiation behavior are very dependent on whether the glasses are "normal" (i.e., non-densifying) or "anomalous" (i.e., densifying) and on the extent of surface mechanical damage. Grinding damage, as opposed to polishing damage, apparently promotes deformation in "normal" glasses, making shear-band formation easier and causing median cracks to form. In contrast, grinding damage in anomalous glasses provides a surface region that apparently deforms, thus reducing residual stresses during unloading; thereby reducing the probability of radial precursor crack formation. These results with the Hertzian indenter are consistent with the results obtained for a similar study that utilized a Vickers indenter (as opposed to the blunt indenter used here) along with acoustic emission testing and fractography [12]. In each case, the severe surface damage due to grinding seemed to promote additional (median type) cracking in the "normal" glass and reduced cracking activity in the "anomalous" glasses. In other words, the grinding damage in the "normal" glass seems to promote shear band formation (causing median cracking) for both Hertzian and sharp indentation; whereas for the "anomalous" glasses, grinding damage appeared to create a disturbed surface region which acts to reduce stresses (and ultimately cracking) by its own deformation for both types of indentation. Based on these results, it is obvious that glass composition, i.e., structure, and surface damage, play a significant role in the cracking behavior of glasses.

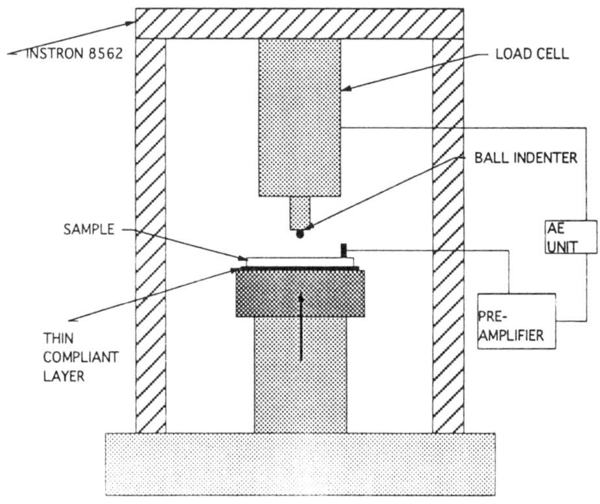

Figure 1. Schematic diagram of experimental set-up for Hertzian indentation cracking tests.

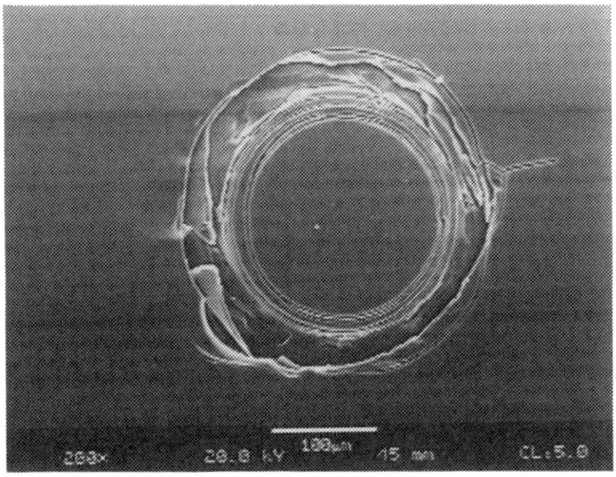

Figure 2. Scanning electron photomicrograph of a Hertzian indentation in "As-Received" soda-lime-silica glass. The white bar represents 100 μm.

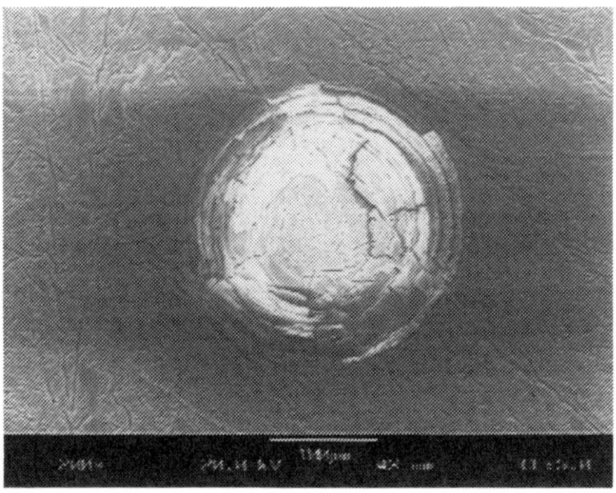

Figure 3. Scanning electron photomicrograph of a Hertzian indentation in (120-grit) "ground" soda-lime-silica. Note the much smaller inner ring crack in comparison with Figure 2. The white bar represents 100 μm.

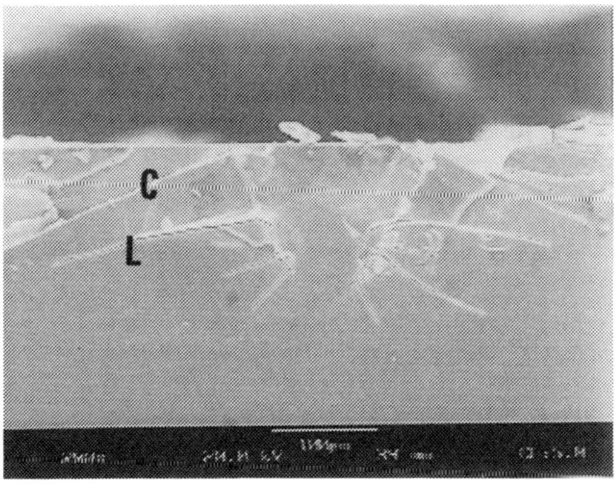

Figure 4.Scanning electron photomicrograph of a cross-section of a Hertzian indentation in "As-Received" soda-lime-silca. "C" indicates "cone" cracking, while "L" shows "lateral" cracking. The white bar represents 100 μm.

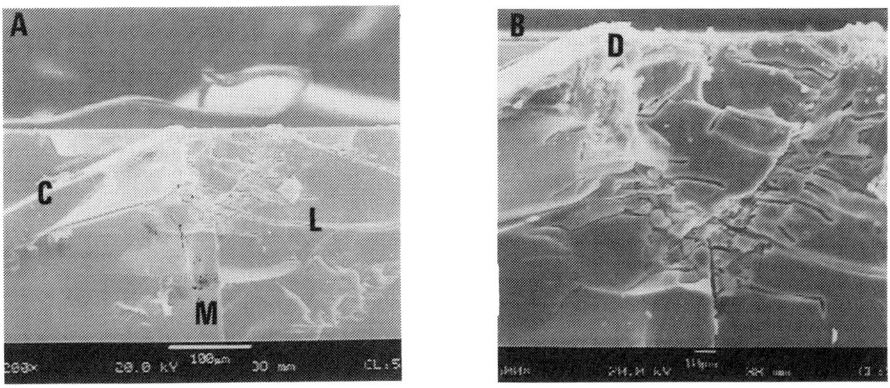

Figure 5 (A and B). Scanning electron photomicrographs of a cross-section of a Hertzian indentation in (120-grit) "ground" soda-lime-silica. (A) "C" indicates "cone" cracking, "M" shows a "median" crack, while "L" designates "lateral" cracking. The white bar represents 100 μm; (B) the same cross-section showing the "displaced" surface material (designated by "D"). Here the white bar represents only 10 μm.

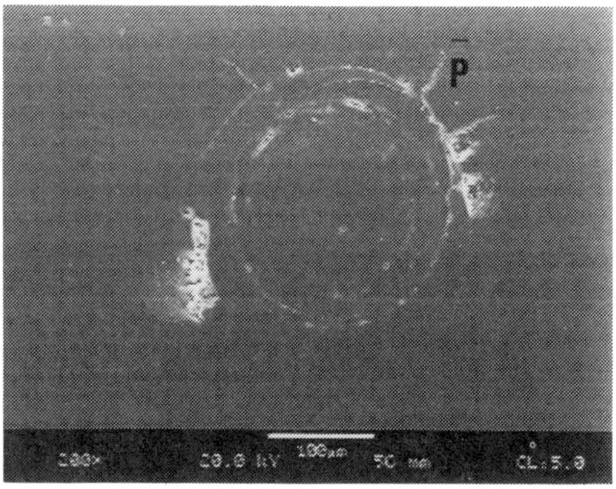

Figure 6.Scanning elctron photomicrograph of a Hertzian indentation of "As-Received" vitreous silica. "P" represents a precursor to "radial" cracking. The white bar represents 100 μm.

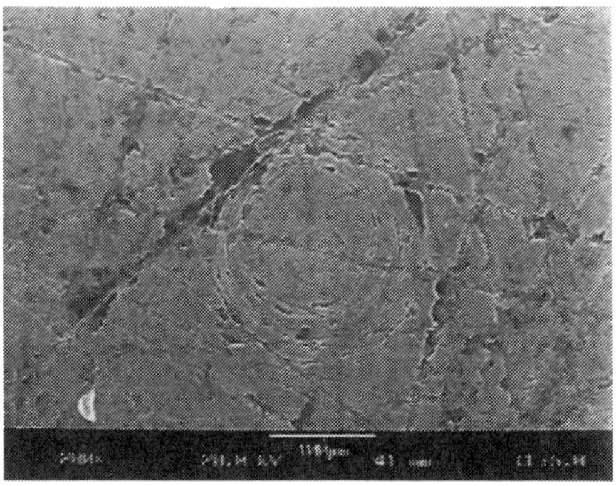

Figure 7.Scanning electon photomicrograph of a Hertzian indentation in (120-grit) "ground" vitreous silica. Note that there is no evidence of "radial" precursor cracks. The white bar represents 100 μm.

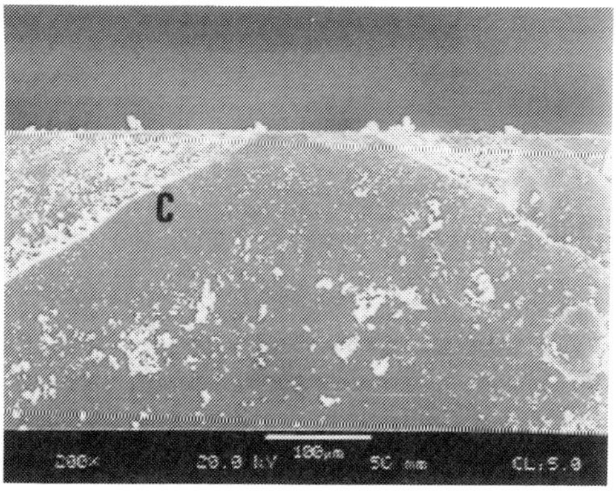

Figure 8.Scanning electron photomicrograph of a cross-section of a Hertzian indentation in "As-Received" vitreous silica. "C" indicates "cone" cracking. The white bar represents 100 μm.

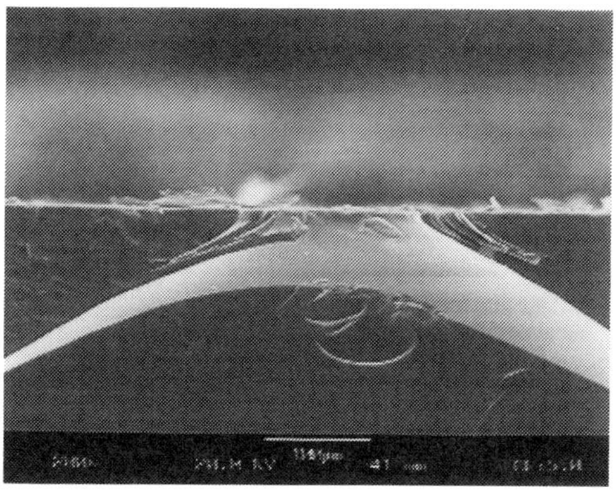

Figure 9.Scanning electron photomicrograph of a cross-section of a Hertzian indentation in (120-grit) "ground" vitreous silica. The white bar represents 100 μm.

V. REFERENCES

[1] Cook, R. and Pharr, G; "Direct Observation and Analysis of Indentation Cracking in Glasses and Ceramics," J. Am. Ceram. Soc. **73** (1990) No. 4, p. 787-817

[2] Chiang, S., Marshall, D., and Evans, A.; "The Response of Solids to Elastic/Plastic Indentation. II. Fracture Initiation," J. Appl. Phys. **53** (1982) p. 312-317

[3] Arora, A., Marshall, D., Lawn, B., and Swain, M.; "Indentation Deformation of Normal and Anomalous Glasses," J. Non-Cryst. Solids **31** (1979) p. 415-428

[4] Peter, K.; "Densification and Flow Phenomena of Glass in Indentation Experiments," J. Non-Cryst. Solids **5** (1970) p. 103-115

[5] Ernsberger, F.; "Role of Densification in Deformation of Glasses Under Point Loading," J. Am. Ceram. Soc. **51** (1968) No. 10, p. 545-547

[6] Argon, A., Hori, Y., Orowan, E.; "Indentation Strength of Glass," J. Am. Ceram. Soc. **43** (1960) No. 2, p. 86-96

[7] Ackakaya, R., "Formation of Inner Ring Cracks During Low-Velocity Thermoelastic Impact on Glass," M.S. Thesis, Alfred University, Alfred, NY (1991)

[8] Knight, C., Swain, M., and Chaudri, M.; "Impact of Small Steel Spheres on Glass Surfaces," J. Mater. Sci. **12** (1977) No. 12, p. 1573-1586

[9] Chaudhri, M., and Walley, S.; "Damage to Glass Surfaces by the Impact of Small Glass and Steel Spheres," Phil. Mag. A **37** (1978) No. 2, p. 153-165

[10] Lankford, J., and Davidson, D., "The Crack Initiation Threshold in Ceramic Materials Subject to Elastic/Plastic Indentation," J. Mater. Sci. **14** (1979) No. 7, p.1662-1668

[11] Jankford, J.; "Threshold Microfracture During Elastic-Plastic Indentation of Ceramics," J. Mater. Sci. **16** (1981) No. 5, p. 1177-1182

[12] Cleary, T., and Varner, J.; "Effects of Surface Treatments on Crack Initiation in Glasses," Glastech. Ber. Glass Sci. Technol. **68**, (1995) No. C1 p. 295-302, (Proc. 3rd Conf. European Soc. of Glass Sci. and Techn., May 22-24, 1995, Würzberg, Germany)

[13] Hagan, J.; "Shear Deformation Under Pyramidal Indentations in Soda-Lime Glass," J. Mater. Sci. **15** (1980) No. 6, p. 1417-1424

VI. ACKNOWLEDGMENTS

The authors would like to express their gratitude to the NSF/Industry and Center for Glass Research at the New York State College of Ceramics at Alfred University for providing the necessary funding to complete this work. Additional appreciation is extended to AFG., Ind., and Corning Incorporated for providing the glasses used in this study.

STRUCTURAL IMPLICATIONS OF ELASTIC PROPERTIES, FRACTURE TOUGHNESS AND FRACTURE SURFACE ENERGIES OF R_2O-Al_2O_3-SiO_2 GLASSES

William C. LaCourse, Walter Mason and Vadim Malot
NYS College of Ceramics
Alfred University
Alfred NY 14802

INTRODUCTION

It is generally accepted that the tensile strengths of silica glass fibers greatly exceed those of most complex commercial fibers. However, there have been few systematic studies of compositional effects on strength, and no formal relationships between composition and strength have been developed as has been done for such properties as thermal expansion, density, and viscosity. Indeed, it would be extremely difficult to develop such a relationship since strengths are strongly influenced by surface flaws, residual surface stresses and environmental conditions. These effects can cause variations of as much as a factor of 100 in a population of "as-received" samples tested in flexure. As a result, any variations in strength due to composition can be lost in the statistics. Abrasion lowers the standard deviation, but introduces the possibility that one is observing different sensitivities to abrasion and/or to slow crack growth, rather than inherent strength.

Clearly, composition is important. Derivations of "theoretical" strength lead to equations involving material parameters such as the elastic constants and the surface energy of the solid.[1] Orowan's simplified energy balance approach leads to

$$\sigma_{max} = \sqrt{\frac{E\,\gamma}{a_o}} \qquad (1)$$

where σ_{max} is the strength in the absence of flaws, E is Young's modulus, γ is the fracture surface energy and a_o is the equilibrium spacing between atomic planes. E and γ are determined by the number and strength of the bonds.

In a completely brittle material the fracture surface energy can be estimated from the concentration and strengths of the interplaner bonds in the material. For SiO_2 glass Doremus[2] estimates 7.9 x 10^{18} Si-O bonds/m^2 with bond strengths of 444 kJ/mole, which leads to a value of $\gamma = 2.9$ J/m^2, and an estimated theoretical strength of about 18 GPa. This is close to values measured at liquid nitrogen temperature for pristine silica fibers. (13 GPa)

Factors influencing E, γ and the theoretical strengths of multicomponent silicate glasses include:

1. Overall network structure.
2. Concentration of non-bridging oxygen.
3. Type of network modifier. (Field strength)
4. Presence of other network ions such as Al or B.
5. Phase separation.

It is nearly impossible to independently study the above effects. Non-bridging oxygen content is usually varied by addition of alkali and/or alkaline earth ions. However, phase separation occurs in the compositional ranges of interest (0-25 %), and glasses are extremely moisture sensitive. Addition of alumina or boron to alkali silicate glasses with compositions outside the two phase region can also be used to vary the NBO content since it is well established that these ions remove NBO in a predictable manner.[3,4] Of course here one must also consider the effects of the Al or B on theoretical strength, and with low Al or B contents the glasses are still moisture sensitive.

The second alternative was used here since effects of phase separation are somewhat more difficult to assess, and earlier measurements in our laboratories[5] indicated an anomalous behavior in the strength of pristine Na_2O - Al_2O_3- SiO_2 fibers, as shown in Figure 1. Strengths of $Na_2O \cdot xAl_2O_3 \cdot (3-x)SiO_2$ fibers increased with the initial Al additions but exhibited a strong maximum at about x = 0.2. An earlier study by Pantano[6] showed that the fracture toughness, measured by indentation fracture strength, showed a maximum at approximately the same composition. Eagan and Swerengen[7] also showed similar effects on fracture toughness. Interestingly, early work in similar glasses showed that the activation energy for Na^+ diffusion also exhibits a maximum in this compositional range.[8] A

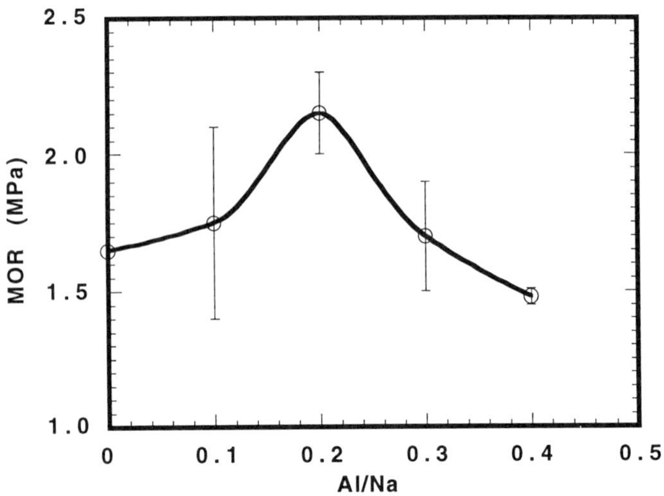

Figure 1. Strength of untouched Na_2O xAl_2O_3 (3-x) SiO_2 glass fibers.

more detailed investigation of the mechanical properties of the aluminosilicates therefore seemed appropriate.

Fracture toughness measurements coupled with independent measurements of E can provide useful information regarding γ. Indentation fracture (crack generation under Vickers loading) is particularly useful for studies of compositional effects in glass since several measurements can be made on a single sample and since complicating factors introduced by grain boundary effects are absent. This technique was used in the present study.

EXPERIMENTAL PROCEDURE

Glass Preparation: Glasses of the following series were prepared. Na_2O-xAl_2O_3-(3-x)SiO_2; Li_2O-xAl_2O_3-(2-x)SiO_2; Na_2O-xAl_2O_3-(2-x)SiO_2; 0.5Na_2O-0.5Li_2O-xAl_2O_3-(3-x)SiO_2; 0.72Na_2O-0.28CaO-xAl_2O_3-(3-x)SiO_2, where x varied from 0.0 to 1.0. The molar percentage of modifier remained constant in each series. Additional systems, including a 33 % mixed alkali and alkali boro- and galliosilicates (Na_2O-xB_2O_3-(3-x)SiO_2; Na_2O-xGa_2O_3-(3-x)SiO_2;) were also studied but will not be discussed in the present paper. Batches yielding 50 grams of glass were prepared using powdered forms of SiO_2, Al_2O_3, and the carbonates of sodium calcium and lithium. Glass compositions with an alumina content greater than 15 mol % contained 0.03 weight percent As_2O_3 as a fining agent. The batches were mixed manually and charged into covered alumina or Pt/Rh crucibles. The crucible was put into an electrical resistance furnace at 1723 K and the batches were fused in air. All batches were first held at 1723K for 3 to 8 hours, depending on composition. Glasses with a high alumina content were subsequently melted in a high temperature furnace at 1923 K for an additional 2 hours.

Bubble free melts were cast into shallow rectangular graphite molds, and were immediately placed in an electrical resistance muffle furnace held 10-15 K above T_g, for 30 minutes and then left to cool in the furnace to room temperature. Wet chemical analyses were performed on randomly selected glasses. The targeted and the actual compositions were typically within 1%. Samples of each composition were wet ground on their two major surfaces to a 600 grit finish using SiC abrasive paper. Next, 1 μm cerium oxide polishing compound was used to polish these surfaces to an optical quality finish. All specimens were reannealed to eliminate residual stresses induced by grinding.

Elastic Properties: The elastic properties were determined at room temperature by the pulse-echo technique in which one measures the transverse and longitudinal sound wave velocities, from which the elastic properties can be calculated. An ultrasonic pulser-receiver equipped with an oscilloscope was used with a 5 MHz longitudinal wave transducer or a 2.25 MHz shear wave transducer, as appropriate, to measure the transit time of a sonic pulse through a plate of known thickness. The velocities so obtained were used together with sample density to calculate Poisson's ratio, shear modulus and Young's modulus. Five repetitions of the measurement of the longitudinal and shear wave velocities on one specimen yielded a standard deviation of 0.5 GPa for Young's modulus and 0.3 GPa for the shear modulus.

Microhardness: The samples used to determine elastic properties were also used to measure Vicker's microhardness The testing was done at room temperature in air. A load of 200 gf was maintained over an interval of 15 seconds. The indentations were checked for cracks emanating from their corners; and, if none was present, the average length of the diagonals of the indentation was measured and used to calculate hardness values. The results are reported as the average of five measurements for each composition, plus and minus one standard deviation.

Toughness and Fracture Surface Energy: A Vicker's diamond indenter was employed for indentation toughness studies, using the same samples used for hardness. A special holder mounted on ball bearings was used to ensure that the specimen was perpendicular to the indenter. Testing was done in an air conditioned laboratory at room temperature. A load of 49 N was applied for 15 seconds. The length of the indentation diagonals (a) and the surface cracks (c) were measured within 1 minute after indentation to minimize any influence of environment-induced post-indentation crack growth. The average values of both quantities were then calculated. Six such measurements were made for each composition and combined to obtain an average crack length and an average half-diagonal length for use in the determination of fracture toughness.

A qualitative evaluation was made of the geometry of the indentation cracks by direct observation of the subsurface cracks as shown in Figure 2. At least one side of each specimen was ground perpendicular to the test surface and polished to optical quality. Cracks were then introduced by indenting near the polished side of the specimen. The indenter was oriented such that one diagonal of the Vicker's diamond was parallel to the edge of the specimen and the other perpendicular to it.

Crack profiles were examined in reflected light using an optical microscope. The objective was focused beneath the surface in order to see the geometry of the cracks in a plane parallel to the side of the specimen. In each composition the cracks were found to be of the Palmqvist geometry so that the appropriate formula for the fracture toughness is[9]

$$K_{IC} = 0.015 \left(\frac{E}{H}\right)^{\frac{3}{2}} \left(\frac{a}{l}\right)^{\frac{1}{2}} \frac{P}{C^{3/2}} \qquad (1)$$

where K_{IC} is the fracture toughness, E = Young's modulus, H = Vicker's microhardness, "a" is half the indentation diameter, "l" the crack length measured from the corner of the indent to the crack tip, c = (a + l) and P is the applied load. Fracture surface energies were calculated assuming the Palmqvist geometry.

Effects of post indentation crack growth was also evaluated for several compositions by carrying out the measurements in an inert environment. While some differences in the measured crack lengths were noted, (decreased length) the trends in the data with composition were duplicated. A comparison of the two types of measurements will be made in a separate paper. Only the data obtained in air are reported here.

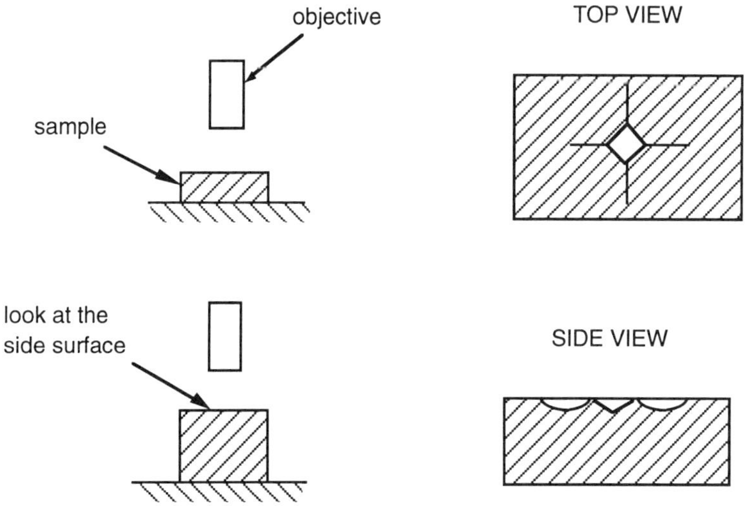

Figure 2. Method for observing subsurface cracking caused by Vicker's indentation. The Palmqvist geometry, observed for all compositions discussed in the present paper is shown.

RESULTS AND DISCUSSION

Data obtained in the present study are summarized in Tables I-V. Figure 3 shows the behavior of Vicker's Hardness (H) for the 25 Na_2O x Al_2O_3 (75-x) SiO_2 glasses. Figure 4 presents the Young's modulus (E) and measured fracture toughness, K_{IC}. Figure 5 shows the calculated fracture surface energies for the same glasses. Values were calculated assuming the Palmqvist geometry. Pantano's data for fracture toughness in liquid nitrogen, measured using an indentation fracture technique exhibits the same compositional dependence, with values slightly higher for each composition.

K_{IC} values measured here and those determined by Pantano show a fairly pronounced maxima at an Al/Na ratio near 0.2. Since E increases continuously in this compositional range the fracture surface energy, γ, also shows a maximum at this same composition, as shown in Figure 5. Recall (Figure 1) that the strength of pristine fibers exhibit a maximum and the activation energy for ionic conductivity shows a pronounced maximum at the same composition.[8]

Table I Properties of Na_2O-xAl_2O_3-$(3-x)SiO_2$ glasses.

% Al	H GPa	E GPa	G GPa	K_{IC} MPa $m^{.5}$	γ J/m^2
0	3.84	58	24	0.51	1.74
2	3.90	60	24	0.54	1.81
4	3.98	60	25	0.55	1.92
6	4.13	61	25	0.55	1.85
8	4.30				
10	4.40	65	26	0.48	1.35
15	4.60	66	27	0.39	0.85
25	4.70	69	28	0.41	0.94

Table II. Properties of $0.5Na_2O$-$0.5Li_2O$-xAl_2O_3-$(3-x)SiO_2$ glasses.

% Al	H GPa	E GPa	G GPa	K_{IC} MPa $m^{.5}$	γ J/m^2
0	5.05	70.	29	0.82	3.8
2	5.15	72	30	0.90	4.5
4	5.20	73	30	0.83	3.7
6	5.30	73	30	0.79	3.4
8	5.45	78	33	0.76	3.0
10	5.60	79	33	0.69	2.4
15	5.70	84	34	0.73	2.5
25	5.90	87	35	0.78	2.6

Table III Properties of $0.72Na_2O$-$0.28CaO$-xAl_2O_3-$(3-x)SiO_2$ glasses.

% Al	H GPa	E GPa	G GPa	K_{IC} MPa $m^{.5}$	γ J/m^2
0	4.45	67	28	0.53	1.7
2	4.75	68	28	0.58	1.9
4	4.80	69	29	0.59	1.9
6	5.00	70	29	0.55	1.7
8	5.10	70	29	0.55	1.7
10	5.25	72	29	0.55	1.6
15	5.30	73	30	0.54	1.6
25	5.35	76	31	0.64	2.1

Table IV Properties of Na_2O-xAl_2O_3-(2-x)SiO_2 glasses.

% Al	H GPa	E GPa	G GPa	K_{IC} MPa $m^{.5}$	γ J/m^2
0	3.95	60	24	0.53	1.75
2	4.05	61	24	0.46	1.30
4	4,25	63	25	0.45	1.20
6	4.30	63	25	0.39	0.85
8	4.30	64	26	0.32	0.60
10	4.40	65.	26	0.29	0.50
20	4.95	70	28	0.36	0.65

Table V. Properties of Li_2O-xAl_2O_3-(2-x)SiO_2.

% Al	H GPa	E GPa	G GPa	K_{IC} MPa $m^{.5}$	γ J/m^2
0	4.80	78	32	0.86	3.75
2	4.80				
4	4.90	79	32	0.63	1.95
6	4.95	82	3	0.57	1.60
8	5.05				
10	5.10	84	33	0.52	1.15
15	5.40	91	36	0.37	0.55
25	5.85	92	37	0.58	1.30

Pantano[6] also found that the stress intensity exponent, ("n" in the crack velocity equation, $V = AK_I^n$) was a maximum (reduced slow crack growth) in the same range. T_g shows no unusual behavior, nor do most other properties.

The relatively smooth increase of T_g with increasing Al_2O_3 content has been explained by a continuous decrease in the concentration of NBO's. However, this alone cannot account for the observed changes in toughness, and activation energy for conduction. Furthermore, a close examination of the behavior of E and H reveals a more complex dependence on alumina content than that predicted from a simple linear decrease in NBO content. Each of the properties increase slowly with increasing Al/Na (R) to approximately 0.2. From R = 0.2 to 0.4 the properties increase more rapidly, and above R = 0.5 - 0.6 the increase again slows. The behavior can be understood on the following basis:

1. The Na_2O 3 SiO_2 glass structure is best described by the "modified random network" structure in which continuous alkali/NBO rich "pathways" surround isolated Si-O-Si rich regions.[10] In such a structure, properties such as T_g, moduli and conductivity are controlled by the alkali rich regions.

2. When Al_2O_3 is added, the Al ions have a choice as to where they enter the structure; in the alkali rich or silica rich regions. It is unlikely that they will enter randomly.

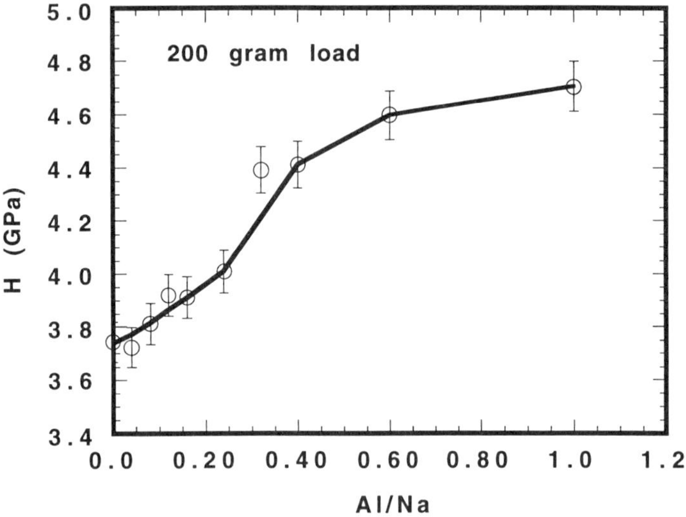

Figure 3. Vickers hardness of 25 % Na_2O xAl_2O_3 (75-x)SiO_2 glasses.

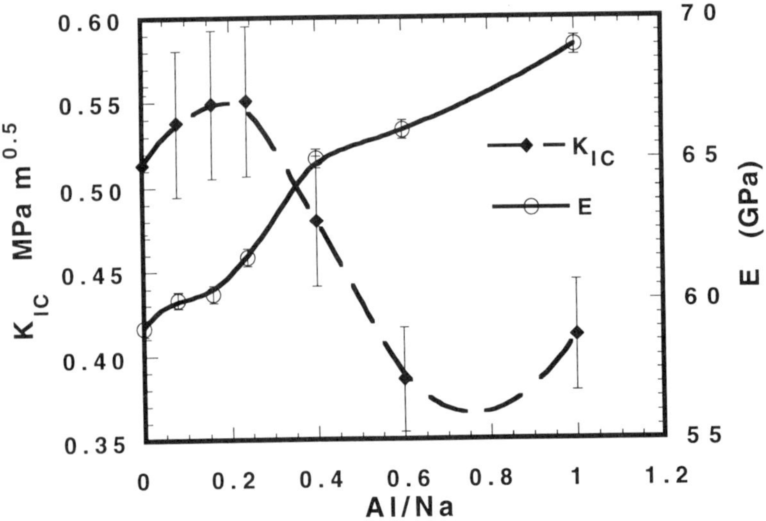

Figure 4. K_{IC} and E for 25 % Na_2O xAl_2O_3 (75-x)SiO_2 glasses.

Pantano suggested that the initial aluminum ions enter the structure preferentially in regions of high non-bridging oxygen content. Thermodynamic considerations and recent molecular dynamic computer simulations of these glasses[11], as well as the compositional dependence of the properties suggest the opposite; i.e. the initial Al^{+3}

enter the silica rich regions. Here, the Al^{+3} assume 4-fold coordination with oxygen, creating a AlO_4 tetrahedron with 4 bridging oxygens. The single negative charge on the tetrahedra, is balanced by moving an Na^+ ion from the continuous alkali rich network to the interior of the Al-silicate cages. Consequences of this are:

1. The isolated Na^+ ions in the vicinity of the AlO_4 tetrahedra do not conduct since the distance between equivalent sites is too large.

2. The Na^+ ions remaining in the alkali rich network (conduction pathways) are less mobile. The pathways shrink in physical size and the concentration of charge carriers in the pathways is reduced. Alkali diffusion decreases and the activation energy increases to that representative of a glass with lower alkali content.

3. Properties which depend on the overall connectivity of the structure (T_g, E, G, H) are less strongly influenced by the initial Al since, at low Al/Na ratios, the strongly bonded, disperse, high-silica cages remain separated by the alkali-rich regions.

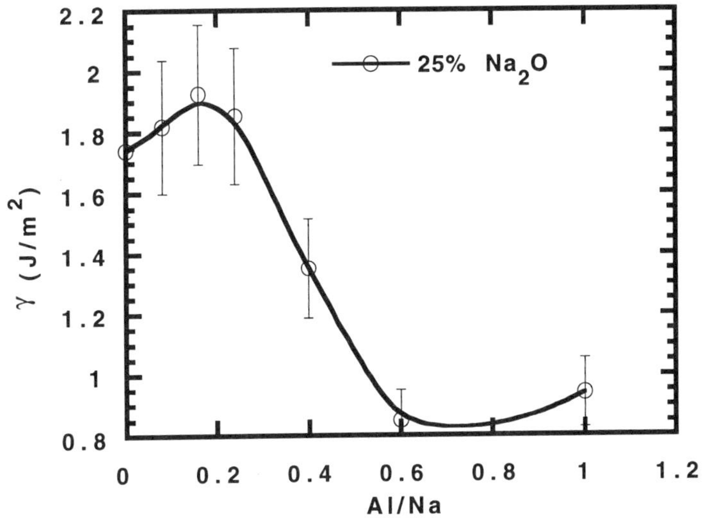

Figure 5. Fracture surface energy of 25 % Na_2O xAl_2O_3 $(75\text{-}x)SiO_2$ glasses.

Thermodynamic arguments for Al entering the silica rich, rather than the alkali rich region are based on the negative charge on the AlO_4 tetrahedra. A NBO would create a second negative charge, and would likely be unstable. Similar arguments have been made to account for the structure of alkali-borate glasses.

K_{IC} and γ increase as the Al enters the silica rich regions, but then decrease rapidly beyond R = 0.3. In this region Al is likely beginning to enter the alkali rich regions, creating strong covalent links between the previoulsly isolated cages. The continuous alkali rich regions become increasingly disrupted by strong Si-O-Si or Al-O-Si bonds between the silica rich regions. Properties determined by network connectivity therefore change rapidly for values of R between 0.2 and 0.4. Work by Cao and Cormack[10] further suggests that in this compositional range the structure begins to convert from the "modified random network" to a random network structure. Thus, while the concentration of non-bridging oxygens continues to decrease, a concurrent total rearrangement of the structure takes place.

The conductivity begins to increase as more Na^+ are bonded to the AlO_4^- tetrahedra. Eventually, new pathways for these ions are setup and the conductivity becomes greater than in the binary alkali silicate at about R = 0.4 since the Na^+ are less tightly bonded to the AlO_4^- tetrahedra than to NBO. E, G, and H increase more rapidly in the same range, (see Fig. 3 and 4) as is expected if the strong Si-O-Si and Al-O-Si bonding becomes extended three dimensionally.

Once the conversion to a three dimensional random network is complete, apparently around R = 0.7 for the 25 % Na_2O glasses, the property changes again slow. The structure of these glasses with R approaching 1.0 is complex. For a 25 % Na_2O glass there are essentially as many AlO_4 tetrahedra as there are SiO_4. Molecular dynamic simulations indicate increasing concentrations of tri-coordinated oxygen and Al-O-Al bonds while the NBO content decreases. It is difficult to predict, a priori, the effect of this combination of structural features on individual properties. Reduction in NBO would increase E, G and T_g, while tri-coordinated oxygens would likely temper this effect. Furthermore, since the transformation to a three dimensional random network has been accomplished, the slower changes in properties with further addition of Al is not unexpected. Changes in K_{IC} and γ also slow and perhaps even reverse as R = 1 is approached. This is probably controlled by the reduction in NBO since the transformation of the network structure is complete by about R = 0.7.

Pantano relates the initial increase in K_{IC} to the decreased mobility of the alkali under stress. Diffusion of a non-equilibrium number of alkali to a crack tip can accelerate bond breakage[12], and could account for the apparent connection between the maximum in "n" values determined in slow fracture experiments and the fracture toughness. We believe other effects may be equally important. Specifically, irreversible local movement (rotation and perhaps limited translation) of the silicate cages should be possible at low Al contents since the alkali rich regions permit continuous bonding during deformation. Total fracture need not occur. The slight increase in K_{IC} and γ with initial Al additions may be due to the increased mass of the "flow" units or the effect of a small concentration of Al in the alkali rich regions which hinder but do not prevent rotation. Beyond R = 0.2 - 0.3 the growing connectivity between silicate rich regions increasingly limits such motion and leads to a rapid decrease in γ, and a decreased K_{IC} despite the increased modulus.

The behavior of the mixed Li-Na aluminosilicate and the Na-Ca aluminosiliicate glasses with total modifier contents of 25%, show similar effects, but the maximum in toughness occurs at lower Al_2O_3 contents. The higher toughness of the mixed alkali aluminosilicates can be accounted for by the increased bonding between silicate rich regions by the Li^+ ions. Stress induced motion of the silicate regions may still occur but the energy absorption is greater. Also, the mobility of the alkali is reduced. In the case of the mixed alkali glasses this may relate to the tendency of Li to remain in the alkali rich regions with Na^+ preferentially associated with the AlO_4^- tetrahedra within the network forming rich regions. The higher fracture toughness, modulus and fracture energies for these glasses and for other Li_2O containing silicates is consistent with this interpretation.

The Na_2O-CaO aluminosilicates show a similar maximum in the low Al_2O_3 region, but at much smaller Al contents, suggesting that the structures of the alumina-free glasses are similar to the simple alkali silicates. The transformation of the structure from a MRN to a RN structure may occur at smaller concentrations.

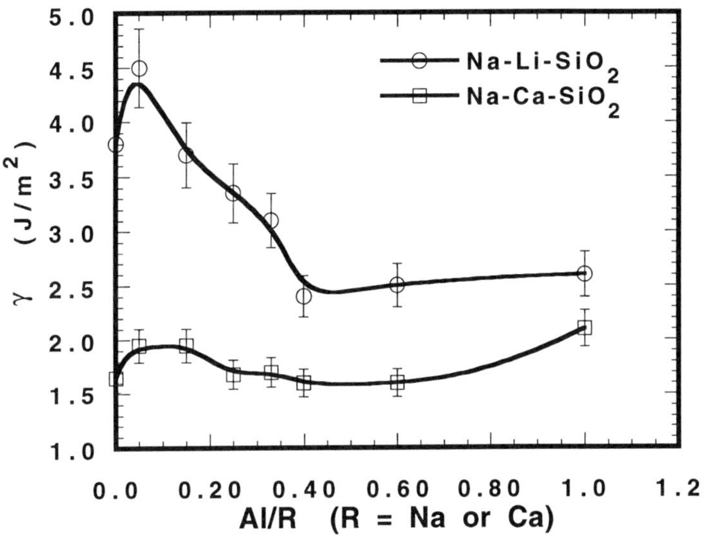

Figure 6. Fracture surface energies for soda-lime aluminosilicate and mixed alkali aluminosilicate glasses.

SUMMARY

The effects of Al_2O_3 on mechanical properties of alkali silicate glasses can be understood on the basis of a total structural rearrangement of the structure that occurs at critical Al_2O_3 contents. For low Al_2O_3 contents a "modified random network" structure exists in which toughness is controlled by energy absorbed when isolated silica rich regions move in a manner similar to interchain motion in polymeric systems. The motion does not require fracture since bonding between silicate units via Na^+ - O^- Na^+ bonds is continuous during this motion. At critical Al_2O_3 contents the isolated silicate rich units begin to link via strong, Si-O-Si or Si-O-Al bonds, and the mobility of the units is decreased leading to rapid decreases in K_{IC} and γ, and to rapid increases in T_g, E and H. Similar structural changes can also account for maxima in the activation energy for conductivity.

ACKNOWLEDGEMENTS:

The authors would like to thank the Center for Glass Research and the Center for Advanced Ceramic Technology at Alfred University for support of this research.

REFERENCES

1. A. Kelly and N.H. Macmillan, Strong Solids, Clarendon Press, Oxford (1986) pp 1-50.

2. R. Doremus, Glass Science Wiley-Interscience, New York 1973.

3. B. M. J. Smets and T. P. A. Lommen, The Incorporation of Aluminium Oxide and Boron Oxide in Sodium Silicate Glasses, Studied by X-ray Photoelectron Spectroscopy, Phys. Chem. Glasses, **22** 158-162 (1981).

4. T. D. Taylor and G. E. Rindone, Properties of Soda Aluminosilicate Glasses: III, Coordination of Aluminum Ions, J. Am. Ceram. Soc., **45** 479-581 (1956).

5. S. Carson, Ph.D. Thesis, NYS College of Ceramics, Alfred University, 1988.

6. C. B. Pantano, Surface Chemistry in Relation to the Strength and Fracture of Silicate Glasses, in Strength of Inorganic Glasses, ed. C. R.Kurkjian, Plenum Press (1985) 37-66.

7. R.J. Eagan and J.C. Swearengen, "Effect of Composition on the Mechacical Properties of Aluminosilicate and Borosilicate Glasses," J. Am. Ceram. Soc., **61** 27-30 (1978).

8. J. O. Isard, "Electrical Conduction in the Aluminosilicate Glasses," J. Soc.Glass Tech., **43,** 113T-123T (1959).

9. M. T. Laugier, Palmqvist Crack Extension and the Center-Loaded Penny Crack Analogy, J. Amer. Ceram. Soc., **68** C51-52 (1985).

10. G. N. Greaves, "EXAFS and the Structure of Glass" J. Non-Cryst. Solids, **71** 203-217 (1985).

11. Y. Cao and A.N. Cormack, A Structural Model for Interpretation of an Anomaly in Alkali Aluminosilicate Glasses at Al/Alkali = 0.2 - 0.4 , pp 137-152 in: Diffusion in Amorphous Materials, eds. H. Jain and D. Gupta, TMS, Warrendale, PA (1994).

12. S.M. Cox, Strength and Ion Mobility, Phys. Chem. Glasses, **10** 226 (1969)

ARC INDUCED FRACTURES FROM LAMPMAKING TEST EQUIPMENT

David Johnston
OSRAM SYLVANIA INC.
Central Research
Beverly, MA

ABSTRACT

Fractography is an invaluable tool for determining the cause of failure as demonstrated in this case history of a fluorescent lamp fracture. Some energy-saving fluorescent lamps require a conductive tin oxide coating on the inner tube surface. This conductive coating is then coated with a fumed alumina dielectric, followed by a fluorescent phosphor coating. Conductive coated energy-saving lamps from various manufacturers were subjected to high magnetic fields. During operation in the magnetic field, small bright sparks were observed forming and extinguishing on the inside surface of the glass. The glass became weak and subject to impact failures at very low levels of energy. Fracture analysis indicated the failures were starting on the inner surface of the glass tube with a thermal origin.

Further analysis revealed the fracture origins were generated when a high energy lamp arc or an ionized plasma in the lamp shorted to the conductive coating through the dielectric coating. The short caused a localized hot spot, fusing alumina and phosphor to the glass, forming microcracks and destroying the conductive nature of the coating.

INTRODUCTION

Fluorescent lamps were developed as lighting products in the 1930's.[1] The lamp is three to four times more efficient than incandescent lamps, and allows for a variety of shapes, colors and spectral outputs that cannot be achieved by other lighting products. Fluorescent lamps operate by ionizing a low pressure gas (usually Ar) which contains Hg vapor. The Hg vapor is excited by an arc and emits UV light, which is converted by a phosphor layer to visible light (see figures 1 & 2).

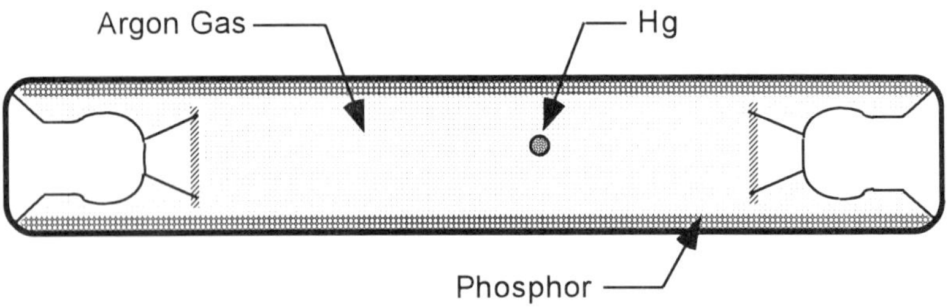

Figure 1 Schematic - fluorescent lamp

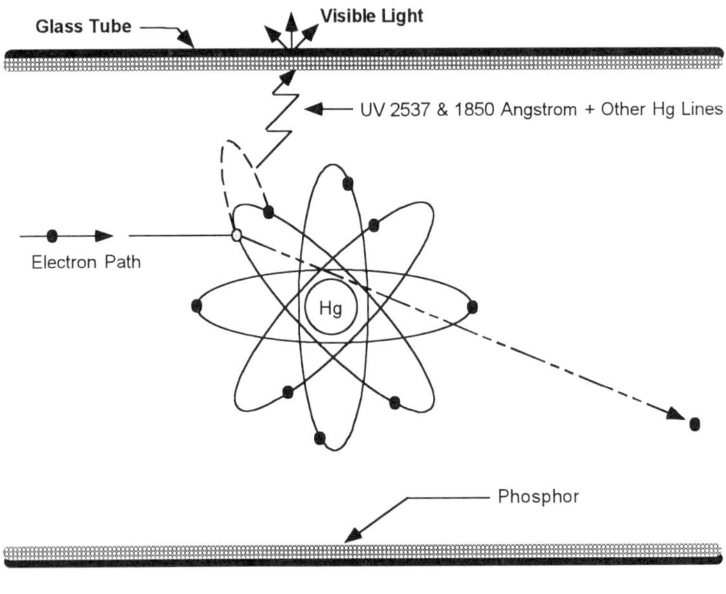

Figure 2 Section - fluorescent lamp

The efficiency of a fluorescent lamp can be increased by substituting Kr for some or all of the Ar gas. This change in gas composition boosts efficiency but also increases the starting voltage of the lamp, making it difficult to start on a

110V circuit.[2] It has long been known that the starting voltage of a fluorescent lamp can be reduced by placing an electrically conductive plane near the lamp. The first energy-saving lamps with a Kr gas mix employed a metal strip on the outside of the lamp to aid starting. This approach gave way to a tin oxide (SnO_{2-x}) conductive coating on the inside surface of the lamp. The conductive coating has advantages in being transparent and inexpensive. The conductive coated glass is coated with a dielectric (fumed alumina) and then with phosphor before being assembled into a lamp (See figure 3). This type of lamp is called an "Energy-saving Lamp".

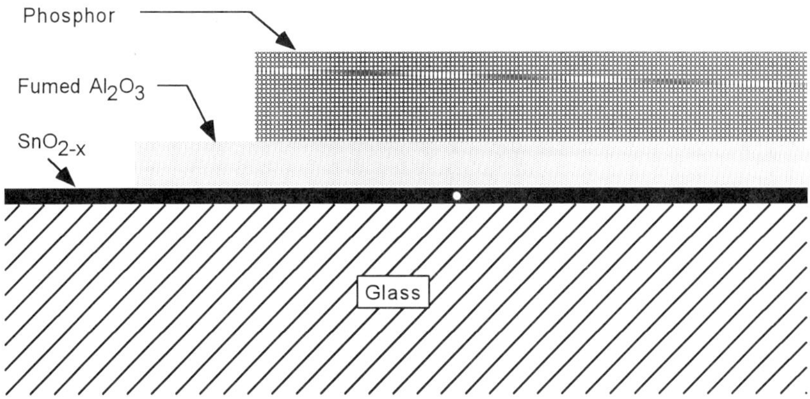

Figure 3 Coatings on energy-saving lamps

The manufacture of lighting products requires handling thin walled glass articles at a variety of temperatures and speeds. The glass is usually coated with various materials, sealed to other glasses, metals, and cements and subjected to intense heat and electrical fields. Test equipment can subject lamps to unusual electrical and physical conditions in the process of determining lamp characteristics. In some cases, testing can produce unanticipated damage to the materials in the lamp.

In a recent test, energy-saving lamps from several manufacturers which employ conductive coated glass were operated in a high magnetic field. During the test on energy-saving lamps, bright sparks or arcs were observed on the inside glass wall. These arcs were small and of short duration. Many arc strikes could be seen over a time period of several seconds. After testing, the glass was extremely weak and subject to impact failures during normal handling.

ANALYSIS

The energy required to cause failure under impact loading was extremely low. There was little evidence of crushed glass, or crack branching that normally occurs when a thin walled tube is fractured by impact. A typical crack of the type under study is shown in figure 4. Careful examination of the outside surface of the lamps revealed a fracture origin that could be seen through the glass wall. Minute burn marks were found on the inside surface of the glass (see figure 4). One such mark was associated with each origin and others were spread throughout the general area of the failure.

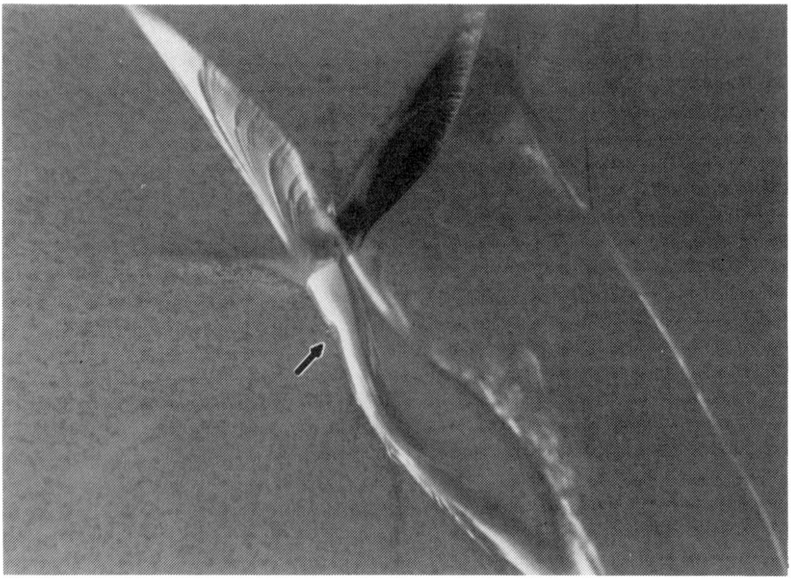

Figure 4 - Planer view of fracture origin as seen from outside surface

Fracture surfaces were carefully broken out and examined. A typical origin is shown in figure 5. The origins were on the inside surface of the tubing and resembled the type of origin that is seen when a large amount of heat is added to or removed from a point on the glass surface. Examination at high magnification revealed adhered material on the surface of the glass that could not be washed off. All of the fracture origins were on the inside surface, directly opposite the suspected point of impact.

Figure 5 - Fracture surface of lamp from figure 4; showing typical origin on inner surface of tube

Several fracture origins were removed from lamps and examined using a Scanning Electron Microscope (SEM) equipped with Energy Dispersive X-Ray Microanalysis (EDX) and Back Scattered Electron Detection (BSED). Analysis showed that fumed alumina and phosphor were fused onto the surface of the glass. There was no evidence of a flux or other foreign materials (those not normally found in a fluorescent lamp) at the point of interest. The typical arc strike appeared as a semi-circular spot 100 μm in diameter. Figure 6 shows the inside surface of the glass at the origin.

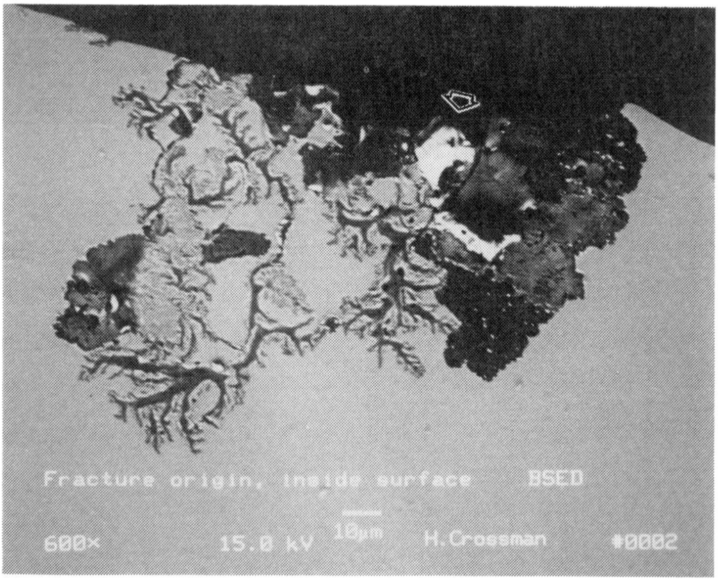

Figure 6 SEM micrograph of the inside surface of a lamp where an arc strike occurred. A fracture originated under the large particle of fused alumina and phosphor shown near the center.

There is evidence of melting, fusion of fumed alumina and phosphor (calcium halophosphate) and holes melted into the glass surface. A distinct pattern of grooves in the surface of the glass trace around the area of the origin. These grooves may be the result of arc wandering around the area of the arc strike. The white spot near the center of the image is a lump of fused phosphor and alumina.

Figure 7 is a micrograph of the fracture origin in the plane of the fracture. Please note the fused particle of alumina and calcium halophosphate. Due to the nature of the fracture origin, it was postulated that the lamp arc was striking the inside wall of the lamp and fusing the fumed alumina and phosphor coatings to the glass.

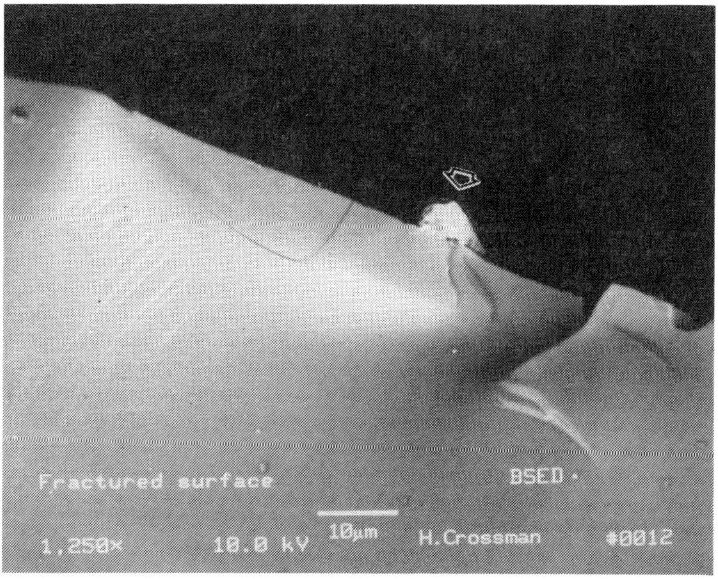

Figure 7 SEM micrograph of the fracture surface from the lamp in figure 6. A large particle of fused alumina and phosphor is located at the origin.

TESTING

Further testing was required to characterize this defect type and to determine if other methods could be used to cause an arc strike. An arc strike is a discharge of the ionized gas in the lamp through the phosphor and dielectric coatings to the conductive coating. Such strikes appear to be self limiting, because the conductive nature of the tin oxide coating is destroyed at high temperature. Testing showed that defects (such as thin spots, asperities, holes and foreign material) in the fumed alumina coating, which acts as a dielectric, could cause an arc strike to occur in the absence of a high magnetic field. Holes and thin-spots were created in the fumed alumina coating of several coated bulbs The area of each hole or thin-spot was marked from the outside of the bulb in a manner that did not influence the failures. These samples were then introduced into normal manufacturing on an automated high speed lampmaking line. Arc strikes with associated fusion of phosphor and dielectric occurred at each place where the dielectric coating had been altered. The electrical energy used to start the lamp during manufacturing combined with the defects that were created in the dielectric, allowing arc strikes to occur. These tests confirmed the hypothesis that arc strikes can also be caused by a discharge through weakness in the dielectric coating.

Glass tubing strength was measured under various conditions using a pendulum impact tester. The following table shows the energy necessary to fracture the glass under the following conditions: 1) phosphor coated; 2) conductive tin oxide, fumed alumina, and phosphor coated which has been hit by an arc strike on the opposite side of the glass from the impact site. Eight to ten samples were tested in each group. The results are shown in joules required to fracture the glass.

IMPACT ENERGY AT FAILURE OF GLASS TUBE

Phosphor Coated	SnO_2, Al_2O_3 & Phosphor Coated with Arc Strike
717×10^{-3} J	8×10^{-3} J

The effect of an arc strike is a reduction in the impact resistance of the conductive coated glass to approximately 1% of that when no arc strike is present. Impact testing shows the dramatic influence such a small defect can have on a thin walled glass tube. In the laboratory, defects similar to arc strikes that were found in test lamps were created by allowing a very short burst from a Tesla coil to strike the glass through the fumed alumina and phosphor coatings. Using this method, similar strength reductions occurred and the failure looked remarkably similar to the ones found on the test equipment.

CONCLUSIONS

Thin walled glass tubes (wall thickness ≤ 0.75 mm) rely on both inside and outside surfaces for strength. The impact resistance of a thin walled tube can be dramatically reduced when a relatively small defect is introduced on the inner surface. In the case of fluorescent lamps using tin oxide conductive coated glass, defects can arise from fusion of the dielectric coating and phosphor to the glass wall when an arc strike occurs between the conductive coating and the ionized gas in the lamp. This type of defect can be created by operating conductive coated lamps in high magnetic fields, by allowing a Tesla coil to discharge to the inside surface of the glass tube, or by purposely creating defects in the dielectric coating.

ACKNOWLEDGMENTS

The author would like to thank the following for their assistance: T. Winner; L. Ungar, H. Crossman & J. Gagne.

REFERENCES

1) W. Allphin, Chapter 6 - Fluorescent Lamps, pp 85-87, The Primer of Lamps and Lighting, Third Edition, Addison-Wesley Publishing Co., Reading MA, 1973

2) W. Elenbaas, J. Funke, TH. Hehenkamp, L. C. Kalff, A. A. Kruithof, J. L. Ouweltjes, L. M. C. Touw, R. Van Der Veen, D. Vermeulen, Chapter VI - Factors affecting the efficiency, pg. 99, Fluorescent Lamps and Lighting, The Macmillan Company, NY, 1959

FAILURE ANALYSIS OF A FIBERGLASS-REINFORCED PLASTIC PRESSURE VESSEL

S. J. Glass, E. K. Beauchamp, M. J. Carr, T. R. Guess, S. L. Monroe, R. J. Moore, A. Slavin, and N. R. Sorenson, Sandia National Laboratories, Albuquerque, NM 87185.

ABSTRACT

A fiberglass-reinforced plastic (FRP) pressure vessel containing sulfuric acid failed catastrophically in service. Preliminary investigations ruled out failure due to sabotage and chemical or mechanical overpressure. Subsequent examination of the fiber fracture surfaces and measurements of mirror radii indicated that fibers failed at stresses significantly below their expected strength. Further examination by scanning electron microscopy and energy dispersive spectroscopy suggested that the glass fibers had been exposed to sulfuric acid, a reagent that corrodes this type of glass and degrades its strength. Finite element analysis predicted axial stresses in the vessel in an exposed region above a supporting steel cradle, that exceeded the fiber-matrix strength of the circular wrap layers of FRP material. Numerous cracks were detected in this region using a vicinal illumination technique. We concluded that vessel failure was due to cracking of the circular wraps under high axial stresses, followed by exposure of the fibers in the cross ply layers to trapped acid, which produced significant strength degradation of these fibers. Progressive rupture of the cross ply fibers, starting at the outer surface of the FRP and extending inwards and laterally, occurred until a crack of critical size was produced.

BACKGROUND

A sulfuric acid vessel, pressurized at the normal operating pressure of 0.31 MPa (45 psi), burst during service. The vessel was one of two identical 200 liter, teflon-lined, fiberglass-reinforced plastic (FRP) vessels used as part of an acid distribution system. The FRP vessel contained several large cracks from which the burst teflon liner bulged as shown in Fig. 1A. The failure event also severely damaged the cabinet and showered the room with debris as shown in Fig. 1B.

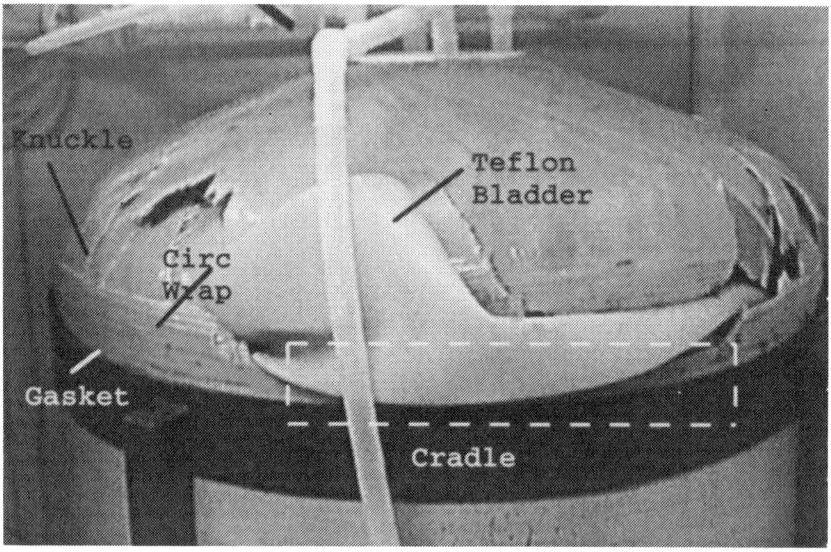

Fig. 1A. The FRP vessel contained several large cracks from which the burst teflon liner bulged. The dashed region was cut from the vessel for fracture surface examination (Fig. 2).

Fig. 1B. The FRP vessel failure also severely damaged the cabinet and showered the room with debris.

Although the room was unoccupied at the time of the failure, a thorough investigation of the event was performed in the interest of equipment reliability and personnel safety. Forensic reconstruction of the event, fractographic analysis of the failed vessel, computer modeling of stresses in the vessel walls, and mechanical testing of sections of the failed vessel provided information about the cause of failure.

PRELIMINARY INVESTIGATION

During the preliminary investigation dozens of failure scenarios were considered, all based on observations or recorded data. At each stage of the investigation, care was taken not to destroy or obscure evidence needed by a competing scenario. The various failure scenarios considered were grouped as follows:

- Sabotage
- Excessive internal pressure
  - Failure at elevated pressure due to chemical reaction, for example sulfuric acid reacts violently with many compounds, including water
  - Failure at elevated pressure due to mechanical over-pressurization
- Failure at normal operating conditions (pressure and temperature)

There was no evidence to suggest that a person had the means or ability to cause this type of failure by sabotage. No evidence was found to uniquely support overpressure by chemical and/or mechanical means. Forensic evidence indicated that the pressure in the failed vessel was not unusually high when it failed. Trajectory analysis of debris, including a level sensor found several meters away, indicated that the failure sequence was consistent with the FRP cracking, followed by the teflon bladder bulging out of the cracked FRP and bursting. A thorough analysis of the gas system also eliminated the possibility of mechanical over-pressurization. It was concluded that the failure occurred under normal operating conditions.

By design, the FRP vessels are expected to be able to withstand 1.4 MPa (200 psi) pressure. That, of course, assumes that the vessel retains its integrity. Thus it was assumed that the vessel integrity must somehow have been compromised and subsequent investigation focused on the fracture surfaces, the mechanical properties of the FRP material, and the stress state in the vessel under normal operating pressure.

FRACTOGRAPHIC ANALYSIS

Following an in situ examination of the ruptured FRP vessel, a 8 cm by 51 cm section (Fig. 1A, section in dashed lines) shown in Fig. 2, containing the primary fractures, was cut from the vessel for detailed examination. The axial location of the primary fractures is near the region on the vessel that was supported/constrained by the metal cradle and rubber gasket (Fig. 1A). The section of composite material containing the primary cracks is constructed from, nominally, seven layers (each ~0.05 cm thick) of fiber-reinforced plastic with alternating + and -32° oriented fibers, termed cross plies. The cross plies are surrounded by a circumferential (circ) wrap (0.23 cm thick) of fiber-reinforced plastic around the cylindrical portion of the vessel. The cross plies support the axial stresses in the vessel and the circ wraps support the hoop stresses.

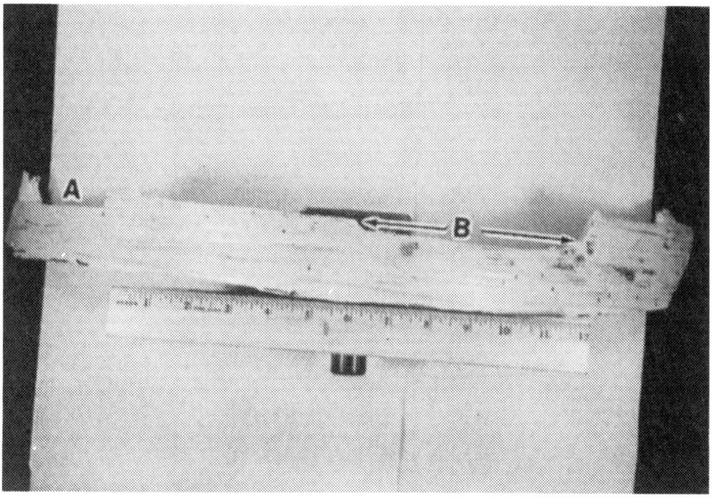

Fig. 2. Following an in situ examination of the ruptured vessel, an 8 cm by 51 cm section (section in dashed lines in Fig. 1), containing the primary fractures, was cut from the vessel for detailed examination. Macroscopic examination of the primary fracture surfaces revealed two areas (Areas A and B) of planar circumferential (or hoop) fracture, normal to the longitudinal axis of the vessel, extending almost completely through the circ wrap and cross plies.

Under normal conditions, overstressed FRP fails by a combination of cracking of the epoxy matrix and glass fibers pulling out of the epoxy matrix. Where fibers are parallel or oriented at some small angle to the principal stress direction, the fracture surface will be rough due to extensive fiber pull-out. If the glass fibers are substantially weakened in the plane where the matrix is cracked, little or no fiber pull-out occurs and the fracture surface will be planar. Where fibers are oriented perpendicular to or almost perpendicular to the principal stress direction, as they are in the circ wrap, the fracture surface will also be planar with no protruding fibers.

Macroscopic examination of the primary fracture surface revealed two areas (Fig. 2, Areas A and B) of planar circumferential (or hoop) fracture, normal to the longitudinal axis of the vessel, extending almost completely through the circ wrap and cross plies as shown in the schematic in Fig. 3A. If the vessel is exposed to sufficiently high axial tensile stresses, this type of planar hoop crack is expected in the circ wrap (a result of interfacial failure between the fibers and the epoxy matrix), but not in the cross plies. Fracture in the cross plies should exhibit extensive fiber pull-out producing a fracture surface that is not planar (Fig. 3B). The planar fracture surface of the cross plies was the first evidence that the vessel had been compromised in some way that weakened the fibers in that plane.

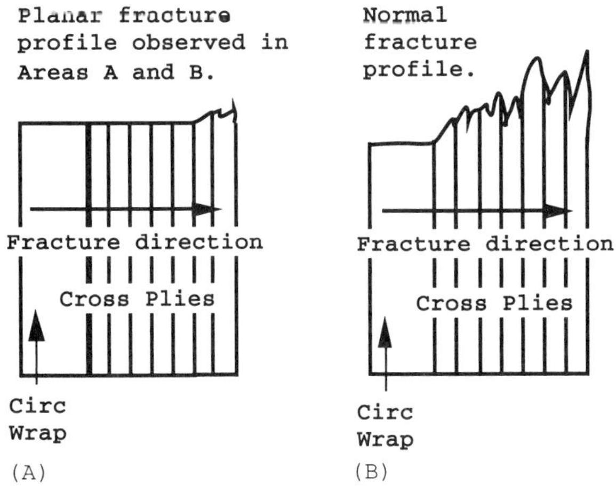

Fig. 3. A) Schematic cross section of the planar fracture surface observed for the circ wrap and cross plies of Area A. B) Schematic cross section of the <u>expected</u> fracture surface topography for the circ wrap and cross plies.

Areas A and B in Fig. 2 were examined using optical microscopy and scanning electron microscopy (SEM). Examination of fiber fracture surfaces from Area A provided evidence that the fracture propagated from the exterior toward the interior of the vessel. Individual fibers, in all of the cross plies, exhibited velocity hackle (Fig. 4) that indicated that all of the failure origins were on the side of the fiber closest to the vessel exterior. This observation suggested that something was degrading the fibers and it was gaining access to the fibers through a matrix crack that grew from the vessel exterior.

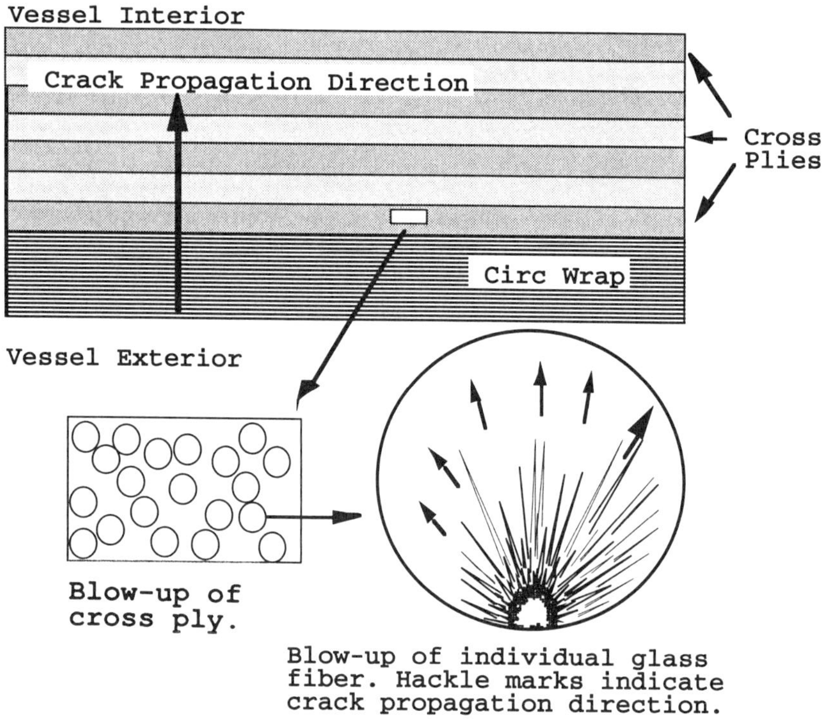

(A)

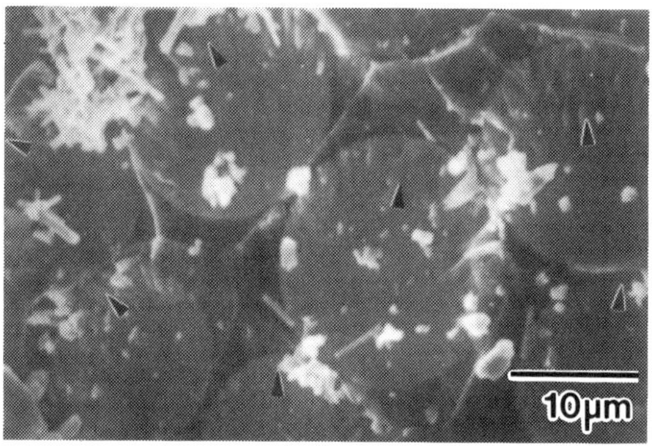

(B)

Fig. 4. (A) Schematic and (B) SEM micrograph of velocity hackle showing direction of crack propagation (note arrows).

The fracture mirror radii provided an estimate of the fiber failure stress.[1-3] Calculated failure stresses of these fibers ranged from 620-1207 MPa (90 to 175 ksi), substantially lower than the nominal 3100 MPa (450 ksi) strength for the E-glass fibers used in the vessel construction.

Planar fracture regions were bounded by fracture regions exhibiting extensive fiber pull-out, which were likely created during the final moment of vessel rupture as the regions of planar fracture linked together. Fibers in these linking regions exhibited mirror radii characteristic of higher strength fibers. These observations suggested that the vessel was fabricated from E-glass fibers with the expected strength level, but that they had been degraded and weakened in some regions.

A portion of Area B was examined using energy dispersive spectroscopy (EDS) to obtain information on fiber composition and on the presence of corrosive agents. Because the fracture surface had been flooded with sulfuric acid during the vessel rupture, and with aqueous solutions during clean-up, unequivocal chemical data could not be obtained from this area. Areas were sought for examination where cracks would not have been exposed to particulate debris, acid, or aqueous solutions. Chemical analyses were also performed on the fracture surfaces of flexure test specimens (described elsewhere in this paper) that were obtained from chemically uncompromised regions of the vessel. One specimen was selected for testing because it was located in the cradle region and contained a hoop crack in the circ wraps. This pre-existing crack did not extend completely through the specimen. The specimen was loaded in flexure such that the exterior surface containing the crack was placed in tension. This caused the pre-existing crack to propagate to completion through the innermost cross plies.

Figure 5 shows the fracture surfaces of a flexure specimen which did not contain a pre-existing crack (A) and the specimen described above that did (B). The direction of crack propagation is from the bottom of the figure to the top. Figure 5A shows no planar fracture in the cross ply regions (Area 2). Figure 5B shows the two types of macroscopic fracture features in the cross plies, i.e., the pre-existing crack exhibits planar fracture (Fig. 5B, Area 1), whereas the fracture surface produced during testing exhibits the expected fiber pull-out (Fig. 5B, Area 2).

SEM examination of the fracture surface of the flexure bar with the pre-existing crack was performed. The pre-existing crack contained planar fracture both in the circ wrap and in some of the cross plies immediately adjacent to the circ wrap. The remainder of the cross plies, which failed during flexure testing, exhibited fiber pull-out (Fig. 6A) and the fibers in this region failed at high stresses. EDS verified the presence of Ca and Al as major constituents of the E-glass fibers (Fig. 6B). This EDS information can be contrasted with the EDS spectrum from material from the planar region of fracture (Fig. 7A). Fibers in this region failed at low stresses (689 MPa). EDS of these fibers (Fig. 7B) indicated Ca and Al depletion, which strongly suggested that fibers had been exposed to a corrosive chemical reagent such as an acid.

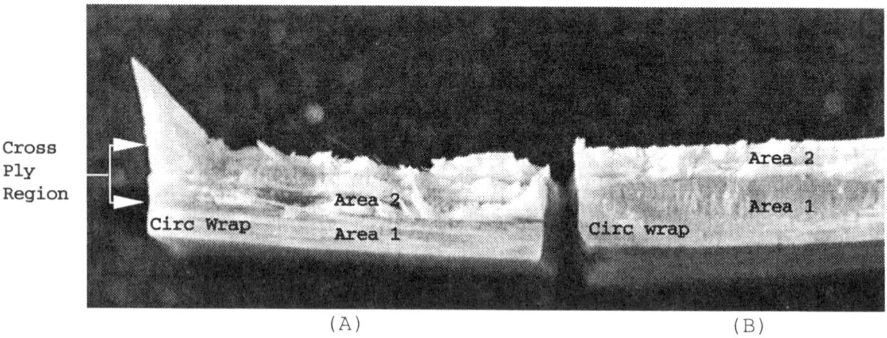

Fig. 5. A) The fracture surface of a flexure specimen that did not contain a pre-existing crack. There is no planar fracture in the cross ply regions (Area 2). B) The fracture surface of a flexure specimen that contained a pre-existing crack in the circ wrap. The direction of crack propagation is from the bottom of the figure to the top. The photo shows the two types of macroscopic fracture features in the cross plies, i.e., the pre-existing crack exhibits planar fracture (Area 1), whereas the fracture surface produced during testing exhibits the expected fiber pull-out (Area 2).

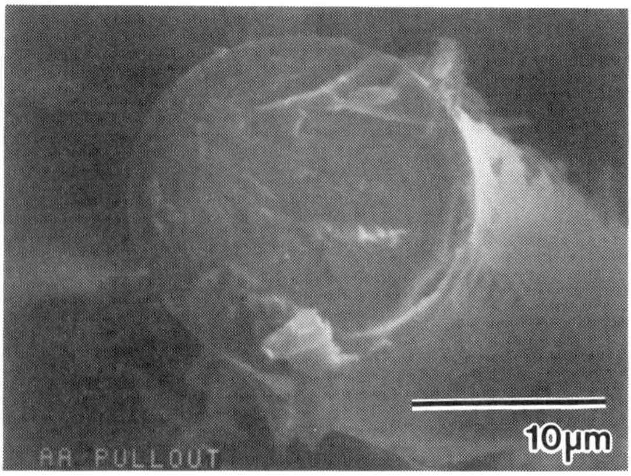

Fig. 6A. SEM micrograph of a fractured fiber from a region of fracture exhibiting extensive fiber pull-out.

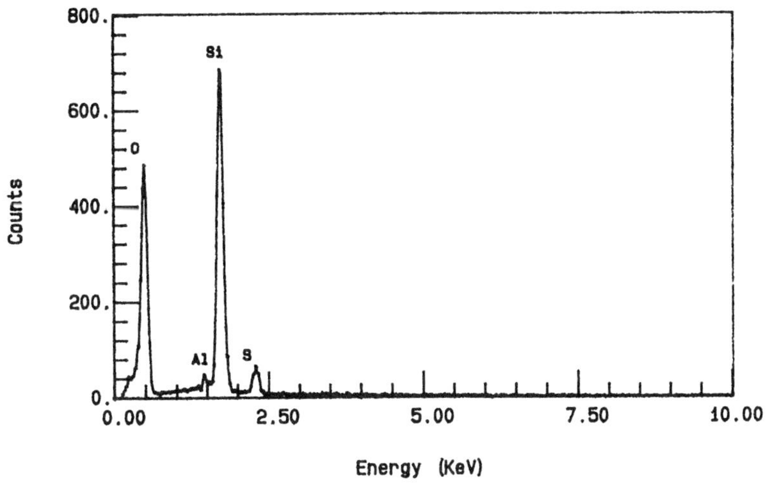

Fig. 7B. EDS of fiber from planar fracture region (Fig. 5B, Area 1) indicated depletion of Ca and Al relative to the EDS spectrum in Fig. 6B.

The source of the acid was identified by examining the normal operation and maintenance of the vessel. The design of the acid vessel incorporates a top cover, which provides support for pipes and valves. During routine valve maintenance sulfuric acid is spilled onto the top of the vessel and subsequently rinsed off with water. The fit of the top cover on the vessel provided a mechanism to trap acid and/or acid rinse-water in the vicinity of the circ wrap. Given this opportunity for exposure of the fibers to sulfuric acid, it is quite probable that Ca and Al were leached from the fibers by the acid.[4,5] Acid dissolution of Ca and Al significantly reduces the tensile strength of E-glass fibers,[4] consistent with our fractographic measurements. Further analysis of deposits on the fracture surface indicated the presence of $Ca(SO_4)_2$, a likely reaction product after prolonged exposure to sulfuric acid.

To examine the question of whether matrix cracking occurred throughout the vessels or preferentially in the vicinity of the failure, the failed vessel was examined for cracks in regions other than the primary fracture surface using a simple optical inspection technique. A fiber optic light source is placed against the surface of the translucent FRP vessel. In this manner, light is internally scattered. With this vicinal illumination cracks oriented normal to the surface scattered the light and were observable as shadows in the halo around the fiber optic light source as shown in Fig. 8. Magnification of 10-20X is helpful, but

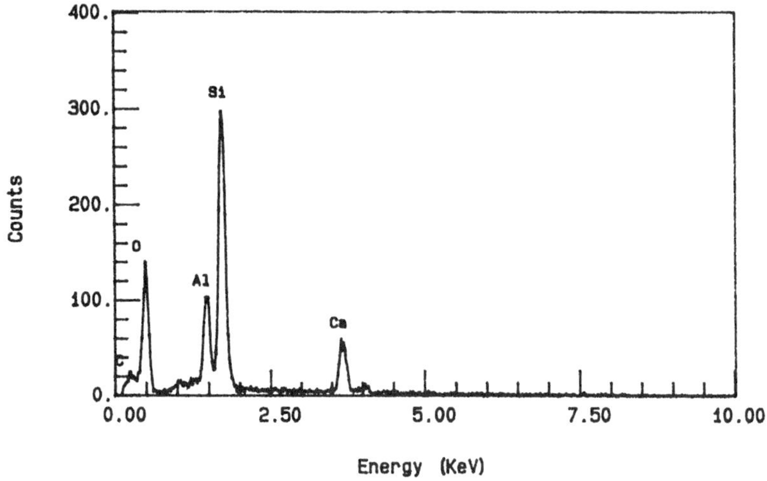

Fig. 6B. EDS verified the presence of Ca and Al as major constituents of E-glass fibers in the region of Fig. 6A.

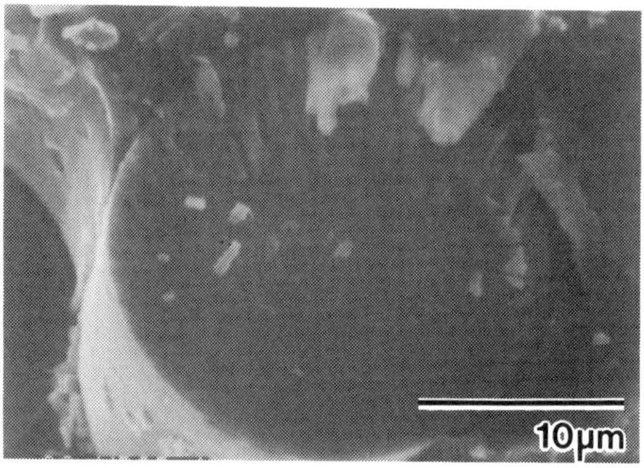

Fig. 7A. SEM micrograph of a fractured fiber from a region of fracture exhibiting planar fracture (Fig. 5B, Area 1).

not necessary for detecting cracks in this manner. Using this technique cracks were located in regions away from the primary crack in the failed vessel and in the companion unfailed vessel.

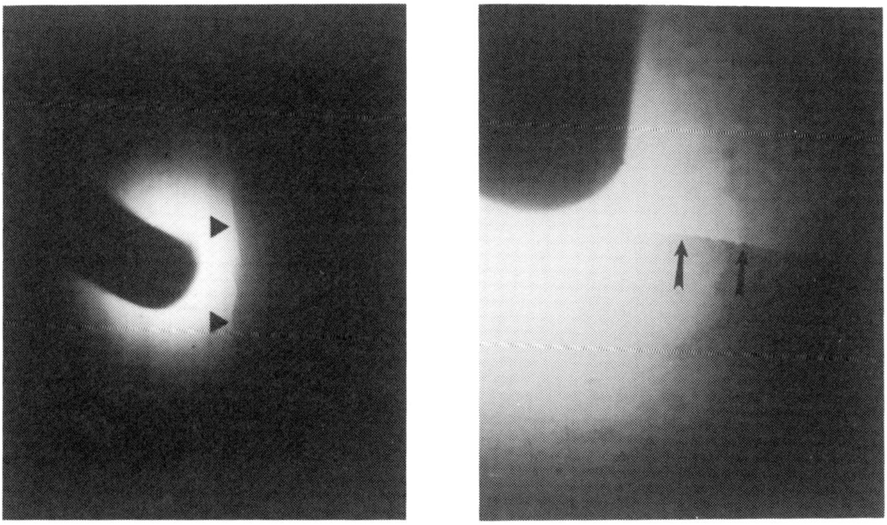

Fig. 8. Under vicinal illumination, cracks oriented normal to the surface scatter the light and are observable as shadows in the halo around the fiber optic light source (note arrows).

Circumferential cracks were found in the circ wraps of both vessels. The cracks were always located in the same axial region; within five cm of the upper "knuckle" transition from the dome top to the cylindrical wall of the vessel (see Fig. 1). This region is very close to, or may include the top of the gasketed upper cradle support band. The cracks varied in length from a fraction of a cm to more than 25 cm in length. There were indications that some cracks had joined with neighbors to form longer cracks. Several cracks intercepted the edge of a cut section and appeared to penetrate the circ wraps. None of these cracks penetrated the entire wall so as to be visible on the interior surface. No hoop cracks were found on the interior of the vessel or at any other location on the exterior surface.

FINITE ELEMENT ANALYSIS

A finite element analysis of the pressure vessel was performed to estimate the stress state prior to failure. The stress estimates, along with experimentally determined material strengths, were essential factors in the determination of the vessel failure scenario. The pressure vessel was modeled as axisymmetric, thereby neglecting any small asymmetries in the area of penetrations in the top of the vessel. 4308 four-noded quadrilateral elements, each with four integration points, were used in the model. To accurately resolve the stresses through the thickness of the vessel and in the failed region, the teflon layer was modeled with two elements through the thickness, the cross plies with three elements through the thickness, and the circ wrap was modeled with four elements through the thickness. Figures 9 and 10 show the analysis model geometry and mesh.

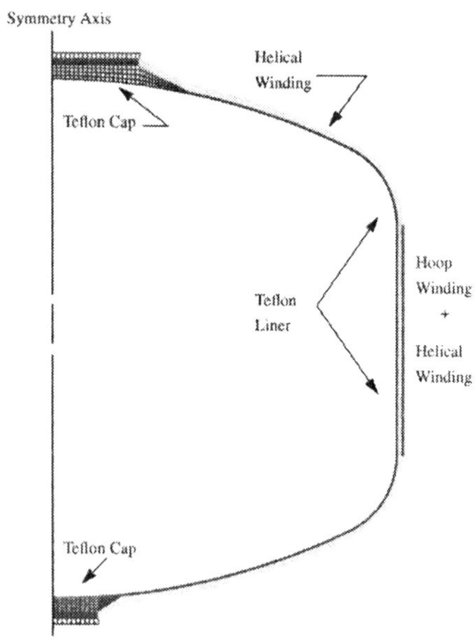

Fig. 9 Finite element model geometry and mesh of the FRP vessel and teflon liner.

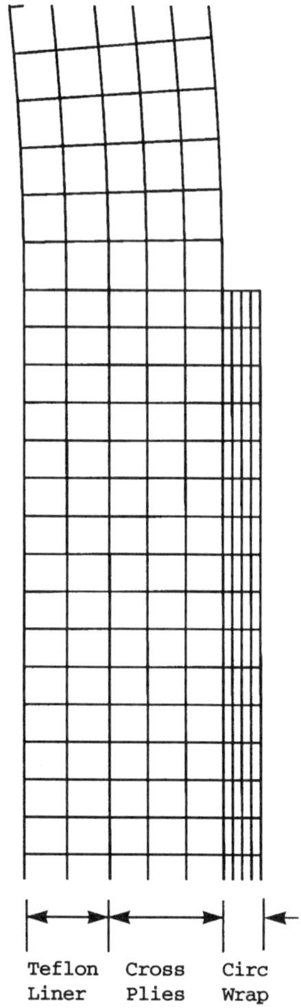

Fig. 10 Detail of the finite element mesh of FRP vessel and teflon liner.

In modeling the orthotropic properties of the FRP, appropriate properties were assigned to both the cross plies and circ wrap. The teflon liner was modeled as an isotropic material. All materials were assumed to be linearly elastic. The cradle was modeled by radially constraining the model in the region of the cradle gasket. The applied loading was a net internal pressure of 0.31 MPa (45 psi). The analysis was run on the CRAY YMP using the commercial finite element analysis code ABAQUS version 4.9.[6]

The analysis shows a region of high axial stress (49 MPa) just above the cradle gasket in the region where cracks were observed. This high stress region is shown in Fig. 11. Such stresses would tend to cause cracking in the circ wrap if its strength is at or below this level.

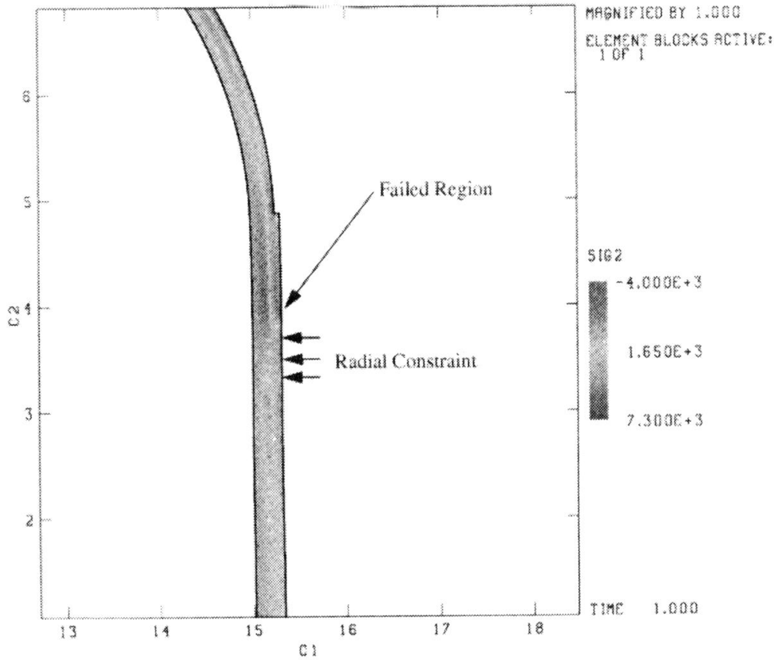

Fig. 11 Finite element analysis model showing stresses in FRP vessel wall. High tensile stresses of 49 MPa (7100 psi) were found above the radial constraint of the cradle, and correspond to the cracked regions of the vessel.

MECHANICAL TESTING

Flexural strength tests (three-point bend) were conducted to obtain the strength of uncracked composite material in the region where elevated tensile stresses were predicted (just above the metal cradle). Flexural strength specimens were also made from cracked material (cracks located by vicinal inspection technique) from this region to compare the fracture surfaces. Specimens were 5.1 mm thick by 22.9 mm wide by 83.8 mm long. Specimens were cut from the vessel and loaded such that the circ wrap layers and the fibers in them were perpendicular to the tensile stress. The span was 69.9 mm and specimens were loaded at 0.13 mm/min.

The initial failure mode was matrix cracking in the circ wrap. Under continued loading, the crack continued to propagate through the matrix of the circ wrap until it reached the circ wrap/cross ply interface. Here the crack either changed direction and produced delamination of the circ wrap/cross ply interface, or continued along a tortuous path through the cross plies, with extensive fiber pull-out. The average tensile stress to initiate matrix cracking in uncracked specimens was 55 ± 9 MPa. This value falls within the reported range of strengths for this material (28-62 MPa). This measured range of strengths is also within the FEA predicted stress level of 49 MPa. Thus the tensile stresses developed under normal operating conditions (0.31 MPa internal pressure), in the region of the circ wraps adjacent to the cradle, are likely to have produced matrix cracking. Such cracks would then provide access to corrosive acid that attacked the glass fibers in the cross plies, degrading their strength. During each pressurization cycle weakened glass fibers failed, producing a slowly growing crack, which provided further access to acid, and produced failure of additional fibers. When the crack reached critical proportions, it propagated catastrophically and produced vessel failure.

As mentioned in the section describing the flexure specimens used for EDS analysis of glass fibers, when specimens from cracked and uncracked regions of the vessel in the vicinity of the circ wrap were examined, two types of fracture surfaces were observed. In the previously uncracked specimen, the fracture surface of the cross ply region exhibited the expected non-planar fracture with extensive fiber debonding and pull-out (Fig. 3B). In the specimens exhibiting pre-cracks in the circ wrap, the fracture surface of the cross ply region exhibited planar fracture with no fiber pull-out adjacent to the circ wrap, but non-planar fracture with extensive fiber pull-out in the remainder of the cross ply (see Fig. 3A). The arrangement (sequence) of these two types of fracture surfaces was similar to that observed on the primary cracks of the failed vessel. The presence of planar fracture in the cross ply material clearly demonstrates the existence of regions of extensive crack growth in the cross plies under normal operating pressures. Thus other cracks existed that could have eventually caused failure; the actual failure was not a fluke or the result of a unique flaw, but rather was simply the first cracked region to reach critical dimensions.

The proposed failure scenario and supporting evidence are as follows:

- There were two primary circumferential through-cracks just above the steel cradle.
- The primary cracks were oriented parallel to the circ wrap direction and the fibers in the circ wrap.
- The fracture surface of the primary crack was planar even in the cross plies, with little evidence of fiber debonding or pull-out.
- The orientation of mirror, mist, and hackle features on glass fiber fracture surfaces from the planar region of the failed cross plies indicated that cracks initiated at the outer surface of the vessel and propagated inward.
- Fractographic evidence revealed that the load-bearing fibers on planar fracture surfaces in the cross ply region failed at twenty to forty percent of their nominal strength, suggesting the possibility that environmental attack and strength degradation had occurred.
- Maintenance was performed on the valves on the top of the vessel that resulted in sulfuric acid being spilled and washed down onto the region of the circ wraps.
- Microanalysis data showed that the Ca and Al contents of the weak E-glass fibers in the planar crack growth area were significantly reduced from that of full strength fibers elsewhere. The depletion levels and resulting strength degradation were similar to those reported in the literature for fibers of similar E-glass compositions after controlled exposure to sulfuric acid.
- Fractographic evidence indicated that fibers that failed during the final overload event (away from the initial planar cracks) and during subsequent laboratory tests failed at full strength.
- Finite element calculations were performed for the vessel pressurized to 0.31 MPa under two conditions: with and without a stiff cradle. The radial constraint produces a tensile stress of approximately 49 MPa in the axial direction, on the outer surface, just above the cradle. This stress was approximately normal to the glass fibers in the circ wrap. Without the constraint of the cradle, a stress of this magnitude does not develop.
- Tensile tests were conducted on undamaged material cut from a region of the failed vessel away from the primary cracks. That material had an average strength of 55 ± 9 MPa, confirming that cracks could form in the circ wrap at the

tensile stress levels produced by the radial constraint of the stiff steel cradle.

- Cracking of the matrix material of the circ wrap at the predicted stress levels provided the acid access to the fibers in the cross plies. Subsequent cracking of the cross plies proceeded non-catastrophically as acid-weakened fibers failed sequentially at the crack tip during subsequent pressurization cycles.
- A sample taken from an area which contained circumferential cracks, but away from the primary fracture, was broken in a three-point bend test. This fracture surface revealed a crack that penetrated about half way through the load-bearing cross plies. The planar character of this partial crack was similar to that observed on the two flat areas of the primary crack.
- One or more of the cracks growing through the cross plies above the cradle reached the critical size for catastrophic crack propagation, resulting in vessel failure at normal operating pressure.

Vessel failure was due to a combination of circumstances. Radial constraint of the steel cradle produced axial tensile stresses in the circ wraps just above the cradle that exceeded the fiber-matrix strength, causing them to crack. Cracks allowed acid, which was spilled onto the vessel during routine maintenance, to be trapped where it could corrode and degrade the strength of load-bearing cross ply fibers. Sequential failure of cross ply fibers occurred until a crack of catastrophic proportions was present.

SUMMARY

A failure investigation was conducted on a ruptured sulfuric acid vessel. Forensic reconstruction of the event, fractographic analysis of the failed vessel, computer modeling of stresses in the vessel walls, and mechanical testing of sections of the failed vessel provided information about the probable cause of failure. It was concluded that the failure occurred at normal operating conditions and was the result of a localized high tensile stress causing cracking of the matrix, leading to acid-exposure and weakening of load-bearing glass fibers in the FRP. Sequential failure of weakened fibers led to crack growth during pressurization cycles until one crack reached the critical dimensions for catastrophic failure. Ceramic fractographic techniques and vicinal illumination were key elements in reconstructing the failure scenario.

REFERENCES

1. J. J. Mecholsky, "Fracture Surface Analysis of Optical Fibers," Engineered Materials Handbook, Vol. 4, Ceramics and Glasses, pp. 663-668, ASME, (1991).
2. A. C. Jaras, B. J. Norman, and S. C. Simmens, "The Measurement of Glass Fibre Strength in Composites from Studies of Their Fracture Surfaces," J. Mater. Sci., 18, pp. 2459-2465 (1983).
3. P. K. Gupta, "Fractography of Fiberglass," pp. 185-206 in Fractography of Glass, Edited by R. C. Bradt and R. E. Tressler, Plenum Press, New York, NY, 1994.
4. B. Das, B. D. Tucker, and J. C. Watson, "Acid Corrosion Analysis of Fibre Glass," J. Mater. Sci. 26 [24] 6606-12 (1991).
5. B. D. Caddock, K. E. Evans, and I. G. Masters, "Diffusion Behaviour of the Core-Sheath Structure in E-Glass Fibres Exposed to Aqueous HCl," J. Mater. Sci. 24 [11] 4100-4105 (1989).
6. ABAQUS Version 4.9, Hibbitt, Karlsson, and Sorensen, Inc.

ACKNOWLEDGMENTS

This work was performed at Sandia National Labs, supported by the U.S. Department of Energy Under Contract #DE-AC04-76DP00789 and as part of Cooperative Research and Development Agreement 01082 with SEMATECH. The authors acknowledge SEMATECH, Intel Corp., and Fluoroware, Inc. for permission to publish and present the results of this investigation. The authors also thank the other Sandia participants in this investigation including Mark Davis, John Geiske, Richard Grant, Fred Greulich, Chuck Gwyn, Will Hareland, Paul Hlava, Alice Kilgo, Bonnie McKenzie, Joe Michael, Gordon Pike, Ken Schuler, and Mark Stavig.

KEYWORD AND AUTHOR INDEX